HAROLD B. KILLEN

University of Houston

MODERN ELECTRONIC COMMUNICATION TECHNIQUES

Macmillan Publishing Company

New York

Collier Macmillan Publishers

London

Acknowledgments

Figs. 5-8, 5-9, 5-11, and 5-12: courtesy of Motorola Semiconductor Products Inc. Fig. 5-14: Quan-Tech Division of Stansbury Industries, Inc. Figs. 10-7, 10-8, 10-9, 10-10, and 10-11: RCA Corporation. Fig. 12-25, 13-1 through 13-19, and 14-31, and Tables 13-1 through 13-4: reprinted with permission from Electronic Design; copyright Hayden Publishing Co., Inc., 1979. Figs. 14-3 through 14-27: reprinted from EDN, 1978, © 1983 Cahners Publishing Co. Fig. 16-6: from William Sinnema, Electronic Transmission Technology: Lines, Waves, and Antennas. © 1979, p. 44. Reprinted by permission of Prentice-Hall, Inc., Englewood Cliffs, N.J. Figs. 17-28, 17-29, 17-30, and 17-34: reprinted with permission from The Bell System Technical Journal. Copyright 1957 by AT&T.

Printed in the United States of America

Macmillan Publishing Company
866 Third Avenue, New York, New York 10022

Collier Macmillan Canada, Inc.

Library of Congress Cataloging in Publication Data

Killen, Harold B.
Modern electronic communication techniques

Includes index.
1. Telecommunication. 2. Electronics. I. Title.
TK5101.K44 1985 621.38'0413 84-5737
ISBN 0-02-363600-9

Printing: 1 2 3 4 5 6 7 8 Year: 5 6 7 8 9 0 1 2

ISBN 0-02-363600-9

PATTERN TYPES

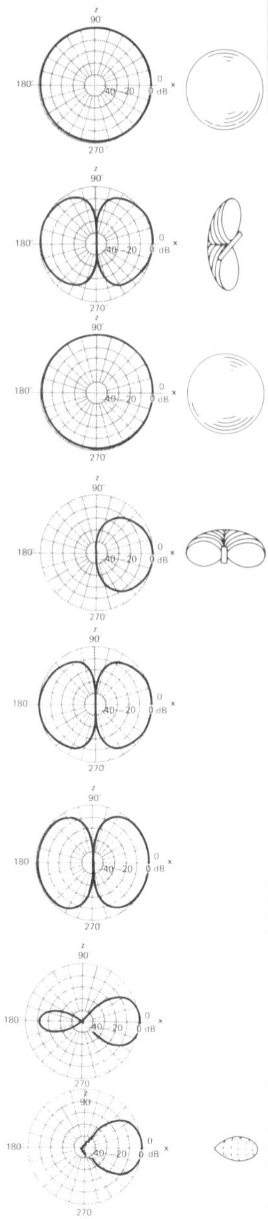

PATTERN #	POLARIZATION	GAIN dB above Dipole	GAIN dB above Isotrope	−3dB BANDWIDTH%	IMPEDANCE Resistive at f_r, R, ohms	CONFIGURATION	TYPE
E	V	0	2.14	40	28	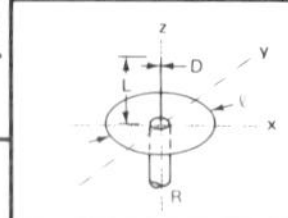	**Dipole over small ground plane** $L = \lambda/4$, $L/D = 53$, $l = 2\lambda$
E	V	0	2.14	45	150	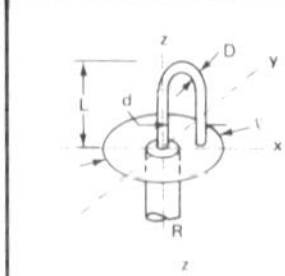	**Folded Unipole over small ground plane** $L = \lambda/4$, $L/D = 53$, $l = 2\lambda$, $L/d = 13$
E	V	0	2.14	16	50	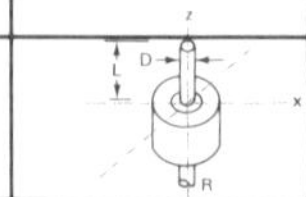	**Coaxial Dipole** $L = \lambda/4$, $L/D = 40$
E	V	0	2.14	200	72	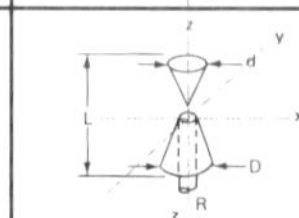	**Biconical Coaxial Dipole** $L = \lambda/2$, $d = \lambda/8$, $D = 3\lambda/8$
E	V	0	2.14	300	50	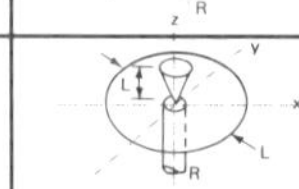	**Disc-Cone or Rod Disc-Cone** $L = \lambda/4$, $l = \lambda$
E	V	12	14.14	25	20	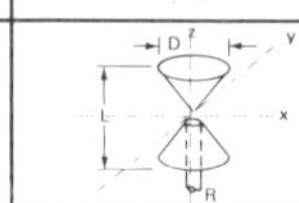	**Biconical Horn** $L = 9\lambda/2$, $D = 14\lambda$
F	H	0	2.14	70	350	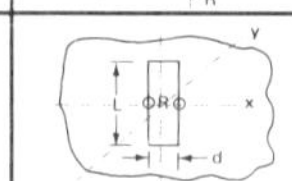	**Slot in Large Ground Plane** $L = \lambda/2$, $l/d = 29$
B	H	1	3.14	13	45	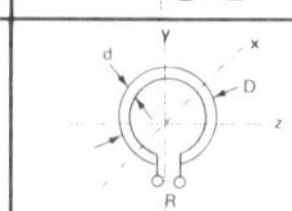	**Vertical Full Wave Loop** $D = \frac{\lambda}{\pi}$, $D/d = 36$
G	Circ.	8	10.14	200	130	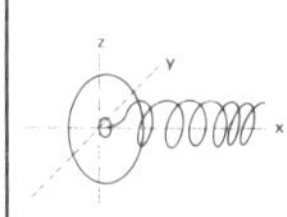	**Helical over reflector screen, tube 6λ long coiled into 6 turns λ/4 apart**
H	H	14.5	16.74	100	600	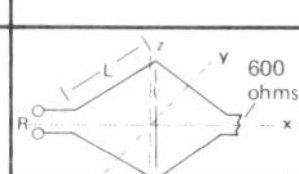	**Rhombic** $L = 9\lambda$, $l = 9\lambda/2$
H	H	12.5	14.74	30	300	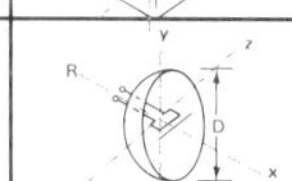	**Parabolic with folded dipole feed (λ/2)** $D = 5\lambda/2$
H	H	13	15.14	35	50	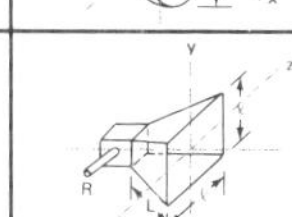	**Horn, coaxial feed** $l = 3\lambda$, $L = 3\lambda$

(Continued on back endpaper.)

To my wife Phyllis and our children
Michelle, Lou Ann, and Cheryl

PREFACE

This book is intended for use as a two-semester course in electronic communications in a two- or four-year technology program. In general, a knowledge of basic ac and dc electrical circuit theory along with algebra is sufficient for the beginning student in this area. An effort was made to avoid the use of advanced mathematics.

To make the book self-contained, the first four chapters include extensive material on circuits, Fourier series, and filters. A treatment of noise is contained in Chapter 5. This introductory material is referred back to repeatedly throughout the text both to aid the student and to provide continuity for the text.

The book is divided into five parts. Chapter 6 begins Part II and introduces the student to communication circuits. This material prepares the student for Part III: Modulation/Demodulation Techniques. The last two parts consists of six systems chapters. These chapters cover fiber optics communications, data communications, and radio-frequency communications. A discussion of satellite communications is included as an integral feature of modern communication systems.

Numerous solved examples are provided throughout the text to aid the student.

Proven practical system design examples are included in Chapter 15 to illustrate key design techniques. A special effort was made throughout the text to include the latest in integrated-circuit techniques, and design nomographs are given to aid in system design, particularly for radio-frequency path design as described in Chapter 17.

The author wishes to express his appreciation to the many reviewers who contributed to the formation of this book and to the organizations who gave permission to use technical material. Special thanks is also due to the staff at Macmillan Publishing Company for their work in producing the text.

HAROLD B. KILLEN

CONTENTS

I FUNDAMENTALS

II COMMUNICATION CIRCUITS

III MODULATION/DEMODULATION TECHNIQUES

PART

FUNDAMENTALS

CHAPTER 1

Introduction

It has been said that the field of communications mirrors the state of development of a technological society. If so, we are witnessing a revolution each four to five years. We have become accustomed to the television revolution, the space revolution, the digital/microprocessor revolution, the consumer revolution, and currently the data communications revolution. This trend is continuing and perhaps even accelerating. Communication via fiber optic data links are becoming commonplace. Remarkably, the areas of interest noted about are bound together by fundamental, easily understood communication principles. To those who are intrigued by deep-space communication systems, CB radios, or automatic garage door openers, I invite you to join in this study of communication techniques.

1-1 TRANSMISSION OF INFORMATION

The purpose of a communication system is to transmit information from one place to another. The performance of the system is measured by the amount of information unavoidably lost during transmission. If the communication system is a wireless one, the information is superimposed (modulated) onto a higher-frequency carrier.

All carrier communication systems have in common the basic functional components illustrated in Figure 1-1. In the modulator, the information to be transmitted is imposed on the carrier signal produced by the signal generator. The transducer (transmitting antenna) couples the modulated signal to the propagating medium, where it propagates as an electromagnetic wave. A small portion of this signal reaches the receiver, where another transducer (receiving antenna) couples the alternating signal plus noise to the selective receiver. The selective receiver filters the desired modulated carrier signal. In the demodulator, the modulating signal containing the information is separated from the carrier and recovered.

The characteristics of a communication system are markedly dependent on that portion of the electromagnetic spectrum chosen for use (Figure 1-2). In radio systems the electromagnetic carrier wave is selected from the radio-frequency (RF) portion of the spectrum. Microwave or millimeter systems have carrier frequencies from those portions of the spectrum. Optical communication systems have carrier frequencies from the infrared, visible, and ultraviolet frequencies.

In the communication system the amount of information that is transmitted is directly related to the bandwidth (frequency spectrum) of the modulated carrier, which is generally some fixed percentage of the carrier frequency itself. Thus, increasing the carrier frequency increases the transmission bandwidth and therefore the overall capacity of the system to transmit information—hence the emergence of optical communication systems as a field of particular interest.

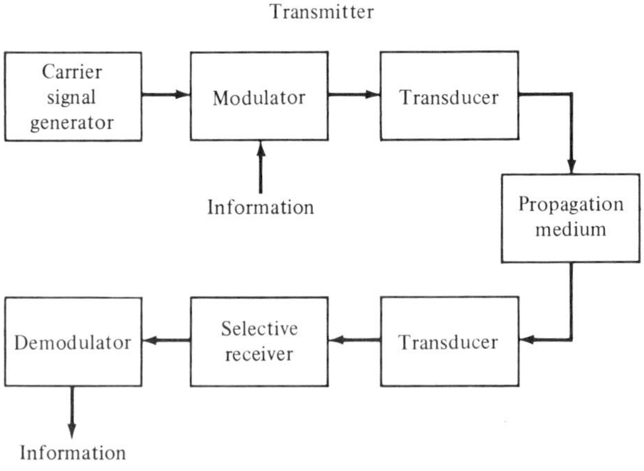

FIGURE 1-1. Basic functional components of a carrier communication system.

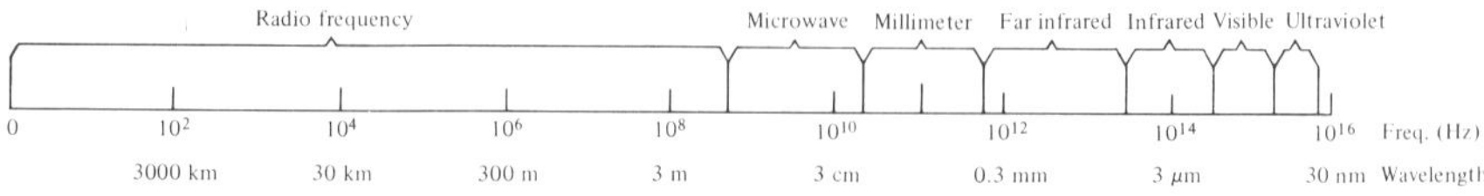

FIGURE 1-2. The electromagnetic spectrum.

1-2 THE COMMUNICATION SYSTEM

Referring to Figure 1-1, the output of the carrier signal generator is a time-varying sinusoidal signal which is described by

$$E_c(t) = A_m \sin(\omega_c t + \phi) \tag{1-1}$$

Note that there are three parameters which may be varied: the amplitude A_m, the frequency ω_c, and the phase ϕ. Using the information signal to vary A_m results in amplitude modulation (AM) of the carrier signal. Varying ω_c or ϕ results in frequency modulation (FM) and phase modulation (PM), respectively. Modulation, then, is the process of using a lower-frequency signal to modulate a higher-frequency signal.

Demodulation is the inverse, that is, removing the higher-frequency signal and recovering (detecting) the lower-frequency information signal. The study of communication theory, then, is how to achieve this operation.

1-3 DESIGN CONSIDERATIONS

There are two basic limitations on the performance of a communication system:

- The bandwidth of frequencies associated with the transmitted signal.
- Electrical noise.

Electrical noise consists of any unwanted voltages or currents that end up appearing in the receiver. Basically, it may be divided into two categories:

- Terrestrial (human-made) and atmospheric.
- Extraterrestrial—noise from outer space.

The troublesome nature of noise may be illustrated quite simply. The communications receiver is a very sensitive instrument. The electrical signal presented to the receiver by the receiving antenna may be no greater than 1 microvolt (μv). This signal is amplified many times to provide a usable output at, for example, a speaker. It does not require much noise voltage to render the signal useless. A fundamental performance consideration, then, is the ratio of signal voltage (power) to noise voltage (power) at the receiver input. As an illustration of signal-to-noise ratio (SNR) requirements in the video field, an acceptable picture on a color television receiver requires an input SNR on the order of 1000 : 1.

With this design requirement established, the frequency bandwidth associated with the transmitted signal may be appreciated. Noise is found to be present over the entire usable electromagnetic spectrum (Figure 1-2). Thus much of this noise spectrum may be filtered by the receiver if the receiver bandwidth is no greater than that necessary to process the received signal. The bandwidth of frequencies associated with the transmitted signal then becomes of extreme importance. A detailed analysis with examples is provided in a later chapter.

1-4 COMPARISON OF RADIO-FREQUENCY AND LIGHT-FREQUENCY COMMUNICATIONS SYSTEMS

Typical radio-frequency and light-frequency communication systems are shown in Figure 1-3. The carrier in both cases [electromagnetic radiation (EMR)] is described by the same physical laws. The difference between radio and light radiation is quantitative only. Light source frequencies are several orders of magnitude higher than those generated by even the highest radio-frequency sources. This frequency difference results in a number of differences in the two systems.

Transmission of radio frequencies involves the electrical modulation of a single-frequency carrier signal that can be described precisely by an amplitude–time relationship [equation (1-1)]. At light frequencies the carrier can be broadband and must then be described in a statistical sense. Modulation is achieved by varying the intensity of the light. In both systems the wavelength, λ, of the EMR is given by

$$\lambda = \frac{300}{f_{\text{MHz}}} \quad \text{meters} \tag{1-2}$$

where f_{MHz} is the frequency in megaherz. We shall see that the physical size of the antenna is directly dependent on the wavelength. This results in different sizes and types of antennas for the same transmission efficiency.

The receiver for the radio communication system filters and amplifies the modulated carrier prior to demodulation. The signal is always electrical. On the

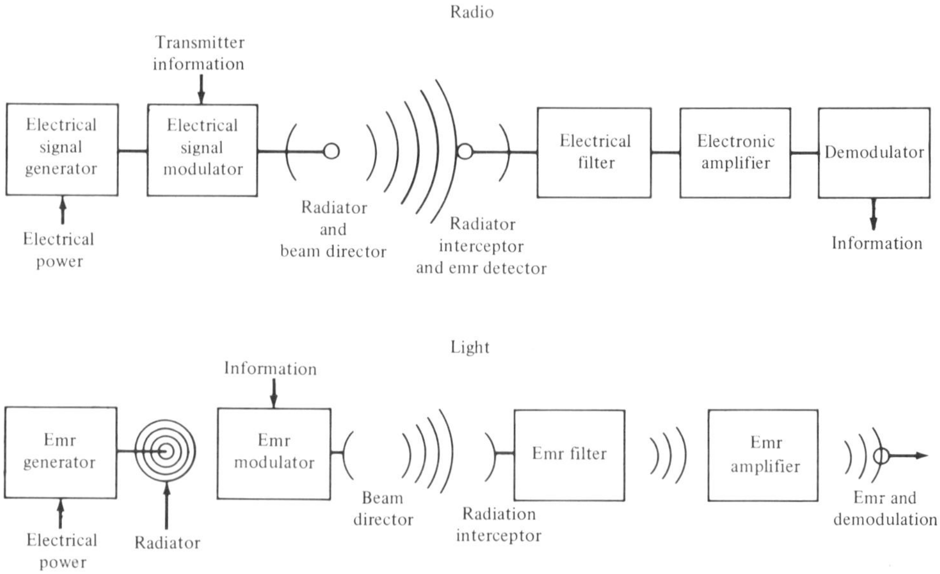

FIGURE 1-3. Comparison of point-to-point radio- and light-frequency communication systems (EMR: electromagnetic radiation).

other hand, the light receiver filters and demodulates a signal that consists of a complex electromagnetic wave. The signal must be demodulated before it can be processed as an electrical signal. Present technology has not yet produced components that will process electrical signals at light frequencies. Because of the difference in wavelength between a radio communication system and a light communication system, the greater practical difference in design is the antenna.

1-5

DEPENDENCE OF THE PHYSICAL CHARACTERISTICS OF ANTENNAS ON FREQUENCY

Any type of EMR transmitting antenna is described basically in terms of its radiation pattern. Referring to Figure 1-3, we see that the antenna for the radio communication system may consist of a small antenna backed by a reflecting paraboloid "dish." The reflector serves to focus the radiation along the path between the transmitting and receiving stations. At the receiver a similar antenna intercepts or collects a portion of the energy in the EMR wave. The "pattern" in which the EMR is directed can be described by the gain and beam angle. The *gain* of an antenna relates to the maximum power density of the radiation pattern. Thus antenna gain may be defined as

$$G_a = \frac{\text{maximum power density of the radiation pattern}}{\text{power density of an isotropic antenna}} \tag{1-3}$$

where an isotropic antenna is one that radiates electromagnetic energy equally in all directions.

If the total transmitter power delivered to the antenna is P_T watts, the isotropic antenna transmits an electromagnetic wave which has a power density of $P_T/4\pi$ watts per unit solid angle. Thus equation (1-3) may be expressed as

$$\begin{aligned} G_a &= \frac{P_M}{P_T/4\pi} \\ &= \frac{4\pi P_m}{P_T} \end{aligned} \tag{1-4}$$

where P_M is the maximum power density of the radiation pattern. For an antenna whose radiation efficiency is not 100%, equation (1-4) is usually expressed as

$$G_a = K\frac{4\pi P_M}{P_T} \tag{1-5}$$

where $K < 1$.

The antenna beam angle, Ω, is defined as the solid angle in steradians, measured from the antenna, in which the maximum power must be concentrated in order to have the same total power (refer to Figure 1-4). Thus

$$G_a \frac{P_T}{4\pi}\Omega = P_T \tag{1-6}$$

and $\Omega = 4\pi/G_a$ steradians. It will be shown in Chapter 17 that for a circular antenna of diameter D, the antenna gain may be expressed as

$$G_a = \left(\frac{4D}{\lambda}\right)^2 \tag{1-7}$$

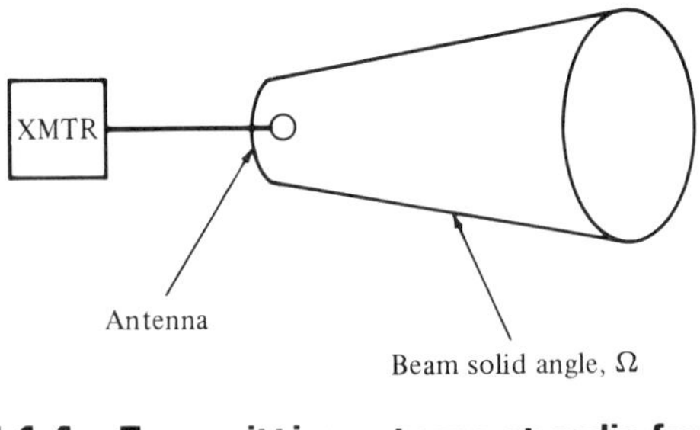

FIGURE 1-4. Transmitting antenna at radio frequencies.

Thus, for a given gain, the diameter of the antenna decreases with the square of the carrier frequency [see equation (1-2)]. If the dimensions of the antenna are held constant, the gain of the antenna increases proportionally with the square of the carrier frequency. This gain, however, can be realized only if all the antenna dimensions are controlled to a small fraction of the wavelength. Consequently, manufacturing tolerances become smaller as the carrier frequency is increased.

At high frequencies, more of the carrier power can be beamed directly at the receiver. The paraboloid antenna can no longer be used because the dimensions become too small and the power density too large. A lens is used as the transmitting antenna (Figure 1-5).

The transmission efficiency remains directly proportional to the square of the frequency for a given diameter and equation (1-7) still applies. Antenna gain and beamwidth versus diameter are shown in Figure 1-6.

1-6 RADIATION CHANNEL

The region between the transmitter and the receiver is referred to as the *radiation channel*. If this region consists of open space, for example, this results in the propagation of unguided waves. On the other other hand, if the wave is constrained to travel in a coaxial cable or fiber optic link, we have a radiation channel in which there are guided waves. Communication system design then requires that the properties of the propagation path be considered. In particular, data communications using guided waves will be discussed in detail.

1-7 SUMMARY

This first chapter has been used to present an overview of communication systems. We see that communication system design and analysis can be divided into four main areas:

- Selection of communication system type: radio or light.
- Design and performance of the receiver.
- Carrier power requirements for a given communication link and receiver.
- Selection of the type of modulation.

The important topics mentioned in this chapter are considered in detail in later chapters, where their impact on design will be emphasized. The background

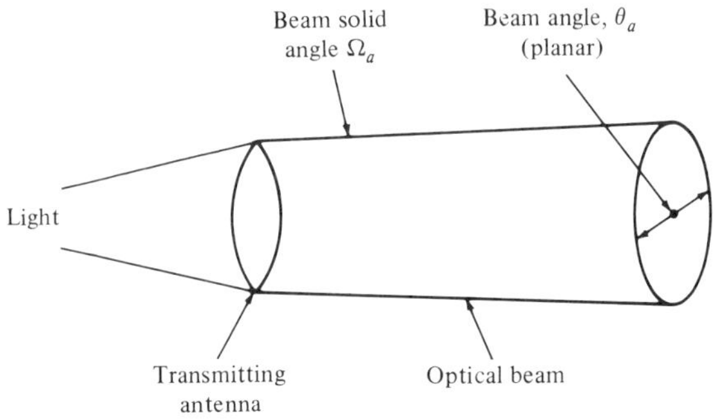

FIGURE 1-5. Transmitting antenna at light frequencies.

information necessary to design communication systems is presented in the first nine chapters. This material is succeeded by system considerations, and finally radio-wave propagation is treated. Sufficient examples are included in each chapter to illustrate problem-solving techniques. The potential impact of communications on each of us is staggering. May this book open up this potential to each of you.

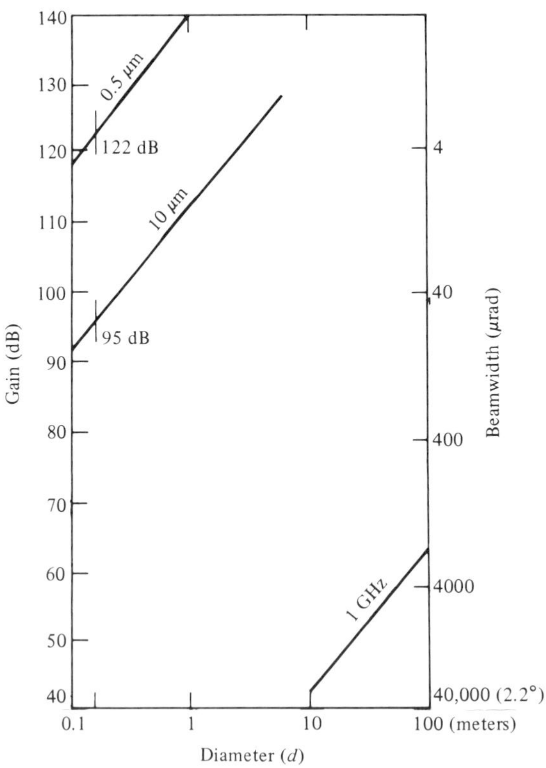

FIGURE 1-6. Antenna gain and beamwidth versus diameter.

CHAPTER 2

Transmission in Electrical Networks

In Chapter 1 we considered the overall concepts of communication theory. We now begin a detailed study of the theory necessary to understand and construct communication systems. Circuits designed to operate at radio frequencies are generally limited to a single frequency or to a specific band of frequencies. To this end, we first consider circuits that are "tuned" to respond to specific frequencies. Following this, the frequency spectra of specific signals are analyzed and the results are correlated with the tuned circuit response.

2-1 SERIES RESONANT CIRCUIT

The *RLC* series circuit of Figure 2-1 is of such interest and importance in communication theory that we begin our study with this circuit. The results achieved in analyzing this circuit will be applied to succeeding circuits. The most striking characteristic of this circuit is that it has a low impedance to current

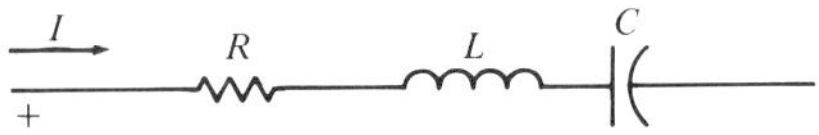

FIGURE 2-1. Series resonant circuit.

flow at one frequency and a relatively high impedance to current flow at all frequencies that differ significantly from this frequency.

The elements of the circuit are in series and the total circuit impedance may be expressed as

$$\begin{aligned} Z &= R + j\omega L + \frac{1}{j\omega C} \\ &= R + j\left(\omega L - \frac{1}{\omega C}\right) \end{aligned} \tag{2-1}$$

where $\omega = 2\pi f$. The reactive terms are in parentheses with the positive term representing inductive reactance and the negative term representing capacitive reactance. Equation (2-1) is often expressed as

$$Z = R + j(X_L - X_C) \tag{2-2}$$

where

$$X_L = \omega L = 2\pi f L$$

and

$$X_C = \frac{1}{\omega C} = \frac{1}{2\pi f C}$$

It is apparent that at some frequency ω_0, $X_L = X_C$ and the reactive term is equal to zero. This frequency is termed the *resonant frequency*. Expressed mathematically, it follows that at resonance

$$\omega_0 L - \frac{1}{\omega_0 C} = 0 \tag{2-3}$$

or

$$\omega_0^2 = \frac{1}{LC} \tag{2-4}$$

From equation (2-4) the resonant frequency, f_0, may be expressed as

$$f_0 = \frac{1}{2\pi\sqrt{LC}} \tag{2-5}$$

At this frequency the net reactance of the circuit is zero and the impedance, Z, of the circuit is a minimum. Thus the current, I, is maximum. Using complex algebra, the impedence may be expressed in terms of a magnitude and angle. That is,

$$Z = \sqrt{R^2 + (X_L - X_C)^2} \;\angle \phi = \tan^{-1} \frac{X_L - X_C}{R} \qquad (2\text{-}6)$$

The impedance of a series resonant circuit with its components is shown graphically in Figure 2-2. When the reactive component of the impedance is zero, the phase angle ϕ is zero and the line current, I, is in phase with the applied voltage.

EXAMPLE 2-1

A series circuit contains an inductor L, a 1-microfarad (μF) capacitor, and a 10-ohm (Ω) resistor. (a) Find the value of L such that the circuit will resonate at 1.0 kilohertz (kHz), (b) find the current I and the voltage across each component at resonance if the applied voltage is 10 V root mean square (rms), and (c) sketch the transform diagram.

SOLUTION

(a) $L = \dfrac{1}{(2\pi f_0)^2 C} = \dfrac{1}{(2\pi \times 1000)^2(10^{-6})} = 25.36$ millihenries (mH)

(b) $I = \dfrac{E}{R} = \dfrac{10\angle 0^\circ}{10} = 1\angle 0^\circ$ amperes (A)

$|V_L| = IX_L = I\omega L$

$= (1)(2\pi \times 1000)(25.36 \times 10^{-3}) = 158.9$ V

$|V_C| = |V_L| = 158.9$ V

$V_R = IR = (1)(10) = 10$ V

(c) $V_L = jX_L I$

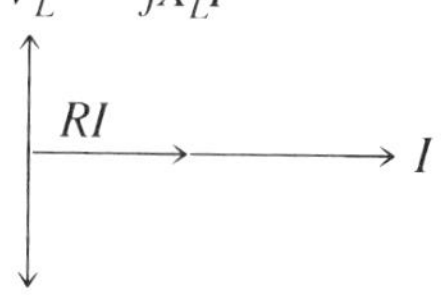

$V_C = -jX_c I$

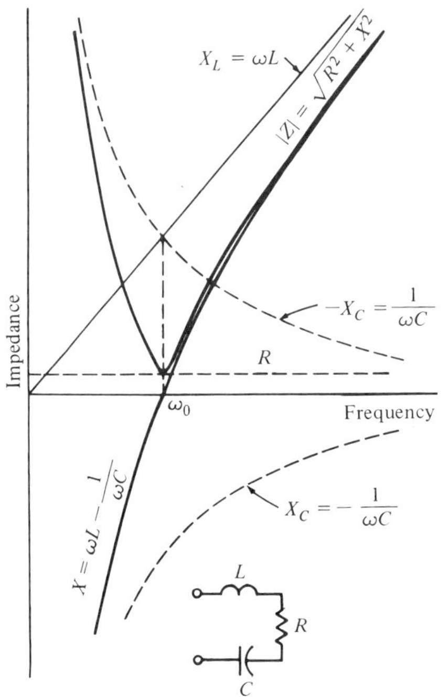

FIGURE 2-2. Impedance of a series resonant circuit and its three components.

2-2

Q RISE IN VOLTAGE

Referring to Example 2-1, we see that the voltage across the inductor and the capacitor is 158.9 V with only 10 V applied. This voltage rise is a characteristic of the circuit at resonance and is large if the reactive element (X_L or X_C) is 10 to 100 times as great as R. This ratio will be defined as the Q or *quality factor* of the circuit. Expressed mathematically,

$$Q = \frac{\omega L}{R} = \frac{1}{\omega CR} \tag{2-7}$$

The voltage across the inductance may be expressed in terms of the Q and the applied voltage. At resonance, the line current, I, is given by

$$I = \frac{E}{R} \tag{2-8}$$

and the voltage across the inductance may be expressed as

$$V_L = IX_L = I\omega L \tag{2-9}$$

Substituting for the current from equation (2-8), we obtain

$$\begin{aligned} V_L &= \frac{E}{R}\omega L = \frac{\omega L}{R} E \\ &= QE \end{aligned} \tag{2-10}$$

Thus the voltage across the inductor at resonance is Q times the applied voltage. Similarly, it can be shown that

$$V_C = QE \tag{2-11}$$

2-3 HALF-POWER POINTS

The preceding analysis and computations were performed at the resonant frequency, f_0. Calculation of the circuit parameters becomes somewhat more complex if the frequency is allowed to change on either side of the resonant frequency. Plotting the response of the circuit versus frequency illustrates the frequency selectivity of the circuit. Since the current is common to all elements of the circuit, it is convenient to graph the current versus the frequency. The general expression for the current, I, is

$$I = \frac{E}{R + j\left(\omega L - \dfrac{1}{\omega C}\right)} \tag{2-12}$$

and the magnitude of the current versus frequency is shown in Figure 2-3. As the frequency is changed to either side of resonance, the reactance increases, causing a decrease in the line current, I. We see, then, that the resonant circuit selects a band of frequencies. Note that at frequencies f_1 and f_2 the current is $0.707I_{max}$, where I_{max} occurs at the resonant frequency.

Recall that only the resistance in a circuit consumes power. Thus at frequencies f_1 and f_2 the power input to the circuit is

$$\begin{aligned} P &= (0.707I_m)^2R = 0.5I_mR \\ &= 0.5P_m \end{aligned} \tag{2-13}$$

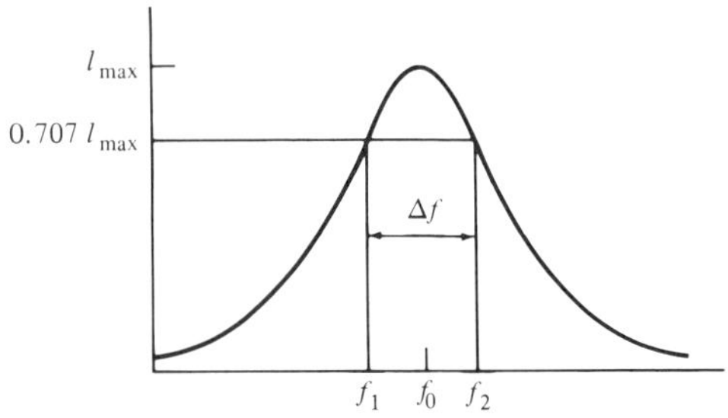

FIGURE 2-3. Bandwith of a series resonant circuit.

Thus the frequencies f_1 and f_2 on the resonance curve are referred to as the *half-power points*. The bandwidth, Δf, of the circuit is

$$\Delta f = f_2 - f_1 \tag{2-14}$$

For the current to drop to $0.707 I_{max}$ at f_1 and f_2 it is necessary that the impedance, Z, be $\sqrt{2}$ times the impedance at resonance. Since the impedance at resonance is $Z = R$, the circuit impedance at the half-power points is $Z = \sqrt{2}R$. That is, the net reactance must equal R. Thus, at frequency f_2 the inductive reactance must have increased by an amount equal to $\frac{1}{2}R$ and the capacitive reactance decreased by an amount equal to $\frac{1}{2}R$. At frequency f_1 the inverse must be true. Expressed mathematically, then,

$$2\pi f_2 L - 2\pi f_1 L = R \tag{2-15}$$

or

$$f_2 - f_1 = \frac{R}{2\pi L}$$

Dividing both sides of this equation by the resonant frequency, f_0, yields

$$\frac{f_2 - f_1}{f_0} = \frac{R}{2\pi f_0 L} = \frac{1}{Q} \tag{2-16}$$

From equation (2-16) the Q of the circuit may be expressed as

$$Q = \frac{f_0}{f_2 - f_1} = \frac{f_0}{\Delta f} \tag{2-17}$$

We see that increasing Q is equivalent to decreasing the bandwidth, Δf, and making the circuit more selective.

We now have three relations for the Q of a series resonant circuit. These are

$$Q = \frac{\omega_0 L}{R} = \frac{1}{\omega_0 CR} \tag{2-18}$$

$$Q = \frac{\omega_0}{\omega_2 - \omega_1} \tag{2-19}$$

$$Q = \left|\frac{V_L}{E}\right| = \left|\frac{V_C}{E}\right| \tag{2-20}$$

An even more basic definition for Q can be shown to be

$$Q = \frac{\text{energy stored}}{\text{energy disipated per cycle}} \tag{2-21}$$

This definition is more basic than the previous three and applies to any resonant system. The admittance, Y, of a series resonant circuit is given by

$$Y = \frac{1}{R + j\left(\omega L - \dfrac{1}{\omega C}\right)} \tag{2-22}$$

and the effect of different values of circuit resistance is indicated in Figure 2-4. The phase angle, ϕ, associated with a series resonant circuit is given by

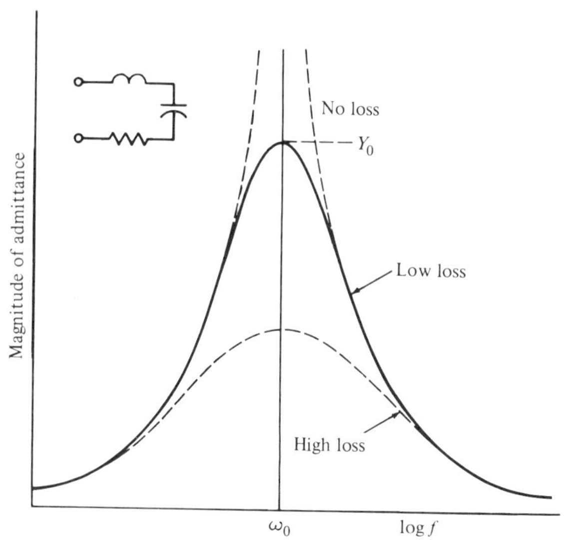

FIGURE 2-4. Admittance of a series resonant circuit.

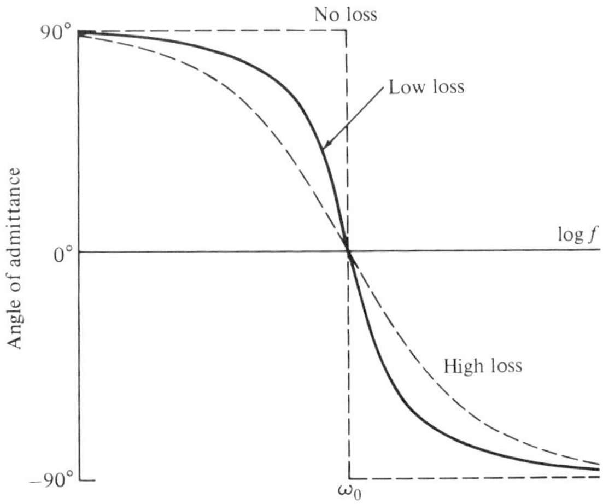

FIGURE 2-5. Phase angle of a series resonant circuit.

$$\phi = \tan^{-1} \frac{X_L - X_C}{R} \tag{2-23}$$

and is shown graphically in Figure 2-5.

EXAMPLE 2-2

A series *RLC* circuit with $Q = 250$ is resonant at 1.5 mHz. (a) Find the frequencies at the half-power points, and (b) find the frequencies the average power to the resonant circuit is one-tenth the power at resonance, input voltage remaining constant.

SOLUTION

(a) $\Delta f = f_2 - f_1 = \dfrac{f_0}{Q} = \dfrac{1.5 \times 10^6}{250} = 6000 \text{ Hz}$

$$f_2 = f_0 + \frac{\Delta f}{2} = 1.5 \times 10^6 + \frac{6000}{2} = 1.503 \text{ MHz}$$

$$f_1 = f_0 - \frac{\Delta f}{2} = 1.5 \times 10^6 - \frac{6000}{2} = 1.497 \text{ MHz}$$

(b) $P = \dfrac{P_0}{10} =$

$$\left|\frac{E}{Z}\right|^2 R = \frac{1}{10}\left(\frac{E}{R}\right)^2 R$$

$$|Z|^2 = 10R^2$$

$$R^2 + \left(\omega L - \frac{1}{\omega C}\right)^2 = 10R^2$$

$$R^2\left[1 + \left(\frac{\omega L}{R} - \frac{1}{\omega RC}\right)^2\right] = R^2\left[1 + \left(\frac{\omega}{\omega_0}\frac{\omega_0 L}{R} - \frac{\omega_0}{\omega}\frac{1}{\omega_0 RC}\right)^2\right]$$

$$R^2\left[1 + \left(\frac{\omega}{\omega_0}Q - \frac{\omega_0}{\omega}Q\right)^2\right] = 10R^2$$

$$\omega^2 - \frac{3\omega_0}{Q}\omega - \omega_0^2 = 0$$

$$\omega_{1,2} = \frac{3\omega_0}{Q} + \sqrt{\left(\frac{3\omega_0}{2Q}\right)^2 + \omega_0^2}$$

$$= \frac{3(2\pi)(1.5 \times 10^6)}{250}$$

$$\pm \sqrt{\left[\frac{3(2\pi)(1.5 \times 10^6}{2(250)}\right]^2 + [(2\pi)(1.5 \times 10^6)]^2}$$

$$\omega_2 = 9{,}476{,}520 \text{ radians (rad)}$$

$$f_2 = \frac{\omega_2}{2\pi} = 1.509 \text{ MHz}$$

$$f_1 = f_0 - 9 \text{ kHz} = 1.491 \text{ MHz}$$

2-4

PARALLEL RESONANT CIRCUIT

The parallel resonant circuit is shown in Figure 2-6. This circuit is also composed of *RLC* elements, but its behavior is in many respects the dual of the series circuit. It has high impedance at resonance, whereas the series circuit has high admittance. The diagrams remain the same, with currents replacing voltages and voltages replacing currents.

The admittance, Y, looking into the terminals of the circuit is

$$Y = G + j\omega C - j\frac{1}{\omega L} = G + j\left(\omega C - \frac{1}{\omega L}\right) \tag{2-24}$$

It is evident that there exists a frequency, ω, such that the susceptance of the

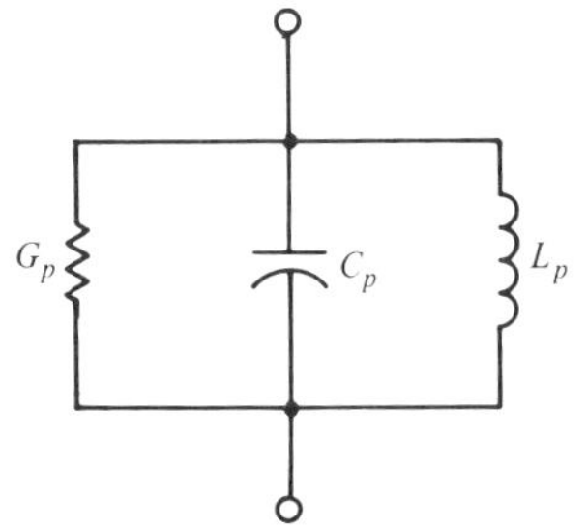

FIGURE 2-6. Parallel resonant circuit.

circuit is zero. That is,

$$\omega C - \frac{1}{\omega L} = 0$$

or

$$\omega_0 = \frac{1}{\sqrt{LC}} \tag{2-25}$$

This is the same formula that defines resonance for the series resonant circuit. Since the admittance is purely real at this frequency it is evident that an applied voltage and resultant terminal current must be in phase.

Let us consider another circuit that exhibits parallel resonance. In Figure 2-7 the resistance is in series with the inductance. The resistance may consist of, for example, the resistance of the windings of the inductor. We shall show later that the two circuits are equivalent at a particular frequency for a given set of elements. If we now apply the condition for resonance to this circuit—that

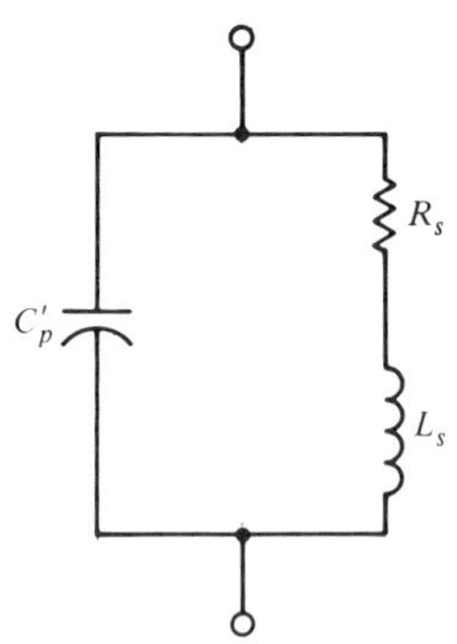

FIGURE 2-7. Parallel resonant circuit.

is, the phase angle between applied voltage and current must be zero—this is equivalent to specifying that the circuit admittance, Y, is purely real. The admittance may be expressed as

$$\begin{aligned} Y &= \frac{1}{R + j\omega L} + j\omega C \\ &= \frac{R}{R^2 + (\omega L)^2} + j\left[\omega C - \frac{\omega L}{R^2 + (\omega L)^2}\right] \end{aligned} \tag{2-26}$$

The resonant frequency, ω_0, is found by setting the reactive part of equation (2-26) equal to zero. That is,

$$\omega C - \frac{\omega L}{R^2 + (\omega L)^2} = 0$$

or (2-27)

$$\omega_0 = \sqrt{\frac{1}{LC} - \left(\frac{R}{L}\right)^2}$$

Note that the series resistance, R, shifts the resonant frequency.

From equation (2-26), at resonance the admittance of the circuit is given by

$$Y_0 = \frac{R}{R^2 + (\omega L)^2} \tag{2-28}$$

From equation (2-27) we have the relation

$$R^2 + (\omega L)^2 = \frac{L}{C} \tag{2-29}$$

Substituting into equation (2-28), the admittance, Y, at resonance is

$$Y_p = \frac{RC}{L} \tag{2-30}$$

The input impedance of the circuit at resonance is

$$Z_p = \frac{L}{RC} \tag{2-31}$$

CIRCUIT *Q* FOR A PARALLEL RESONANT CIRCUIT

The Q of a parallel resonant circuit can be approached in terms of the fundamental definition defined by equation (2-21). It is easy to relate the previous ideas if the physical action of the circuit is considered. Assume that a sinusoidal voltage is applied to the circuit in Figure 2-7. If we let the inductive current, i_L, be $i_L = I_m \sin \omega t$, the voltage v_L, will be $v_L = L(d_i/d_t) = \omega L I_m \cos \omega t$. The instantaneous power to the inductance is

$$\begin{aligned} P &= vi = \omega L I_m^2 \sin \omega t \cos \omega t \\ &= \omega L I_m^2 (\tfrac{1}{2} \sin 2\omega t) \end{aligned} \tag{2-32}$$

Curves relating power, energy, and current are shown in Figure 2-8a. Similar curves for power to a capacitance are shown in Figure 2-8b. From Figure 2-8 it is evident that power is alternatively delivered to the inductor and capacitor. During the time an element is not receiving energy, it is returning its energy to the circuit. This results in a current that oscillates in the *RLC* loop. This storage feature of the circuit results in its often being described as a *tank circuit*. The oscillating current sees the *RLC* elements as a series circuit and thus the previous

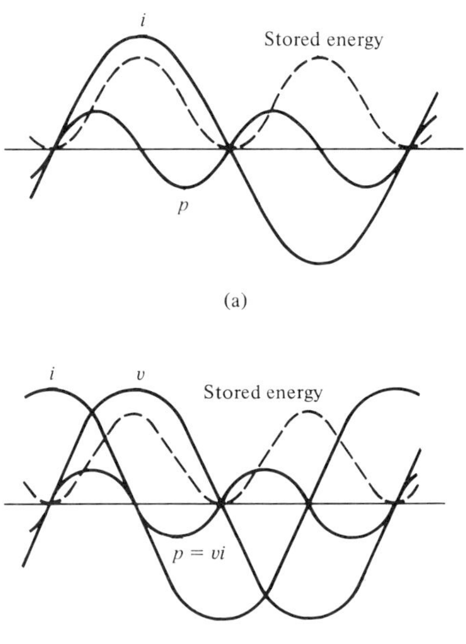

FIGURE 2-8. (a) Power and energy to inductance; (b) power and energy to capacitance.

definition of Q is applicable. That is,

$$Q = \frac{\omega L}{R} = \frac{1}{\omega RC} \tag{2-33}$$

The only current from the source is that necessary to compensate for losses in the circuit. If $R = 0$, the tank circuit is lossless.

EXAMPLE 2-3

A parallel circuit has a capacitor of 50 picofarads (pF) in one branch and an inductance of 100 μH with 10 Ω resistance in the second branch. The applied voltage is 50 V rms. Find (a) f_0, I_L, I_C, Q, and (b) z_0, Z_{max}, I_T.

SOLUTION

(a) $$f_0 = \frac{1}{2\pi}\sqrt{\frac{1}{LC} - \left(\frac{R}{L}\right)^2}$$

$$= \frac{1}{2\pi}\sqrt{\frac{1}{(100 \times 10^{-6})(50 \times 10^{-12})} - \left(\frac{10}{100 \times 10^{-6}}\right)^2}$$

$$= 2.248 \text{ MHz}$$

$$I_L = \frac{E}{R + j\omega L} = \frac{50\angle 0^\circ}{10 + j(2\pi)(2.248 \times 10^6)(100 \times 10^{-6})}$$

$$= 35.3 \angle -89.55^\circ \text{ mA}$$

$$I_C = \frac{E}{1/j\omega C} = j\omega CE$$

$$= j(2\pi)(2.248 \times 10^6)(50 \times 10^{-12})50\angle 0^\circ$$

$$= 35.3\angle +90^\circ$$

$$Q = \frac{\omega L}{R} = \frac{(2\pi)(2.248 \times 10^6)(100 \times 10^{-6})}{10}$$

$$= 141.1$$

(b) $$Z_p = \frac{L}{RC} = \frac{100 \times 10^{-6}}{10(50 \times 10^{-12}} = 200 \text{ k}\Omega$$

$$Z = \frac{R + j\omega L\left(\frac{1}{j\omega C}\right)}{R + j\left(\omega L - \frac{1}{\omega C}\right)}$$

$$|Z_{max}| = \left|\frac{(R + j\omega L)(1/j\omega C)}{R}\right|$$

$$= \left|\frac{-j\sqrt{\dfrac{L}{C}}\left(R + j\sqrt{\dfrac{L}{C}}\right)}{10}\right|$$

$$= \left|\frac{-j\sqrt{\dfrac{100 \times 10^{-6}}{50 \times 10^{-12}}}\left(10 + j\sqrt{\dfrac{100 \times 10^{-6}}{50 \times 10^{-12}}}\right)}{10}\right|$$

$$= 201.02\text{k}\Omega$$

$$I_T = \frac{E}{Z_p} = \frac{50\angle 0^\circ}{2 \times 10^5} = 0.25 \text{ mA}$$

Note from Example 2-3 that if we compare the ratio of oscillating current to terminal current, we see a Q rise in current at the resonant frequency. That is,

$$Q = \frac{I_L}{I_T} = \frac{35.2}{0.25} = 141.2 \quad (2\text{-}34)$$

Thus there is a Q rise in the tank current at resonance. Plots of the total current and circuit impedance versus frequency are shown in Figure 2-9.

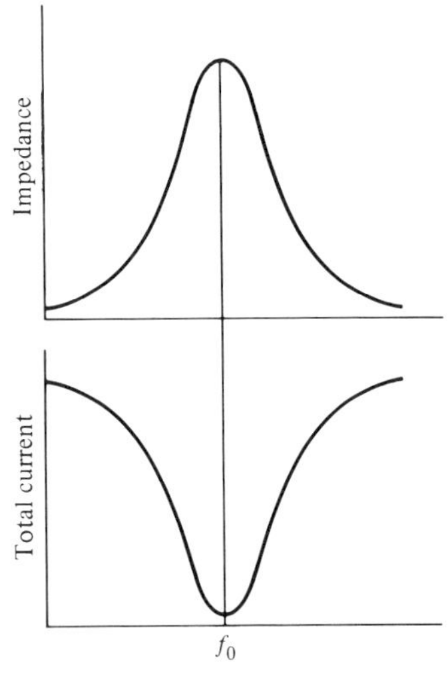

FIGURE 2-9. Parallel resonance curves.

The bandwidth of a parallel resonant circuit is the range of frequencies where the total impedance drops to 0.707 of the parallel resistance. This results in a bandwidth of

$$\Delta f = \frac{f_0}{Q} \tag{2-35}$$

which is identical to the expression for a series resonant circuit.

2-6 EQUIVALENCE OF SERIES AND PARALLEL RESONANT CIRCUITS

It was stated previously that the circuit in Figure 2-7 could be shown to be equivalent to the circuit in Figure 2-6. The input admittance for the circuit in Figure 2-6 is

$$y_{\text{in}} = G_p + j\omega C_p - j\frac{1}{\omega L_p} \tag{2-36}$$

Similarly, for the circuit in Figure 2-7,

$$\begin{aligned} y_{\text{in}} &= j\omega C'_p + \frac{1}{R_s + j\omega L_s} \\ &= \frac{R_s}{R_s^2 + (\omega L_s)^2} + j\omega C'_p - j\frac{\omega L_s}{R_s^2 + (\omega L_s)^2} \end{aligned} \tag{2-37}$$

If we now set equations (2-36) and (2-37) equal and equate corresponding terms, we have the following:

$$G_p = \frac{R_s}{R_s^2 + (\omega L_s)^2} \tag{2-38}$$

$$C_p = C'_p \tag{2-39}$$

$$B_L = \frac{1}{\omega L_p} = \frac{\omega L_s}{R_s^2 + (\omega L_s)^2} \tag{2-40}$$

That is, for the two circuits to have the same admittance, the series inductance and resistance translate as a different parallel value. The capacitor is unchanged by the translation.

EXAMPLE 2-4

Determine the equivalent parallel form for the circuit in Example 2-3 at the resonant frequency.

SOLUTION

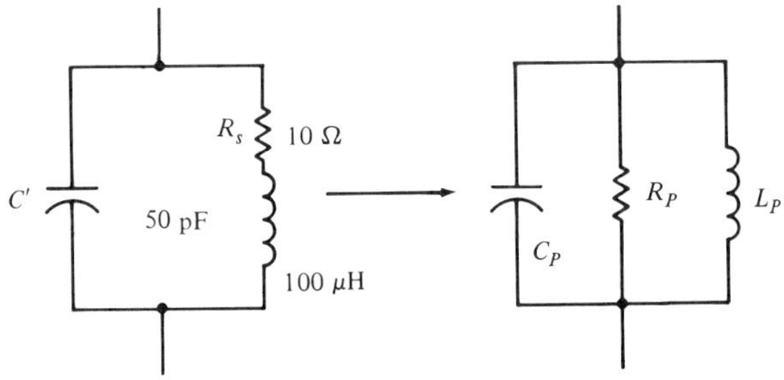

$$C_p = C_p' = 50 \text{ pF}$$

$$R_p = \frac{1}{G_p} = \frac{R_s^2 + (\omega L_s)^2}{R_s} = \frac{10^2 + [(2\pi)(2.248 \times 10^6)(100 \times 10^{-6})]^2}{10}$$

$$= 199{,}312.11\ \Omega$$

$$L_p = \frac{R_s^2 + (\omega L_s)^2}{\omega^2 L_s} = \frac{10^2 + [(2\pi)(2.248 \times 10^6)]^2}{[(2\pi)(2.248 \times 10^6)]^2(100 \times 10^{-6})}$$

$$= 100\ \mu\text{H}$$

A similar procedure may be followed to obtain an equivalent series circuit from a parallel circuit. The equivalent series resistance and inductance are

$$R_s = \frac{G_p}{G_p^2 + (\omega L_p)^2} \tag{2-41}$$

$$L_s = \frac{\omega L_p}{G_p^2 + (\omega L_p)^2} \tag{2-42}$$

Proof is left as an exercise for the student.

2.7 EQUIVALENT *Q* OF A LOADED PARALLEL RESONANT CIRCUIT

It is evident from equation (2-41) that if a load is connected to a parallel resonant circuit, the equivalent series resistance is changed and thus the Q must be altered.

This can be illustrated simply by an example. We see that the bandwidth of a loaded tuned circuit is related to the effective Q_e of the circuit.

EXAMPLE 2-5

A 100-kΩ load is connected across the parallel circuit described in Example 2-3. Determine (a) the Q and bandwidth without the load and (b) the Q and bandwidth with the load.

SOLUTION

(a) For the unloaded case:

$$Q = \frac{\omega_0 L_s}{R_s} = \frac{(2\pi)(2.248 \times 10^6)(100 \times 10^{-6})}{10} = 141.17$$

$$\text{BW} = \frac{f_0}{Q} = \frac{2.248 \times 10^6}{141.17} = 15.92 \text{ kHz}$$

(b) For the loaded case:

$$Z_e = Z_p \| R_L = \frac{(2 \times 10^5)(10^5)}{2 \times 10^5 + 10^5} = 66.67 \text{ k}\Omega$$

$$= \frac{L}{R_e C} \rightarrow R_e = \frac{L}{Z_e C}$$

$$R_e = \frac{100 \times 10^{-6}}{(66.67 \times 10^3)(50 \times 10^{-12})} = 29.999\ \Omega$$

$$Q_e = \frac{\omega_0 L_s}{29.999} = 47.06$$

$$\text{BW} = \frac{f_0}{Q_e} = \frac{2.248 \times 10^6}{47.06} = 47.77 \text{ kHz}$$

2-8 COUPLED CIRCUITS

When energy is transferred between circuits some form of coupling must be employed. For maximum voltage transfer the load must look like a high impedance. Maximum power transfer dictates that the load must be matched to the internal impedance of the source. Magnetically coupled circuits are often used in communication circuits because the foregoing conditions are easily met.

2-9

MAGNETIC COUPLING

Magnetic circuits are defined to be circuits in which one circuit affects another by a mutual magnetic field. A number of factors must be considered when dealing with magnetically coupled circuits. These are:

1. The polarity of the magnetic effect on each circuit must be considered.
2. The degree of mutual coupling must be specified, including a measure of the closeness of coupling between the circuits.
3. The common magnetic field may induce unequal voltages in the circuits because of unequal number of turns of wire in the coupled coils.
4. An equivalent circuit is often used for analysis and design instead of the actual circuit.

2-10

MUTUAL INDUCTANCE

Mutual inductance is defined as the ratio of induced voltage in one circuit to the rate of change of current in another. Specifically, a transformer presents an example of a magnetic circuit with mutual inductance.

Refer to Figure 2-10. The voltages V_p and V_s may be expressed as

$$V_p = L_{11}\frac{di_1}{dt} \pm L_{21}\frac{di_2}{dt} \tag{2-43}$$

$$V_s = \pm L_{21}\frac{di_1}{dt} + L_{22}\frac{di_2}{dt} \tag{2-44}$$

where L_{11} and L_{22} is the self-inductance of the coils and L_{12} and L_{21} is the mutual inductance. In all cases $L_{12} = L_{21} = M$. The convention followed for the $\pm$

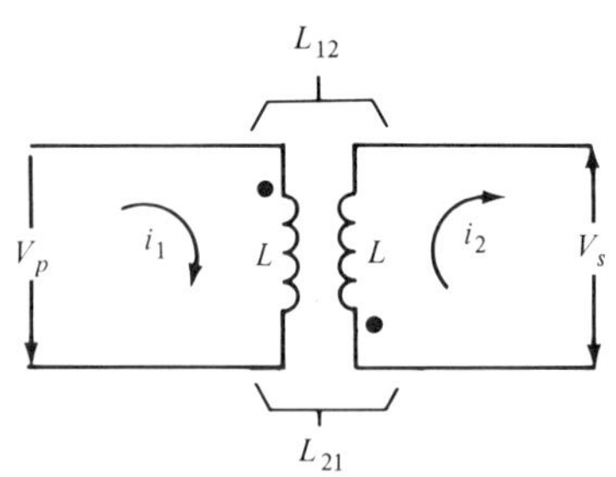

FIGURE 2-10. Transformer coupling.

signs preceding the induced voltages is as follows. If both currents enter the dotted or undotted ends of the coils, the sign is positive. If one current enters a dotted end and the other an undotted end, the sign is negative. For sinusoidal currents, equations (2-43) and (2-44) become

$$V_p = j\omega(L_{11}I_1 \pm MI_2) \tag{2-45}$$

$$V_s = j\omega(\pm MI_1 + L_{22}I_2) \tag{2-46}$$

2-11 COEFFICIENT OF COUPLING

The degree of coupling or interaction between two circuits is measured by a coupling coefficient, k. It is defined as the ratio of the mutual impedance, Z_m, to the square root of the total impedance of the two circuits. This definition may be expressed as

$$k = \frac{Z_m}{\sqrt{Z_1 Z_2}} \tag{2-47}$$

For the circuit in Figure 2-10, this expression becomes

$$k = \frac{M}{\sqrt{L_{11}L_{22}}} \tag{2-48}$$

If the circuits are separated or are oriented in such a manner that there is little mutual magnetic flux, they are said to be "loosely" coupled and k is a small number. For example, air-core transformers may have a coefficient of coupling, k, on the order of 0.01 to 0.1. For transformers with iron cores, k may have values on the order of 0.5 to 0.6. The maximum value of k is 1.0 when there is 100% coupling. Power transformers with steel cores have nearly 100% coupling.

2-12 IDEAL LOW-FREQUENCY TRANSFORMER

In the ideal low-frequency transformer all of the flux established by the primary ampere-turns must link the secondary. According to Kirchhoff's voltage law, a voltage must appear across the primary that is exactly equal to the applied voltage. Furthermore, by Faraday's law the induced voltage is proportional to the rate of change of flux. Expressed mathematically, this is

$$V_p = N_p \frac{d\phi}{dt} \tag{2-49}$$

Since all of the flux links the secondary, a voltage will be induced in the secondary winding,

$$E_s = N_s \frac{d\phi}{dt} \tag{2-50}$$

Dividing equation (2-49) by (2-50) gives

$$\frac{E_p}{E_s} = \frac{N_p}{N_s} = n \tag{2-51}$$

where $n = N_p/N_s$ is the turns ratio. This ratio, n, is referred to as the *transformation ratio*.

Refer to Figure 2-10. If a load impedance is now connected to the secondary, the voltage induced in the secondary will cause current to flow in the secondary circuit. This secondary current will establish a magnetomotive force equal to N_sI_s ampere-turns, which, by Lenz's law, opposes the flux in the core of the transformer. However, since the transformer is lossless, this flux must not change since it is responsible for inducing a voltage equal to the applied voltage in the primary [equation (2-49)]. Thus, when the secondary is loaded, additional primary current must flow so that sufficient primary ampere-turns are available to exactly cancel the secondary ampere-turns. The primary current is shown in Figure 2-11. In this figure I_m is the small magnetizing current necessary to establish the flux when the secondary is unloaded. Since the circuit appears inductive, this current lags the applied voltage by 90°. The additional primary current, I'_p, is in phase with the applied voltage. This would indicate a resistive component in the primary impedance when the secondary is loaded with a resistive load. It was indicated previously that the additional primary ampere-turns must exactly equal the secondary ampere-turns. This relationship may be expressed as

$$N_pI'_p = N_sI_s$$

or (2-52)

$$\frac{I'_p}{I_s} = \frac{N_s}{N_p} = \frac{1}{n}$$

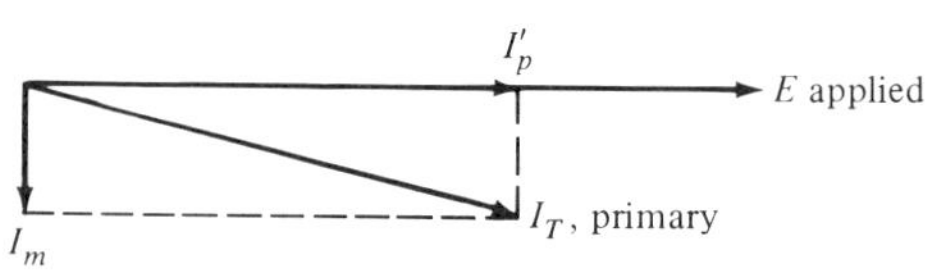

FIGURE 2-11. Primary current with a resistive secondary load.

For conservation of power, the transformation ratio is the inverse of that for voltage.

2-13

IMPEDANCE TRANSFORMATION

In the preceding section, the mechanism in which secondary load current increases primary current was discussed. For a constant applied voltage a change in primary current indicates a change in secondary impedance. Refer to Figure 2-12.

Based on the preceding discussion the effective primary impedance associated with I'_p may be expressed as

$$Z_p = \frac{E_p}{I'_p} \tag{2-53}$$

Substituting for E_p and I'_p from equations (2-51) and (2-52), this expression becomes

$$Z_p = \frac{E_s}{I_s}\left(\frac{N_p}{N_s}\right)^2 = n^2 Z_s \tag{2-54}$$

where Z_s is the impedance of the secondary plus the load. For a good transformer the impedance of the secondary will be small relative to the load and the primary impedance is generally expressed as

$$Z_p = n^2 R_L \tag{2-55}$$

where Z_p is the reflected impedance. That is, the reflected impedance is the square of the turns ratio times the load impedance. It is evident that transformer coupling provides an excellent vehicle for impedance matching.

EXAMPLE 2-6

An audio transformer is listed as 10 kΩ primary reflected impedance with a 10-Ω load. What load must be connected to the transformer to provide an impedance match for an amplifier with a 15-kΩ output impedance?

SOLUTION

$$Z_p = n^2 R_L \rightarrow n^2 = \frac{Z_p}{R_L} = \frac{10{,}000}{10} = 1000 : 1$$

$$R_{L2} = \frac{Z_p}{n^2} = \frac{15{,}000}{1000} = 15\ \Omega$$

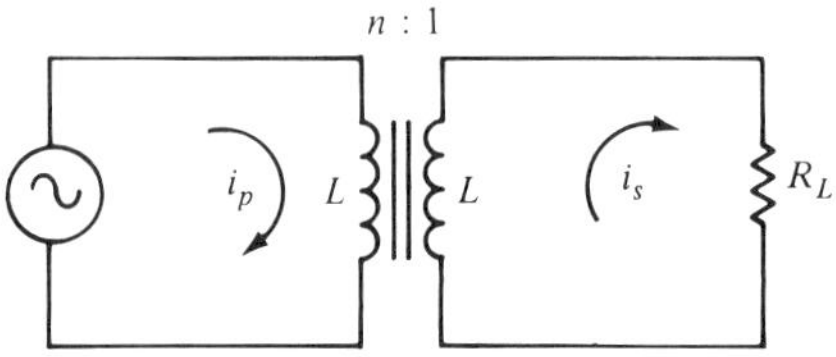

FIGURE 2-12. Transformer with resistive secondary load.

2-14

HIGH-FREQUENCY TRANSFORMER

It was stated previously that an equivalent circuit was often used when working with coupled circuits. This technique provides an aid in analyzing the circuit and also in visualizing circuit operation. We have seen that transformer coupling can be used to provide an impedance match. Also, from the preceding material on tuned circuits we can see that if either the primary or secondary of the transformer is tuned, the bandwidth may be controlled and voltage amplification obtained as well. If both coils are tuned, the tuning process becomes more complicated and the transformer is usually used for a fixed frequency band only.

With all of the advantages listed above it is obviously to our advantage to understand tuned RF transformers thoroughly. The circuit of Figure 2-10 is repeated in Figure 2-13 with the resistance of the windings included. The loop equations for this circuit are

$$V_p = (R_1 + j\omega L_{11})I_1 - j\omega MI_2 \tag{2-56}$$

$$V_s = -j\omega MI_1 + (R_2 + j\omega L_{22})I_2 \tag{2-57}$$

$$V_p = (R_1 + Z_p)I_1 - Z_m I_2 \tag{2-58}$$

$$V_s = -Z_m I_1 + Z_s I_2 \tag{2-59}$$

Equations (2-58) and (2-59) may also be obtained identically from the equivalent circuit in Figure 2-14, in which

$$Z_p = R_1 + j\omega L_1 \tag{2-60}$$

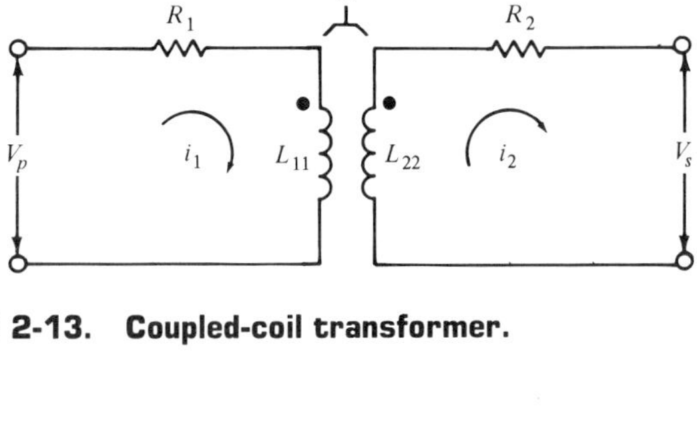

FIGURE 2-13. Coupled-coil transformer.

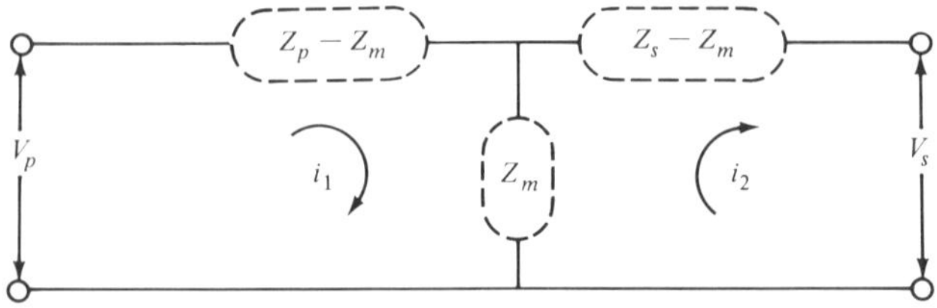

FIGURE 2-14. Coupled-coil transformer equivalent circuit.

$$Z_s = R_2 + j\omega L_2 \tag{2-61}$$

$$Z_m = j\omega M \tag{2-62}$$

2-15

TUNED PRIMARY RF TRANSFORMER

A tuned primary RF transformer is shown in Figure 2-15. To simplify the analysis, stray capacitance associated with both the primary and secondary windings is neglected. At very high frequencies this would cause serious error. The transformer polarity is assumed to be as shown by the dots, and the loop equations are

$$V_p = I_1 Z_p - I_2 Z_m \tag{2-63a}$$

$$0 = I_1 Z_m + I_2(Z_s + Z_L) \tag{2-63b}$$

Solving for the primary current yields

$$I_1 = \frac{V_p(Z_p + Z_L)}{Z_p(Z_s + Z_L) - Z_m^2} \tag{2-64}$$

The ratio V_p/I_1 is called the *driving-point impedance* and may be expressed as

$$\begin{aligned} Z_p' &= Z_p - \frac{Z_m^2}{Z_s + Z_L} \\ &= R_p + j\omega L_p + \frac{\omega^2 M^2}{Z_s + Z_L} \end{aligned} \tag{2-65}$$

Examining equation (2-65) it is evident that there is an impedance, Z_r, reflected into the primary from the secondary given by

$$Z_r = \frac{\omega^2 M^2}{Z_s + Z_L} \tag{2-66}$$

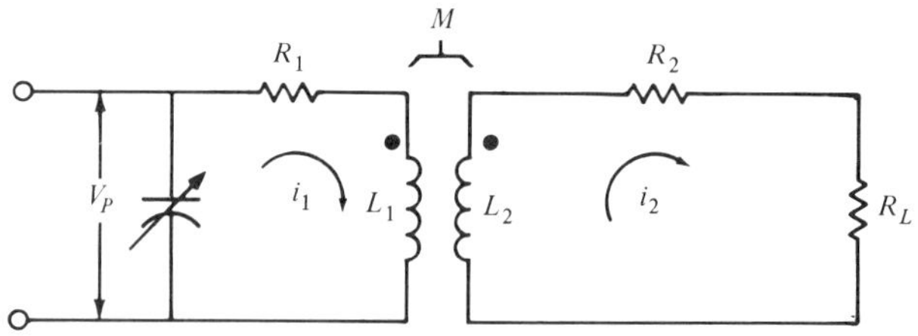

FIGURE 2-15. Tuned primary RF transformer.

if $|Z_L| >> Z_s$, this expression reduces to

$$Z_r = \frac{\omega^2 M^2}{Z_L} \tag{2-67}$$

Referring to equation (2-66), if $Z_s + Z_L$ is purely real, the reflected component is purely resistive. If $Z_s + Z_L$ has a capacitive (inductive) component, this reflects into the primary as an inductive (capacitive) component. A tuned circuit, therefore, can be detuned by the reactive impedance reflected into it.

EXAMPLE 2-7

For the circuit in Figure 2.15, find (a) the reflected impedance at the resonant frequency, (b) the input impedance looking into the primary, and, (c) the bandwidth of the circuit. $R_1 = R_2 = 5\Omega$, $L_1 = L_2 = 100\ \mu\text{H}$, $C = 200\ pF$, $k = 0.1$, $R_L = 10\ K\Omega$.

SOLUTION

$$f_0 = \frac{1}{2\pi\sqrt{LC}} = \frac{1}{2\pi\sqrt{(100 \times 10^{-6})(200 \times 10^{-12})}}$$

$$= 1.125 \text{ MHz}$$

$$Z_r = \frac{\omega_0^2 M^2}{Z_L} = \frac{[(2\pi)(1.125 \times 10^6)]^2[(0.1)\sqrt{(100 \times 10^{-6})^2}]}{10{,}000}$$

$$= 0.005\ \Omega$$

$$Z_p = \frac{L}{RC} = \frac{100 \times 10^{-6}}{(5 + 0.005)(200 \times 10^{-12})} = 99.9\ K\Omega$$

$$Q_e = \frac{\omega L_1}{R_e} = \frac{(2\pi)(1.125 \times 10^6)(100 \times 10^{-6})}{5 + 0.005} = 141.16$$

$$\text{BW} = \frac{f_0}{Q_e} = \frac{1.125 \times 10^6}{141.16} = 7.97 \text{ kHz}$$

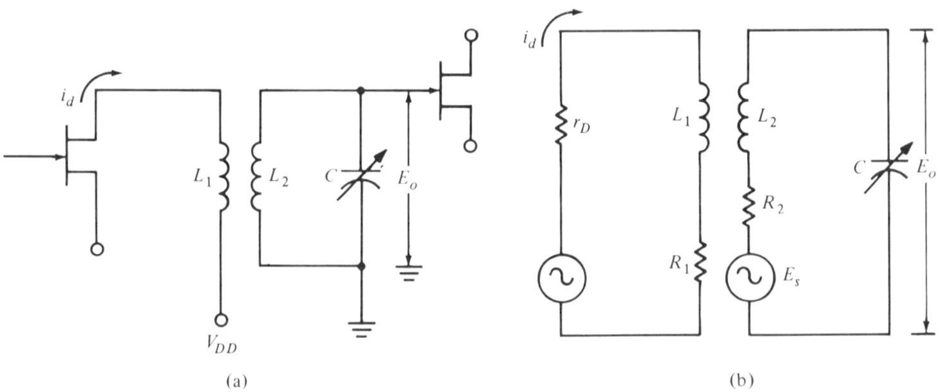

FIGURE 2-16. (a) Tuned secondary RF amplifier; (b) equivalent circuit.

2-16

TUNED SECONDARY RF AMPLIFIER

The circuit for a tuned secondary amplifier is shown in both schematic and equivalent circuit forms in Figure 2-16. Because of the high drain resistance the effect of reflected impedance into the primary from the secondary has a negligible effect on the drain current, i_p. Also, for small changes in frequency on either side of the resonance, the primary impedance, Z_p', is relatively constant. Thus i_p is relatively constant with frequency. The induced voltage ($E_s = \omega M i_p$) is then relatively constant. As far as the induced voltage E_s is concerned, it sees a series circuit consisting of R_2, L_2, and C. The input impedance of the second field-effect transistor (FET) is large and there is little loading effect on the secondary. Thus the output voltage, E_0, at resonance is $E_0 = Q_s E_s$. Amplification due to the Q of the primary is negligible. The primary Q_p is approximately

$$Q_p \approx \frac{\omega_0 L_1}{r_d} \tag{2-68}$$

The large value of the drain resistance makes this a very small number.

EXAMPLE 2-8

For the circuit in Figure 2-16, $L_1 = 50\ \mu\text{H}$, $L_2 = 200\ \mu\text{H}$, $r_d = 50\ \text{k}\Omega$, $k = 0.6$, and $Q_s = 150$. Calculate Q_p and E_0 if the secondary is tuned to 796 kHz and $i_d = 2\ \mu\text{A}$.

SOLUTION

$$Q_p \approx \frac{\omega_0 L_1}{r_d} = \frac{(2\pi)(7.96 \times 10^5)(50 \times 10^{-6})}{5 \times 10^5} = 0.0005$$

$$M = k\sqrt{L_1 L_2} = (0.6)(50 \times 10^{-6})(200 \times 10^{-6}) = 60 \ \mu\text{H}$$

$$R_2 = \frac{\omega_0 L_2}{Q_s} = \frac{(2\pi)(7.98 \times 10^5)(200 \times 10^{-6})}{150} = 6.68 \ \Omega$$

$$R_2' = R_2 + \frac{\omega_0^2 M^2}{r_d} = 6.62 + \frac{[(2\pi)(7.98 \times 10^5)]^2(60 \times 10^{-6})^2}{5 \times 10^5}$$

$$= 6.86 \ \Omega$$

$$Q_e = \frac{\omega_0 L_2}{R_2'} = \frac{(2\pi)(7.98 \times 10^5)(200 \times 10^{-6})}{6.86} = 146.18$$

$$E_o = Q_e E_s$$

$$= (146.18)(2\pi)(7.98 \times 10^5)(60 \times 10^{-6})(2 \times 10^{-6})$$

$$= 87.9 \text{ mV}$$

In this example the bandwidth is BW = 796 kHz/146.18 = 5.459 kHz. This bandwidth would probably be too narrow and a secondary coil with a lower Q would need to be chosen. The penalty is lower gain for the circuit.

2-17

DOUBLE-TUNED TRANSFORMER-COUPLED RF AMPLIFIERS

The circuit for a double-tuned RF amplifier is shown in Figure 2-17. This circuit may have a response quite different from the two examined previously. For example, suppose that the primary and secondary circuits are initially separated physically and that they are each tuned to resonate at the same frequency, f_0. The coils are then moved closer together so that a small degree of coupling exists. With a primary current, i_t, flowing, an output voltage, E_o, will appear. If the frequency of the signal is varied through f_0, the primary current, the secondary current, and voltages E_s and E_o all pass through a peak at f_0. Next, move the coils much closer together. Impedance from the secondary reflects into the primary. At a frequency above resonance the secondary is inductive

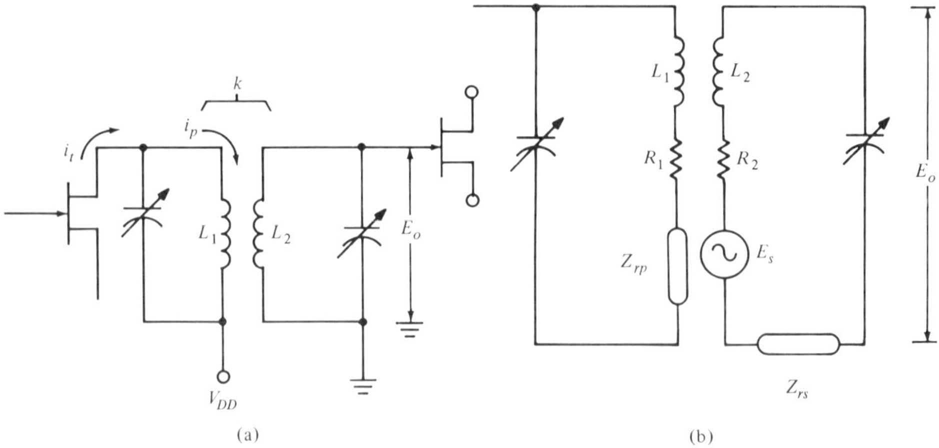

FIGURE 2-17. (a) Double-tuned RF amplifier; (b) equivalent circuit.

and the impedance coupled into the primary is capacitive. This effectively alters the primary impedance to make it less inductive. Thus the resonant frequency of the primary is raised, which changes the frequency for maximum response. The same explanation can be applied for a frequency below f_0. Thus the frequency response curve may contain two peaks with a dip between them.

Figure 2-18 shows this response for different degrees of coupling. Amplifiers of the double-tuned type are generally used in the intermediate-frequency (IF) stages of superheterodyne receivers, where a response curve with steep sides and suitable width and height may be obtained by careful adjustment.

2-17.1 Critical Coupling

If the coupling is increased, ωM increases and the reflected impedance increases. At resonance the reflected impedance is purely real and causes the Q of the

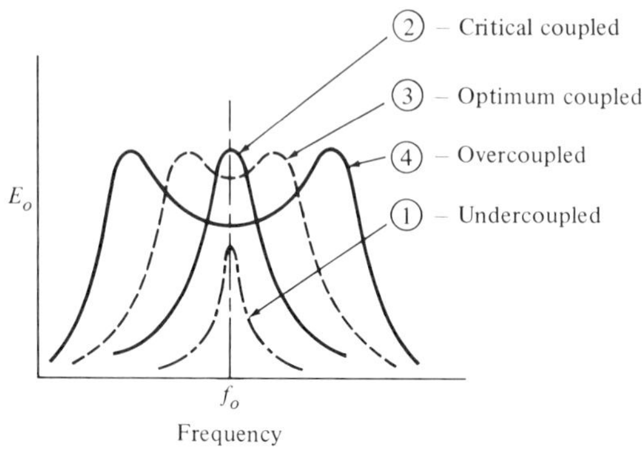

FIGURE 2-18 Effect of coupling on the response curve of a double-tuned inductive coupled RF transformer.

primary to decrease. If the coupling continues to increase, a point is reached where the resistance, R_r, reflected into the primary is equal to the primary resistance, R_1. The effective Q of the primary is decreased to one-half its original value. This point is referred to as *critical coupling*. The decrease in i_p is exactly offset by the increase in ωM and the induced voltage E_s reaches its maximum possible value. Thus $E_o = QE$ is maximum. As a result of the decrease in Q the bandwidth is increased.

2-17.2 Overcoupling

If we again increase the coupling, this time to a value greater than critical coupling, k_c, this results in overcoupling. The best response curve is obtained with a coupling coefficient 50% greater than critical, or

$$k_{\text{opt}} = 1.5k_c \tag{2-69}$$

The effect of continuing to increase the coupling is reflected in Figure 2-18. The coefficient of coupling and the bandwidth of a double-tuned circuit are given by

$$\text{BW} = kf_0 \tag{2-70}$$

2-17.3 Effect of *Q* on Coupling

At critical coupling, the reflected resistance is equal to the primary resistance. Expressed mathematically,

$$R_1 = R_r = \frac{(\omega M)^2}{R_2}$$

or (2-71)

$$\omega M = \sqrt{R_1 R_2}$$

At resonance, considering each circuit by itself (no reflected resistance), we have

$$R_1 = \frac{\omega L_1}{Q_p} \qquad \text{and} \qquad R_2 = \frac{\omega L_2}{Q_s}$$

After substituting these relations into equation (2-71) and dividing both sides of the equation by ω, we have

$$M = \sqrt{\frac{L_1}{Q_p} \times \frac{L_2}{Q_s}} \tag{2-72}$$

However, by definition,

$$M = k\sqrt{L_1 L_2} \tag{2-73}$$

Equating equations (2-72) and (2-73) and dividing both sides by $\sqrt{L_1 L_2}$ results in

$$k_c = \frac{1}{\sqrt{Q_p Q_s}} \tag{2-74}$$

This relation provides a means of determining the critical coefficient of coupling, k_c, in terms of the uncoupled Q's of the primary and secondary circuits.

EXAMPLE 2-9

For the circuit of Figure 2-17, the following values apply: $L_1 = L_2 = 200\ \mu\text{H}$; $R_1 = R_2 = 10\ \Omega$; $C_1 = C_2 = 250$ pF; $k = 0.02$, and $I_t = 2\ \mu\text{A}$. Find the voltage at the gate of the second transistor.

SOLUTION

The current in the primary winding will be $Q_p \times I_t$:

$$Q_p = \frac{\omega L_1}{R_1 + R_r}$$

$$\omega = 2\pi \frac{1}{2\pi\sqrt{L_1 C_1}} = \frac{1}{\sqrt{L_1 C_1}}$$

$$= \frac{1}{\sqrt{(200 \times 10^{-6})(250 \times 10^{-12})}} = 4.472 \times 10^6$$

$$M = k\sqrt{L_1 L_2} = 0.02\sqrt{(200 \times 10^{-6})^2} = 4\ \mu\text{H}$$

$$R_r = \frac{(\omega M)^2}{R_s} = \frac{[(14.472 \times 10^6)(4 \times 10^{-6})]^2}{10} = 32\ \Omega$$

$$Q_p = \frac{(4.472 \times 10^6)(200 \times 10^{-6})}{10 + 32} = 27.95$$

$$I_{L1} = QI_t = 27.95 \times 2 \times 10^{-6} = 55.9\ \mu\text{A}$$

The induced voltage in the tank circuit L_2C_2 is

$$E_s = \omega M I_{L1} = (4.472 \times 10^6)(4 \times 10^{-6})(55.9 \times 10^{-6})$$

$$= 1\ \text{mV}$$

The voltage at the gate of the second transistor is

$$E_{C2} = Q_s E_s$$

$$Q_s = \frac{\omega L_2}{R_s + R_r}$$

$$R_r = \frac{(\omega M)^2}{R_p} = \frac{[(4.472 \times 10^6)(4 \times 10^{-6})]^2}{10} = 32\ \Omega$$

$$Q_s = \frac{(4.472 \times 10^6)(200 \times 10^{-6})}{10 + 32} = 27.95$$

$$E_{C2} = (27.95)(1 \times 10^{-3}) = 27.95\ \text{mV}$$

2-18

OTHER IMPEDANCE-MATCHING NETWORKS

We have seen how transformers are used to match load impedance to a given source impedance. When maximum power transfer is not required, inductive or capacitive tapping may be employed quite effectively. Typical circuits are shown in Figure 2-19. Essentially, both the inductive tap (Figure 2-19a); and the capacitive tape (Figure 2-19b) of these circuits act like voltage dividers. That is,

$$V_{out} = \frac{X_2}{X_1 + X_2} V_{in} \tag{2-75}$$

where X represents the reactance of the respective elements.

The circuits shown in Figure 2-19 are generally used to couple a load impedance to a tuned circuit while seeking to minimize the effect of the load R_L on the Q of the circuit. This requires that $X_2 << R_L$. The equivalent circuits, referred to the input, for the circuits in Figure 2-19 are given in Figure 2-20.

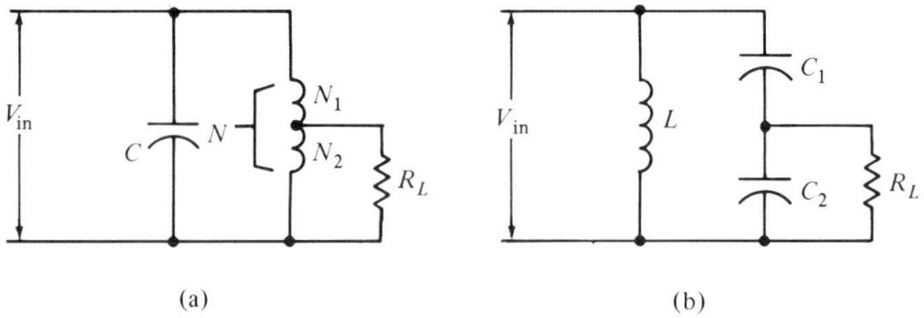

FIGURE 2-19. (a) Inductive tap; (b) capacitive tap.

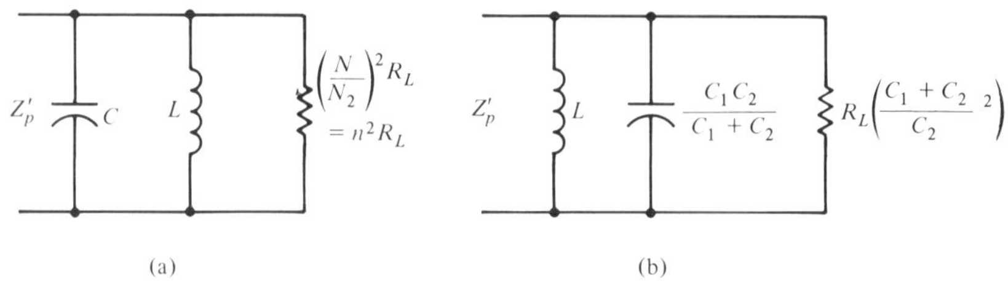

FIGURE 2-20. Equivalent circuits: (a) inductive tap; (b) capacitive tap.

Referring to these circuits, it is easily seen that the previous analysis may be applied to the tapped circuits.

EXAMPLE 2-10

For the circuit of Figure 2-19a the following values apply: $L = 200\ \mu\text{H}$, $C = 250$ pF, and the unloaded $Q = 100$. A 10-kΩ load is taped into the coil in the ratio 1 : 10. Find the equivalent Q of the circuit and the input impedance, Z'_p.

SOLUTION

The unloaded impedance, Z_p, of the tank circuit is

$$Z_p = \frac{L}{RC}$$

$$\omega_0 = \frac{1}{\sqrt{LC}} = \frac{1}{\sqrt{(200 \times 10^{-6})(250 \times 10^{-12})}} = 4.472 \times 10^6$$

$$R = \frac{\omega L_1}{Q} = \frac{(4.472 \times 10^6)(200 \times 10^{-6})}{100} = 8.94\ \Omega$$

Then

$$Z_p = \frac{200 \times 10^{-6}}{(8.94)(250 \times 10^{-12})} = 89.48\ \text{k}\Omega$$

$$Z'_p = Z_p \| n^2 R_L = 89.48\text{k}\Omega \| (10)^2(10{,}000) = 82.130\ \text{k}\Omega$$

$$R' = \frac{L}{Z'_p C} = \frac{200 \times 10^{-6}}{(82{,}130)(250 \times 10^{-12})} = 9.7\ \Omega$$

$$Q_e = \frac{\omega L}{R'} = \frac{(4.472 \times 10^6)(200 \times 10^{-6})}{9.7} = 92.2$$

In this circuit the effective input impedance with the load at resonance is 82.130 kΩ and the Q is reduced to 92.2. When it is desired to reduce the effect of the source impedance on the load, the tap is made on the input side. A circuit for accomplishing this is shown in Figure 2-21.

2-18.1 L Network

The L network shown in Figure 2-22 is the simplest of matching networks. The input impedance, Z_p, may be expressed as

$$\begin{aligned} Z_p &= j\omega L + \frac{(1/j\omega C)R_L}{R_L + 1/j\omega C} \\ &= \frac{R_L}{1 + (\omega R_L C)^2} + j\omega L - \frac{j\omega C R_L^2}{1 + (\omega R_L C)^2} \end{aligned} \tag{2-76}$$

At resonance the reactive part of the impedance must equate to zero. Thus

$$\omega L - \frac{\omega C R_L^2}{1 + (\omega R_L C)^2} = 0 \tag{2-77}$$

Solving for the resonant frequency, ω_0,

$$\omega_0 = \sqrt{\frac{1}{LC} - \frac{1}{R^2 C^2}} \tag{2-78}$$

The real part of the input impedance is

$$R = \frac{R_L}{1 + (\omega R_L C)^2} \tag{2-79}$$

Using results from equation (2-77), the impedance at resonance may be expressed as

$$R_p = \frac{L}{RC} \tag{2-80}$$

We remember these results from tuned circuits.

Next, consider the effect of a source impedance as shown in Figure 2-23. Considering the network in Figure 2-23b, it is evident that when $R_p = R_L$, the equivalent Q of the network will be exactly halved. The form of the equivalent circuit makes it clear that the analytical material on tuned circuits applies to this circuit as well.

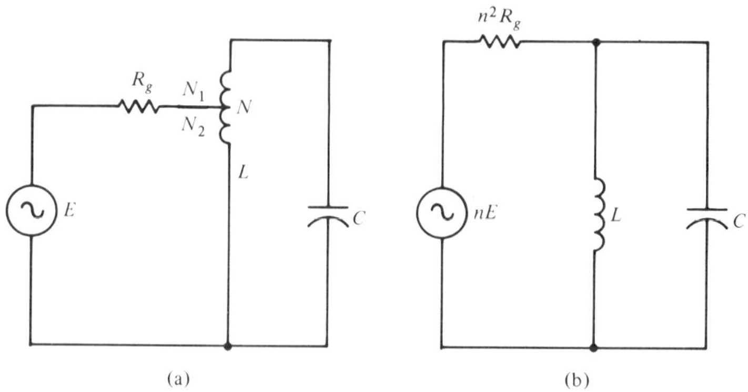

FIGURE 2-21. (a) Tapped on primary; (b) equivalent circuit.

EXAMPLE 2-11

Design an L network to match a 1000-Ω load to a 50-Ω source at 3 MHz.

SOLUTION

$$R_s = R_p = \frac{L}{R_L C} = 50$$

$$\frac{L}{C} = R_L R_s = (1000)(50) = 5 \times 10^4$$

$$\omega_0 = \sqrt{\frac{1}{LC} - \frac{1}{R_L^2 C^2}}$$

$$C = \frac{R_L^2 - 5 \times 10^4}{L R_L^2 \omega_0^2}$$

Using $L = 5 \times 10^4 C$, we have

$$C = \sqrt{\frac{R_L^2 - 5 \times 10^4}{(5 \times 10^4)(R_L^2)(\omega_0)^2}} = \sqrt{\frac{(10^3)^2 - 5 \times 10^4}{(5 \times 10^4)(10^3)^2(2\pi)^2(3 \times 10^6)^2}}$$

$$= 594 \text{ pF}$$

$$L = 5 \times 10^4 C = (5 \times 10^4)(594 \times 10^{-12}) = 29.7\ \mu\text{H}$$

Note from equation (2-80) that both L and C may be made variable to effect an impedance match. The L network may also be configured with L and C interchanged.

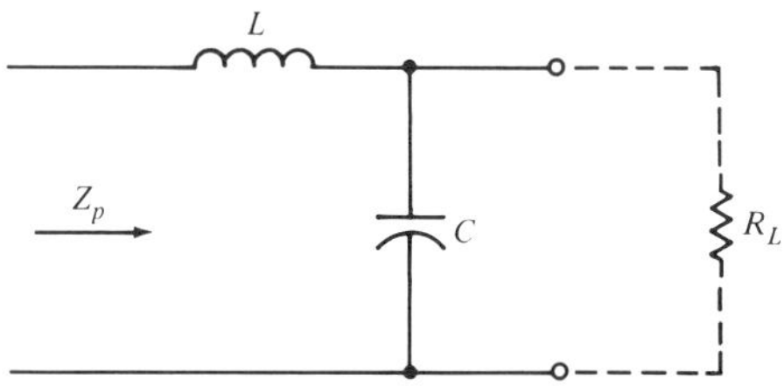

FIGURE 2-22. L network.

2-18.2 Π Network

This circuit is shown in Figure 2-24. There are many impedance-matching networks, with the Π network being one of the more popular ones.

For a given value of input and output resistance there are many networks that will give a particular match. We have seen in Section 2-18.1 that only two reactive elements are needed to effect an impedance match. Adding a third element introduces a redundancy. Thus one element can be assigned an arbitrary value and the other two elements found.

For example, suppose that C_2 is set to a particular value. The parallel combination of C_2 and R_L can be transformed to an equivalent series circuit. Then L can be found by breaking it down into two components, L_1 and L_2. This is indicated in Figure 2-25. The component L_2 tunes out the reactance of the equivalent capacitor, C_e. That is,

$$X_{L2} = X_{Ce} = \frac{X_{C2}}{1 + (X_{C2}/R_L)^2} \tag{2-81}$$

The circuit now reduces to that shown in Figure 2-26. This network is recognized as an equivalent L network. Generally speaking, the inductance, L, is fixed and

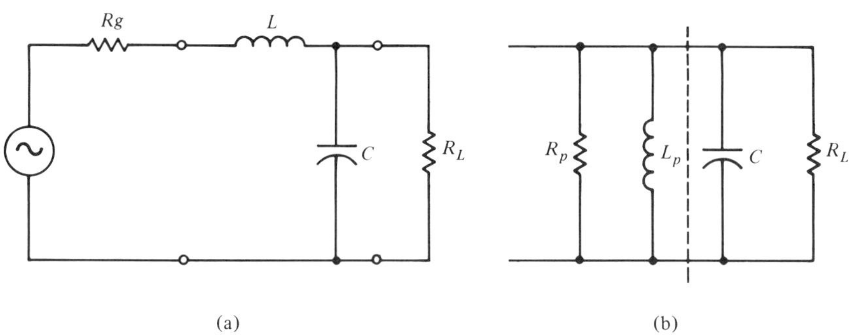

FIGURE 2-23. (a) L network with source and load impedance; (b) Equivalent network for L and R_g.

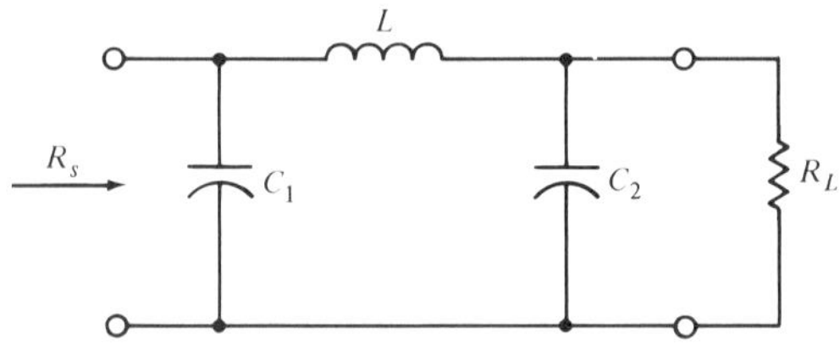

FIGURE 2-24. π **network.**

C_2 provides the loading on the source. The capacitance C_1 is then used for tuning.

2-19

ATTENUATORS

It is often desirable to introduce a fixed amount of attenuation between the source and the load. A popular T pad which is shown in Figure 2-27 is used to match a 50-Ω source and load. The attenuation or insertion loss is defined as the ratio of load current with the pad in place to the load current with the pad removed. Thus the load current without pad is

$$I = \frac{E_s}{R_s + R_L} \tag{2-82}$$

With the pad inserted, loop equations or any other suitable circuit analysis technique may be used to find the load current. The loop equations for this circuit are

$$E = (R_s + R_1 + R_3)I_1 - R_3I_2 \tag{2-83a}$$

$$0 = -R_3I_1 + (R_2 + R_3 + R_L)I_2 \tag{2-83b}$$

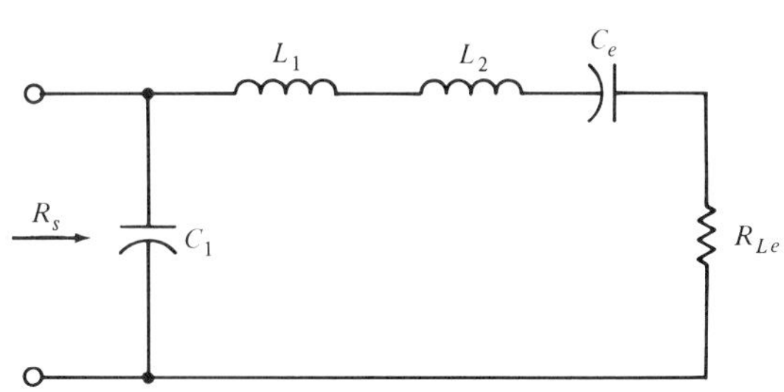

FIGURE 2-25. Equivalent π **network.**

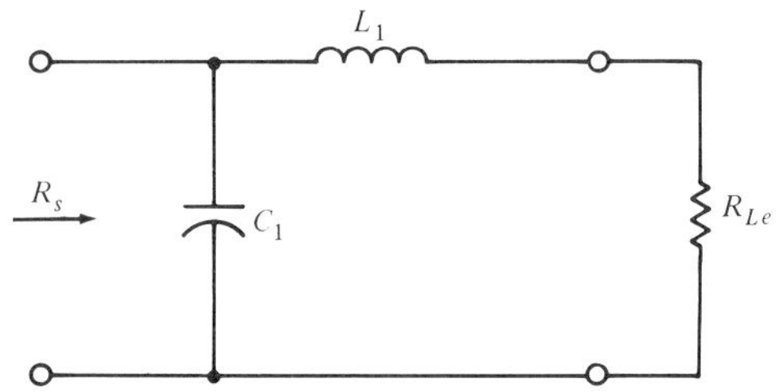

FIGURE 2-26. Equivalent L network.

The load current I_2 and is equal to

$$I_2 = \frac{ER_3}{(R_s + R_1 + R_3)(R_2 + R_3 + R_L) - R_3^2} \tag{2-84}$$

and the attenuation I_2/I is

$$\frac{I_2}{I} = \frac{R_3(R_s + R_L)}{(R_s + R_1 + R_3)(R_2 + R_3 + R_L) - R_3^2}$$

$$= \frac{25(50 + 50)}{(50 + 31 + 25)(31 + 25 + 50) - (25)^2} = 0.236$$

Expressed in decibels, the attenuation is

$$20 \log_{10} \frac{I_2}{I_1} = 20 \log_{10} 0.236 = -12.56 \text{ dB}$$

An L pad is often used to match a source to a load. Two configurations, shown in Figure 2-28, are applicable depending on the relative magnitudes of the source and load impedances. Analysis of the circuit proceeds by writing expressions for the input and output impedance with the source and load connected. For the circuit in Figure 2-28a,

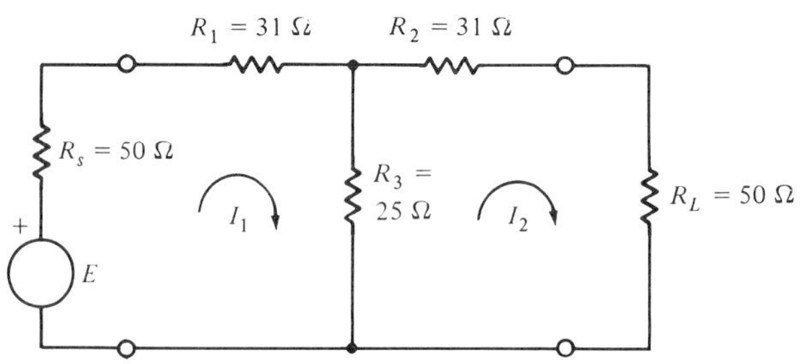

FIGURE 2-27. T-pad attentuator for 50 Ω match.

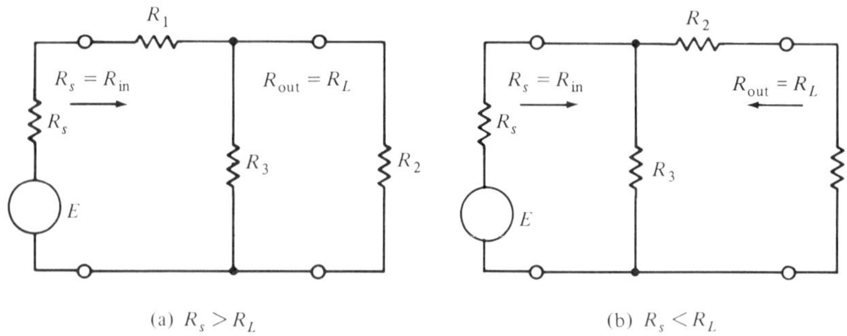

FIGURE 2-28. L pad attenuator.

$$R_{in} = R_s = R_1 + \frac{R_3 R_L}{R_3 + R_L}$$
$$R_o = R_L = \frac{R_3(R_1 + R_s)}{R_1 + R_3 + R_s} \qquad (2\text{-}85)$$

Solving for R_3 in equation (2-85), we have

$$R_3 = \frac{R_L(R_s - R_1)}{R_L - R_s + R_1}$$
$$R_3 = \frac{R_L(R_s + R_1)}{R_1 + R_s - R_L} \qquad (2\text{-}86)$$

After equating the two relations in (2-86), R_1 may be expressed as

$$R_1 = \sqrt{R_s(R_s - R_L)} \qquad (2\text{-}87)$$

A value for R_3 may now be found from equation (2-86). Performing a similar analysis for the circuit in Figure 2-28b, an expression for R_2 is

$$R_2 = \sqrt{R_L(R_L - R_s)} \qquad (2\text{-}88)$$

EXAMPLE 2-12

Design an L network to match a 100-Ω source to a 72-Ω load.

SOLUTION

$$R_1 = \sqrt{R_s(R_s - R_L)} = \sqrt{100(100 - 72)} = 52.91\ \Omega$$

$$R_3 = \frac{R_L(R_s + R_1)}{R_1 + R_s - R_L} = \frac{72(100 + 52.91)}{52.9 + 100 - 72} = 136.07\ \Omega$$

We see from this example that L pads are easy to design and convenient to use. Two precautions are in order when using resistive attenuators:

1. The load and source impedances must be real if an impedance match is to be realized.
2. The network will be flat with frequency only until a high enough frequency is reached where the resistor no longer looks like a pure resistance.

PROBLEMS

2-1. A series circuit consists of $L = 100\ \mu H$, $C = 400$ pF, and $R = 20\ \Omega$. Find the resonant frequency.

2-2. A series circuit has the following circuit constants: $C = 300$ pF, $L = 220\ \mu H$, and $R = 20\ \Omega$. The supply voltage is 60 V. Find (a) the resonant frequency, (b) the impedance at resonance, (c) the line current at resonance, (d) the circuit Q, and (e) the voltage across each component.

2-3. A series circuit, as in Figure 2-1, contains a 1-μF capacitor. It is resonant at 1000 Hz. Plot curves as in Figure 2-2, the value of R being (a) 75 Ω and (b) 15 Ω.

2-4. Plot curves as in Figures 2-4 and 2-5 for the admittance (magnitude and angle) of the circuit described in Problem 2-1. Plot for R equal to (a) 75 Ω, (b) 15 Ω, and (c) zero.

2-5. In a series circuit, L = 65 μH, $C = 1.56$ nanofarads (nF; 10^{-9} farad), $R = 5.10\ \Omega$. Compute (a) the resonant frequency, (b) the quality factor Q, and (c) the impedance of the circuit at a frequency of 1% above resonance.

2-6. It is required that a series circuit of L, C, and R be resonant at 1 MHz per second(s). Its bandwidth (between half-power points) is to be 5000 Hz and its input impedance at resonance is to be 50 Ω. Find L, C, and R (at 1 MHz/s).

2-7. A series circuit contains $L = 159\ \mu H$, $C = 159$ pF, and $R = 20\ \Omega$. The supply voltage is 10 V. (a) Calculate the circuit impedance Z at resonance and for frequencies at 25, 50, 100, and 200 kHz above and below resonance. Indicate whether these impedances are resistive, inductive, or capacitive. (b) Plot a curve of Z versus f. (c) Calculate the line current for the frequencies above. Plot a curve of I versus f on a second sheet. (d) Calculate the voltage across the capacitor E_c for the same frequencies. Plot this curve on a third sheet.

2-8. A capacitor of 100 pF is connected in parallel with a coil of 253 μH and 8.2-Ω resistance. Find the resonant frequency.

2-9. Applied voltage is held constant at the terminals of a resonant circuit. Consider (a) the series resonant circuit, (b) the three-branch parallel circuit, and (c) the two-branch parallel circuit (Figures 2-1, 2-6, and 2-7, respectively). P_0 is power received by the circuit at resonance; P_1 is power received when detuned to make $Q\delta = \frac{1}{2}$. Find P_1/P_0 for each. $\delta = (\omega - \omega_0)/\omega_0$.

2-10. In the circuit shown, at 100 Hz, $Z_L = j100$, $Z_C = -j100$, and $R = 200\ \Omega$. (a) Find the input admittance at 1000 Hz and draw voltage and current transforms. (b) Find the resonant circuit transforms. (c) Find the resonant frequency (how do you define resonance?) and the input admittance at resonant frequency.

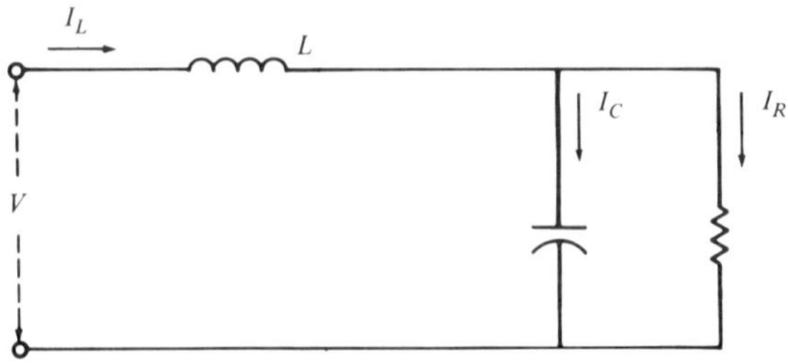

PROBLEM 2-10

2-11. A parallel resonant circuit has a series resistance of 10 Ω. At resonance, the inductive reactance is 1000 Ω and the line current is 0.5 mA. What is the value of the circulating current?

2-12. The tank circuit of Problem 2-8 ($L = 253$ μH, $C = 100$ pF, $R_s = 82\ \Omega$) is shunted with a load resistance R_L of 0.5 MΩ. Find the new (loaded) value of (a) circuit impedance, (b) circuit Q, and (c) bandwidth. (d) What resistance in series with the coil would have the same effect as this shunt resistance?

2-13. An inductive-coupled circuit using an untuned primary and tuned secondary has the following values: $L_1 = 250$ μH; $L_2 = 200$ μH; effective resistances of the windings = 10 Ω and 6 Ω, respectively; secondary capacitor = 200 pF; coefficient of coupling = 0.042. Calculate the value of the reflected impedance and the total primary impedance at (a) the resonant frequency of the secondary, (b) 770 kHz, and (c) 880 kHz.

2-14. A FET ($r_d = 50$ kΩ) is coupled to a second FET via an untuned-primary, tuned-secondary transformer. The alternating-current (ac) component of drain current in the first FET is 20 μA. The coupling values are $L_1 = 10$ mH, $L_2 = 150$ μH and 9.42 Ω, $k = 0.02$, and $C_2 = 169$ pF. Find (a) the voltage at the gate of the second FET at the resonant frequency and (b) the bandwidth of the resonance curve. (c) Repeat part (a) for 20 kHz below resonance.

2-15. A double-tuned circuit has a primary Q of 83 and a secondary Q of 120. Both windings are tuned to 1700 kHz, and their coupling coefficient is 0.013. Is the circuit overcoupled, critical, or undercoupled?

2-16. An RF double-tuned transformer is designed for use at an intermediate frequency of 450 kHz. Primary and secondary are alike: $L = 2$ mH and $Q = 100$. (a) What value of coupling should be used for a 18-kHz bandwidth? (b) What should the Q of the coils be for best response? (c) What should the effective resistance of the coils be for this value of Q? (d) What are the resistances of the given coils? (e) What value of series resistance should be added? (f) What value of shunt resistance would produce the same effect?

2-17. An impedance-coupled amplifier has the following characteristics: a metal-oxide semiconductor field-effect transistor (MOSFET) with $g_{os} = 1/r_d = 6.67$ microsiemens (μS) and $g_{fs} = 10{,}000$ S; a tank circuit with $L = 30$ μH and 12.5 Ω, and $C = 20$ pF; gate resistor = 220 kΩ. Find (a) the effective load impedance, (b) the Q of the coupling network, (c) the resonant frequency and bandwidth of this circuit, (d) the gain of the stage, and (e) the output voltage for an input signal of 5 mV.

2-18. An RF amplifier using MOSFETs ($g_{fs} = 6100$) is coupled to the next stage by a double-tuned transformer whose constants are $L_1 = 159$ μH and 10 Ω, $C_1 = 159$ pF, L_2 and C_2 values the same as for the primary, and coupling coefficient = 0.01. The signal voltage applied to the first gate is 0.82 mV. Find (a) the drain current (ac component), (b) the primary coil current, (c) the induced secondary voltage, (d) the gate signal at the next FET, (e) the gain (gate to gate), and (f) the resonant frequency and bandwidth of this stage.

2-19. Consider the series resonant circuit and the parallel resonant circuits with three or two branches (Figures 2-1, 2-6, and 2-7, respectively). If the resistance in each is R, what is the impedance at resonance of each?

2-20. A series circuit consists of an impedance coil having a resistance of 55 Ω, an inductance of 0.281 H, and a 25-μF capacitor. (a) Calculate the resonant frequency. For the condition of resonance and an impressed voltage of 110 V, determine (b) the total current and power and (c) the voltage drops across the impedance coil, the capacitor, and the resistor.

2-21. Calculate the resonant frequency of a two-branch parallel circuit given the following parameters: $R_L = 30\ \Omega$, $R_C = 30\ \Omega$, $L = 0.106$ H, and $C = 365\ \mu$F.

2-22. An impedance coil having a resistance of 20 Ω and an inductance of 4 mH is connected in parallel with a 0.4-μF capacitor. Calculate the resonant frequency (a) taking the resistance into account and (b) neglecting the resistance.

2-23. A coil with a Q of 90 and a capacitor form a parallel resonant circuit tuned to 4.5 MHz. The total impedance of the circuit at resonance is 60 kΩ. Find (a) the inductance of the coil and (b) the capacitance of the capacitor.

2-24. A transmitter operating on an assigned frequency of 1850 kHz delivers maximum power to an antenna when the antenna is tuned to represent a load consisting of a 200-pF capacitance in parallel with a 160-Ω resistance. Express the internal impedance "seen" at the output terminals of the transmitter in rectangular coordinates.

2-25. A tuned circuit in a radio transmitter is composed of a tank coil with an inductance of 80 μH tuned by an 800-pF capacitor. The RF voltage across this tank circuit is 1200 V rms. (a) If the overall Q of the circuit is 30, calculate the tank current. (b) At what rate must RF energy be supplied to this circuit to maintain the amplitude of the voltage?

2-26. A parallel LC tank circuit has $L = 30\ mh$, $C = 0.50\ \mu f$ and a coil resistance of 6 ohms. Find f_0, Q, BW, and Z_{max}.

2-27. A parallel LC tank circuit has a Q of 60 and a coil winding resistance of 3 ohms. Find the circuit's resistance at resonance.

CHAPTER 3

Filters

In Chapter 2 we looked at a variety of networks commonly employed in communication circuits. In particular we observed that tuned circuits had associated with them a particular band of frequencies which along with their other properties could be used to perform filtering. Since one of the major design problems in implementing communication systems is to be able to select the frequency band of interest, it is desirable to be able to design filter networks quickly and accurately. For the purpose of discussion, we may divide filters into two broad groups:

- Passive filters.
- Active filters.

Passive filters contain no active device such as a transistor, operational amplifier, and so on. We may further subdivide both groups of filters into four functional categories:

- Low pass.
- High pass.

- Bandpass.
- Band-reject.

At this point most of us have some familiarity with filters. In the following we wish to expand this knowledge with a discussion of filter design techniques.

3-1

RC FILTERS

The simple but important *RC* filter will be discussed first in order to introduce terminology. This filter is shown in Figure 3-1 in the low-pass configuration. The ratio of output voltage to input voltage as a function of frequency is defined to be the transfer function of the filter. Thus, treating the network as a voltage divider, the transfer function is

$$\frac{E_o(j\omega)}{E_i(j\omega)} = H(j\omega) = \frac{1/j\omega C}{R + 1/j\omega C} = \frac{1}{1 + j\omega RC} \qquad (3\text{-}1)$$

Equation (3-1) is a complex number and may be written in polar form as follows:

$$H(j\omega) = \frac{1}{\sqrt{1 + (\omega RC)^2}} \underline{/\phi = -\tan^{-1} \omega RC} \qquad (3\text{-}2)$$

where ϕ is the phase shift associated with the network and the remainder of the expression relates to the magnitude of the voltage transfer function. That is,

$$|H(j\omega)| = \frac{1}{\sqrt{1 + (\omega RC)^2}} \qquad (3\text{-}3a)$$

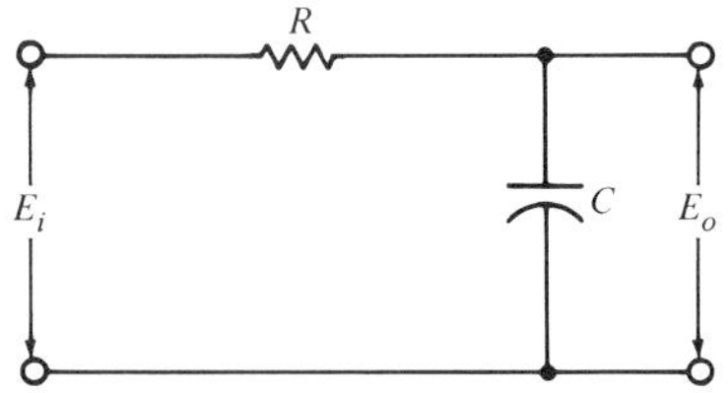

FIGURE 3-1. *RC* low-pass filter.

and

$$\phi = -\tan^{-1} \omega RC \tag{3-3b}$$

The characteristics of the expressions in equations (3-3) are most easily observed in graphical form. Common practice is to convert $|H(j\omega)|$ to decibels and then plot the magnitude versus frequency. This is accomplished by taking $20 \log_{10} |H(j\omega)|$. Thus

$$\begin{aligned} 20 \log_{10} |H(j\omega)| &= 20 \log_{10} \frac{1}{\sqrt{1 + (\omega RC)^2}} \\ &= -20 \log_{10} \sqrt{1 + (\omega RC)^2} \quad \text{dB} \end{aligned} \tag{3-4}$$

The magnitude and phase angle versus frequency are shown in Figure 3-2. Note in Figure 3-2a that when $\omega_1 = 1/RC$ the output is -3 dB below the input. This is easily seen from equation (3-4) when $\omega RC = 1$. That is,

$$20 \log_{10} |H(j\omega)| = -20 \log_{10} \sqrt{1 + (1)^2} = -3 \text{ dB}$$

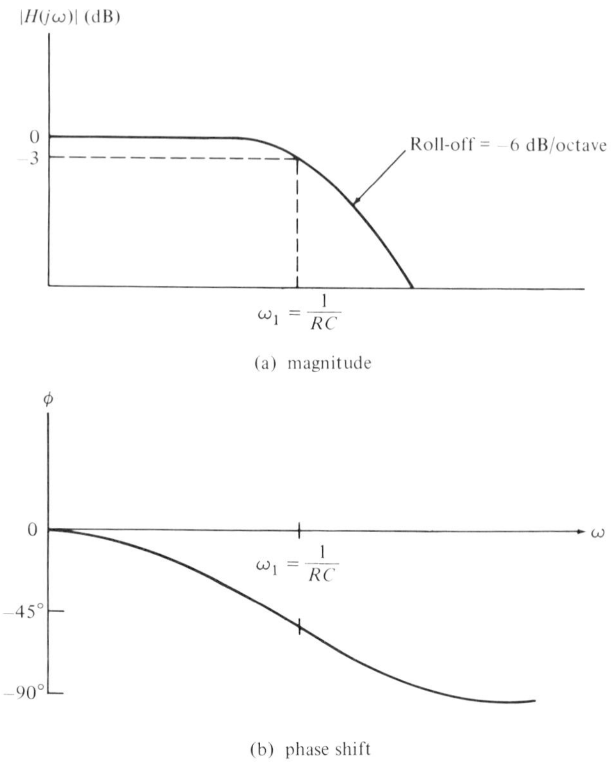

FIGURE 3-2. *RC* low-pass filter characteristics.

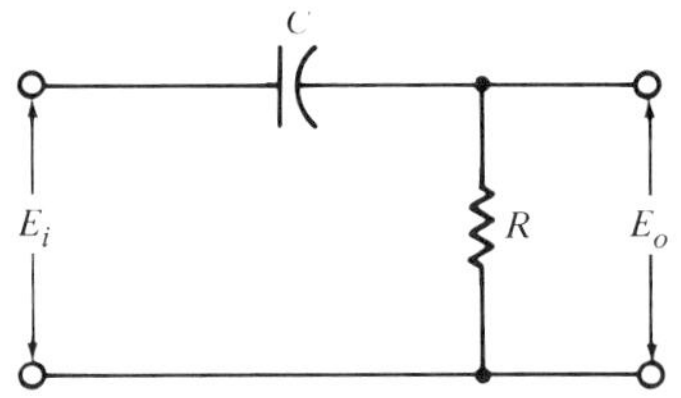

FIGURE 3-3. ***RC*** **high-pass filter.**

The -3-dB frequency is usually taken to define the bandwidth. It may be observed from Figure 3-2b that at this frequency the output voltage lags the input voltage in phase by 45°.

Another characteristic of interest related to the filter is its roll-off rate, shown to be -6 dB/octave. Again this is easily seen from equation (3-4). Consider a frequency ω such that $\omega RC >> 1$ and also that $\omega = 2\omega_1$. Then from equation (3-4)

$$\begin{aligned} 20 \log_{10} |H(j\omega)| &= -20 \log_{10} (\omega RC)^2 \\ &= -20 \log 2\omega_1 RC \\ &= -20 \log 2 - 20 \log_{10} \omega_1 RC \\ &= -6 - 20 \log_{10} \omega_1 RC \end{aligned}$$

Thus in the roll-off region when the frequency is doubled, the filter response changes by -6 dB. This indicates a characteristic of *RC* filters: namely, a pair of elements, one *R* and one *C*, is necessary to achieve a 6-dB/octave change in filter response.

If we now reverse the *R* and *C* elements as in Figure 3-3, a high-pass filtering effect is achieved. The transfer function, $H(j\omega)$, for this filter is given by

$$\begin{aligned} H(j\omega) &= \frac{R}{R + 1/j\omega C} = \frac{j\omega RC}{1 + j\omega RC} \\ &= \frac{\omega RC}{\sqrt{1 + (\omega RC)^2}} \underline{/ \phi = \frac{\pi}{2} - \tan^{-1} \omega RC} \end{aligned} \tag{3-5}$$

The magnitude and phase angle for this filter are graphed in Figure 3-4. Note that this filter exhibits $+6$-dB/octave roll-off and has a positive phase shift. In Chapter 4 the roll-off characteristics of these networks will be reexamined in terms of potential uses as integrators and differentiators.

EXAMPLE 3-1

Design an *RC* low-pass filter with a cutoff frequency, ω_c, equal to 1000 rad/s. Determine the −3-dB cutoff frequency if the filter is loaded with a 1-kΩ load.

SOLUTION

$$H(j\omega) = \frac{1}{1 + j\omega RC}$$

$$\omega RC = 1 \rightarrow R = \frac{1}{\omega C} = \frac{1}{1000C}$$

Choose $C = 1\ \mu f$.

$$R = \frac{1}{(1000)(10^{-6})} = 1\ \text{k}\Omega$$

$$R_L = 1\ \text{k}\Omega$$

$$H(j\omega) = \frac{Z_p}{R + Z_p}$$

$$Z_p = X_C \| R_L = \frac{R_L}{1 + j\omega CR_L}$$

$$H(j\omega) = \frac{R}{(R + R_L) + j\omega CRR_L} \rightarrow \omega CRR_L = R + R_L$$

$$\omega_c = \frac{R + R_L}{RR_LC} = \frac{1000 + 1000}{(1000)(1000)(10^{-6})} = 2000\ \text{rad/s}$$

3-2

CLASSICAL PASSIVE FILTER DESIGN

One of the classical uses of tuned *RLC* circuits is in filtering. Two techniques can be used to attenuate undesired signal frequencies. These are (1) connection of a high-impedance path in series with the signal and (2) connection of a low-impedance path in shunt with the undesired signal. Three basic filter configurations applying this technique are shown in Figure 3-5.

We see that these configurations consists of the familiar L type, the T type, and the Π type. Variations of these configurations are used to form a number of filter types. The circuit components that produce the specific impedances Z_1

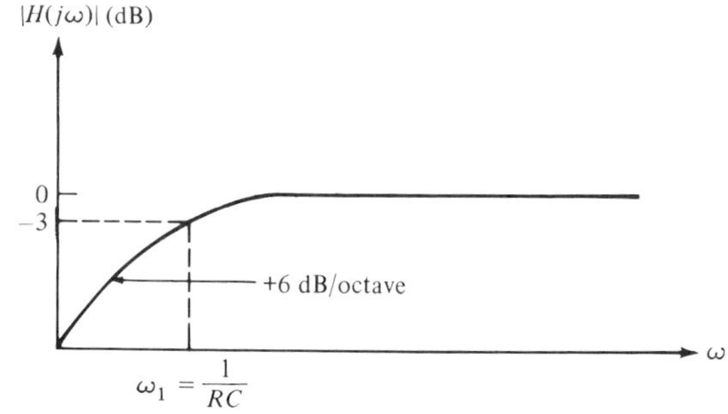

(a) magnitude

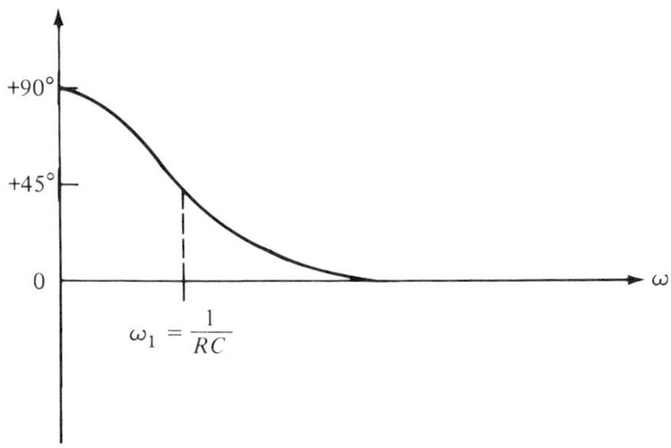

(b) phase shift

FIGURE 3-4. ***RC* high-pass filter characteristics.**

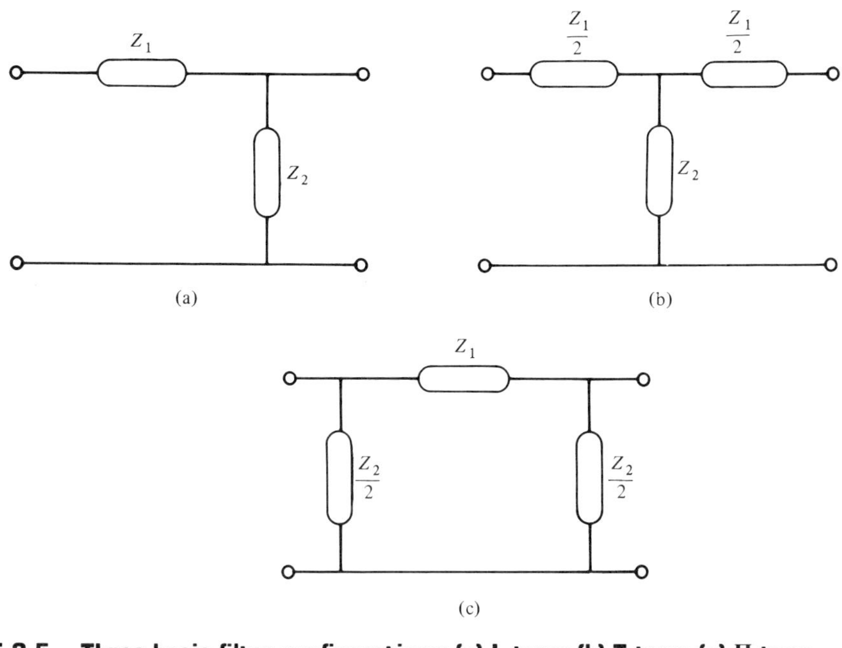

FIGURE 3-5. Three basic filter configurations: (a) L type; (b) T type; (c) Π type.

or Z_2 depend on the frequency requirements. For example, if the filter is high pass, the series element is a capacitor and the shunt element an inductor. For a low-pass filter the reverse is true: the series element is an inductor and the shunt element is a capacitor.

When a passive filter is inserted in a circuit it is necessary that both the input and output impedances of the filter match the circuit impedance. For the filter, this impedance is determined by the L and C values. Specifically, the *surge* or *characteristic impedance* of the filter is given by

$$Z_0 = \sqrt{\frac{L}{C}} \tag{3-6}$$

If the filter is terminated in an impedance equal to the characteristic impedance of the filter, this makes the composite circuit appear as though the LC sections are repeated out to infinity. This principle is examined further in Chapter 16. Design equations for the filter sections in Figure 3-5 are given below in terms of the cutoff frequency, f_c.

Low pass:
(a) L type

$$L = \frac{Z_0}{\pi f_c}, \qquad C = \frac{1}{\pi f_c Z_0} \tag{3-7}$$

(b) T type

$$L_1 = L_2 = \frac{Z_0}{2\pi f_c} \qquad C = \frac{1}{\pi f_c Z_0} \tag{3-8}$$

(c) Π type

$$L = \frac{Z_0}{\pi f_c} \qquad C_1 = C_2 = \frac{2}{\pi f_c Z_0} \tag{3-9}$$

High pass:
(a) L type

$$L = \frac{Z_0}{4\pi f_c} \qquad C = \frac{1}{4\pi f_c Z_0} \tag{3-10}$$

(b) T type

$$L = \frac{Z_0}{4\pi f_c} \qquad C_1 = C_2 = \frac{2}{4\pi f_c Z_0} \tag{3-11}$$

(c) Π type

$$L_1 = L_2 = \frac{2Z_0}{4\pi f_c} \qquad C = \frac{1}{4\pi f_c Z_0} \tag{3-12}$$

If the product of the filter series impedance, Z_1, and the filter shunt impedance, Z_2, is constant at all frequencies, these filters are referred to as *constant-k* filters.

EXAMPLE 3-2

A constant-k filter is to be terminated with a 50-Ω load and is to have a cutoff frequency of 1 MHz. Calculate L and C for the T configuration.

SOLUTION

$$L_1 = L_2 = \frac{Z_0}{2\pi f_c} = \frac{50}{(2\pi)(10^6)} = 7.96\ \mu\text{H}$$

$$C = \frac{1}{\pi f_c Z_0} = \frac{1}{\pi(10^6)(50)} = 0.0064\ \mu\text{F}$$

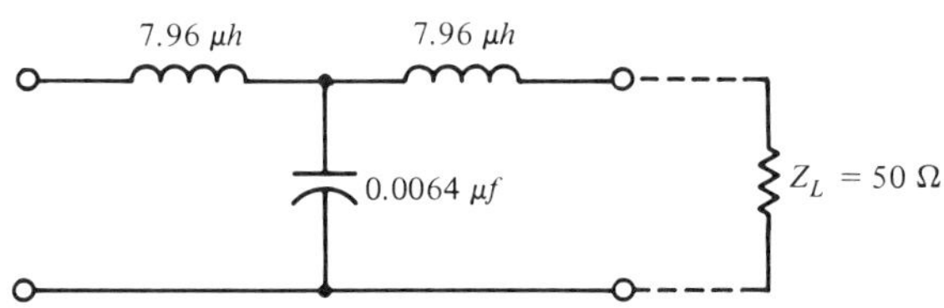

3-2.1 Constant-*k* Bandpass Filter

The configuration and frequency response for the L, T, and π constant-k bandpass filters are shown in Figure 3-6. The elements in the series leg are series resonant while the shunt elements are parallel resonant. For a broad frequency response range, the filter is made up of multiple sections usually using a combination of L, T, and Π networks. The design equations for the constant-k bandpass are as follows:

$$L_1 = \frac{Z_0}{\pi(f_2 - f_1)} \tag{3-13}$$

$$L_2 = \frac{(f_2 - f_1)Z_0}{4\pi f_1 f_2} \tag{3-14}$$

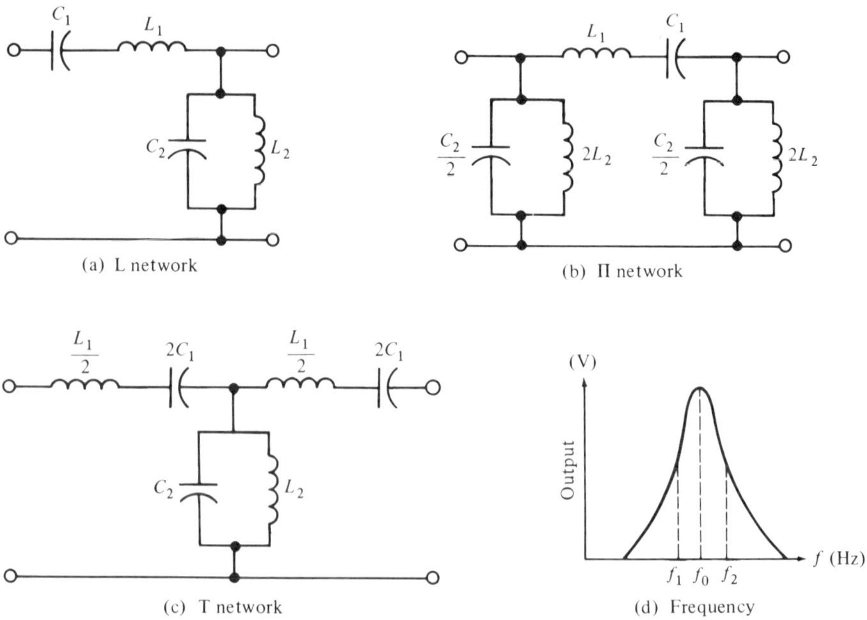

FIGURE 3-6. Bandpass constant-*k* filter: (a) L network; (b) Π network; (c) T network; (d) frequency response.

$$C_1 = \frac{f_2 - f_1}{4\pi f_1 f_2 Z_0} \tag{3-15}$$

$$C_2 = \frac{1}{\pi(f_2 - f_1)Z_0} \tag{3-16}$$

EXAMPLE 3-3

A constant-*k* filter is desired that will pass the band of frequencies from 1 to 1.6 MHz. The filter is terminated by a 50-Ω load. Use the L network.

SOLUTION

$$L_1 = \frac{Z_0}{\pi(f_2 - f_1)} = \frac{50}{\pi(1.6 \times 10^6 - 10^6)} = 26.52\ \mu\text{H}$$

$$L_2 = \frac{(f_2 - f_1)Z_0}{4\pi f_1 f_2} = \frac{(0.6 \times 10^6)(50)}{4\pi(10^6)(1.6 \times 10^6)} = 1.49\ \mu\text{H}$$

$$C_1 = \frac{f_2 - f_1}{4\pi f_1 f_2 Z_0} = \frac{0.6 \times 10^6}{4\pi(10^6)(1.6 \times 10^6)(50)} = 597\ \text{pF}$$

$$C_2 = \frac{1}{\pi(f_2 - f_2)Z_0} = \frac{1}{\pi(0.6 \times 10^6)(50)} = 0.01\ \mu\text{F}$$

3-2.2 Constant-*k* Band-Reject Filter

The configuration and filter response for a constant-*k* band-reject filter are shown in Figure 3-7. Again these filters may be made up of many sections. The design equations for the constant-*k* band-reject filter are as follows:

$$L_1 = \frac{(f_2 - f_1)Z_0}{\pi f_1 f_2} \tag{3-17}$$

$$L_2 = \frac{Z_0}{4\pi(f_2 - f_1)} \tag{3-18}$$

$$C_1 = \frac{1}{4\pi(f_2 - f_1)Z_0} \tag{3-19}$$

$$C_2 = \frac{f_2 - f_1}{\pi Z_0 f_1 f_2} \tag{3-20}$$

3-2.3 *m*-Derived Filter

The characteristics of the constant-*k* filter can be improved by adding an element to the section that provides infinite attenuation at a frequency designated as f_∞.

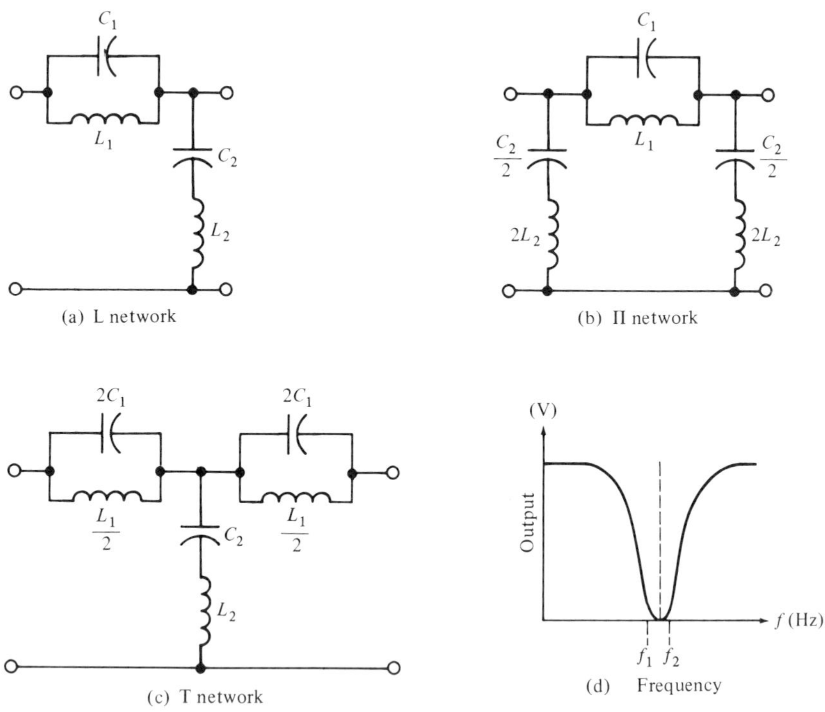

FIGURE 3-7. Band-reject filter: (a) L network; (b) Π network; (c) T network; (d) frequency response.

Constant-k networks have moderate frequency cutoff with a high rejection factor for unwanted frequencies. The m-derived networks have sharper cutoff characteristics but do not have higher rejection of unwanted frequencies. The circuit configuration with frequency response for a low-pass m-derived network is shown in Figure 3-8. Note that the m-derived network has one resonant network per element. The term *m-derived* comes from a constant value which is the ratio of the reactance of the resonant element in the m-derived network to the same element in the comparable constant-k filter. The constant m is also the ratio of the cutoff frequency, f_c, of a high-pass or low-pass filter to the frequency of highest rejection. A typical value is $m = 0.6$. Specific values may be calculated by equation (3-21).

$$m = \sqrt{1 - \left(\frac{f_c}{f_\infty}\right)^2} \tag{3-21}$$

The design equations are as follows:

$$L_1 = \frac{mZ_0}{\pi f_c} \tag{3-22}$$

$$C_1 = \frac{m}{\pi f_c Z_0} \tag{3-23}$$

$$L_2 = \frac{m}{\pi f_c Z_0} \tag{3-24}$$

$$C_2 = \frac{1 - m^2}{4m} \frac{1}{\pi f_c Z_0} \tag{3-25}$$

The configuration for a high-pass m-derived filter is shown in Figure 3-9. The applicable design equations are as follows:

$$m = \sqrt{1 - \left(\frac{f_\infty}{f_c}\right)^2} \tag{3-26}$$

$$L_1 = \frac{Z_0}{4\pi f_c m} \tag{3-27}$$

$$C_1 = \frac{1}{4\pi f_c m Z_0} \tag{3-28}$$

$$L_2 = \frac{mZ_0}{(1 - m^2)\pi f_c} \tag{3-29}$$

$$C_2 = \frac{m}{(1 - m^2)\pi f_c Z_0} \tag{3-30}$$

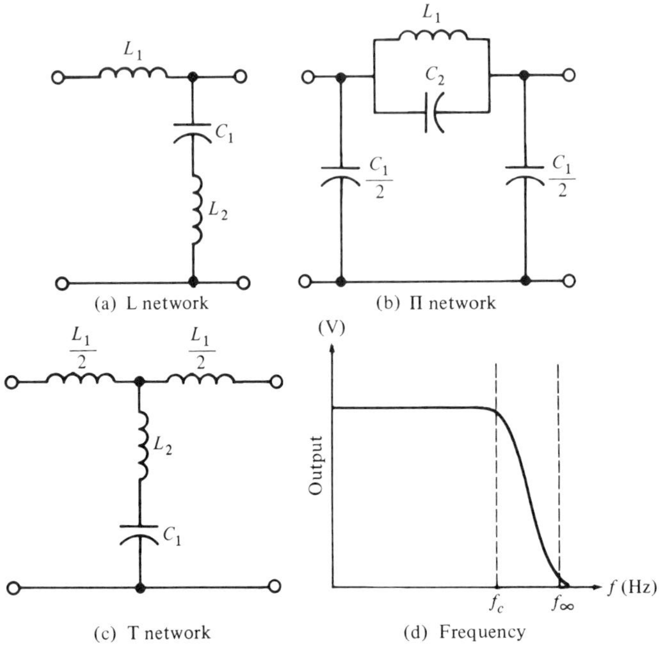

FIGURE 3-8. Low-pass *m*-derived filter: (a) L network; (b) Π network; (c) T network; (d) frequency response.

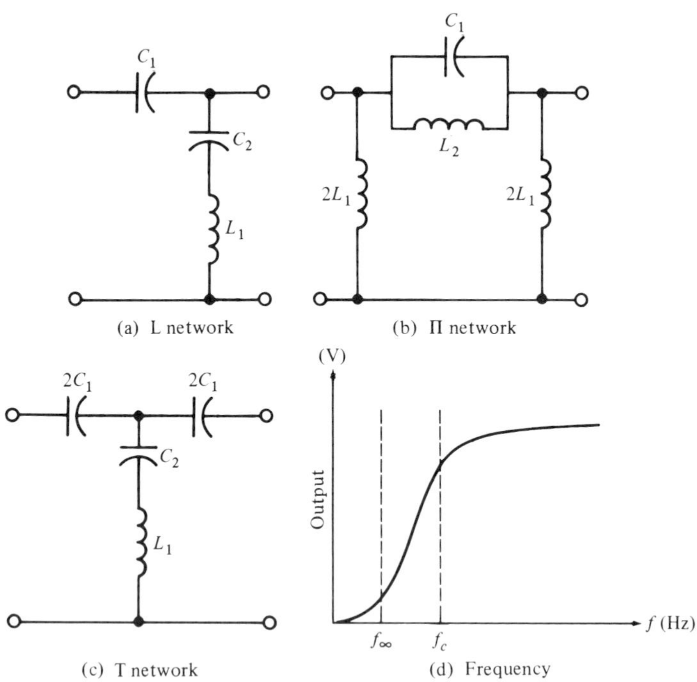

FIGURE 3-9. High-pass *m*-derived filter: (a) L network; (b) Π network; (c) T network; (d) frequency response.

EXAMPLE 3-4

Given an m-derived Π-network high-pass filter with a cutoff frequency of 6 MHz and 50 termination. Assuming that $m = 0.6$, calculate f_∞, C_1, L_1, and L_2.

SOLUTION

$$f_\infty = f_c\sqrt{\frac{1}{1 - m^2}} = 6 \times 10^6\sqrt{\frac{1}{1 - 0.6^2}} = 7.5 \text{ mHz}$$

$$L_1 = \frac{Z_0}{4\pi f_c m} = \frac{50}{4\pi(6 \times 10^6)(0.6)} = 1.105\ \mu\text{H}$$

$$L_2 = \frac{mZ_0}{(1 - m^2)\pi f_c} = \frac{(0.6)(50)}{(1 - 0.6^2)(\pi)(6 \times 10^6)} = 2.49\ \mu\text{H}$$

$$C_1 = \frac{1}{4\pi f_c m Z_0} = \frac{1}{4\pi(6 \times 10^6)(0.6)(50)} = 442 \text{ pF}$$

3-3 ACTIVE FILTER DESIGN

Chebyshev, Butterworth, and Bessel are not names to be found on road maps, but when considered in terms of modern filter design they serve a similar purpose—steering the filter designer or user in the right direction. Modern filter design, with its emphasis on computer solution, makes possible a better performance filter design with fewer components than the multisection "classical" filters. In particular, the above-mentioned filters are easily implemented using an active device. This makes low-frequency design particularly attractive since inductors can be eliminated entirely, thus eliminating large inductors at low frequency.

Modern filters are described by *poles* and *zeros* that identify the resonant frequencies over the frequency band. A zero represents a zero-impedance frequency and a pole represents a high-impedance frequency. For example, a parallel *LC* network connected in shunt with the signal would represent a pole (maximum impedance to ground at the network's resonant frequency). Conversely, a series *LC* network connected in shunt with the signal would represent a zero (minimum impedance to ground) at the network's resonant frequency. In terms of the filter transfer function, the roots of the numerator represent zeros and the roots of the denominator represent poles. The pole–zero notation originates with the Laplace transform and may be a real or complex number. It is usually denoted as $s = \sigma + j\omega$. Thus the pole–zero locations may be plotted in the s plane.

For example, we may rewrite equation (3-1) as

$$H(s) = \frac{1}{1 + sRC} \tag{3-31}$$

Setting the denominator equal to zero and solving for the roots yields

$$s = -\frac{1}{RC} \tag{3-32}$$

Thus the RC low-pass filter has a single root located on the negative real axis (Figure 3-10a).

The pole–zero locations for a high-pass RC filter are shown in Figure 3-10b. It is easily shown from the transfer function that an LC network has poles that fall on the $j\omega$ axis. Thus if we can vary the pole–zero locations, we can achieve the effect of different networks. An active device makes this possible.

The s-plane pole–zero locations for Bessel, Chebyshev, and Butterworth filters are shown in Figure 3-11. The characteristic of these three basic filter responses are:

- *Chebyshev:* maximum roll-off rate.
- *Butterworth:* passband amplitude flatness.
- *Bessel:* time-delay flatness.

Note that, be it Bessel, Butterworth, or Chebyshev, only one pole is on the negative real axis. The other poles occur in complex-conjugate pairs. There are several networks that can be made to have the pole locations of Figure 3-11. One popular configuration is shown in Figure 3-12.

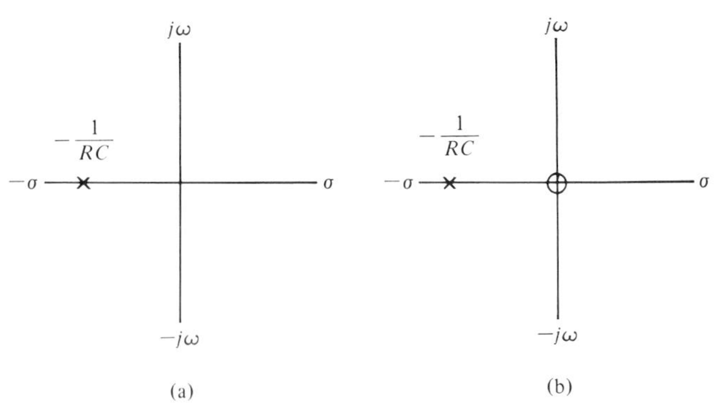

FIGURE 3-10. ***s* plane: (a) low pass; (b) high pass.**

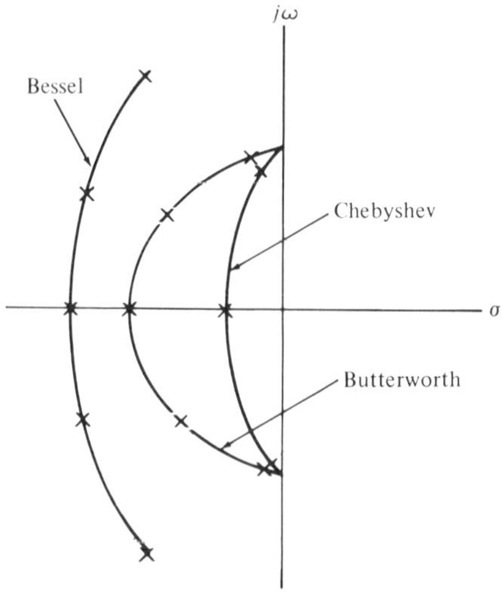

FIGURE 3-11. *s* plane response.

The transfer function for this network is

$$H(s) = \frac{1}{s^2C_1C_2R_1R_2 + s[C_2(R_1 + R_2) + C_1R_1(1 - A)] + 1} \tag{3-33}$$

which reduces to the transfer function of the network itself when $A = 0$. The pole locations are at

$$s = \frac{-b \pm \sqrt{b^2 - 4ac}}{2a} \tag{3-34}$$

where $a = C_1C_2R_1R_2$, $b = C_1R_2(1 - A) + C_2(R_1 + R_2)$, and $C = 1$. We see from equation (3-34) that as the gain, A, changes, the pole locations change. This movement of the pole locations is plotted in Figure 3-13 as the amplifier

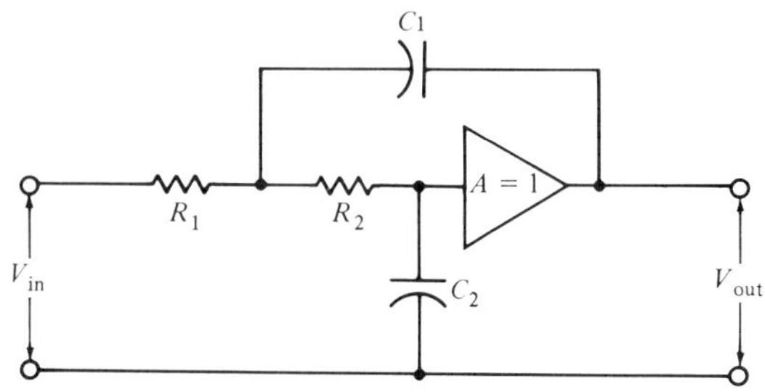

FIGURE 3-12. Active filter with unity-gain amplifier.

TABLE 3-1 Computer Solutions for Normalized Filter Components

Pole	C_1	C_2	C_3	C_1	C_2	C_3
	Bessel			**Butterworth**		
2	9.066 −1	6.799 −1		1.414 +0	7.071 −1	
3	1.423 +0	9.880 −1	2.538 −1	3.546 +0	1.392 +0	2.024 −1
4	7.351 −1	6.746 −1		1.082 +0	9.241 −1	
	1.012 +0	3.900 −1		2.613 +0	3.825 −1	
5	1.009 +0	8.712 −1	3.095 −1	1.753 +0	1.354 +0	4.214 −1
	1.041 +0	3.098 −1		3.235 +0	3.089 −1	
6	6.352 −1	6.098 −1		1.035 +0	9.660 −1	
	7.225 −1	4.835 −1		1.414 +0	7.071 −1	
	1.073 +0	2.561 −1		3.863 +0	2.588 −1	
7	8.532 −1	7.792 −1	3.027 −1	1.531 +0	1.336 +0	4.885 −1
	7.250 −1	4.151 −1		1.604 +0	6.235 −1	
	1.098 +0	2.164 −1		4.493 +0	2.225 −1	
8	5.673 −1	5.539 −1		1.019 +0	9.809 −1	
	6.090 −1	4.861 −1		1.202 +0	8.313 −1	
	7.257 −1	3.590 −1		1.800 +0	5.557 −1	
	1.116 +0	1.857 −1		5.125 +0	1.950 −1	
9	7.564 −1	7.070 −1	2.851 −1	1.455 +0	1.327 +0	5.169 −1
	6.048 −1	4.352 −1		1.305 +0	7.661 −1	
	7.307 −1	3.157 −1		1.999 +0	4.999 −1	
	1.137 +0	1.628 −1		5.758 +0	1.736 −1	
10	5.172 −1	5.092 −1		1.012 +0	9.874 −1	
	5.412 −1	4.682 −1		1.122 +0	8.908 −1	
	5.999 −1	3.896 −1		1.414 +0	7.071 −1	
	7.326 −1	2.792 −1		2.202 +0	4.540 −1	
	1.151 +0	1.437 −1		6.389 +0	1.563 −1	
	0.1-dB Chebyshev			**0.25-dB Chebyshev**		
2	1.638 +0	6.955 −1		1.778 +0	6.789 −1	
3	6.653 +0	1.825 +0	1.345 −1	8.551 +0	2.018 +0	1.109 −1
4	1.901 +0	1.241 +0		2.221 +0	1.285 +0	
	4.592 +0	2.409 −1		5.363 +0	2.084 −1	
5	4.446 +0	2.520 +0	3.804 −1	5.543 +0	2.892 +0	3.425 −1
	6.810 +0	1.580 −1		8.061 +0	1.341 −1	
6	2.553 +0	1.776 +0		3.044 +0	1.875 +0	
	3.487 +0	4.917 −1		4.159 +0	4.296 −1	
	9.531 +0	1.110 −1		1.136 +1	9.323 −2	
7	5.175 +0	3.322 +0	5.693 −1	6.471 +0	3.876 +0	5.223 −1
	4.546 +0	3.331 −1		5.448 +0	2.839 −1	
	1.273 +1	8.194 −2		1.526 +1	6.844 −2	
8	3.270 +0	2.323 +0		3.932 +0	2.474 +0	
	3.857 +0	6.890 −1		4.638 +0	6.062 −1	
	5.773 +0	2.398 −1		6.942 +0	2.019 −1	
	1.644 +1	6.292 −2		1.976 +1	5.234 −2	
9	6.194 +0	4.161 +0	7.483 −1	7.766 +0	4.891 +0	6.919 −1
	4.678 +0	4.655 −1		5.637 +0	3.983 −1	
	7.169 +0	1.812 −1		8.639 +0	1.514 −1	
	2.064 +1	4.980 −2		2.487 +1	4.131 −2	
10	4.011 +0	2.877 +0		4.843 +0	3.075 +0	
	4.447 +0	8.756 −1		5.368 +0	7.725 −1	
	5.603 +0	3.353 −1		6.766 +0	2.830 −1	
	8.727 +0	1.419 −1		1.053 +1	1.181 −1	
	2.532 +1	4.037 −2		3.057 +1	3.344 −2	

TABLE 3-1 (continued)

Pole	C_1	C_2	C_3	C_1	C_2	C_3
	0.5-dB Chebyshev			**1-dB Chebyshev**		
2	1.949 +0	6.533 −1		2.218 +0	6.061 −1	
3	1.123 +1	2.250 +0	8.950 −2	1.618 +1	2.567 +0	6.428 −2
4	2.582 +0	1.298 +0		3.125 +0	1.269 +0	
	6.233 +0	1.802 −1		7.546 +0	1.489 −1	
5	6.842 +0	3.317 +0	3.033 −1	8.884 +0	3.935 +0	2.540 −1
	9.462 +0	1.144 −1		1.155 +1	9.355 −2	
6	3.592 +0	1.921 +0		4.410 +	1.904 +0	
	4.907 +0	3.743 −1		0	3.117 −1	
	1.340 +1	7.902 −2		6.024 +0	6.425 −2	
				1.646 +1		
7	7.973 +0	4.483 +0	4.698 −1	1.029 +1	5.382 +0	4.012 −1
	6.446 +0	2.429 −1		7.941 +0	1.993 −1	
	1.807 +1	5.778 −2		2.225 +1	4.684 −2	
8	4.665 +0	2.547 +0		5.756 +0	2.538 +0	
	5.502 +0	5.303 −1		6.792 +0	4.435 −1	
	8.237 +0	1.714 −1		1.015 +1	1.395 −1	
	2.345 +1	4.409 −2		2.894 +1	3.568 −2	
9	9.563 +0	5.680 +0	6.259 −1	1.233 +1	6.853 +0	5.382 −1
	6.697 +0	3.419 −1		8.281 +0	2.813 −1	
	1.026 +1	1.279 −1		1.268 +1	1.038 −1	
	2.954 +1	3.475 −2		3.651 +1	2.808 −2	
10	5.760 +0	3.175 +0		7.125 +0	3.170 +0	
	6.383 +0	6.773 −1		7.897 +0	5.680 −1	
	8.048 +0	2.406 −1		9.952 +0	1.962 −1	
	1.253 +1	9.952 −2		1.550 +1	8.054 −2	
	3.636 +1	2.810 −2		4.498 +1	2.269 −2	
	2-dB Chebyshev			**3-dB Chebyshev**		
2	2.672 +0	5.246 −1		3.103 +0	4.558 −1	
3	2.782 +1	3.113 +0	3.892 −2	4.342 +1	3.629 −0	
4	4.021 +0	1.103 +0		4.863 +0	1.049 +0	
	9.707 +0	1.150 −1		1.174 +1	9.432 −2	
5	1.240 +1	4.953 +0	1.963 −1	1.579 +1	5.917 +0	1.600 −1
	1.499 +1	7.169 −2		1.822 +1	5.860 −2	
6	5.750 +0	1.769 +0		7.012 +0	1.607 +0	
	7.853 +0	2.426 −1		9.575 +0	2.002 −1	
	2.146 +1	4.902 −2		2.615 +1	4.004 −2	
7	1.421 +1	6.851 +0	3.169 −1	1.793 +1	8.236 +0	2.621 −1
	1.038 +1	1.532 −1		1.267 +1	1.257 −1	
	2.910 +1	3.568 −2		3.553 +1	2.912 −2	
8	7.539 +0	2.367 +0		9.217 +0	2.156 +0	
	8.896 +0	3.467 −1		1.087 +1	2.866 −1	
	1.331 +1	1.067 −1		1.627 +1	8.733 −2	
	3.791 +1	2.714 −2		4.634 +1	2.215 −2	
9	1.696 +1	8.763 +0	4.285 −1	2.135 +1	1.056 +1	3.561 −1
	1.086 +1	2.168 −1		1.328 +1	1.779 −1	
	1.663 +1	7.916 −2		2.035 +1	6.467 −2	
	4.790 +1	2.135 −2		5.860 +1	1.742 −2	
10	9.355 +0	2.964 +0		1.144 +1	2.704 +0	
	1.037 +1	4.451 −1		1.269 +1	3.681 −1	
	1.306 +1	1.502 −1		1.599 +1	1.231 −1	
	2.035 +1	6.133 −2		2.491 +1	5.007 −2	
	5.908 +1	1.724 −2		7.231 +1	1.407 −2	

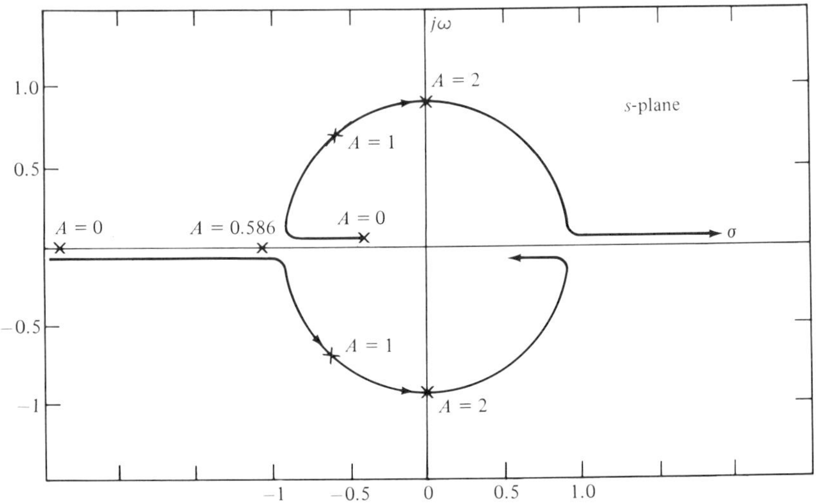

FIGURE 3-13. Pole–zero plot for variable amplifier gain, A.

gain is varied. In accordance with standard circuit theory, only poles in the left-half s plane represent stable networks.

3-3.1 Butterworth Response

Referring to Figure 3-11, we see that the poles for the Butterworth response lie on a circle. In particular, if $A = 1$, then from Figure 3-13, the poles are located at $s = -0.707 \pm j0.707$, which is the same as a two-pole Butterworth low-pass filter. Thus the unity-gain amplifier causes a two-pole RC network to behave in the same manner as a two-pole LC network. Another approach is to fix the gain of the amplifier at unity and then vary pole locations by varying the RC elements. For example, an operational amplifier can be connected as a voltage follower. This has the advantage of providing low output impedance and allows sections to be cascaded without increasing the overall gain of the filter. The remainder of this discussion follows this approach.

3-3.2 Filter Tables

Computer solutions for the normalized filter components are shown in Table 3-1. In using the tables only one three-pole circuit configuration may be used per filter since only one pole may occur on the real axis. Thus, if the required number of poles is N, the total number of active circuits needed is $N/2$ for N even and $(N - 1)/2$ for N odd. In the tables, the values given are all normalized for a 3-dB cutoff frequency of 1 rad/s and are expressed in farads and ohms. The filter sections that may be used are shown in Figure 3-14. The filter sections comprising the filter should be driven from a voltage source, that is, a low-

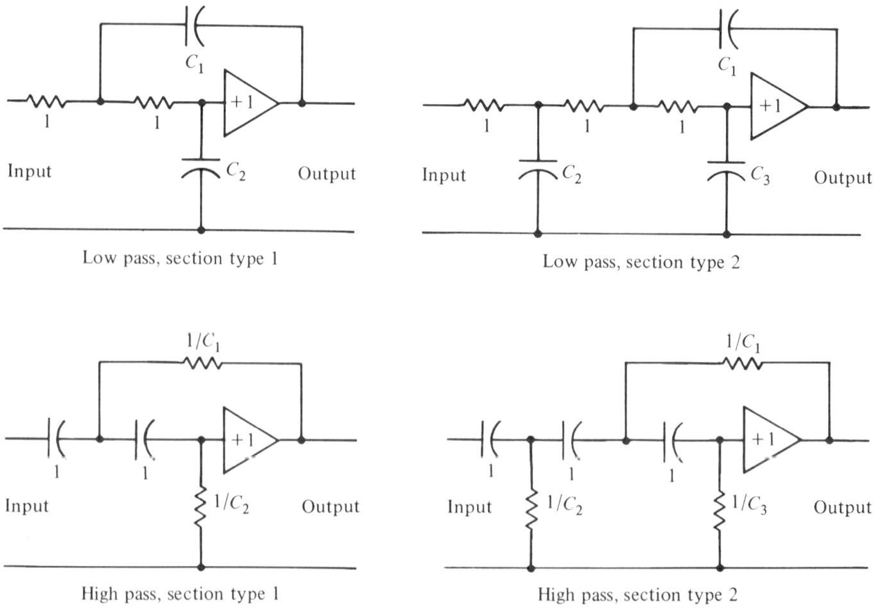

FIGURE 3-14. Filter section type.

impedance source. The load impedance may be any value that does not overload the output-section amplifier.

3-3.3 Filter Design

When designing a low-pass filter, for example, the designer must first select the type of filter desired and the number of poles required. Remember that for an *RC* combination the roll-off is -6 dB/octave. The procedure, then, is to look up the appropriate table values and divide all capacitor values by $2\pi f_c$ to achieve frequency scaling, where f_c is the -3-dB cutoff frequency. Next, resistor values are multiplied by M and capacitor values are divided by M. This permits convenient element values to be selected. The quantity M is an arbitrarily selected impedance scaling factor. An example will illustrate the procedure.

EXAMPLE 3-5

Design a five-pole 1-dB ripple Chebyshev low-pass filter with a -3-dB cutoff frequency of 100 Hz. Capacitor values are desired to be 0.1 μF or less.

SOLUTION

The solution is shown in step sequence in Figure 3-15.

1. In Figure 3-15a, component values are selected from the table.
2. In Figure 3-15b, frequency scaling is accomplished by dividing all capacitor values by $\omega_c = (2\pi)(100) = 200\pi$.
3. In Figure 3-15c, impedance scaling is accomplished by dividing each capacitor value by M and multiplying each resistor value by $M = 1{,}400$. The value of M may differ for each section.

High-pass filter design follows a similar procedure. The designer selects the filter type and number of poles required and then calculates the normalized element values by finding the reciprocal of the table values. All capacitor values

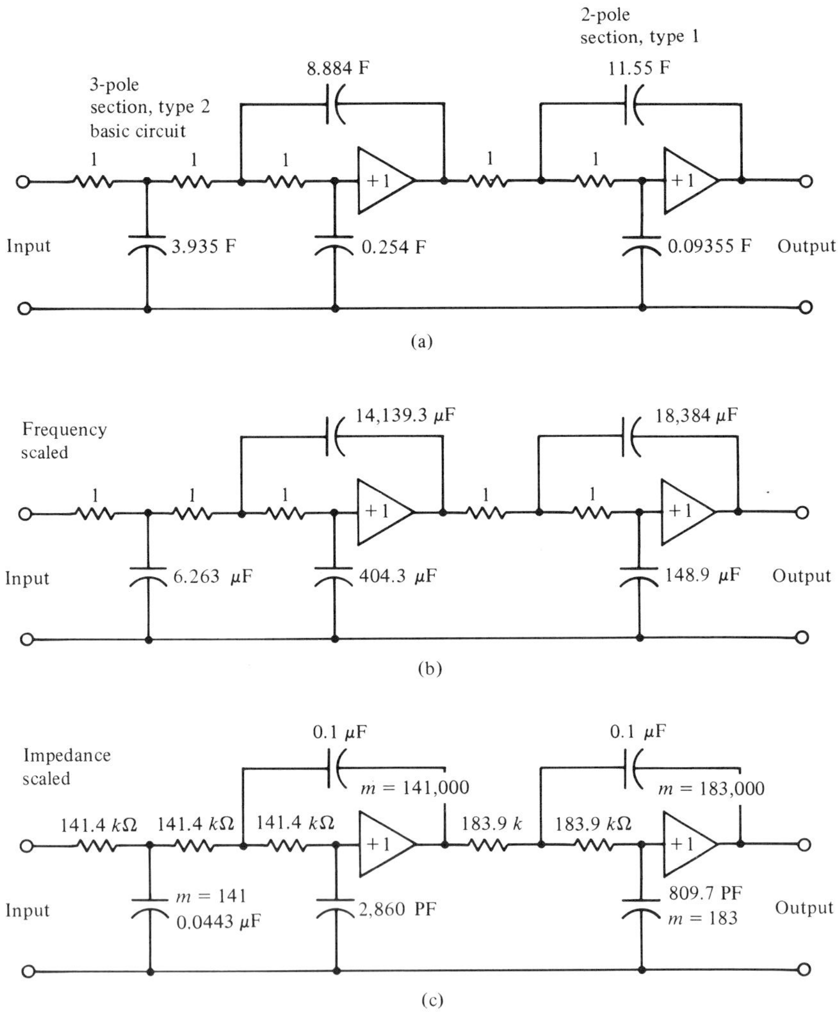

FIGURE 3-15. Filter design. Component values for a five pole Chebyshev (a) are selected from the tables, frequency scaled in (b) and impedance scaled in (c).

are then divided by $2\pi f_c$ and by the impedance scaling factor M. Resistor values are multiplied by M. Again the value of M may differ for each section.

3-3.4 Cascading Different Sections

A cascaded high-pass and low-pass configuration may be used to achieve a bandpass filter (see Figure 3-16a). The roll-off rate may be varied by varying the number of poles. For very narrowband filters, $0.1f_c$ or less, this approach may not be satisfactory, since any loss will be the sum of the loss for the two filters. A band-reject filter may be realized by a parallel configuration (Figure 3-16b).

In using the tables it is necessary that the order of the filter sections be the same as the order in which they occur in the tables. That is, the active low-pass tables are arranged so that the input section has a gradual roll-off, while following sections have increasingly sharp peaks in their responses. This combination of responses produces the desired flat passband and sharp cutoff. If a sharply peaked section were placed at the input, the unity-gain amplifier could be easily overdriven, since peak gains of 20 dB or more are not uncommon.

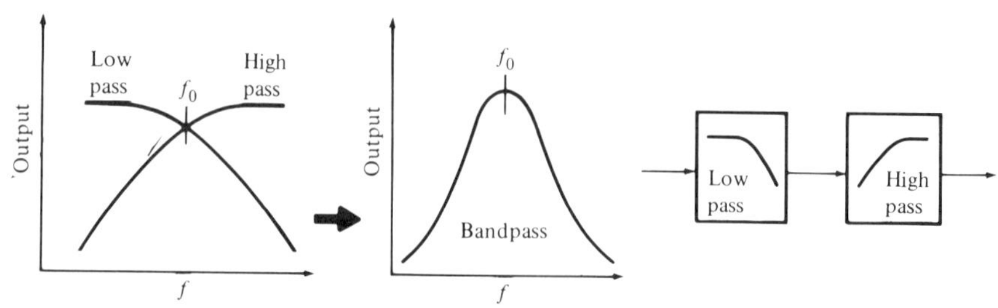

(a)

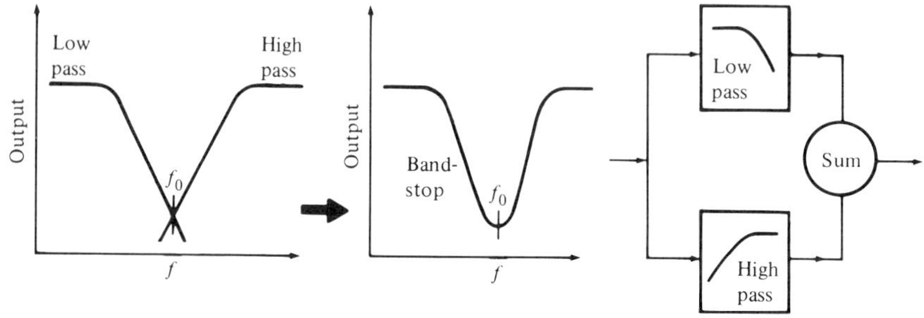

(b)

FIGURE 3-16. Filter configurations: (a) bandpass; (b) band-reject.

3-1. Design a high-pass *RC* filter with a cutoff frequency, ω_c, equal to 2000 rad/s. Determine the -3-dB frequency if the filter is loaded with a 2-kΩ load.

3-2. Design a constant-*k* bandpass filter that will pass the band of frequencies from 3 to 3.2 MHz. The filter is terminated in a 100-Ω load. Use a Π network in your design.

3-3. Given an *m*-derived T-network low-pass filter with a cutoff frequency of 30 MHz and 100-Ω termination. If $m = 0.6$, calculate f_∞, L_1, and C_1.

3-4. Design a low-pass Chebyshev 0.1-dB ripple filter with a cutoff frequency, ω_c, equal to 100,000 rad/s and a roll-off rate of -18 dB/octave.

3-5. Repeat Problem 3-4 for a maximally flat response filter.

3-6. Design a bandpass filter using Butterworth sections with a center frequency of 10,000 rad/s and a passband of 2000 rad/s. The roll-off rate desired is 12 dB/octave.

3-7. Using an operational-amplifier adder circuit, design a band-reject filter with a center frequency of 100,000 rad/s. The reject band is to be 5000 rad/s, with roll-off at 12 dB/octave. Use Butterworth sections.

3-8. Repeat Problem 3-4 for a high-pass filter.

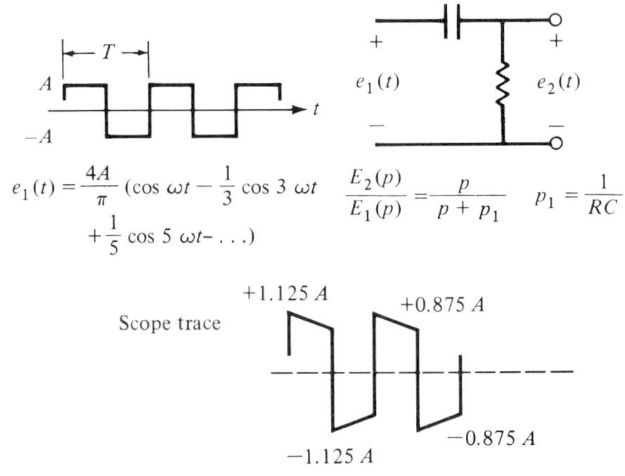

PROBLEM 3-9

3-9. Two students were asked to explain the effects of passing a square wave through a high-pass *RC* filter (see the figure). Student A reasoned that since the fundamental frequency of the square wave was four octaves above the −3-dB roll-off frequency of the filter, its attenuation and that of all of the harmonics should be negligible. The output should look like the input. Student B disagreed and to prove her point, set up an experiment and obtained the oscilloscope trace for the output shown in the lower part of the figure. Explain the fallacy in student A's argument. Describe the characteristics of a network to be connected in tandem with the *RC* network so that no overall distortion will result.

CHAPTER 4

Waveform Spectra

The previous chapters have dealt with networks whose response is frequency dependent. In particular we were able to calculate the network's response for a particular sinusoidal frequency. The question now arises: What is the response if the stimulus is not sinusoidal? We shall see that any waveform can be resolved into a sum of sinusoidal waves of different frequencies. Each component frequency can then be treated by methods and concepts that are familiar to us. The individual results can then be recombined into a nonsinusoidal waveform which will be the correct answer for the problem. This result, then, is the fundamental reason for developing circuit analysis in terms of sinusoidal frequencies. In communication systems we must nearly always process nonsinusoidal signals. Thus it is necessary to acquire techniques for determining the frequency content or spectrum of nonsinusoidal waves.

FOURIER SERIES

In the preceding section we indicated that a nonsinusoidal repetitive wave could be resolved into a sum of sinusoidal waves. Expressed mathematically, this statement is

$$f(t) = \tfrac{1}{2}a_0 + a_1 \cos \omega t + a_2 \cos 2\omega t + a_3 \cos 3\omega t + \cdots \qquad (4\text{-}1)$$
$$b_1 \sin \omega t + b_2 \sin 2\omega t + b_3 \sin 3\omega t + \cdots$$

Equation (4-1) is referred to as the *Fourier series*. The first term in the series, $\frac{1}{2}a_0$, represents the *dc component* of the wave if the waveform is a current or voltage waveform. The terms with a_1 and b_1 coefficients represent the fundamental frequency of the wave. The second harmonic is represented by the terms with a_2 and b_2 coefficients, and the higher harmonics are represented accordingly.

In order that the series represent the complex waveform, it is necessary that the series converge. That is, as more and more terms (higher harmonics) are added to the series, it must continually become a better approximation to the waveform. To find the series that represents a particular function, $f(t)$, we need only find the coefficients a_0, a_1, b_1, a_2, b_2, and so on.

The process for doing this is based on the fact that the sine and cosine functions are orthogonal. That is, the average of the cross-products is zero. In terms of an integral this may be expressed as

$$\int_0^{2\pi} \sin mx \cos nx \, dx = 0 \qquad (4\text{-}2)$$

where m and n are any integer. Those unfamiliar with the calculus may skip to Section 4-2 and simply use the results.

The following additional definite integrals will be used in developing formulas that may be used to calculate the a and b coefficients:

$$\int_0^{2\pi} \sin mx \, dx = 0 \qquad \int_0^{2\pi} \cos mx \, dx = 0 \qquad (4\text{-}3)$$

$$\int_0^{2\pi} (\sin mx)^2 \, dx = \pi \qquad \int_0^{2\pi} (\cos mx)^2 \, dx = \pi \qquad (4\text{-}4)$$

$$\int_0^{2\pi} \cos mx \cos nx \, dx = 0 \qquad (4\text{-}5)$$

Using these integrals, let us first find coefficient a_2. The process is to multiply equation (4-1) through by $\cos 2\omega t \, d(\omega t)$ and integrate from 0 to 2π. Thus

$$\begin{aligned}\int_0^{2\pi} f(t) \cos 2\omega t \, d(\omega t) &= \int_0^{2\pi} \tfrac{1}{2}a_0 \cos 2\omega t \, d(\omega t) \\ &+ \int_0^{2\pi} a_1 \cos \omega t \cos 2\omega t \, d(\omega t) \\ &+ \int_0^{2\pi} a_2 (\cos 2\omega t)^2 \, d(\omega t) + \cdots \qquad (4\text{-}6) \\ &+ \int_0^{2\pi} b_1 \sin \omega t \cos 2\omega t \, d(\omega t) \\ &+ \int_0^{2\pi} b_2 \sin 2\omega t \cos 2\omega t \, d(\omega t) + \cdots\end{aligned}$$

Using equations (4-2) through (4-5), we see that all of the terms are zero except the first and fourth. Equation (4-6) reduces to

$$\int_0^{2\pi} f(t) \cos 2\omega t \, d(\omega t) = \int_0^{2\pi} a_2 (\cos 2\,\omega t)^2 \, d\omega t = a_2\pi \qquad (4\text{-}7)$$

Thus

$$a_2 = \frac{1}{\pi} \int_0^{2\pi} f(t) \cos 2\omega t \, d(\omega t) \qquad (4\text{-}8)$$

Similar development can be used to find the other a coefficients. We may generalize the results for any n as follows:

$$a_n = \frac{1}{\pi} \int_0^{2\pi} f(t) \cos n\omega t \, d(\omega t) \qquad (4\text{-}9)$$

A similar development for the sine terms gives

$$b_n = \frac{1}{\pi} \int_0^{2\pi} f(t) \sin n\omega t \, d(\omega t) \qquad (4\text{-}10)$$

Thus equations (4-9) and (4-10) may be used to determine completely the coefficients for the series represented by equation (4-1).

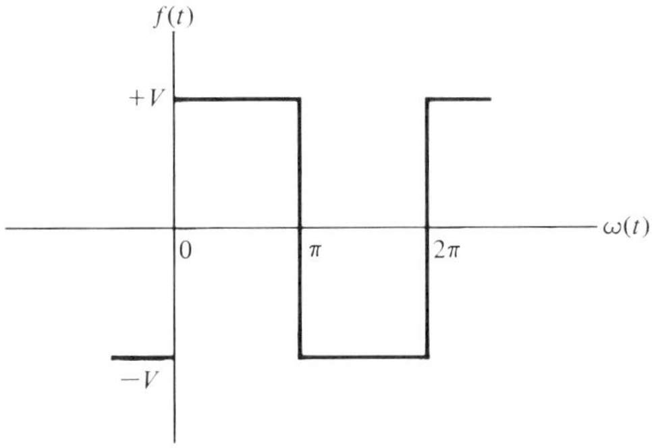

FIGURE 4-1. Square-wave function.

EXAMPLE 4-1

(a) Find the Fourier series for the square wave in Figure 4-1, (b) sketch the spectrum, and, (c) sketch the sum of the first two components.

SOLUTION

$$\text{(a) } a_n = \frac{1}{\pi}\int_0^{\pi} (+V)\cos nx\, dx + \frac{1}{\pi}\int_{\pi}^{2\pi} (-V)\cos nx\, dx$$

$$= \frac{V}{n\pi}(2\sin n\pi - \sin n\,2\pi) = 0 \qquad n = 1, 2, 3, \ldots$$

$$a_0 = \frac{1}{\pi}\int_0^{\pi} (+V)\, dx + \frac{1}{\pi}\int_{\pi}^{2\pi} (-V)\, dx = \frac{V}{\pi}(\pi - 0 - 2\pi + \pi) = 0$$

$$b_n = \frac{1}{\pi}\int_0^{\pi} (+V)\sin nx\, dx + \int_{\pi}^{2\pi} (-V)\sin nx\, dx$$

$$= \frac{V}{n\pi}(1 - 2\cos n\pi + \cos n\,2\pi)$$

From these results

$$b_1 = \frac{4V}{\pi},\quad b_2 = 0,\quad b_3 = \frac{4V}{3\pi},\quad b_4 = 0,\quad b_5 = \frac{4V}{5\pi},\quad \ldots$$

$$f(t) = \frac{4V}{\pi}\left(\sin \omega t + \frac{1}{3}\sin 3\omega t + \frac{1}{5}\sin \omega t + \cdots\right)$$

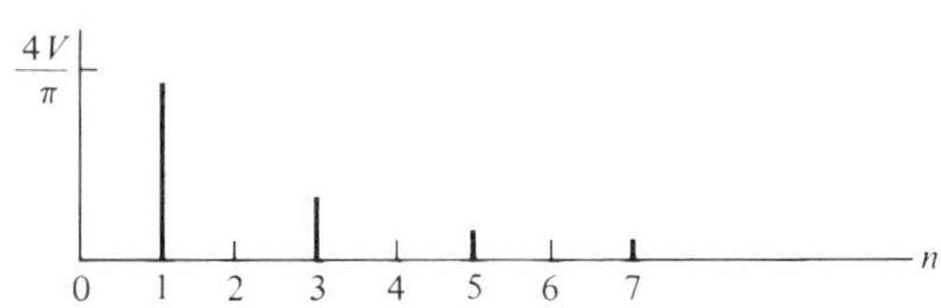

(b)

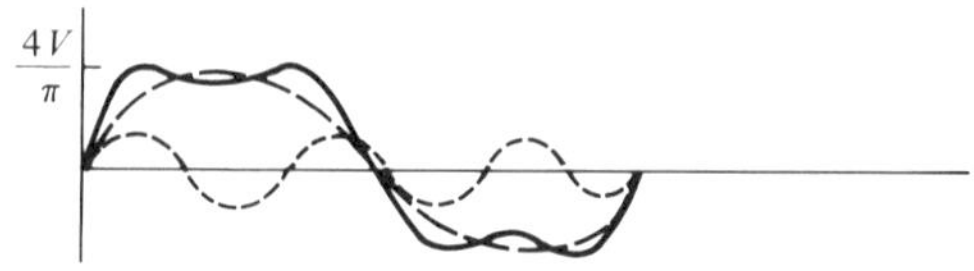

(c)

We see that the higher harmonics are needed to square up the wave. Therefore, we may associate the higher frequencies in a waveform with the discontinuties.

4-1.1 Choice of Axis

The choice of axis will determine the form of the Fourier series. A complete discussion of this topic is beyond the scope of this text. However, we may state the following facts briefly. If the function meets the following test:

$$f(x) = -f(-x) \tag{4-11}$$

it is an *odd function* and the Fourier series contains only sine terms. This is the case for the square wave in Example 4-1. If the function meets the following test:

$$f(x) = f(-x) \tag{4-12}$$

it is an an *even function* and the series contains only cosine terms. Refer to Figure 4-2, which is an even function. The Fourier series for this function is

$$f(t) = \frac{4V}{\pi}\left(\cos \omega t - \frac{1}{3}\cos 3\omega t + \frac{1}{5}\cos 5\ \omega t + \cdots\right) \tag{4-13}$$

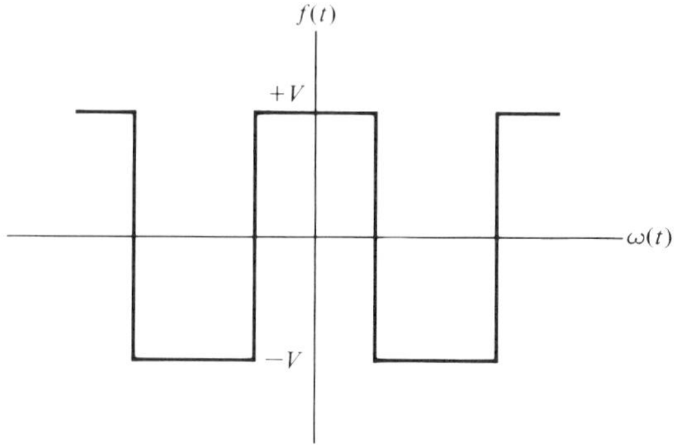

FIGURE 4-2. Square-wave function: even.

TABLE 4-1 Fourier Series for Common Waveforms

Function	Series
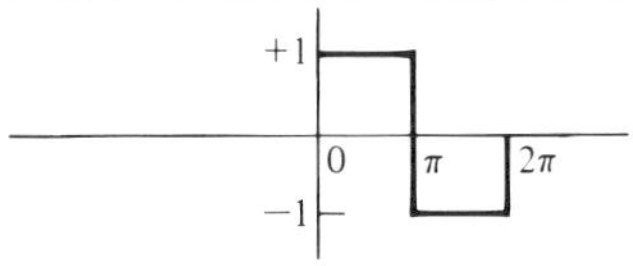	$f(t) = \frac{4}{\pi} \Sigma \frac{1}{n} \sin n\omega t$ $- n$ odd
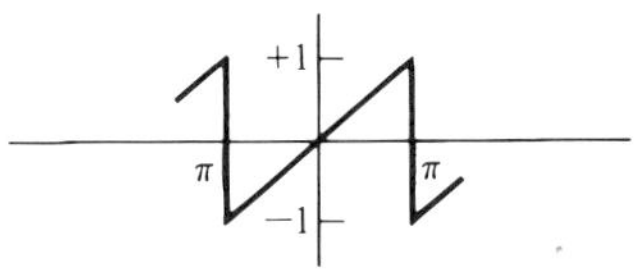	$f(t) = \frac{2}{\pi} \Sigma(-1)^{n-1} \sin n\omega t$ $-$ integer n
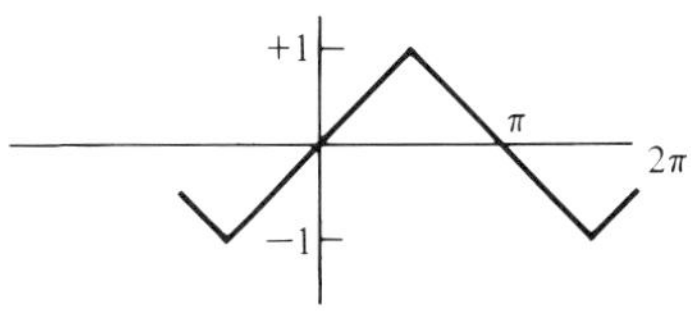	$f(t) = \frac{8}{\pi^2} \Sigma(-1)^{\left(\frac{n-1}{2}\right)} \frac{1}{n^2} \sin n\omega t$ $- n$ odd
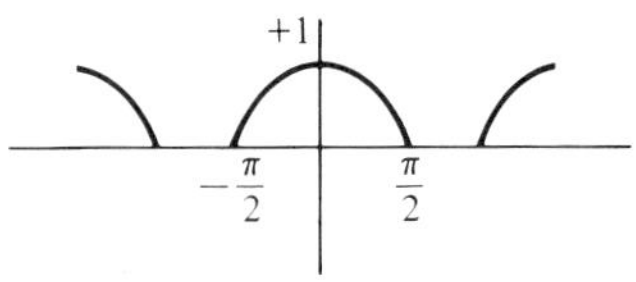	$f(t) = \frac{1}{\pi}(1 + \frac{1}{2} \cos \omega t + \frac{2}{3} \cos 2\omega t - \frac{2}{15} \cos 4\omega t + \frac{2}{35} \cos 6\omega t + \ldots)$
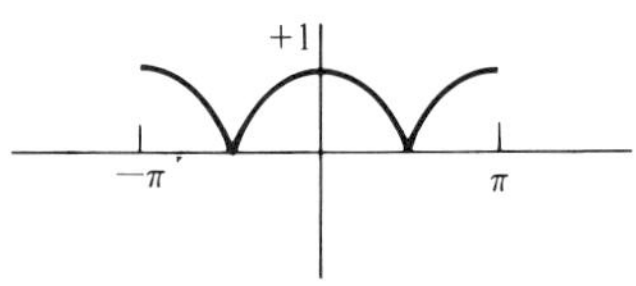	$f(t) = \frac{2}{\pi} (1 + \frac{2}{3} \cos 2\omega t - \frac{2}{15} \cos 4\omega t + \frac{2}{35} \cos 6\omega t + \ldots)$
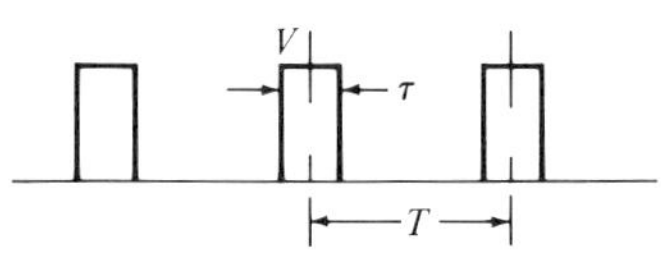	$f(t) = \frac{V\tau}{T} + \sum_{n=1}^{\infty} \frac{2V\tau}{T} \frac{\sin n \frac{\pi\tau}{T}}{n \frac{\tau}{T}} \cos n\omega t$
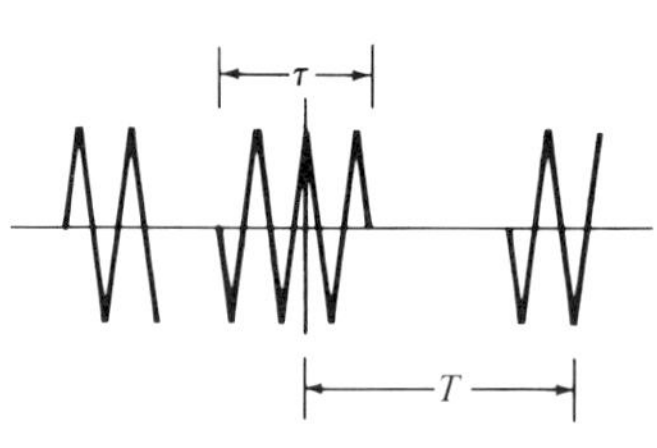	$f(t) = V \frac{\tau}{2T} \sum_{n=1}^{\infty} \left[\frac{\sin(n\omega_0 - \omega_C) \frac{\tau}{2}}{(n\omega_0 - \omega_c)\, \tau/2} + \frac{\sin(n\omega_0 + \omega_c) \frac{\tau}{2}}{(n\omega_0 - \omega_c)\, \tau/2} \right] \cos n\omega_0 t$

If the function is neither even nor odd, the Fourier series will contain both sine and cosine terms. In calculating the series, we are usually free to place the vertical axis wherever we wish.

4-2

FOURIER SERIES FOR COMMON WAVEFORMS

Certain waveforms occur in use with such frequency that the series is tabulated. Some of the more common ones are contained in Table 4-1.

4-3

USE IN CIRCUIT ANALYSIS

The uses of the Fourier series are varied and find many applications. For example, from Example 4-1b we see that most of the energy in the square wave is contained in the fundamental and harmonics out to the seventh harmonic. Therefore, we would expect that a filter or amplifier would need a bandwidth of at least $BW = 7\omega_1$ to pass a square wave with minimal distortion. Example 4-2 illustrates the use of Fourier series in a filter problem.

EXAMPLE 4-2

An *LC* filter is used to filter the output of a full-wave rectifier as shown below. Calculate the filter output voltage.

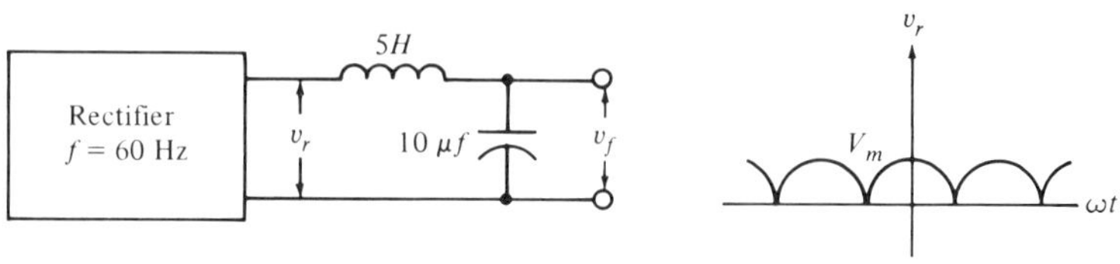

SOLUTION

From Table 4-1,

$$V_r(t) = \frac{2}{\pi}\left(1 + \frac{2}{3}\cos 2\omega t - \frac{2}{15}\cos 4\omega t + \frac{2}{35}\cos 6\omega t + \cdots\right)$$

The transfer function for the filter is

$$H(j\omega) = \frac{1/j\omega_n C}{j\omega_n L + 1/j\omega_n C} = \frac{1}{1 - \omega_n^2 LC}$$

$$= \frac{1}{1 - \omega_n^2 (5 \times 10^{-5})} \qquad \omega_n = n\omega = n(377)$$

Thus

$$E_o(j\omega) = H(j\omega)E_{\text{in}}(j\omega)$$

$n = 0$:

$$E_o = \frac{1}{1 - 0}\left(\frac{2}{\pi}\right) = \frac{2}{\pi}$$

$n = 2$:

$$E_o = \frac{1}{1 - (2 \times 377)^2(5 \times 10^{-5})}\left(\frac{2}{3}\right) \cos 2\omega t$$

$$= 0.0243 \cos 2\omega t = 0.0243 \cos (2\omega t + \pi)$$

$n = 4$:

$$E_o = \frac{1}{1 - (4 \times 377)^2(5 \times 10^{-5})}\left(\frac{2}{15}\right) \cos 4\omega t$$

$$= +0.00117 \cos 4\omega t$$

$n = 6$:

$$E_o = \frac{1}{1 - (6 \times 377)^2(5 \times 10^{-5})}\left(\frac{2}{35}\right) \cos 6\omega t$$

$$= -0.00022 \cos 6\omega t = 0.00022 \cos (6\omega t + \pi)$$

$$v_f = \frac{2}{\pi}[1 + 0.0243 \cos (2\omega t + \pi) + 0.00117 \cos 4\omega t$$

$$+ 0.00022 \cos (6\omega t + \pi) + \cdots]$$

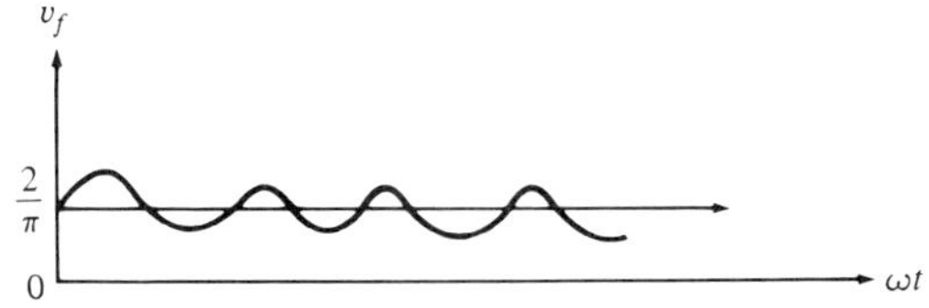

Thus, knowing the filter transfer function and the Fourier series for the input complex wave, it is rather easy to calculate the voltage output from the filter. In many cases we may not make the actual computations but are simply guided in our work by the Fourier series concept. Instruments known as spectrum analyzers are available for measuring the spectrum of complex waveforms.

4-4 RMS VALUE AND POWER

When a complex waveform is applied across a resistor, the energy is expended as heat. We know from circuit theory how to calculate the rms value of a sine wave and thus the power. The Fourier series, then, allows us to calculate the rms value of a complex waveform. In fact, the *rms value of a complex wave* is the square root of the sum of the squares of the rms values of the components.

To express this in mathematical form, let the Fourier series for a complex wave be

$$E(t) = E_0 + E_{1m} \sin(\omega_1 t - \theta_1) + E_{2m} \sin(2\omega_1 t - \theta_2) + \cdots \quad (4\text{-}14)$$

The rms value of the voltage is

$$E_{\text{rms}} = \sqrt{E_0^2 + E_1^2 + E_2^2 + \cdots} \quad (4\text{-}15)$$

where E_0 is the dc component and E_1, E_2, . . . , are the rms values of the harmonics. The rms values for the harmonics are expressed as

$$E_1 = \frac{E_{1m}}{\sqrt{2}}, \qquad E_2 = \frac{E_{2m}}{\sqrt{2}} \quad (4\text{-}16)$$

EXAMPLE 4-3

Find the rms value for the filter output voltage in Example 4-2.

SOLUTION

$$E_f = \sqrt{E_0^2 + \frac{1}{2}E_{2m}^2 + \frac{1}{2}E_{4m}^2 + \frac{1}{2}E_{6m}^2 + \cdots}$$

$$= \sqrt{\left(\frac{2}{\pi}\right)^2 + \left[\left(\frac{2}{\pi}\right)\left(\frac{1}{2}\right)(0.0243)\right]^2 + \left[\left(\frac{2}{\pi}\right)\left(\frac{1}{2}\right)(0.00117)\right]^2 + \left[\left(\frac{2}{\pi}\right)\left(\frac{1}{2}\right)(0.00022)\right]^2}$$

$$= 0.6367 \text{ V}$$

EFFECT OF FILTERS ON COMPLEX WAVEFORMS

Without explicitly stating the case, we have already approached this problem to some extent. In Example 4-2, if we compare the Fourier series representing the input and output of the filter, respectively, we see that both the phase and amplitude of the waveform component are modified. The discussion of low-pass and high-pass *RC* filters in Chapter 3 adds additional information on this subject. Thus, when considering the effect of filtering in the frequency domain, the results are well documented. At this point less has been said about how these effects appear in the time domain. For example, if a square wave is passed through an *RC* low-pass filter, how do we relate the frequency-domain filtering effects to the actual appearance of the output waveform? A consideration of this topic enables us to think effectively in either the frequency domain or the time domain.

In considering an application of filtering we would expect that as long as the signal of interest had frequency components that fell in the passband of the filter, the output should look like the input with possibly a phase shift. Such is essentially the case. Some filters possess this characteristic better than others, such as the Bessel filter. Of interest, then, is what happens to frequencies that fall in the roll-off region of the filter.

Consider a single-frequency sinusoidal wave; that is,

$$e(t) = E_m \cos \omega t = \mathrm{Re}(E_{m}e^{j\omega t})$$

If we take the integral of $e(t)$, the following result occurs:

$$\begin{aligned} \int e(t)\,dt &= \mathrm{Re}\int (E_{m}e^{j\omega t}\,dt) \\ &= \mathrm{Re}\left(\frac{1}{j\omega} E_{m}e^{j\omega t}\right) = \frac{1}{j\omega}\,e(t) \end{aligned} \qquad (4\text{-}17)$$

That is, multiplying by $1/j\omega$ in the frequency domain is the same as integration in the time domain. In symbol form, this is

$$\frac{1}{j\omega} \rightarrow \int dt$$

Now $1/j\omega = (1/\omega)e^{-j(\pi/2)}$. Thus integration causes the waveform to be divided by the amplitude of the frequency and the phase is shifted by $-\pi/2$ radians. It is instructive to convert $|1/\omega|$ to decibels and plot this function as shown in Figure 4-3. We see that the function which represents integration changes at a

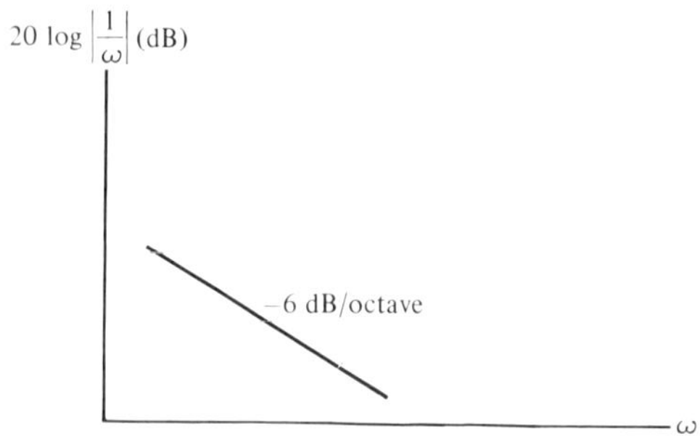

FIGURE 4-3. 1/ω in decibels versus ω.

−6 dB/octave rate. This is precisely the roll-off rate of the *RC* filter, and thus frequencies that fall in this roll-off region will be integrated. Considering the square wave, we remember that the higher harmonics are associated with the discontinuities or sharp corners of the square wave. If the higher frequencies fall in the roll-off region of a low-pass filter, we would expect the filter output to display an integration effect (see Figure 4-4).

In communication circuits we must often perform integration. We see from the discussion above that the circuit in Figure 4-4 will perform integration if the *RC* time constant is set such that all of the signal frequencies fall in the −6 dB/octave roll-off region. We see this from the filter transfer function, $F(j\omega)$.

$$F(j\omega) = \frac{1}{1 + j\omega RC} \tag{4-18}$$

For $\omega RC >> 1$, equation (4-18) reduces to

$$F(j\omega) \; \frac{1}{j\omega RC} \tag{4-19}$$

Taking $20 \log |F(j\omega)|$ and plotting versus ω, the integration action of the *RC* network is easily displayed. Thus from Figure 4-5 we see that all of the signal frequencies must be greater than $\omega_1 = 1/RC$.

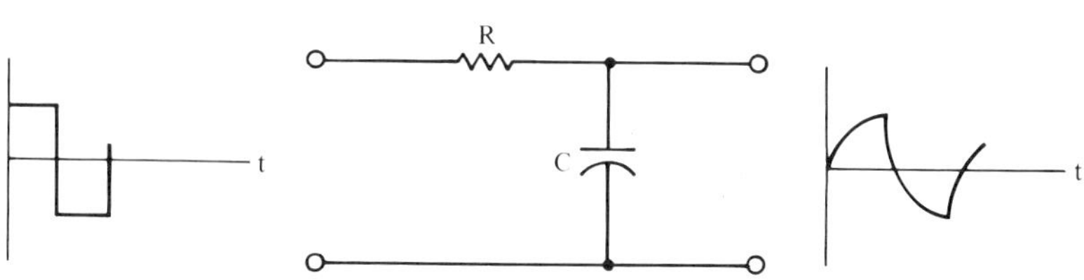

FIGURE 4-4. *RC* filter with insufficient bandwidth.

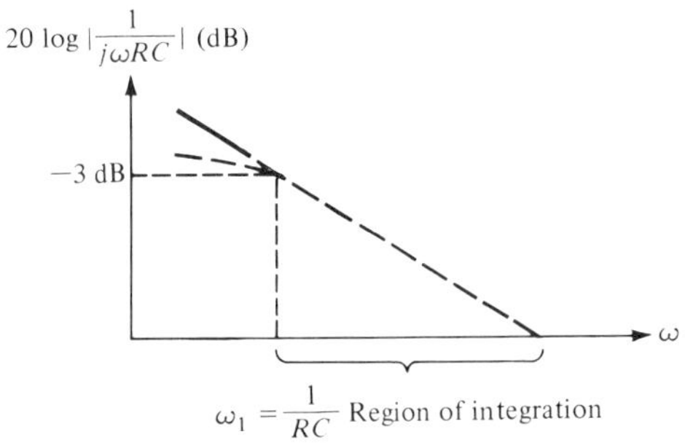

FIGURE 4-5. ***RC* network region of integration.**

The *RC* high-pass filter can be used to perform differentiation. In equation (4-17), if we take the derivative instead of the integral, we note that multiplying by $j\omega$ in the frequency domain is equivalent to differentiating in the time domain; that is,

$$j\omega \rightarrow \frac{d}{dt}$$

Now $j\omega = \omega e^{j(\pi/2)}$; thus differentiation causes the waveform to be multiplied by $|\omega|$ and the phase of each frequency is advanced by $\pi/2$ radians. The magnitude of $j\omega$ in decibels versus ω is plotted in Figure 4-6. We see that a function which increases at +6 dB/octave represents differentiation. This is precisely the slope of the high-pass *RC* network. This circuit is often used as a differentiator in communication circuits.

4-6 INFINITE PULSE TRAIN

The periodic pulse train shown in Figure 4-7 is of particular interest in communications theory and design. Its Fourier series is given in Table 4-1. From

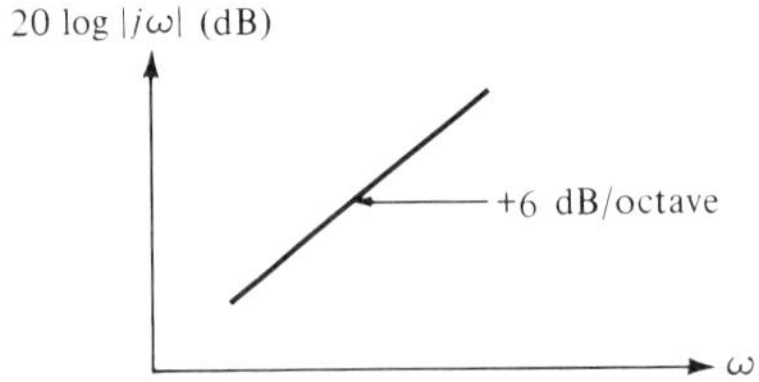

FIGURE 4-6. **$|j\omega|$ (dB) versus ω.**

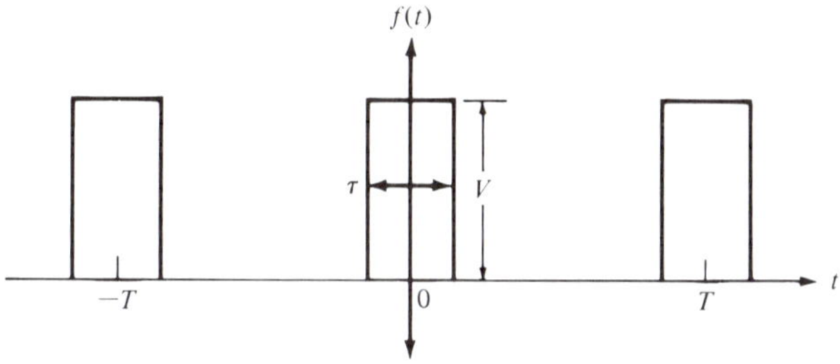

FIGURE 4-7. Infinite pulse train.

this series, the envelope of the frequency spectrum is given by

$$c_n = V\frac{\tau}{T}\frac{\sin (n\omega_0)\tau/2}{n\omega_0\tau/2} \tag{4-20}$$

which has the familiar sin x/x form, where $x = n\omega_0\tau/2$. The envelope is shown in Figure 4-8, where the spacing between successive lines in the spectrum is $\omega_0 = 2\pi/T$. The fundamental angular frequency, ω_0, and the zero crossings are separated by $2\pi/\tau$. Note that as the period T increases, the discrete spectral lines move closer together. As T decreases, the lines move farther apart. If we let the period, T, increase indefinitely, the lines crowd together and ultimately form a continuous frequency spectrum. For design purposes, the bandwidth of this pulse is often taken to be the distance between the first zero crossings. That is,

$$\text{BW} \approx \frac{2\pi}{\tau} \tag{4-21}$$

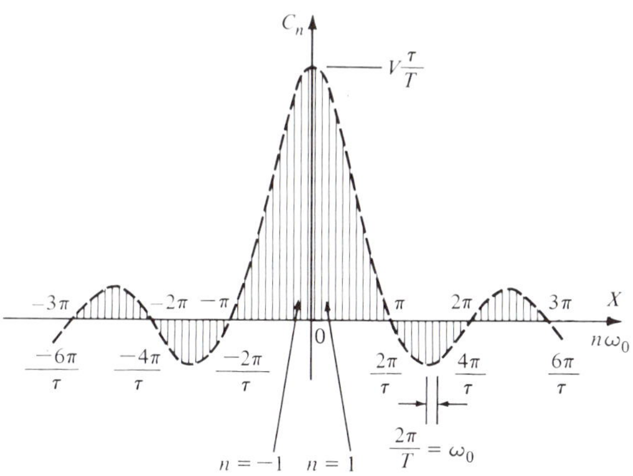

FIGURE 4-8. Frequency spectrum of the periodic pulse train.

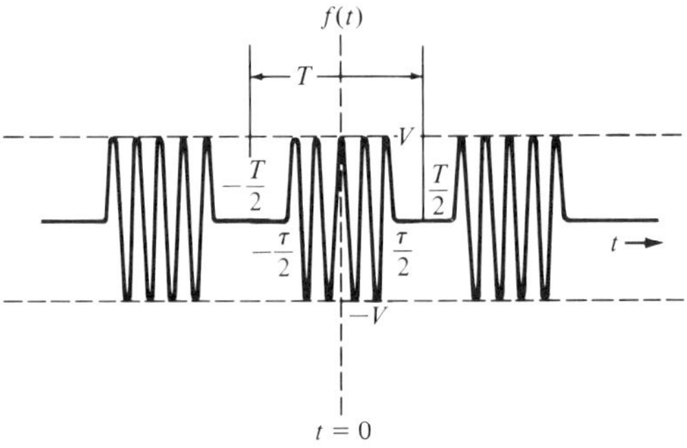

FIGURE 4-9. Pulsed RF wave.

4-7

PULSED RF WAVE

Consider now an application where the pulse train is used to turn an RF transmitter off and on, resulting in a pulsed RF wave (Figure 4-9). The Fourier series for this waveform is given in Table 4-1. The envelope of the spectrum is given by

$$c_n = V\frac{\tau}{2T}\left[\frac{\sin(n\omega_0 - \omega_c)\tau/2}{(n\omega_0 - \omega_c)\tau/2} + \frac{\sin(n\omega_0 + \omega_c)\tau/2}{(n\omega_0 + \omega_c)\tau/2}\right] \tag{4-22}$$

and the frequency spectrum is shown in Figure 4-10. We see that equation (4-22) is identical to (4-20) except for a symmetrical translation of the frequency components $n\omega_0$ to both sides of the origin. That is, $(n\omega_0 - \omega_c)$ and $(n\omega_0 + \omega_c)$, where ω_c is the frequency of the RF wave. The bandwidth is approximately $4\pi/\tau$ and is centered about the RF frequency, ω_c.

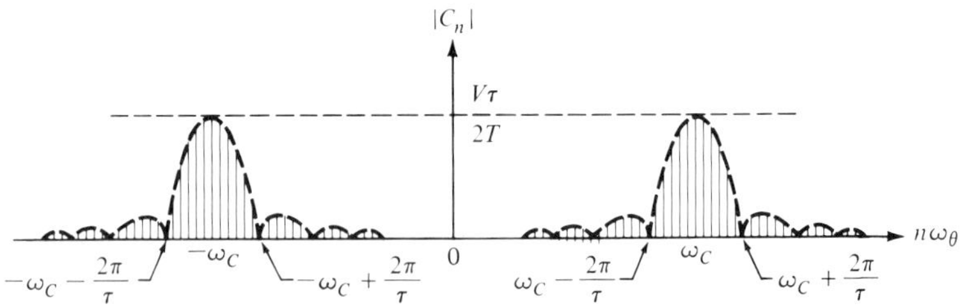

FIGURE 4-10. Frequency spectrum of pulsed RF wave.

EXAMPLE 4-4

A pulsed RF radar is needed which will be capable of resolving targets separated a distance of 10 m. Assume an RF frequency of 1 GHz and that the targets are located in a radial line from the radar site. (a) Calculate the pulse duration, τ, and (b) determine the bandwidth of the RF input amplifier of the radar, assuming that the reflected signal is a time-shifted version of the signal transmitted.

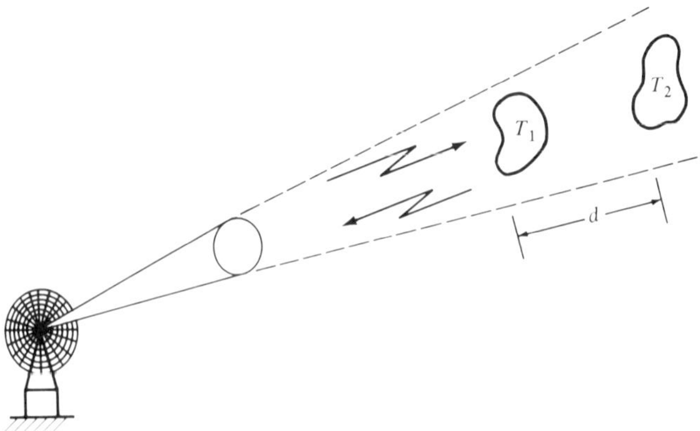

SOLUTION

Referring to the figure, a common antenna is used for both transmitting and receiving. During the pulse period, τ, energy is radiated, and during the off period, $T - \tau$, the antenna receives energy (target reflections) and directs this to the radar receiver, which "looks" for a target return.

(a) $d = C\tau = 10 \text{ m}$

$$\tau = \frac{10 \text{ m}}{3 \times 10^8 \text{ m/s}} = 3.33 \times 10^{-8}$$

(b) $$\text{BW} = \frac{4\pi}{\tau} = \frac{4\pi}{3.33 \times 10^{-8}} = 3.774 \times 10^8 \text{ Hz}$$

$$= 377 \text{ MHz}$$

We see that increased radar resolution can be achieved only by an increase in the RF bandwidth.

From the previous discussion we see that a periodic function will have a power spectrum that has a series of discrete lines at harmonically related frequencies. A Fourier series can also be used to represent nonperiodic functions over any selected time period, but the series will represent only the nonperiodic function in the selected interval of time, and not elsewhere.

THE FOURIER INTEGRAL AND CONTINUOUS FREQUENCY SPECTRA

The Fourier integral relates to continuous frequency spectra. Our previous discussion on Fourier series and the infinite pulse train leads directly into this subject. Equation (4-1) may be expressed as

$$f(\omega t) = \tfrac{1}{2}a_0 + \sum_{n=1}^{\infty} (a_n \cos n\omega t + b_n \sin n\omega t) \tag{4-23}$$

Now the sine and cosine terms may be expressed as exponentials. That is,

$$\begin{aligned} \cos \omega t &= \frac{1}{2}(e^{j\omega t} + e^{-j\omega t}) \\ \sin \omega t &= \frac{1}{2j}(e^{j\omega t} - e^{-j\omega t}) \end{aligned} \tag{4-24}$$

Substituting these equations into (4-23) results in

$$\begin{aligned} f(\omega t) &= \frac{1}{2}a_0 + \sum_{n=1}^{\infty}\left(a_n \frac{e^{jn\omega t} + e^{-jn\omega t}}{2} + b_n \frac{e^{jn\omega t} - e^{-jn\omega t}}{2j}\right) \\ &= \frac{1}{2}a_0 + \sum_{n=1}^{\infty}\left(\frac{a_n - jb_n}{2} e^{jn\omega t} + \frac{a_n + jb_n}{2} e^{-jn\omega t}\right) \end{aligned} \tag{4-25}$$

To simplify this expression, let

$$c_n = \frac{a_n - jb_n}{2} \qquad c_{-n} = \frac{a_n + jb_n}{2} \qquad c_0 = \mathrm{J}a_0$$

Equation (4-25) now has the form

$$f(\omega t) = c_0 + \sum_{n=1}^{\infty} (c_n e^{jn\omega t} + c_{-n} e^{-jn\omega t}) \tag{4-26}$$

Next if we let n range from $-\infty$ to $+\infty$, equation (4-26) reduces to

$$f(\omega t) = \sum_{n=-\infty}^{\infty} c_n e^{jn\omega t} \tag{4-27}$$

The coefficients c_n can now be evaluated in terms of a_n and b_n, which we already know. That is,

$$
\begin{aligned}
c_n &= \frac{1}{2\pi}\int_0^{2\pi} f(\omega t)\cos n\omega t \, d(\omega t) - \frac{j}{2\pi}\int_0^{2\pi} f(\omega t)\sin n\omega t \, d(\omega t) \\
&= \frac{1}{2\pi}\int_0^{2\pi} f(\omega t)(\cos n\omega t - j\sin n\omega t)\, d(\omega t) \qquad (4\text{-}28) \\
&= \frac{1}{2\pi}\int_0^{2\pi} f(\omega t) e^{-jn\omega t}\, d(\omega t)
\end{aligned}
$$

Thus equation (4-27) is the exponential form of the Fourier series and the coefficients c_n are found using equation (4-28). We see from equation (4-27) that we have both positive and negative frequencies. It takes a positive and a negative-frequency combination to form a physical frequency.

Now, how does this relate to the Fourier integral? The infinite pulse train is shown in Figure 4-7 and the Fourier coefficients c_n are computed using equation (4-28).

$$
\begin{aligned}
c_n &= \frac{1}{2\pi}\int_{-\pi}^{\pi} f(\omega_0 t) e^{-jn\omega_0 t}\, d(\omega_0 t) \\
&= \frac{1}{2\pi}\int_{-\tau/2}^{\tau/2} V e^{-jn\omega_0 t}\, d(\omega_0 t) \qquad (4\text{-}29) \\
&= V\frac{\tau}{T}\frac{\sin(n\omega_0\tau/2)}{n\omega_0\tau/2}
\end{aligned}
$$

This result was already given in equation (4-20). For a particular waveform, the ratio τ/T is constant and c_n varies as $\sin x/x$ (Figure 4-8).

The situation we are interested in is what happens if the period becomes infinite, leaving us with a single pulse. We have already observed that as the period increases, the harmonic components crowd closer together. The limiting situation, then, is if T goes to ∞. To accomplish this mathematically, begin with equation (4-27) and substitute equation (4-28) for c_n. Thus

$$
f(\omega_0 t) = \sum_{n=-\infty}^{\infty}\left[\frac{1}{2\pi}\int_{-\pi}^{\pi} f(\omega_0 t) e^{-jn\omega_0 t} d\omega_0 t\right] e^{jn\omega_0 t} \qquad (4\text{-}30)
$$

Net, let $\omega_0 = \Delta\omega$ and $T \rightarrow \infty$. Introduce a new variable $n\omega_0 = \omega$ into equation (4-30) and change the variable of integration to dt instead of $d(\omega t)$. After making these changes, equation (4-30) becomes

$$f(t) = \sum_{n=-\infty}^{\infty} \left[\frac{1}{2\pi} \int_{-\pi/\Delta\omega}^{\pi/\Delta\omega} f(t)e^{-j\omega t} dt \right] e^{j\omega t}\, \Delta\omega \tag{4-31}$$

Note that the limits on the integral are changed to reflect the change of integration variables. Next, take the limit as $\Delta\omega \rightarrow 0$. In the limit the summation process becomes an integration process and equation (4-31) becomes

$$f(t) = \int_{-\infty}^{\infty} \left[\frac{1}{2\pi} \int_{-\infty}^{\infty} f(t)e^{-j\omega t}\, dt \right] e^{j\omega t}\, d\omega \tag{4-32}$$

This equation is the *Fourier integral*. Let

$$g(\omega) = \frac{1}{2\pi} \int_{-\infty}^{\infty} f(t)e^{-j\omega t}\, dt \tag{4-33}$$

Using this expression, equation (4-32) may be written as

$$f(t) = \int_{-\infty}^{\infty} g(\omega)e^{j\omega t}\, d\omega \tag{4-34}$$

Equations (4-33) and (4-34) are referred to in the literature as the *Fourier transform pair*.

EXAMPLE 4-5

Find the Fourier transform for the single plus shown.

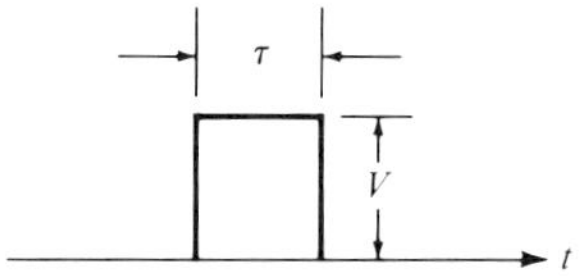

SOLUTION

$$g(\omega) = \frac{1}{2\pi} \int_{-\infty}^{\infty} f(t)e^{-j\omega t}\, d\omega$$

$$= \frac{1}{2\pi} \int_{-\tau/2}^{\tau/2} (V)e^{-j\omega t}\, d\omega = \frac{V\tau}{2\pi} \frac{\sin(\omega\tau/2)}{\omega\tau/2}$$

We see from the expression for $g(\omega)$ that the spectrum for the single pulse is a continuous spectrum and that it has the familiar sin x/x envelope. Note from equation (4-34) that if we know the spectrum $g(\omega)$, we can perform the inverse function and find $f(t)$. We will come into contact with the Fourier transform again in Chapter 12. In digital applications the digital version is known as the *discrete Fourier transform* (DFT).

PROBLEMS

4-1. The first five terms of the output of a half-wave rectifier are given in Table 4-1. Plot these five component terms and add them. Does it seem likely that the infinite series will converge to the rectified half-wave? Find the rms value.

4-2. Current is 1 A for 1 ms, from $t = -0.5$ to $+0.5$ ms. Current is then zero for 3 ms, from $t = 0.5$ to 3.5 ms. The current pulse is then repeated, and continues to be repeated with a 4-ms period. (a) Sketch the wave of current, and (b) find its analysis in the form of equation (4-1). Give terms through the fifth harmonic.

4-3. Voltage is 10 V for 3 ms and then zero for 1 ms repeated periodically. Sketch the wave and analyze it in the form of equation (4-1), selecting zero time at such a point that the series will contain no sine terms. Give the coefficients of all terms through the fifth harmonic.

4-4. Considering symmetry, what terms will not appear in the Fourier series [form of equation (4-1)] for the function shown? Determine the remaining terms and give the series through the third harmonic.

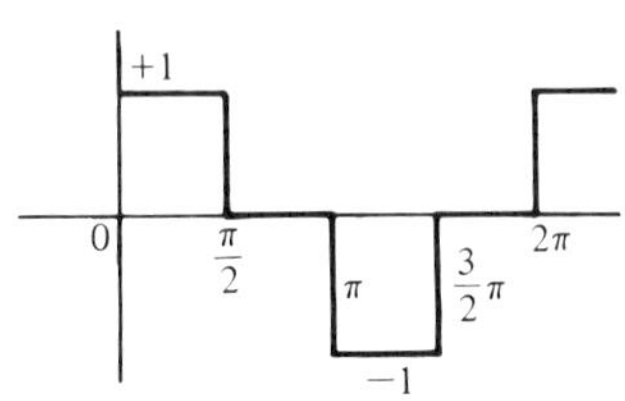

PROBLEM 4-4

4-5. Find the Fourier series (through the fifth harmonic) for the clipped sine wave shown.

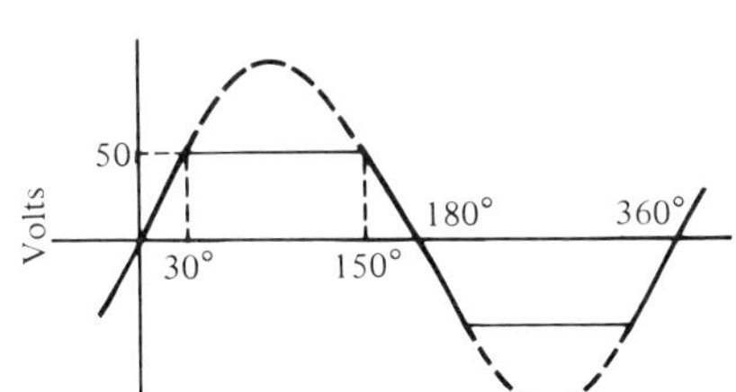

PROBLEM 4-5

4-6. Find the Fourier series (through the fifth harmonic) for a wave similar to that shown with Problem 4-5, but with only the positive crests clipped, the negative half-cycles remaining sinusoidal.

4-7. A 1-V square wave (Figure 4-1) of 100-Hz frequency is applied to a circuit of R and C in series. $R = 16 \times 10^3\ \Omega$, $C = 10^{-7}$ F. Express voltage across the resistance as a Fourier series (through the seventh harmonic).

4.8. (a) Find the Fourier series representation of each of the pulse trains in the figure shown. Choose the time origin so that a cosine series is obtained in each case. (b) Plot the first 10 Fourier coefficients versus frequency for each pulse train. (c) Compare the plot of part (a) of the figure with those for parts (b) and (c), paying particular attention to the rate of decrease of the higher-frequency components.

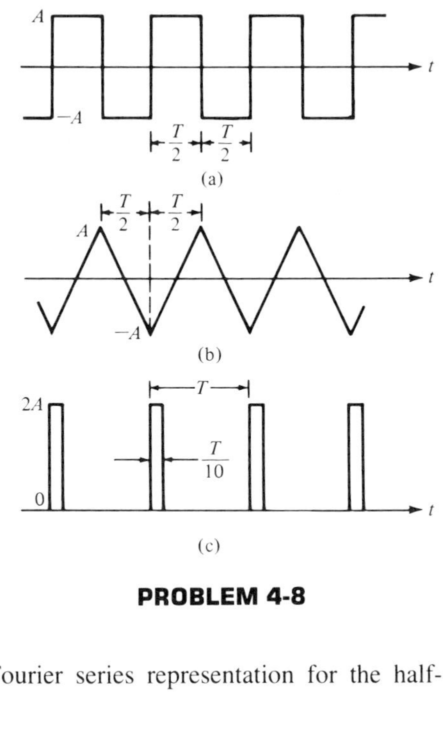

PROBLEM 4-8

4-9. Find the cosine Fourier series representation for the half-wave-rectified sine wave shown.

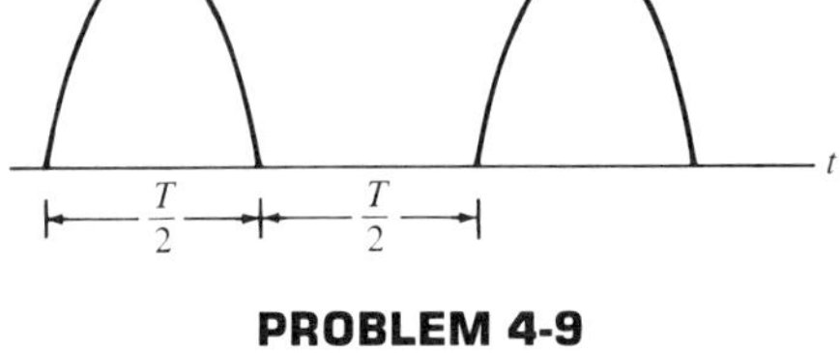

PROBLEM 4-9

4-10. Each of the pulse trains shown in the figure represents a voltage $v(t)$ appearing across a resistor of 1 Ω. (a) Find the total average power dissipated in the resistor by each voltage. (b) Find the percentage of total average power contributed by the fundamental of each voltage. (c) Find the percentage of the total power due to the first 10 harmonics of each voltage. (d) Find the percentage of the total power due to those harmonics of the voltage within the first zero-crossing interval of the amplitude spectrum.

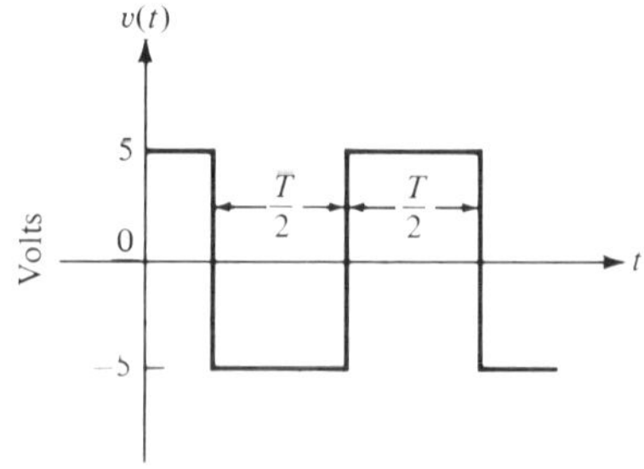

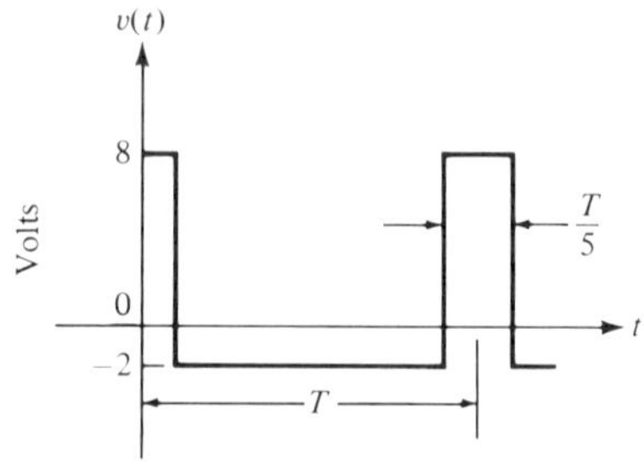

PROBLEM 4-10

4-11. A teletype printer uses a code consisting of seven time intervals per character (letter). This includes start and stop pulses during the first and last time intervals corresponding to each letter. During the remaining five intervals the signal may be either a logic 1 or a logic 0. The average character has the form shown in the figure. Assuming that the average word in the English language contains five letters and that one character is needed to transmit a space between words, there are on the average six characters per word. The rate of transmission is 25 words per minute.

(a) Find the length of one time interval.

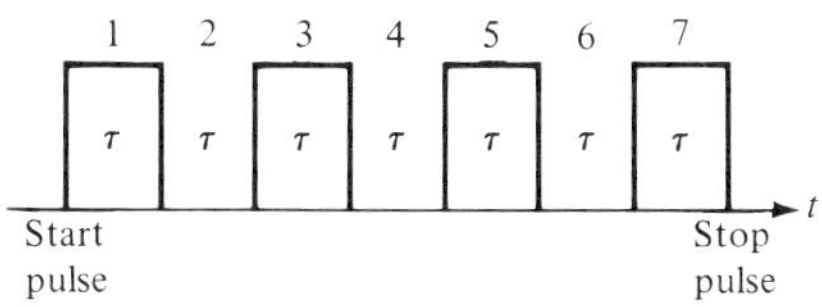

PROBLEM 4-11

(b) Find the bandwidth needed to transmit the first five harmonics of the pulse train shown. What bandwidth would be needed if the rate of transmission were increased to 100 words per minute?

4-12. (a) Sketch the frequency spectra of the following periodic pulse trains. Complete out to the first three zero crossings.
(b) How does the required transmission bandwidth relate to the pulse duration and the zero crossings?

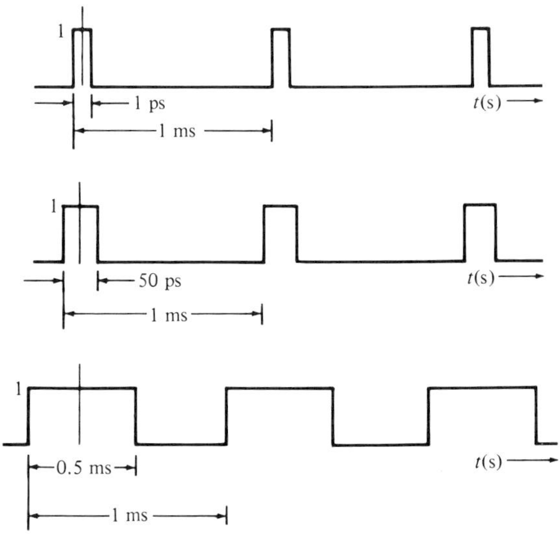

PROBLEM 4-12

4-13. A 3 KHz square wave is passed through a low pass filter with a −3 dB cutoff frequency of 12 KHz. Sketch the output signal and explain why distortion occurs.

CHAPTER 5

Noise

Communication systems often must process extremely small signals, and thus it is necessary to consider the noise aspects of the system. The noise of the system in conjunction with the inherent noisiness of the signal source determines the smallest signal that can usefully be processed. We are all familiar with the noise in a television picture. The "snow" present in the picture is caused primarily by noise sources in the set and partially by noise received from the antenna.

5-1 DESCRIPTION OF NOISE

Electrical noise may be defined as random electrical fluctuations. If we look at a noise voltage, we find that we cannot predict the amplitude of the noise voltage at any specific instant of time. An oscilloscope display of a noise voltage may appear as shown in Figure 5-1. By using probability theory, it can be shown that there are two things that can be determined about the noise voltage: (1) the

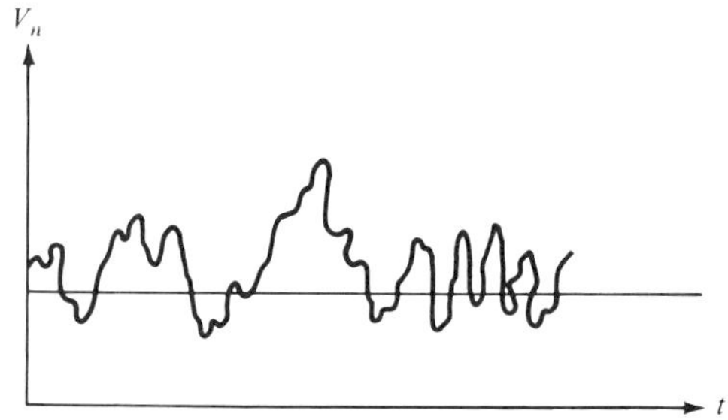

FIGURE 5-1. Random noise voltage.

average energy (or power) contained in the noise waveform, and (2) the power spectrum. We are already familiar with the concept of average power. For example, this is analogous to speaking of the rms value of a sinusoidal wave. We do not consider the time variation, only the average effect. The concept of a power spectrum was discussed at length in Chapter 4. To extend our understanding of electrical noise, then, we need to be aware of the different types of electrical noise.

5-2 TYPES OF NOISE

The following discussion refers to noise voltages generated internally in electrical/electronic components. It does not include atmospheric noise, ignition noise, and so on (i.e., external noise sources). An understanding of noise produced by components is necessary since with careful design it is possible to bring noise levels down to nearly theoretical levels.

For practical purposes, electrical noise may be divided into five basic categories:

- Thermal noise.
- Shot noise.
- Excess noise.
- Avalanche noise.
- Multistate noise.

These will be discussed sequentially.

5-2.1 Thermal Noise

Thermal noise results from the random thermal motion of the carriers within the conducting material. In 1928, Nyquist showed that the average noise power available for any conductor is

$$\text{available noise power} = kTB$$

where

T = temperature of conductor (degrees Kelvin: 273.1°C)
k = Boltzmann's constant: 1.38×10^{-23} w-s/deg
B = bandwidth over which the noise is measured (i.e., BW = $f_H - F_L$)

From basic circuit theory we remember that available power is obtained when the load is matched to the source. Since the available power is $E^2/4R$, we can write

$$\frac{\overline{E_n^2}}{4R} = kTB$$

or (5-1)

$$E_n = \sqrt{4kTBR}$$

where $\overline{E_n^2}$ is the rms value of the noise voltage squared. Equation (5-1) specifies the noise voltage that appears across any resistance R. It can be seen from this expression that the thermal noise for a given bandwidth is determined by the temperature and resistance. Thermal noise has a *flat* or *white* frequency spectrum and in terms of statistics has a Gaussian amplitude distribution.

When working with the various noise sources in a circuit, it is useful to have an equivalent circuit of the source. Since the noise voltage appears across the resistor, we may indicate the equivalent circuit as a noise generator in series with a noiseless resistor (Figure 5-2).

5-2.2 Shot Noise

Shot noise is generated as a result of the random passage of discrete carriers across a barrier or discontinuity such as a semiconductor function. This, then, is a characteristic of diodes and transistors. The basic equation for shot noise is

$$i_n = \sqrt{2qI_{dc}B} \tag{5-2}$$

where

i_n = shot noise (rms amperes)
q = electron charge: 1.59×10^{-19} coulomb (c)
I_{dc} = diode direct current (amperes)
B = equivalent bandwidth (hertz)

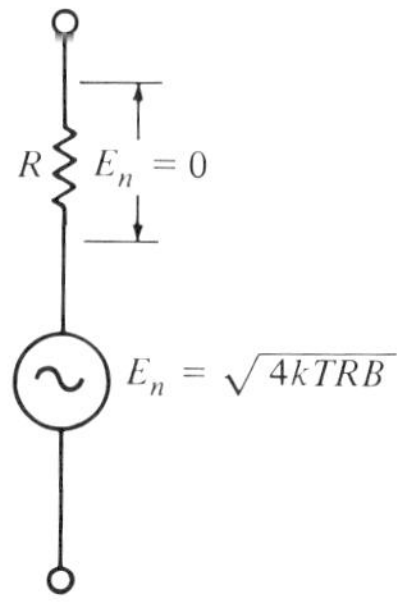

FIGURE 5-2. Equivalent circuit of a resistor.

Like thermal noise, shot noise has a flat frequency spectrum and a Gaussian amplitude distribution. This characteristic allows shot noise to be modeled as an equivalent resistance.

5-2.3 Excess Noise

This noise is sometimes referred to by such terms as current noise, bulk noise, and $1/f$ noise. Excess noise results from the passage of current through a semiconductor material. The exact mechanisms responsible for excess noise are not clearly understood, but the noise is essentially given by the following expression:

$$(E_e^2)^{1/2} = \left[\int_{f_1}^{f_2} (kI^{\alpha}f^{-\gamma})\, df\right]^{1/2} \tag{5-3}$$

where k, α, and γ varies with the specific component. Excess noise is a characteristic of transistors and diodes.

5-2.4 Avalanche Noise

Avalanche noise, like shot noise, is a junction phenomenon. The mechanism is as follows. Carriers in a high-voltage gradient develop sufficient energy to dislodge additional carriers through physical impact. These dislodged carriers, in turn, have sufficient energy to dislodge still more carriers. This process continues in cascade fashion. The random passage of these ''bundles'' of carriers across the junction generates avalanche noise. All transistors and diodes break into an avalanche condition if sufficient voltage is applied. A problem develops if this occurs within rated condition. Avalanche noise has a Gaussian distribution.

5-2.5 Multistate Noise

Multistate noise consists of erratic switching that is generated within the device at various sharply defined levels of applied current. It is found occasionally in transistors and somewhat more often in diodes. At present, little is known of the origin of multistate noise. Research indicates that it is related to a surface phenomenon.

5-3

TRANSISTOR NOISE SOURCES

Transistors generate thermal and shot noise which may be computed from the previous formulas. In addition, they generate excess or $1/f$ noise. Since both the thermal and shot noise are "white" noise, we may use two fictitious noise sources in circuit analysis. Transistor noise power versus frequency is shown in Figure 5-3. The white noise voltage may be computed using the Nyquist equation; that is,

$$\overline{E_n^2} = 4kTBR_{eq} \qquad (5\text{-}4)$$

where R_{eq} is the equivalent resistance that would produce E_n^2. The $1/f$ noise voltage may be computed using equation (5-3) if the constants are known. Generally, α is in the neighborhood of 2 and γ is near 1. k is a constant of the material. The usual procedure is to measure this noise.

5-4

REACTANCE AND EQUIVALENT NOISE BANDWIDTH

Inductive and capacitive reactance, with the exception of electrolytic capacitors, do not produce any thermal noise. This follows because a pure reactance does not dissipate power. The effect of reactance must be considered, however, be-

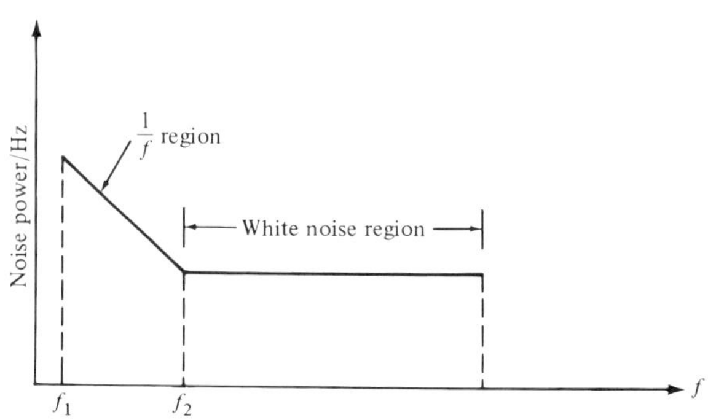

FIGURE 5-3. Transistor noise power versus frequency.

cause of its effect on the spectrum. Consider, for example, the RC low-pass filter. We know that the transfer function $H(j\omega)$ is

$$|H(j\omega)| = \frac{1}{\sqrt{1 + (\omega RC)^2}} \tag{5-5}$$

If an input power of $N(f) = kT$ watts per hertz is input to the filter, the output noise power is

$$\begin{aligned} N_o(f) &= |H(j\omega)|^2 kT \\ &= \frac{kT}{1 + (\omega RC)^2} \end{aligned} \tag{5-6}$$

If we integrate equation (5-6) from zero to infinity, the following results:

$$P_{no} = \int_0^\infty N_o(f)\,df = \int_0^\infty \frac{kT}{1 + (\omega RC)^2}\,df = \frac{kT}{4RC} \tag{5-7}$$

The total noise power out is given by $kT/4RC$. Comparing this to equation (5-1), we see that an equivalent noise bandwidth is given by

$$B_{\text{eff}} = \frac{1}{4RC} \tag{5-8}$$

The concept of an equivalent noise bandwidth is often used in the analysis of communication systems.

Electrolytic capacitors (including tantalum) comprise semiconductors of very large junction area. Thus electrolytic capacitors are subject to essentially the same type of noise problems as are other semiconductors. For example, the presence of substantial amounts of excess noise indicates leakage paths either at the terminations or within the capacitor. A particular problem in electrolytic capacitors is momentary avalanche noise. This indicates the tendency of a capacitor to ''punch through.''

EXAMPLE 5-1

Refer to the circuit shown. An input white noise $N(f) = 1\ \mu\text{W/Hz}$ is present at the input. (a) Compute the output noise power and (b) sketch the equivalent square-noise bandwidth versus the actual filter bandwidth.

SOLUTION

(a) $P_{no} = \dfrac{kT}{4RC} = \dfrac{10^{-6}}{(4)(100)(10^{-6})} = 2.5\text{ mW}$

$$(b)\ f_{-3\text{dB}} = \frac{1}{2\pi RC} = \frac{1}{2\pi(100)(10^{-6})} = 1590 \text{ Hz}$$

$$B_{\text{eff}} = \frac{1}{4RC} = \frac{1}{(4)(100)(10^{-6})} = 2500 \text{ Hz}$$

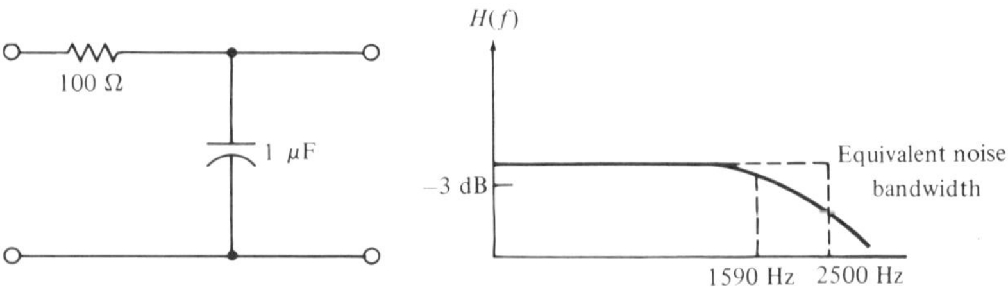

5-5

RESISTORS IN SERIES AND IN PARALLEL

Consider two resistors in series, as shown in Figure 5-4. The total noise voltage squared between points a and b is

$$\begin{aligned} E_{nT}^2 &= E_{n1}^2 + E_{n2}^2 = 4kTR_1B + 4kTR_2B \\ &= 4kTB(R_1 + R_2) \end{aligned} \tag{5-9}$$

or

$$E_{nT} = \sqrt{4kTB(R_1 + R_2)} \tag{5-10}$$

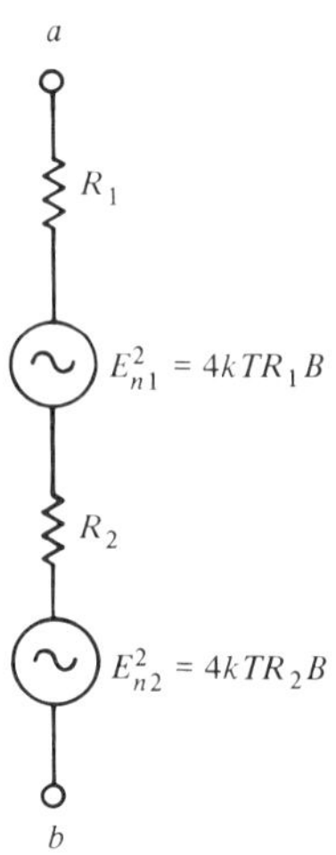

FIGURE 5-4. Equivalent circuit of two resistors in series.

Thus the total noise voltage may be computed by summing the resistors in series. This result may be extended to N resistors in series.

A similar result may be shown for resistors in parallel. Thévenin's equivalent circuit for the noise source in Figure 5-1 may be replaced by its Norton's equivalent circuit. That is,

$$I_n^2 = \frac{4kTB}{R} = 4GkTB \tag{5-11}$$

Then for two resistors in parallel,

$$\begin{aligned} I_n^2 &= I_{n1}^2 + I_{n2}^2 = 4kTB(G_1 + G_2) \\ &= 4kTB\,\frac{R_1 + R_2}{R_1R_2} \end{aligned} \tag{5-12}$$

The parallel combination may be replaced by its parallel equivalent.

5-6 EQUIVALENT NOISE RESISTANCE

The noise that originates in a device is often represented by a fictitious resistance, referenced to the input of the device and the device itself is then assumed to be noiseless. Thus an amplifier may have an actual input resistance R_i and a fictitious resistance R_n referred to the input. Thus in terms of noise produced the input resistance is $R_i + R_n$.

EXAMPLE 5-2

An amplifier at room temperature has a bandwidth of 7 MHz and an equivalent shot noise resistance of 300 Ω. The actual input resistance is 200 Ω. Assuming a gain of 150, what is the output noise?

SOLUTION

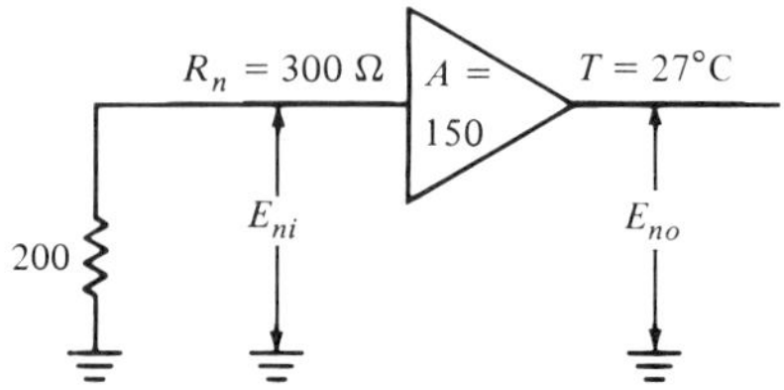

$$\begin{aligned} E_{no} &= AE_{ni} \\ &= 150[\sqrt{4kT(R_1 + R_n)\Delta f}] \\ &= 150[\sqrt{(4)(1.38 \times 10^{-23})(273 + 27)(200 + 300)(7 \times 10^6)}] \\ &= 1.142 \text{ mV} \end{aligned}$$

5-7

EQUIVALENT NOISE RESISTANCE FOR CASCADED AMPLIFIERS

Several amplifiers in cascade are shown in Figure 5-5. As mentioned previously, it is convenient to refer all of the noise-producing entities back to the amplifier input as an equivalent resistance. The noise voltage at the output due to R_3 is

$$E_{n3} = \sqrt{4kT \, \Delta f R_3} \tag{5-13}$$

This same voltage would be present at the output of the amplifier if there were no R_3 but there were an R_3' present at the input of the second amplifier. That is,

$$\begin{aligned} E'_{n3} &= \frac{E_{n3}}{A_2} = \frac{\sqrt{4kT \, \Delta f R_3}}{A_2} = \sqrt{4kT \, \Delta f \frac{R_3}{A_2^2}} \\ &= \sqrt{4kT \, \Delta f R_3'} \end{aligned} \tag{5-14}$$

Thus the equivalent resistance R_3 is given by

$$R_3' = \frac{R_3}{A_2^2} \tag{5-15}$$

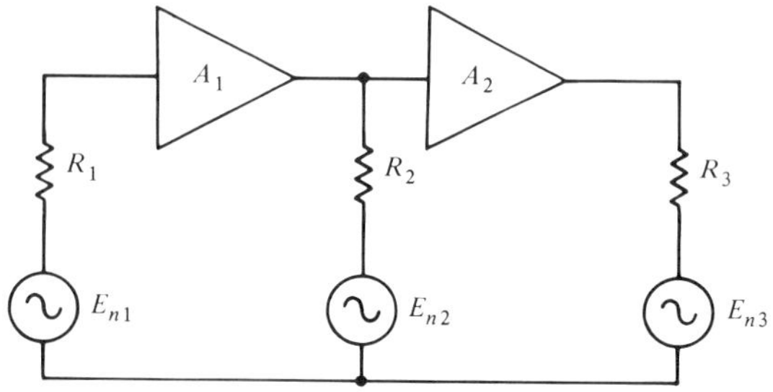

FIGURE 5-5. Cascaded amplifiers.

The actual input resistance at the second stage is R_2, but the equivalent noise resistance R'_{n2} is

$$R'_{n2} = R_2 + R'_3 = R_2 + \frac{R_3}{A_2^2} \tag{5-16}$$

Similarly, at the input to the first amplifier the noise resistance R'_{n2} referred to the input is

$$\begin{aligned} R'_n &= \frac{R'_{n2}}{A_1^2} = \frac{R_2 + R_3/A_2^2}{A_1^2} \\ &= \frac{R_2}{A_1^2} + \frac{R_3}{A_1^2 A_2^2} \end{aligned} \tag{5-17}$$

The total equivalent noise resistance for the cascaded amplifier is

$$R_{eq} = R_1 + \frac{R_2}{A_1^2} + \frac{R_3}{A_1^2 A_2^2} \tag{5-18}$$

EXAMPLE 5-3

The amplifier in Figure 5-5 has the following values: $R_1 = 500\ \Omega$, $R_2 = 30\ k\Omega$, $R_3 = 1\ M\Omega$, $A_1 = 10$, and $A_2 = 20$. In addition, there is an equivalent shot noise resistance at the inputs of amplifiers 1 and 2 of 2 kΩ and 10 kΩ, respectively. The amplifier has a bandwidth of 10 kHz and operates at room temperature. Calculate (a) R_{eq}, (b) E_{ni}, and (c) E_{no}.

SOLUTION

(a) $$R_{eq} = 500 + 2000 + \frac{30{,}000 + 10{,}000}{10^2} + \frac{1{,}000{,}000}{(10^2)(20)^2}$$

$$= 2500 + 400 + 25 = 2925\ \Omega$$

(b) $$E_{ni} = \sqrt{4kTR_{eq}\,\Delta f}$$

$$= \sqrt{(4)(1.38 \times 10^{-21})(300)(2925)(10{,}000)}$$

$$= 696\ \mu V$$

(c) $$E_{no} = A_1 A_2 E_{ni}$$

$$= (10)(20)(696)(10^{-6}) = 139.2\ mV$$

Note from this example that the primary noise-producing mechanisms are centered around the first stage. Thus we see the necessity for designing a low-noise amplifier for the first RF stage in a receiver.

5-8 NOISE FIGURE

A receiver or amplifier that is "ideal" adds no noise of its own to the signal being amplified. In practice, of course, this is not possible. A measure of the noise produced by a practical receiver or amplifier is the *noise figure* or *noise factor*. The noise figure, F, of a linear system may be defined as

$$F = \frac{(S/N)_i}{(S/N)_o} \tag{5-19}$$

where

$(S/N)_i$ = ratio of input signal power to input noise power
$(S/N)_o$ = ratio of output signal power to output noise power

The term S/N is read "signal-to-noise ratio" or SNR.

5-8.1 Calculation of Noise Figure

Consider the amplifier equivalent circuit shown in Figure 5-6. The signal appearing across the terminals of the amplifier is expressed as

$$E_{si} = \frac{R_T}{R_a + R_T} E_s \tag{5-20}$$

and the input signal power is

$$P_{si} = \frac{E_{si}^2}{R_T} = \frac{R_T}{(R_a + R_T)^2} E_s^2 \tag{5-21}$$

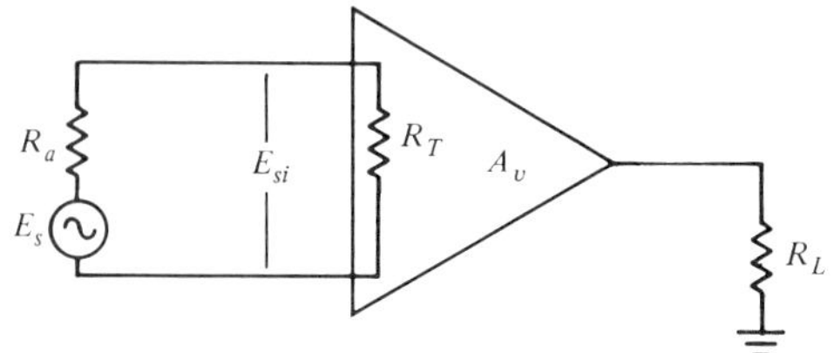

FIGURE 5-6. Linear amplifier.

The input noise voltage is produced by the parallel combination of R_a and R_T. That is,

$$E_{ni}^2 = 4kT\,\Delta f \frac{R_a R_T}{R_a + R_T} \tag{5-22}$$

and the noise power in is

$$P_{ni} = \frac{E_{ni}^2}{R_T} = \frac{4kT\,\Delta f R_a}{R_a + R_T} \tag{5-23}$$

Using equations (5-21) and (5-23), the input SNR is

$$(S/N)_i = \frac{R_T E_s^2}{4kT\,\Delta f R_a (R_a + R_T)} \tag{5-24}$$

Referring to equation (5-19), we see that if we can find the output SNR, we will know the noise figure. The signal voltage out is the input signal amplified by the voltage gain. Thus

$$E_{so} = AE_{si}$$

and the output signal power is

$$P_{so} = \frac{E_{so}^2}{R_L} = \frac{(AE_{si})^2}{R_L} = \frac{A^2 R_T^2 E_s^2}{(R_a + R_T)^2 R_L} \tag{5-25}$$

where E_{si} is given by equation (5-20). The output SNR may now be expressed as

$$(S/N)_o = \frac{P_{so}}{P_{no}} = \frac{A^2 R_T^2 E_s^2}{(R_a + R_T)^2 R_L P_{no}} \tag{5-26}$$

where P_{no} is the output noise power. The noise figure, then, is the ratio of equations (5-24) and (5-26). That is,

$$F = \frac{(S/N)_i}{(S/N)_o} = \frac{R_L (R_a + R_T) P_{no}}{4kT\,\Delta f A^2 R_a R_T} \tag{5-27}$$

To make this a usable expression for the noise figure, it remains to find an expression for the noise power, P_{no}. The noise power out may be computed from the equivalent noise resistance. Referring to Figure 5-6 and equation

(5-18), we see that R_{eq} is

$$R_{eq} = R_a \| R_T + R'_n \tag{5-28}$$

where R'_n is as in equation (5-17). The noise voltage in may be expressed as

$$E_n = \sqrt{4kTR_{eq}\,\Delta f} \tag{5-29}$$

and the noise power out is then given by

$$P_{no} = \frac{E_{no}^2}{R_L} = \frac{(AE_{ni})^2}{R_L} = \frac{4kT\,\Delta f R_{eq} A^2}{R_L} \tag{5-30}$$

Substituting this expression into equation (5-27) results in

$$F = R_{eq}\frac{R_a + R_T}{R_a R_T} \tag{5-31}$$

Now by using equation (5-28), this expression becomes

$$F = \left(\frac{R_a R_T}{R_a + R_T} + R'_n\right)\frac{R_a + R_T}{R_a R_T} = 1 + \frac{R'_n(R_a + R_T)}{R_a R_T} \tag{5-32}$$

If $R_T >> R_a$, this expression reduces to

$$F = 1 + \frac{R'_n}{R_a} \tag{5-33}$$

This is often referred to as the *mismatched* case.

EXAMPLE 5-4

A single-stage amplifier has an input impedance of 75 Ω and is driven by a 75-Ω source with a 1-mV signal. The equivalent shot noise resistance at the input of the amplifier is 100 Ω and the load on the amplifier is 1 kΩ. The amplifier has a voltage gain of 100, a bandwidth of 20 kHz, and operates at room temperature. Calculate (a) the output noise voltage and (b) the amplifier noise figure.

SOLUTION

(a) $R_{eq} = R_a \| R_T + R_n + \dfrac{R_L}{A^2}$

$$= \frac{(75)(75)}{75 + 75} + 100 + \frac{1000}{(100)^2} = 137.6\ \Omega$$

$$E_{no} = AE_{ni} = 100\sqrt{4kTR_{\text{eq}}\,\Delta f}$$

$$= 100\sqrt{(4)(1.38 \times 10^{-23})(273 + 27)(137.6)(2 \times 10^5)}$$

$$= 67.51\ \mu\text{V}$$

(b) $$F = 1 + \frac{R_n'(R_a + R_T)}{R_a R_T} = 1 + \frac{(100.1)(150)}{75^2}$$

$$= 3.66 \text{ or } 5.63 \text{ dB}$$

5-8.2 Noise Figure of Networks in Cascade

The noise figure for a network is expressed by equation (5-19). That is,

$$F = \frac{(S/N)_i}{(S/N)_o} = \frac{S_i}{S_o}\frac{N_o}{N_i} = \frac{N_o}{GN_i} \tag{5-34}$$

where G is the gain of the network. Then using equations (5-34) and (5-1), the noise out, N_o, may be expressed as

$$N_o = FGN_i = FGkTB \tag{5-35}$$

If we refer this noise to the input, the equivalent input noise, N_i, is

$$N_i = FkTB \tag{5-36}$$

The source contributes noise equal to kTB. Thus the network must contribute the difference. Thus the amplifier contribution is

$$\begin{aligned} P_{na} &= N_i - N_s = FkTB - kT_B \\ &= (F - 1)kTB \end{aligned} \tag{5-37}$$

Now consider two networks in cascade, as shown in Figure 5-7. From equations (5-36) and (5-37), the noise input to network 2 is

$$N_{o1} = G_1F_1kTB_n + (F_2 - 1)kTB \tag{5-38}$$

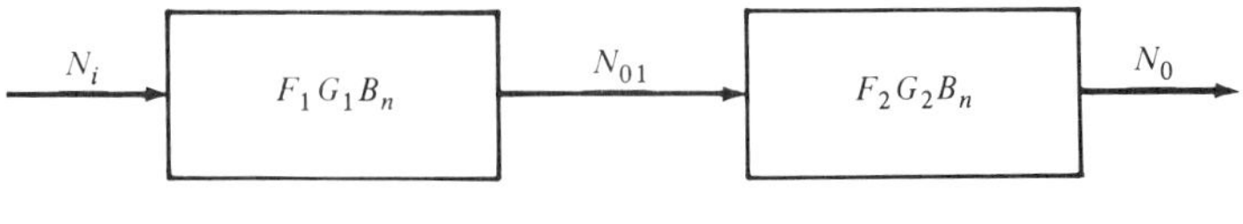

FIGURE 5-7. Two networks in cascade.

and the noise out is the gain G_2 times the noise in, N_{o1}.

$$N_o = G_2 N_{o1} = G_2 G_1 F_1 kTB_n + G_2(F_2 - 1)kT_B \tag{5-39}$$

Using equations (5-34) and (5-39), the noise figure may be written as

$$\begin{aligned} F &= \frac{N_o}{G_T N_i} = \frac{N_o}{G_1 G_2 N_i} = \frac{G_1 G_2 F_1 kTB_n + G_2(F_2 - 1)kTB}{G_1 G_2 kTB_n} \\ &= F_1 + \frac{F_2 - 1}{G_1} \end{aligned} \tag{5-40}$$

This result may be extended for N networks in cascade and is

$$F = F_1 + \frac{F_2 - 1}{G_1} + \frac{F_3 - 1}{G_1 G_2} + \cdots + \frac{F_{N-1}}{G_1 G_2 \cdots G_{N-1}} \tag{5-41}$$

A similar expression may be derived for the overall noise figure when the bandwidths and/or the temperature of the individual networks are not the same.

EXAMPLE 5-5

A television set has an RF front-end amplifier with a gain of 20 dB and a noise figure of 7 dB. It is followed by a mixer with a gain of 15 and a noise figure of 10 dB. Calculate the overall noise figure.

SOLUTION

$$F_1 = 7 \text{ dB} \rightarrow 5.01$$

$$F_2 = 10 \text{ dB} \rightarrow 10$$

$$G_1 = 20 \text{ dB} \rightarrow 100$$

$$F = F_1 + \frac{F_2 - 1}{G_1} = 5.01 + \frac{10 - 1}{100} = 5.1$$

$$F_{\text{dB}} = 10 \log_{10} 5.1 = 7.07 \text{ dB}$$

5-8.3 Effective Noise Temperature

The noise figure concept was originally formulated to describe the performance of relatively noisy networks. The use of the noise figure with a standard temperature $T_0 = 290°\text{K}$ is not as convenient with low-noise devices as is the *effective noise figure*. The amplifier noise contribution is expressed by equation

(5-37). We can represent this noise in terms of a fictitious noise temperature, T_e, as follows:

$$kT_eB = (F - 1)kT_0B$$

or

$$T_e = (F - 1)T_0$$

Thus the noise figure can be represented by a temperature, T_e, at the input of the network, which would account for the noise at the output. We should emphasize that the effective noise temperature of a network is not necessarily equal to the physical noise temperature at the input terminals. The effective noise temperature of N networks in cascade is

$$T_e = T_1 + \frac{T_2}{G_1} + \frac{T_3}{G_1G_2} + \cdots + \frac{T_N}{G_1G_2 \cdots G_{N-1}} \tag{5-42}$$

The effective noise temperature is preferred for describing low-noise devices.

EXAMPLE 5-6

A receiver has a noise figure of 3 dB measured with respect to a standard noise temperature of $T_0 = 290°K$. What is the effective noise temperature, T_e?

SOLUTION

$$T_e = (F - 1)T_0 = (2 - 1)(290°K) = 290°K$$

Note that the effective noise temperature is not dependent on the actual choice of the standard temperature, T_0, as long as the noise figure is measured with respect to the same standard.

5-8.4 Equivalent Antenna Noise Temperature

Noise is one of the limiting factors of any communication system. In estimating the noise due to external sources, it is necessary to consider the gain and orientation of the antenna. Since the available noise is expressed in terms of a bandwidth, it may be expressed as an effective noise temperature, T_a. That is,

$$P_n = kT_aB$$

or (5-43)

$$T_a = \frac{P_n}{kB}$$

where

P_n = noise power available from an equivalent loss-free antenna
T_a = equivalent antenna noise temperature

If we know the antenna temperature and the equivalent noise temperature of the receiver input, the total noise temperature at the amplifier input is

$$T = T_a + T_e \qquad (5\text{-}44)$$

The total noise power, then, is

$$P_n = k(T_a + T_e)B \qquad (5\text{-}45)$$

5-9 SEMICONDUCTOR NOISE FIGURE

The sensitivity of an amplifier is limited by the signal-to-noise ratio available at the antenna input. We saw from the discussion on noise figure that if the systems adds noise, it degrades this ratio. The choice of semiconductor device to be used as an amplifier, mixer, and so on, is thus of extreme importance. From a practical standpoint, a distinction needs to be made between two types of noise figure: spot noise figure and average noise figure. *Spot noise figure* is the noise figure measured in a narrow frequency band. For example, it is useful for specifying the noise figure at individual frequencies. *Average noise figure* is measured over a band of frequencies such as the bandwidth of an RF amplifier and is useful in specifying system performance.

Noise figure variation versus frequency for all types of semiconductor devices is a complex subject. Fortunately, a study of transistor noise figure versus frequency covers several characteristics that are common with other devices. We have already touched on this subject in Section 5-3. The calculation of noise figure for a transistor utilizes an equivalent circuit involving device parameters and noise sources. The complexity of the resultant equation for noise figure will depend on the complexity of the equivalent circuit. An expression for transistor noise figure is given by Nielsen's equation:

$$F = 1 + \frac{r_e}{2R_g} + \frac{r_b'}{R_g} + \frac{(R_g + r_e + r_b')^2}{2\beta_0 R_g r_e}\left[1 + (\beta_0 + 1)\left(\frac{f}{f_{ab}}\right)^2\right] \tag{5-46}$$

where

$$r_e = \frac{26}{I_E(\text{mA})} \quad \text{ohms}$$

r_b' = base spreading resistance (ohms)

R_g = source resistance (ohms)

β_0 = low-frequency common-emitter current gain

f = frequency (hertz)

f_{ab} = common-base cutoff frequency (hertz)

Figure 5-8 gives the measured and calculated curves of noise figures versus frequency for the 2N4957 transistor. Note that the curve is relatively flat at midband and increases approximately 6 dB/octave at high frequencies. The discrepancy between the two curves at midband could be due to the equivalent circuit used or in the assumed noise sources.

5-9.1 Transistor Noise Figure

The transistor noise figure at low frequencies ($f < 1$ kHz) is not predicted by Neilsen's equation. The major noise source in this region is excess noise, discussed in Section 5-2.3. A typical noise figure versus frequency curve for the 2N4952 is shown in Figure 5-9. Note that in the $1/f$ region the slope is ap-

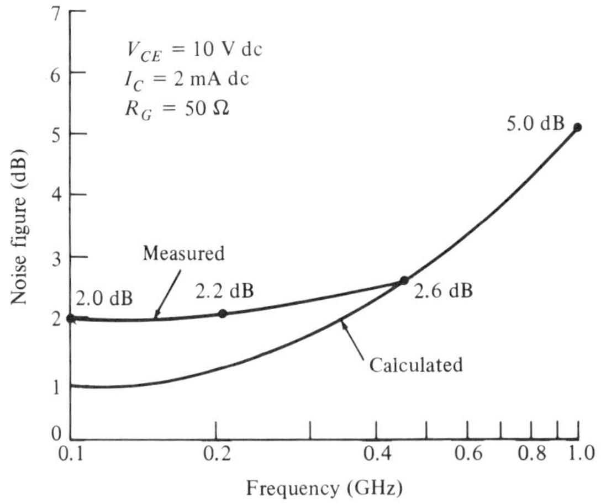

FIGURE 5-8. Typical spot noise figure versus frequency (calculated and measured) for 2N4957 PNP transistor.

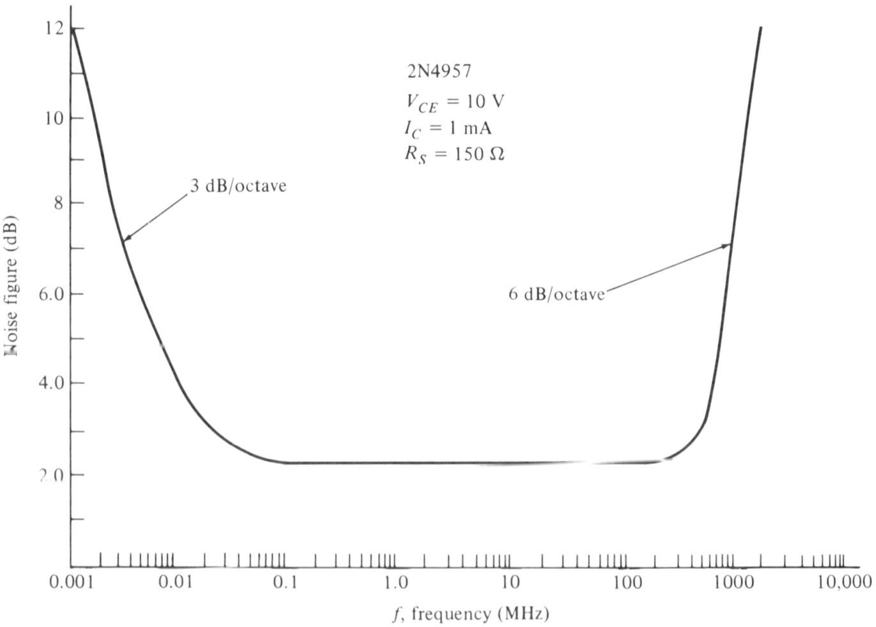

FIGURE 5-9. Noise figure asymptotes.

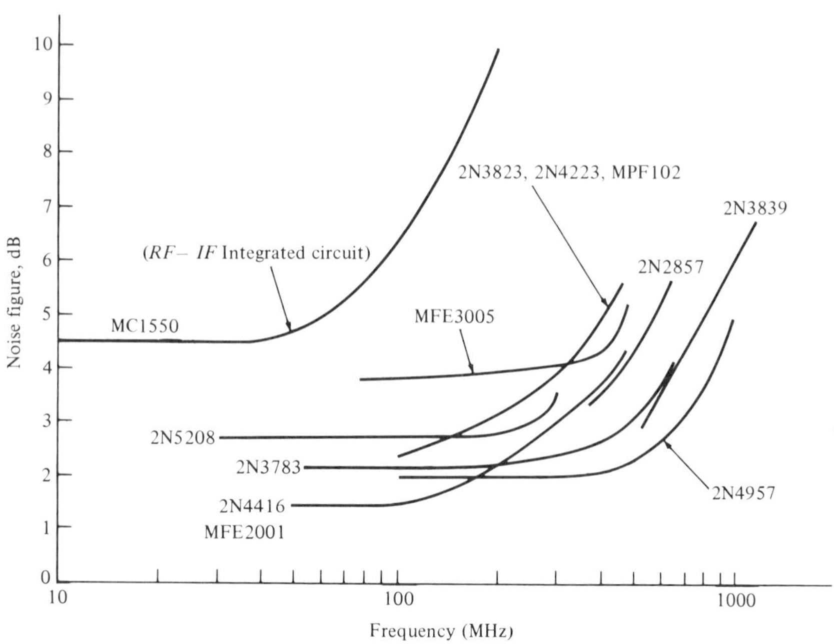

FIGURE 5-10. Noise figure versus frequency for various RF semiconductor devices.

proximately -3 dB/octave. The noise figure curves for various semiconductor devices is shown in Figure 5-10.

Equivalent noise circuits vary depending on device type, such as junction transistors or MOSFETs, but the same general noise figure characteristics prevail (see Figures 5-11 and 5-12). Note from these figures that by a proper selection of source resistance, the noise figure can be minimized. Also, it can be seen that a similar dependence exists between noise figure and bias current.

An additional design variable of importance that is not shown by equation (5-46) is the source susceptance. For many high-frequency devices, lower noise figures are attained by mistuning the input circuit rather than tuning for maximum gain—that is, the base (common-emitter configuration) sees a resistance shunted by a susceptance (usually capacitive). The value of this susceptance is found by empirical methods. Again, the considerations are true with junction transistors and MOSFETs. That is, the noise figure, in general, is optimum for a particular bias point and source admittance.

5-10 NOISE MEASUREMENTS

A number of methods are available for making noise figure measurements. The technique used often depends on the equipment available and the frequency region of interest.

5-10.1 Low-Frequency Noise Figure Measurements Using a Signal Generator

This method is one of the simplest for measuring the noise figure at low frequencies. The test setup is shown in Figure 5-13. The application of this test is to fix the input signal-to-noise ratio and then measure the output signal-to-noise ratio. The noise figure then, by equation (5-19), is

$$F(\text{dB}) = 20 \log_{10} \frac{E_{si}}{E_{ni}} - 20 \log_{10} \frac{E_{so}}{E_{no}} \tag{5-47}$$

where

E_{si} = signal input voltage
E_{ni} = noise input voltage
E_{so} = signal output voltage
E_{no} = noise output voltage

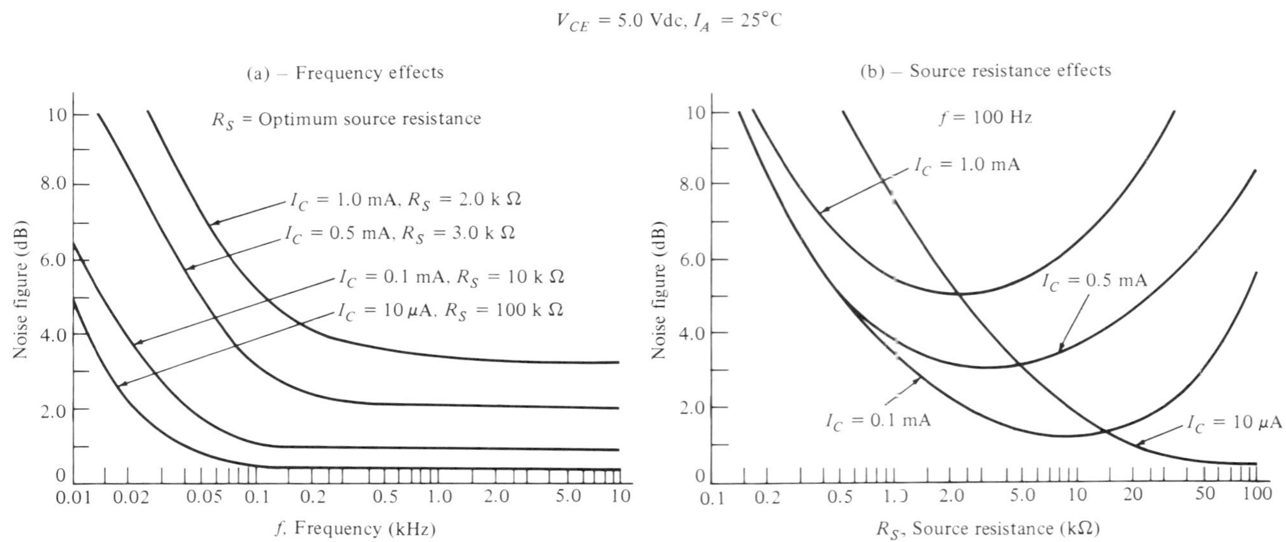

FIGURE 5-12. Noise figures versus frequency for the 2N4220 J-FET.

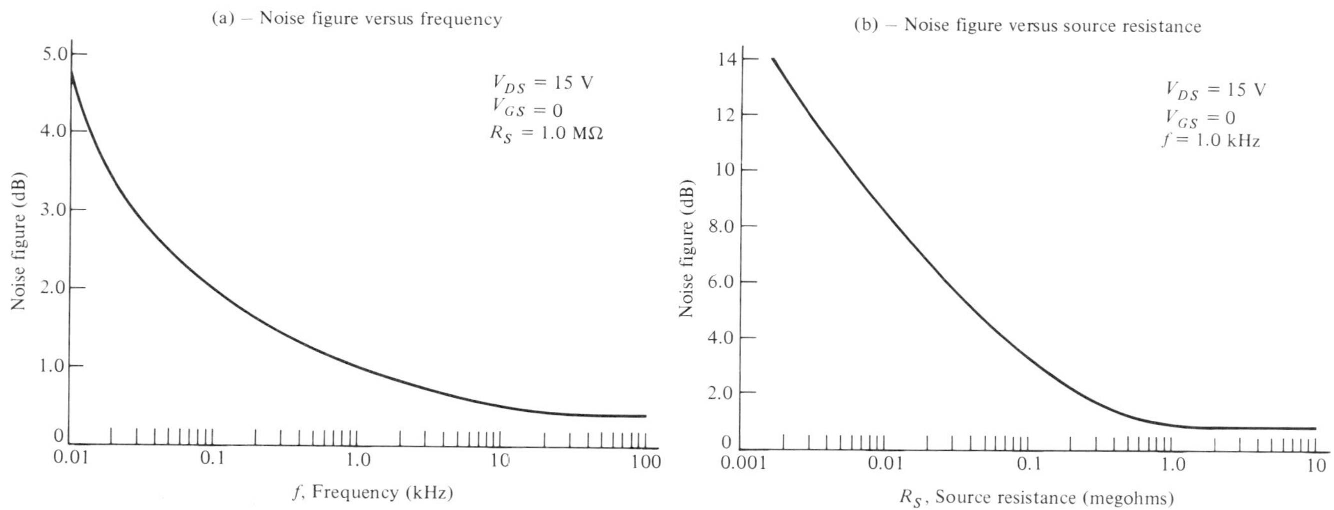

FIGURE 5-11. Noise figure considerations for the 2N5088 and 2N5089 silicon NPN transistors.

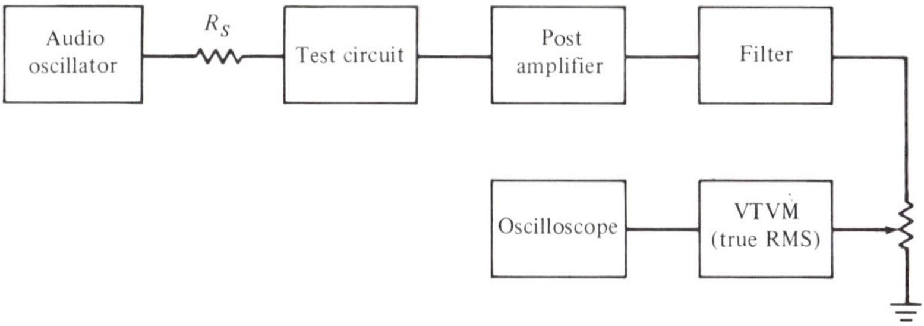

FIGURE 5-13. Test setup for low-frequency noise figure measurement: signal generator method.

If the source resistance, R_s, is known, the input noise voltage, E_{ni}, may be calculated by equation (5-1) The signal voltage, E_{ni}, is then adjusted so that the input signal-to-noise ratio is a convenient value, such as 20 dB. Thus equation (5-47) becomes

$$F(\text{dB}) = 20 \text{ dB} - 20 \log_{10} \frac{E_{so}}{E_{no}} \tag{5-48}$$

The procedure is as follows. First adjust the input voltage signal-to-noise ratio to 20 dB. Next, adjust the variable attenuator so that the voltmeter indication of E_{so} is full scale. Following this step, reduce the input signal to zero and note the drop in output voltage (decibels). Since the output voltage without a signal is E_{no}, the actual drop in output in decibels represents the ratio of E_{so} and E_{no}. Substitute this ratio in equation (5-48) and compute the noise figure.

Note that the measured output signal also includes E_{no}. Thus the actual signal measured is $(E_{so}^2 + E_{no}^2)^{1/2}$. This will result in the inclusion of an error in the measured noise figure. However, the error is less than 0.5 dB for a device having a noise figure up to 10 dB if the input signal-to-noise ratio is 20 dB.

5-10.2 Constant-Gain Method for Low-Frequency Noise Figure Measurements

This method is based on maintaining a constant known voltage gain and measuring the noise voltage out, E_{no}. Equation (5-34) may be expressed in terms of voltages as

$$F = \frac{E_{no}^2}{E_{ni}^2(G_v)^2} \tag{5-49}$$

where G_v represents the voltage gain. A commercial instrument employing this principle is the Quan-Tech noise analyzer, model 2173C-2181 (see Figure

5-14). This instrument measures noise at five frequencies simultaneously and is an industry standard for low-frequency noise measurements.

5-10.3 Output-Doubling Method for High-Frequency Noise Measurements

One of the simplest procedures for measuring high-frequency noise is the noise-doubling method. This method involves increasing the noise generator output until the noise power in the output is doubled. A typical test setup is shown in Figure 5-15. One of the most commonly used sources of noise is the saturated thermionic diode, which is a vacuum diode with a tungsten filament and is operated so that the anode current is dependent on the filament temperature. When operated in this fashion, there is no ''space-charge smoothing'' of the noise output from the filament and the available noise due to the random nature of the emission of electrons (shot noise).

When the noise power produced by the noise generator is equal to that produced by the circuit under test, the following relation holds:

$$P_{no} = GP_{ng} \tag{5-50}$$

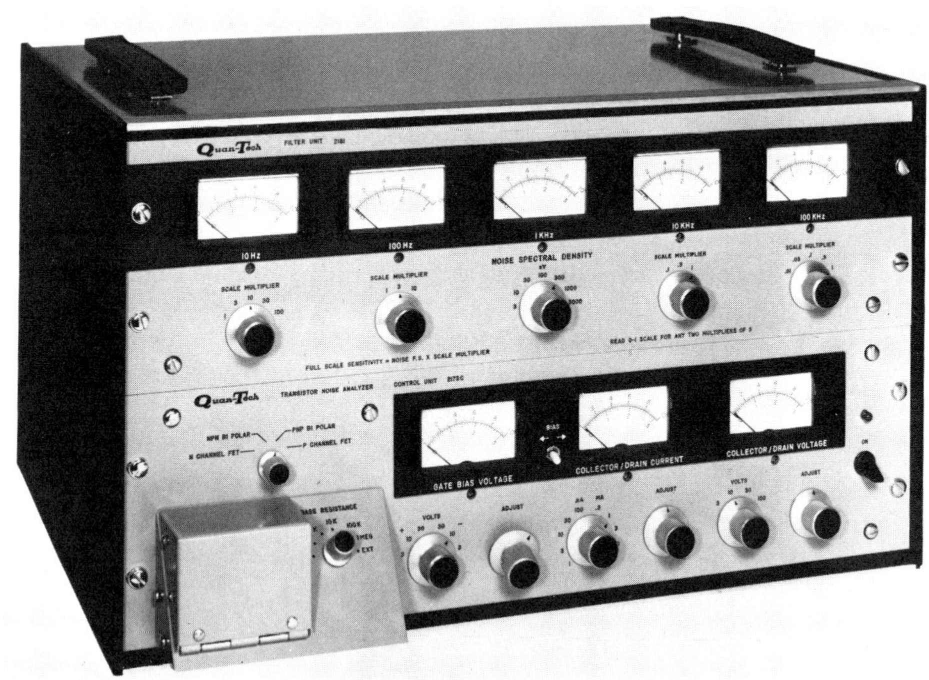

FIGURE 5-14. Quan-Tech noise analyzer.

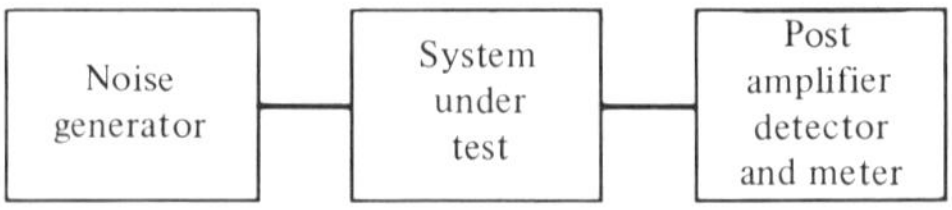

FIGURE 5-15. Noise figure system using a postamplifier.

where

P_{no} = noise power out from source and the device
G − power gain
P_{ng} = noise generator power output

Substituting equation (5-50) into equation (5-34) yields for the noise figure:

$$F = \frac{GP_{ng}}{GN_i} = \frac{P_{ng}}{P_{ni}} \tag{5-51}$$

The actual values of P_{ng} and P_{ni} are functions of both the source resistance and the input impedance of the device under test. The power terms in equation (5-51) may be defined on an available power basis. Thus

$$F = \frac{i_{ng}^2 R_g/4}{e_{ng}^2 R_g/4} = \frac{2qI_{dc}BR_g/4}{kTB} = \frac{qI_{dc}R_g}{2kT} \tag{5-52}$$

where equations (5-1) and (5-2) are used. If we assume that the temperature of the source is 290°K and $R_g = 50$, this expression reduces to

$$F = I_{dc} \qquad \text{mA} \tag{5-53}$$

The actual technique used to double the noise output power involves more than connecting an RF voltmeter to the output of the test circuit and noting when the twice power output point is reached. First, the noise power levels are very low and a postamplifier is required to increase these levels to the capability of the RF voltmeter. In addition, this power amplifier (if spot noise is desired) must have a bandwidth less than the test amplifier and must not overload for the signal levels encountered. Finally, the actual RF voltmeter used must be linear and must respond to the rms voltage.

The contribution of noise by the postamplifier is taken into account by equation (5-41). That is,

$$F = F_1 + \frac{F_2 - 1}{G_1} + \frac{F_3 - 1}{G_1 G_2} + \cdots + \frac{F_{N-1}}{G_1 G_2 \cdots G_{N-1}}$$

where

F = overall noise figure (power ratio)
F_1 = noise figure of the first stage (power ratio)
F_2 = noise figure of the second stage (power ratio)
F_3 = noise figure of the third stage (power ratio)
G_1 = power gain of the first stage (ratio)
G_2 = power gain of the second stage (ratio)

This same technique may be used to measure the noise figure of, for example, a superheterodyne receiver. The noise generator is connected to the input terminals of the receiver. With the noise generator turned off, the IF and AF controls are adjusted to give a convenient reading on the output meter. The diode noise generator is then turned on and the noise generator output is adjusted until the receiver output or diode current increases by $\sqrt{2}$, which means that the noise power output has doubled. The noise figure is then given by equation (5-53). If the temperature is other than 290°K, it may be corrected by the following:

$$F(\text{at } 290°\text{K}) = F(\text{at different temp.}) \times \frac{290 \times T}{290} \tag{5-54}$$

where T is the different temperature in °K.

The effect of image frequency response must be considered if the superheterodyne receiver has significant response at the image frequency. Image frequency will be discussed in Chapter 10. Essentially, if there is significant response at the image frequency, an artificially low noise figure will be measured since the receiver output will be increased by noise injected at the image frequency. The noise figure as measured must therefore be increased by a numerical amount corresponding to the ratio of the sum of signal and image frequency response to signal frequency response. If there is no image rejection, the correction will be a factor of 2 (3 dB).

5-10.4 Commercial Equipment Technique

More sophisticated noise measurement techniques generally use a noise generator with two modes of operation. One mode uses the diode noise output power and the second mode utilizes the noise output from the diode source resistance at room temperature. A typical setup is shown in Figure 5-16.

The general equation relating noise figure and the two output levels is

$$F = 10 \log_{10} \frac{T_2 - T_0}{T_0} - 10 \log_{10} \frac{P_{no2}}{P_{no1}} - 1 \tag{5-55}$$

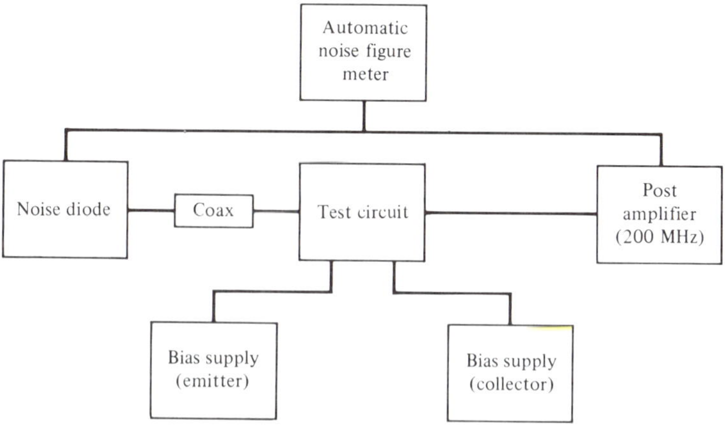

FIGURE 5-16. Typical test setup for 200-MH$_z$ noise figure measurement.

where the expression $[(T_2 - T_0)/T_0]$ is a measure of the relative excess noise power from the noise generator and P_{no1} and P_{no2} are the two resultant power output levels. These levels correspond respectively to the "fired" and "unfired" modes of operation of the noise source.

The relative excess noise power is usually specified by the manufacturer. A typical value for an argon gas tube is 15.2 dB. However, for accurate measurements, each source must be individually checked at each frequency of intended use.

Equation (5-55) is the basis for several of the most popular manual and automatic commercial noise characterization systems. Automatic noise measuring equipment, in general, gate the noise diode "on" and "off" at a fixed rate and continuously record the output power ratio. Methods of reducing this ratio to a continuous noise figure reading vary with the manufacturer.

5-10.5 Typical VHF Noise Figure Measurements Using Automatic Equipment

A typical 200-MHz noise figure test setup using automatic noise measuring equipment is shown in Figure 5-16. The noise diode is connected to the test circuit through a special coaxial cable which transforms the 50-Ω noise diode impedance to a higher impedance level (such as 100 Ω) with negligible circuit loss. Input circuit losses are important since each decibel of input circuit loss adds directly to the noise figure. The special coaxial cable length is trimmed to exactly one-quarter of a 200-MHz wavelength, such that when one end of the cable is terminated in 50 Ω, the resultant impedance level at the other end is the desired resistive value with minimum shunting capacitance (0.2 pF maximum). The cable characteristic impedance necessary to give the desired source

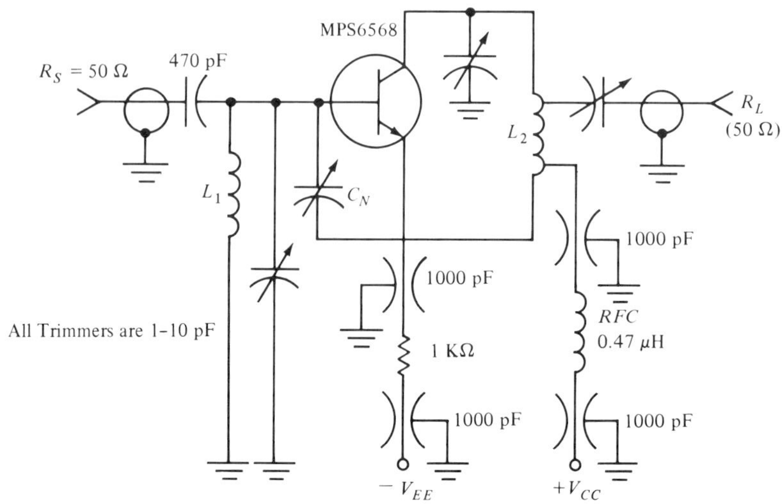

L_1 = 3 turns no. 18 tinned wire, 1/4 inch I.D.
Air wound – length is 3/16 inch.

L_2 = 5 1/2 turns no. 16 tinned wire, 3/8 inch I.D.
Air wound – length is 1/2 inch, V_{CC} feeds tap 2 1/2 turns from collector end. Output tap, 1/2 turn from collector end.

FIGURE 5-17. Typical 200 MH$_Z$ noise test circuit.

resistance is calculated from

$$R_0^2 = R_S R_L$$

where

R_0 = characteristic impedance of cable (ohms)
R_L = resistance of noise diode (50 Ω)
R_S = desired source resistance (ohms)

Several values of source resistance values are readily attained by using coaxial cable with standard values of characteristic impedance. Additional resistance values can be attained by varying the center-conductor wire size of the cable.

Directly following the test amplifier is the postamplifier. This amplifier provides the additional gain necessary to operate the automatic noise figure equipment. Although the required gain is also a function of the test circuit gain, a typical power gain is 50 dB. Since system bandwidth is determined by the automatic noise equipment, no special bandwidth restrictions are required.

Any contribution of the postamplifier to the total measured noise figure may be calculated with equation (5-41). A postamplifier with a noise figure low

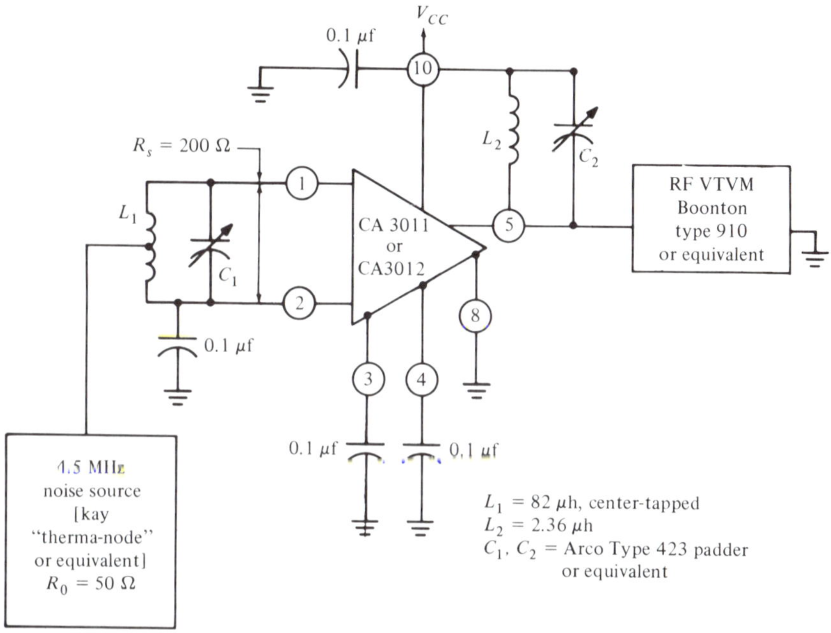

FIGURE 5-18. Noise figure test setup.

enough to avoid any significant contribution to the total noise figure is highly desirable.

The input impedance of the postamplifier should be 50 Ω since this impedance level will provide the load for the test circuit. If the device under test is mismatched at the output, the resultant source resistance seen by the postamplifier can increase to several hundred ohms and may result in instability problems with the postamplifier. Consequently, the first stage of the postamplifier should be stable regardless of the source resistance.

A typical 200-MHz noise figure test circuit is shown in Figure 5-17. Here a common-emitter neutralized configuration is used with two bias supplies. The desired source impedance seen by the device under test is provided by the special coaxial cable discussed previously, and the input tank circuit provides the necessary tuning. The coil (L_1) must be carefully constructed to minimize input circuit loss; not only should the unloaded Q be high, but the actual inductance value should be as large as possible and still maintain tuning.

In general, all lead lengths should be minimum and circuit layout should be such that the neutralizing capacitor (C_N) is provided the shortest path between the base and L_2, through a small hole in the shield. Piston-type capacitors are recommended for C_N.

The immediate area should be carefully examined for unknown noise generators. A screen room is certainly a "must." As a general statement, each piece of equipment in the screen room must be examined as a potential noise source. For instance, it is not uncommon to find that the signal generator used to measure power gain generates a considerable amount of noise power even with the generator output power level reduced.

In general, when making a noise figure test on integrated circuits, it is best to consult the manufacturer's literature for the test setup to be used. For example, the test setup for the RCA CA3011 or CA3012 wide-band amplifier is shown in Figure 5-18. This information is typical of that provided by semiconductor manufacturers.

PROBLEMS

5-1. Calculate the rms noise voltage appearing across a 1-kΩ resistor at room temperature and at 50°C. Assume a noise bandwidth of 5 kHz for both cases.

5-2. Three resistors—5, 10, and 15 kΩ—are connected in series. For room temperature, calculate the rms noise voltage across each resistor and across the series combination. Calculate the rms voltage across the three resistors connected in parallel.

5-3. For Problem 5-1 calculate the Norton's equivalent circuit.

5-4. A low-pass *RC* filter has a −3-dB cutoff frequency of 159 Hz. $R = 1$ kΩ. Calculate the effective noise bandwidth.

5-5. A parallel circuit is resonant at 100 MHZ and has a Q of 75. C = 25 pF (a) Calculate the effective noise bandwidth of the circuit at resonance. (b) Calculate the noise voltage appearing across the circuit at room temperature.

5-6. An amplifier has a actual input resistance of 1 kΩ and a shot noise resistance of 500 Ω referred to the input. (a) For an amplifier voltage gain of 100 V, calculate the rms output noise voltage at room temperature. (b) Repeat part (a) with a 1-kΩ load at the output of the amplifier.

5-7. A signal generator of internal resistance 100 Ω with an internal electromotive force (EMF) = 20 V is connected to the amplifier of Problem 5-6. (a) Calculate the S/N ratio at the input for a noise bandwidth of 10 kHz at room temperature. (b) Calculate the noise figure.

5-8. A mixer circuit having a noise figure of 16 dB is preceded by an amplifier having a 9 dB noise figure and an available power gain of 25 dB. What is the overall noise figure of the combination?

5-9. The mixer stage of a microwave receiver has a noise figure of 11 dB and is preceded by a low-noise amplifier having an available power gain of 20 dB and an equivalent noise temperature of 33°K. (a) Calculate the effective noise temperature of the combination referred to the input of the amplifier. (b) Calculate the signal power required at the input in order that the S/N ratio be 40 dB, for a bandwidth of 5 MHz, and an antenna noise temperature of 50°K.

5-10. An amplifier consists of two cascaded stages at room temperature. The first stage has a gain of 10 dB, an actual input resistance of 500 Ω, and an equivalent shot noise resistance of 50 Ω. It is driven by a source with 75 Ω impedance and an internal EMF of 3 V. The interstage network between the amplifiers consists of a 10-kΩ resistor and a 50-kΩ resistor in parallel. The second stage has an actual input resistance of 1 kΩ and an equivalent shot noise resistance of 200 Ω. The output load is 2 kΩ. Calculate (a) the equivalent input noise resistance, (b) the noise figure for the amplifier, (c) the input and output signal-to-noise ratios, and (d) the effective noise temperature.

5-11. The noise produced by a resistor is amplified by a perfect (noiseless) amplifier with a voltage gain of 100 and a BW of 200 KHz. A meter at the output reads 200 μV rms. Assuming room temperature, calculate the resistor's resistance. If the BW is reduced to 50 KHz, determine the meter reading.

5-12. Calculate the S/N ratio for a receiver output signal of 7 volts and 0.76 volts noise. Calculate both as a ratio and in dB.

5-13. An amplifier has a BW of 1.5 MHz and an input resistance of 1 K ohms referred to the input. Calculate the rms noise voltage at the amplifier's input. Assume room temperature

5-14. A single stage amplifier has a 150 KHz BW and a voltage gain of 125. A 2 mv signal is applied to the amplifiers input. Calculate the output S/N ratio if the amplifier has a 6 dB NF and the input noise is generated by a 2 K ohm resistor. Assume a temperature of 45°K.

5-15. Calculate the noise power at the input to a receiver with an equivalent noise temperature of 50°K. The receiver receives a signal from an antenna with a 30°K equivalent noise temperature and has a 4 MHz BW. Assume an operating temperature of 40°C.

PART II

COMMUNICATION CIRCUITS

CHAPTER 6

Integrated-Circuit RF and Broadband Amplifiers

This chapter begins Part II of the book. Part I presented the necessary fundamentals needed as background for an understanding of communication theory. In Part II we seek to develop an understanding and awareness of the building blocks needed to realize or design a communication system. Modulation theory, which was discussed briefly in Chapter 1, will be discussed in depth in Part III.

Amplifiers are dear to the heart of anyone who works with communication systems. In communication systems we are often confronted with extremely weak signals in the presence of noise. Thus immediately following the antenna, a need exists for a tuned RF amplifier. In the following sections we examine some of our options. The profusion of linear integrated circuits has done for amplifier design what digital integrated circuits did for transistor logic design. In this chapter we develop the background and techniques for using integrated-circuit amplifiers. For those situations where transistor amplifier design is in vogue, the reader is referred to the numerous transistor texts available.

6-1

RF SMALL-SIGNAL DESIGN USING ADMITTANCE PARAMETERS

There are basically two methods by which integrated circuits may be characterized. The first method is to form a model or equivalent circuit of the internal components that will approximate the actual circuit. The accuracy of the results depends on the accuracy of the model. For integrated circuits that contain more than one or two active devices combined with distributed substrate capacitance and resistor parasitic capacitance, this method of analysis becomes increasingly inaccurate. The second method involves measurement of two-port parameters.

The two-port method characterizes the linear integrated circuit as a black box. The accuracy of this method is limited only by the accuracy of the measurements. Also, an additional advantage is that there are well-established design techniques based around two-port network theory. We are all familiar with two-port network parameters. For example, transistor h parameters/y parameters are measured based on treating the transistor as a two-port network. The y parameters for a linear integrated-circuit amplifier can be measured with commercial instruments such as the Boontoon R-X meter or the General Radio 1607A immittance bridge. A discussion of y parameters follows prior to considering a design example.

Design of a solid-state amplifier in terms of familiar transistor design will be discussed first to introduce admittance parameters. The results transfer directly to linear integrated circuits, where we have access only to the terminals of the device. RF amplifier design using two-port parameters is a systematic, mathematical procedure with an exact solution available for the complete design.

6-2

GENERAL DESIGN CONSIDERATIONS

Design of an RF small-signal amplifier is generally based on a requirement for a specified power gain at a given frequency. Additional specifications will include the bandwidth, stability, and input–output isolation. The procedure is first to select the basic circuit type and then solve the applicable design equations.

The basic circuit types for amplifiers may be categorized according to the following:

- *Feedback:* neutralization, unilateralization, and no feedback.
- *Matching at the transistor terminals:* circuit admittances either matched or unmatched to the transistor input and output admittances.

In the following sections each of these categories is discussed. The considerations leading to the selection of a particular circuit type are discussed together with the design equations.

6-2.1 Transistor Stability

The amplifier must be stable. In particular the potential stability of the transistor must be considered. The Linvill stability factor, C, of the transistor may be computed from equation (6-1). That is,

$$C = \frac{|y_{12}y_{21}|}{2g_{11}g_{12} - \text{Re}(y_{12}y_{21})} \tag{6-1}$$

where

y_{12} = transistor reverse transfer admittance
y_{21} = transistor forward transfer admittance
g_{11} = real part of the input admittance, y_{11}
g_{12} = real part of y_{12}
$\text{Re}(y_{12}y_{21})$ = real part of $y_{12}y_{21}$

When C is less than 1, the transistor is unconditionally stable. If C is greater than 1, it is potentially unstable. When there is no feedback in the circuit, an unconditionally stable transistor will not oscillate with any combination of load and source impedance. A potentially unstable transistor will oscillate with a particular combination of load and source impedance.

Since the load and source impedance can affect the stability of a practical amplifier, it is preferable to include these components in the stability test. A stability factor, k, is defined for this purpose.

$$k = \frac{2(g_{11} + G_s)(g_{22} + G_L)}{|y_{12}y_{21}| + \text{Re}(y_{12}y_{21})} \tag{6-2}$$

where

G_s = real part of the complex source admittance
G_L = real part of the complex load admittance

For this stability factor, if k is greater than 1, the circuit will be stable. If k is less than 1, it is potentially unstable and is likely to oscillate at some frequency. The stability factor is discussed further for different circuit types in the following sections.

GENERAL DESIGN EQUATIONS

There are a number of design equations that are applicable to most types of amplifiers. These equations are discussed next.

6-3.1 Power Gain

The general expression for power gain is

$$G_p = \frac{|y_{21}|^2 \operatorname{Re}(y_L)}{|y_L + y_{22}|^2 \operatorname{Re}\left(y_{11} - \dfrac{y_{12}y_{21}}{y_{22} + y_L}\right)} \tag{6-3}$$

where y_L is the complex load admittance. This expression applies to circuits where there is no external feedback. It also applies to a circuit with feedback if the composite y parameters, y_c, are used. The composite parameters are arrived at by considering the transistor and the feedback network as consisting of two "black boxes" in parallel (see Figure 6-1).

The composite network in Figure 6-1 may be considered as a single block box if the respective admittance parameters are combined. That is,

$$\begin{aligned} y_{11c} &= y_{11t} + y_{11f} \\ y_{12c} &= y_{12t} + y_{12f} \\ y_{21c} &= y_{21t} + y_{21f} \\ y_{22c} &= y_{22t} + y_{22f} \end{aligned} \tag{6-4}$$

where t refers to the transistor and f to the feedback network. Since this approach treats the composite network as a single black box, the composite y parameters may be substituted into any of the design equations applicable to a linear, active, two-port network.

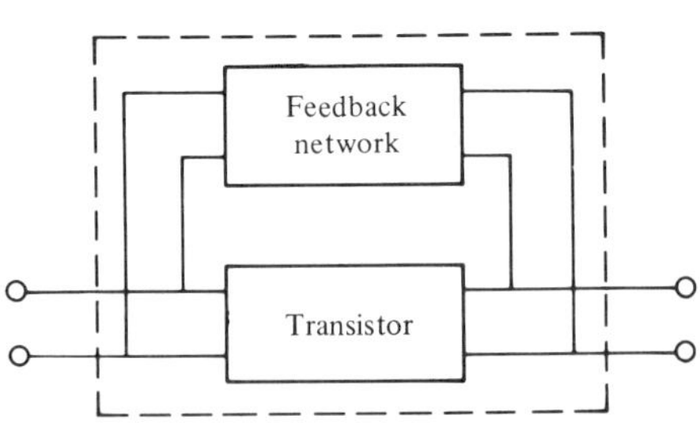

FIGURE 6-1. Transistor with feedback.

The neutralized and unilateralized amplifiers are special cases of this concept. Specifically, employing an external feedback such that $y_{12c} = 0$ is an example of unilateralization. Neutralization consists of employing a feedback network such that y_{12} is some value other than zero. This is discussed in more detail in a later section.

Equation (6-3) provides a solution for power gain of the linear active network only. The input and output networks are considered to be part of the source and load, respectively. Two important points should be kept in mind.

1. The power gain computed from equation (6-3) does not take into account network losses. The input network loss reduces power delivered to the transistor and output network loss is computed as useful power output. The actual power gain is the gain of the transistor less the network losses.
2. Power gain is independent of source admittance. An input mismatch results in less power being delivered to the transistor. Note that equation (6-3) does not contain the term y_s.

In some cases it is desirable to include the effects of input matching in power-gain computations. An expression that includes this parameter is the transducer gain, G_T. The transducer gain is defined as the output power delivered to the load by the transistor divided by the maximum input power available from the source. The equation for transducer gain is

$$G_T = \frac{4\text{Re}(y_s)\ \text{Re}(y_L)|y_{21}|^2}{|(y_{11} + y_s)(y_{22} + y_L) - y_{12}y_{21}|^2} \tag{6-5}$$

The quantity y_L is the composite transistor load admittance composed of both the output network and its load. Also, y_s is the composite transistor source admittance composed of both the input network and its source. Therefore, transducer gain includes the effects of the degree of admittance match of the transistor input terminals but does not take into account input and output network losses. As in equation (6-4), composite y parameters may be substituted.

An additional design equation is that involving the *maximum available gain* (MAG). The MAG is the theoretical power gain of a transistor with its reverse transfer admittance y_{12} set equal to zero and its source and load admittances conjugately matched to y_{11} and y_{22}, respectively. Expressed mathematically, these conditions are

$$\begin{aligned} y_{12} &= 0 \\ y_L &= y_{22}^* \\ y_s &= y_{11}^* \end{aligned} \tag{6-6}$$

Substituting these relations into equation (6-3) yields

$$\text{MAG} = \frac{|y_{21}|^2}{4\text{Re}(y_{22})\,\text{Re}(y_{11})} \tag{6-7}$$

The MAG is a figure of merit only since it is physically impossible to make $y_{12} = 0$ without changing the other parameters. An external feedback may be used to make $y_{12} = 0$, but then the other parameters change.

6-4

TRANSISTOR INPUT AND OUTPUT ADMITTANCES

The input admittance of a transistor is given by

$$y_{\text{in}} = y_{11} - \frac{y_{12}y_{21}}{y_{22} + y_L} \tag{6-8}$$

and the expression for the output admittance is

$$y_{\text{out}} = y_{22} - \frac{y_{12}y_{21}}{y_{11} + y_s} \tag{6-9}$$

If the feedback parameter y_{12} is not zero, y_{in} is dependent on the load admittance and y_{out} is dependent on the source admittance.

6-5

AMPLIFIER STABILITY

Stability of the amplifier can be assured by including stability computations and considering stability in all design decisions relating to feedback and transistor source and load admittances. The potential stability of the transistor should first be computed using equation (6-1). Next, the various alternatives regarding input–output matching and neutralization/unilateralization should be investigated. These alternatives will now be discussed for both the unconditionally stable transistor and the potentially unstable transistor.

It was noted previously that if the Linvill stability factor, C, is less than 1, the transistor is unconditionally stable for any combination of load and source impedance. Thus stability is eliminated as a factor in design, and complete freedom is possible in regard to matching and neutralization to optimize amplifier performance. An amplifier with no feedback is a logical choice for this situation since it offers the advantage of fewer components and a simple tuning procedure. The source and load admittances may be selected for maximum gain or to satisfy other performance requirements. The power gain and transducer gain may be computed using equations (6-3) and (6-5). The input and output admittance may be computed using equations (6-8) and (6-9). Amplifier stability may be computed using equation (6-2), but this is simply for added insurance, since stability is guaranteed by the transistor.

6-5.1 Unilateralized Amplifier

Unilateralization consists of employing an external feedback network to achieve a composite y_{12} of zero. This technique is most often used to achieve stability with a potentially unstable transistor. It is also used, however, when input–output isolation is needed. Design equations are obtained by first obtaining the composite y parameters of the transistor–feedback network combination and then substituting the composite parameters in the general design equations.

The basic condition is that $y_{12} = 0$. For a passive feedback network the following relation is true:

$$y_{11f} = y_{22f} = -y_{12f} = -y_{21f} \tag{6-10}$$

Also,

$$y_{12c} = y_{12t} + y_{12f} = 0$$

or (6-11)

$$y_{12t} = -y_{12f}$$

Thus using equation (6-10), we have

$$y_{12t} = -y_{12f} = y_{11f} = y_{22f} = -y_{21f} \tag{6-12}$$

Substituting these relations into equation (6-4) yields the following set of composite equations:

$$\begin{aligned} y_{11c} &= y_{11t} + y_{12t} \\ y_{22c} &= y_{22t} + y_{12t} \\ y_{12c} &= y_{12t} - y_{12t} = 0 \\ y_{21c} &= y_{21t} - y_{12t} \end{aligned} \tag{6-13}$$

Substituting these composite equations into equations (6-3), (6-5), (6-8), and (6-9) yields

1. Unilateralized power gain

$$G_{pu} = \frac{|y_{21} - y_{12}|^2 \, \text{Re}(y_L)}{|y_L + y_{22} + y_{12}|^2 \, \text{Re}(y_{11})} \tag{6-14}$$

2. Unilateralized transducer gain

$$G_{Tu} = \frac{4\text{Re}(y_s) \, \text{Re}(y_L)|y_{21} - y_{12}|^2}{|(y_{11} + y_{12} + y_s)(y_{22} + y_{12} + y_L)|^2} \tag{6-15}$$

3. Unilateralized input admittance

$$y_{\text{in}} = y_{11} + y_{12} \tag{6-16}$$

4. Unilateralized output admittance

$$y_{\text{out}} = y_{22} + y_{12} \tag{6-17}$$

If y_L is conjugately matched to y_{out}, the unilateralized power gain is

$$G_u = \frac{|y_{21} - y_{12}|^2}{4\text{Re}(y_{11} + y_{12}) \, \text{Re}(y_{22} - y_{12})} \tag{6-18}$$

Note that equations (6-14) through (6-18) are given entirely in transistor parameters. We can see from equations (6-16) and (6-17) that unilateralization results in isolation between the input and the output. That is, y_{in} is independent of y_L and y_{out} is independent of y_s. This means that in a single- or multistage amplifier using unilateralized stages, tuning of any one network will not affect tuning in other parts of the circuit. Thus it is not necessary to repeak an entire amplifier following a change in tuning at a single point.

6-5.2 Neutralized Amplifier

When we say that an amplifier employs neutralization, this implies a feedback network that reduces y_{12} to some value other than zero. Neutralization is used for the same purpose as unilateralization but provides less than complete cancellation of the feedback parameter. A typical example would be a feedback network that sets the reactive part of y_{12c} equal to zero while having only a negligible effect on $g_{12t} = \text{Re}(y_{12t})$. The equations for a particular neutralized case are developed in the same manner as those for the unilateralized case. Since

y_{12} can be any value other than zero, an infinite number of possibilities are available. A particular case is left as an exercise for the student.

6-5.3 Potentially Unstable Transistor

If the Linvill stability factor, C, is greater than 1, the transistor is potentially unstable. Thus certain combinations of source and load impedance will cause oscillations unless a feedback network is used. Stability is achieved by one or both of the following methods:

- Using a feedback network that reduces the composite y_{12} to a value that ensures stability.
- Choosing a source and load admittance combination that provides stability.

These methods are discussed next.

Using Feedback to Achieve Stability. If we use feedback to achieve stability, then either unilateralization or neutralization may be employed. Remember that for unilateralization $y_{12c} = 0$. Thus the numerator of equation (6-1) is zero and the amplifier is unconditionally stable. Power gains and input–output admittances may be computed using equations (6-14) through (6-18). If neutralization is used, the Linvill stability factor, C, can be computed for any transistor–neutralization network combination. Since $y_{12c} \neq 0$, C will have a value other than zero.

Stability Without Feedback. A stable design may be arrived at with a potentially unstable transistor by proper choice of load and source admittances. This can be seen by referring to equation (6-2). G_s and or/G_L may be made large enough to ensure that k is greater than 1. Thus we may choose a value for k and then use equation (6-2) to arrive at values for G_s and G_L. For a particular value of k, the designer may choose any of the following combinations of matching or mismatching of G_s and G_L to the transistor input and output conductances, respectively. That is:

- G_s matched and G_L mismatched.
- G_L matched and G_s mismatched.
- Both G_s and G_L mismatched.

Once G_s and G_L are chosen, the remainder of the design may be completed using the relationships that apply to an amplifier without feedback. Power gain and input–output admittances may be computed using equations (6-3), (6-5), (6-8), and (6-9).

APPLICATIONS

The preceding sections considered amplifier design in terms of admittance parameters. In particular, we inferred that the principles transferred to integrated circuits. In the next section we treat the design of a single-stage 60-MHz integrated-circuit amplifier.

6-6.1 Narrowband Tuned Amplifier (CE-CB)

A typical circuit for design of a single stage is shown in Figure 6-2 and the two-port parameters for this circuit are shown in Figures 6-3 through 6-6. The

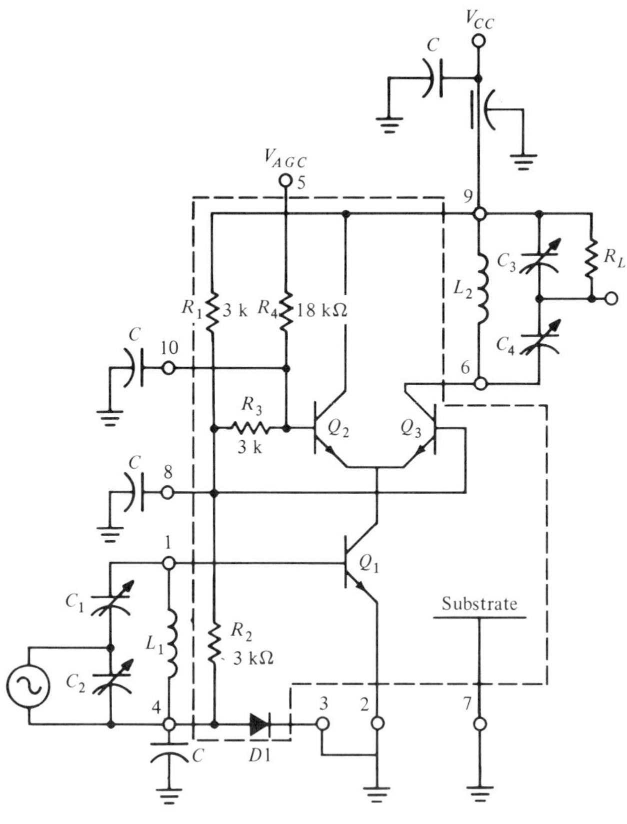

L_1 = 0.23 μH
C_1 = 36 pF
C_2 = 65 pF
C (Bypass) = 1000 pF

L_2 = 0.26 μH
C_3 = 200 pF
C_4 = 27 pF
R_L = 50 Ω

FIGURE 6-2. 60-MHZ tuned amplifier.

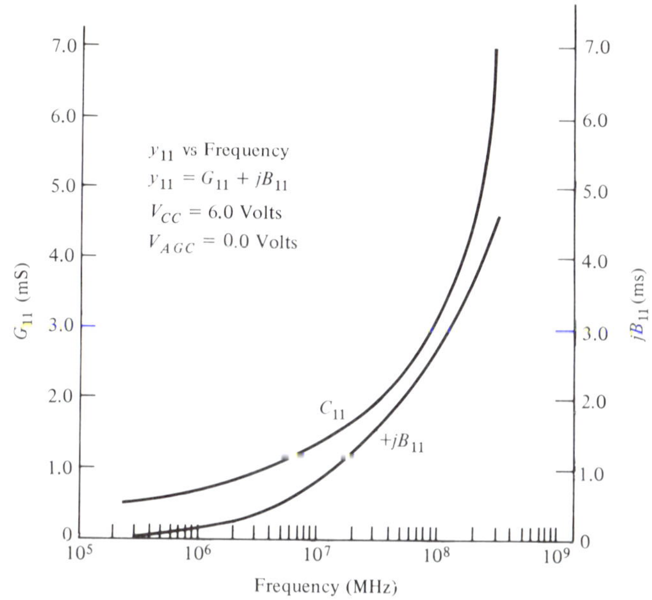

FIGURE 6-3. Input admittance versus frequency.

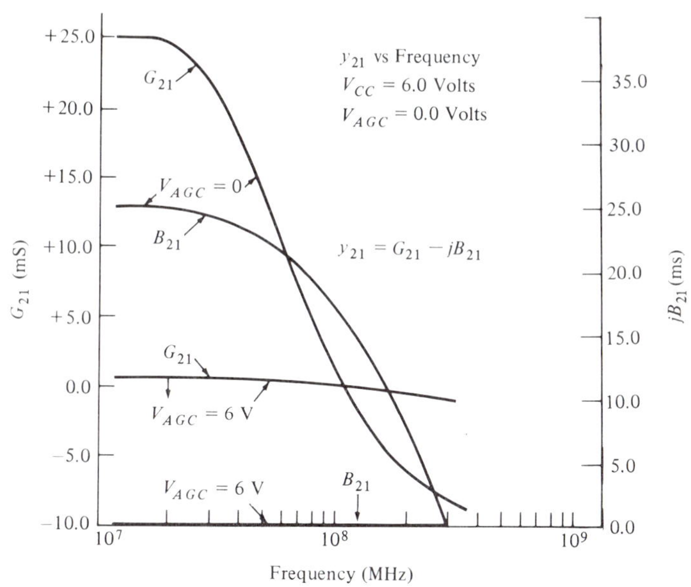

FIGURE 6-4. Forward transfer admittance versus frequency.

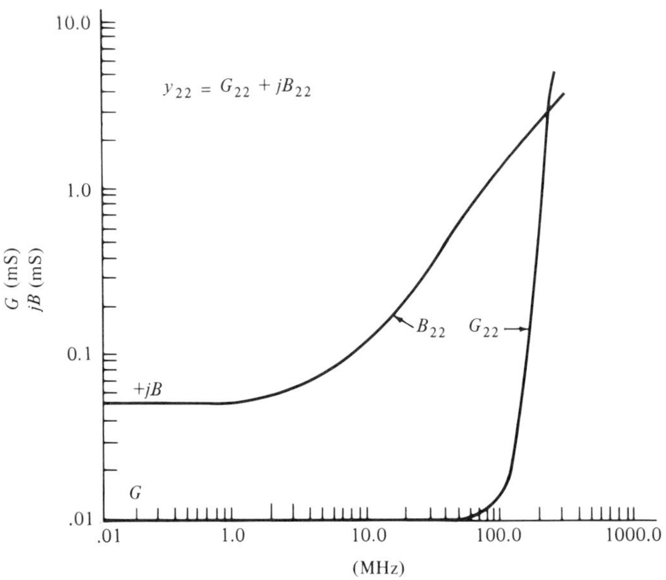

FIGURE 6-5. Output admittance versus frequency.

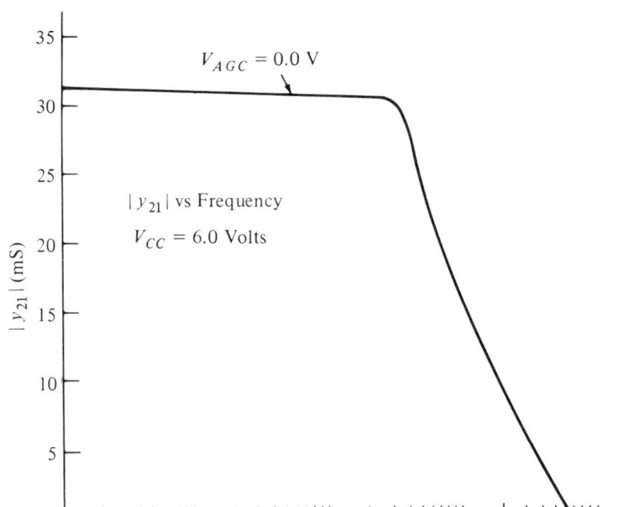

FIGURE 6-6. Magnitude of forward transfer admittance versus frequency.

circuit within the dashed lines is the Motorola 1550 silicon integrated circuit. The design problem, then, consists of determining values for the external components. Before considering a design procedure it is well to examine the operation of the integrated circuit itself.

6-6.2 Circuit Operation

Referring to Figure 6-2, the signal is applied to the base of Q_1 and the output signal is developed across the collector load of Q_3. Thus the combination Q_1–Q_3 acts as a common emitter–common base (CE-CB) pair. Recall from transistor theory that this is the familiar cascode amplifier connection. The Miller capacitance is reduced at the input, and wide bandwidth with gain is achieved with the common-base connection. In addition, the internal feedback, y_{12}, is reduced by two orders of magnitude compared to a single transistor. For this particular circuit y_{12} has a measured value of less than 0.001 mS over the frequency band of the amplifier. A General Radio 1607A immittance bridge is useful for making this measurement.

Note that the collector current from Q_1 splits between Q_2 and Q_3. Thus the gain can be controlled by applying a bias (automatic gain control) to the base of Q_2. The base–emitter voltage on Q_2 need only be 114 mV to switch the major portion of the current through Q_2. This results in minimum signal out at the collector of Q_3. Resistors R_3 and R_4 serve to widen the automatic gain control (AGC) voltage from 114 mV to 0.86 V. This feature makes the AGC line less susceptible to noise. From transistor theory we know that changing the collector load has very little effect on the input impedance of a transistor. Thus the AGC line does not detune the input circuitry.

EXAMPLE 6-1

Design a single-stage 60-MHz amplifier using the circuit of Figure 6-2. Assume that $V_{AGC} = 0$ and $V_{CC} = 6.0$ V dc.

SOLUTION

The following two-port y parameters are obtained from Figures 6-3 through 6-6 at 60 MHz.

$$y_{11} = 2.50 + j2.23 = 3.24 \angle 42.6^\circ \text{ mS}$$

$$y_{12} = \text{(negligible)}$$

$$y_{21} = 10.0 - j19.0 = 21.5 \angle -62.3^\circ \text{ mS}$$

$$y_{22} = 0.01 + j0.80 = 0.80 \angle 90^\circ \text{ mS}$$

Since $y_{12} \approx 0$, we may assume that the maximum available gain (MAG) exists. Thus from equation (6-7),

$$\text{MAG} = \frac{\text{power available from output}}{\text{power available from source}}$$

$$= \frac{|y_{21}|^2}{4\text{Re}(y_{22})\ \text{Re}(y_{11})} = \frac{|21.5 \times 10^{-3}|^2}{(4)(0.01 \times 10^{-3})(2.5 \times 10^{-3})}$$

$$= 4620 \rightarrow 10_{\log} 4620 = 36.7 \text{ dB}$$

Thus the maximum theoretical gain for this circuit is 36.7 dB.

The transducer gain, G_T, is given by equation (6-5). That is,

$$G_T = \frac{\text{power delivered to the load}}{\text{power available from source}}$$

$$= \frac{4|y_{21}|^2\text{Re}(y_s)\ \text{Re}(y_L)}{|(y_{11} + y_s)(y_{22} + y_L) - y_{12}y_{21}|^2}$$

Maximum power gain through a linear two-port network occurs when $y_s = y_{11}^*$ and $y_L = y_{22}^*$. Thus using equations (6-8) and (6-9) with $y_{12} = 0$, we have the following:

$$y_{s(\text{opt})} = y_{11} = 2.50 - j2.30 \text{ mS}$$

$$y_{L(\text{opt})} = y_{22} = 0.01 - j0.80 \text{ mS}$$

The transducer gain is

$$G_T = \frac{4|21.5 \times 10^{-3}|^2(2.5 \times 10^{-3})(0.01 \times 10^{-3})}{|2(2.5 \times 10^{-3})2(0.01 \times 10^{-3}) - 0|^2} = 4620$$

$$10 \log_{10} 4620 = 36.7 \text{ dB}$$

Since $y_{12} \approx 0$, maximum available gain is equal to the transducer gain. In practice, it is difficult to achieve the theoretical value of transducer gain because of losses in the inductors. The finite Q of the collector load results in an equivalent resistance that both loads the output and restricts the gain and bandwidth. Since this resistance also changes the output admittance, it must be considered in designing the output network.

The input and output matching networks are shown in Figure 6-7. Obviously, many different designs may be effected for these networks. Considering the input network first, the approach taken here will be to broadband the input network and design the matching network to conjugate match y_{11}. Thus the

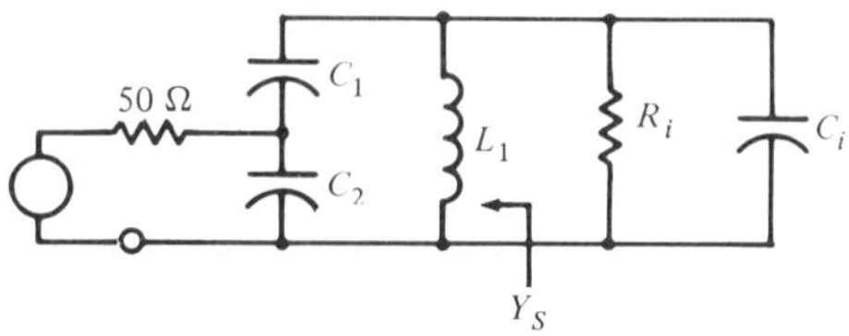

(a) Input matching network

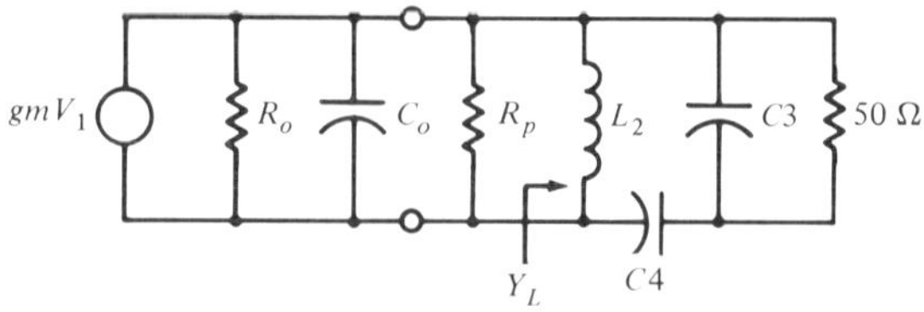

(b) Output matching network

FIGURE 6-7. Input and output matching network.

matching network must transform the 50-Ω source into y_{11}^*. The optimum source admittance is

$$y_s(\text{opt}) = y_{11}^* = 2.50 - j2.30 \text{ mS}$$

A graphical procedure is useful for designing the input network. It is well known that reactive L structures will transform any impedance Z_L at a particular frequency into any desired input impedance. An example will suffice. Referring to Figure 6-8, a single L structure is shown with load Z_L. The L structure operates by first modifying the susceptance of Z_L (or y_L) through the susceptance B_1 to obtain y_a, Z_a. The reactance of Z_a is modified by X_2 to obtain Z_{in}.

An impedance plane or a Smith chart with two sets of coordinates is useful for this design (Figure 6-9). Using the impedance plane, we can follow the change in impedance as we proceed from the network output back to the input. We see that one set of coordinates correspond to the real and imaginary parts of the impedance and the other set to the admittance.

As an example of its use, consider an example where $Z_L = 1.5\ \Omega$. Adding positive susceptance B_1 implies moving clockwise on a constant-G circle. Add-

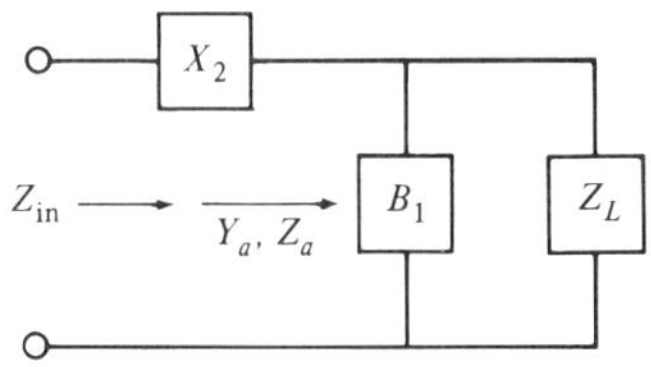

FIGURE 6-8. Reactive L structure.

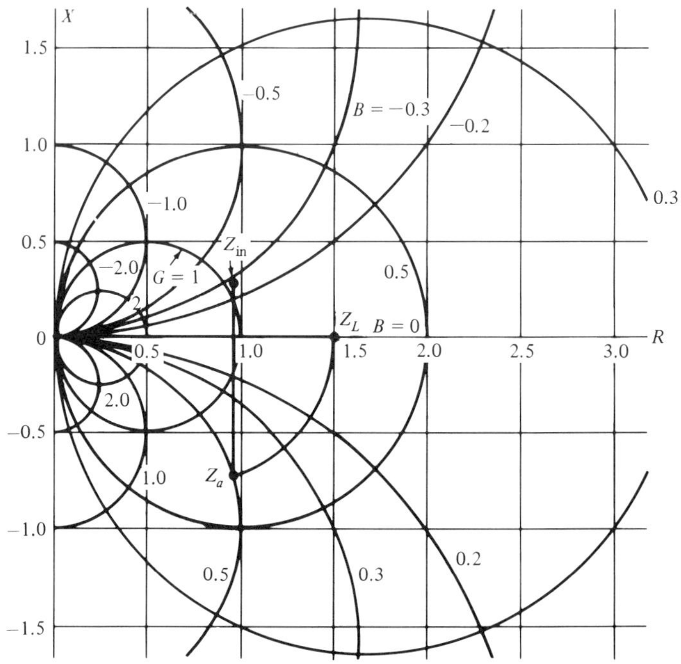

$$Z = R + jX = \frac{1}{Y} = \frac{1}{G + jB}$$

FIGURE 6-9. Plane and impedance and admittance coordinates.

ing series inductance implies moving vertically upward on a constant-*R* line. In this example $B_1 = 0.5$ S and $x_2 = 1\ \Omega$. The input impedance is seen to be $0.95 + j0.28\ \Omega$. In moving from the output port to the input port, note that we alternately move along LOCI of constant conductance and constant resistance. Note that for a given input and output impedance more than one solution is available. In addition, the technique may be used with different configurations for the L structure. Using this technique for the present problem, values for L_1, C_1, and C_2 are calculated to be

$$L_1 = 0.23\ \mu\text{H}$$

$$C_1 = 36\ \text{pF}$$

$$C_2 = 65\ \text{pF}$$

To complete the design, values for the output-matching network must be calculated. The bandwidth may be calculated using equation (2-35). The bandwidth, Δf, is

$$\Delta f = \frac{f_0}{Q} = \frac{f_0}{w_0 R_T C_T} = \frac{1}{2\pi R_T C_T}$$

where

R_T = total R at the output
C_T = total C at the output

In Figure 6-7, $R_p = 1/G_p$ is the equivalent resistance across the output caused by the finite unloaded Q of L_2 [see equation (2-38)]. For this design example, the measured value of R_p at 60 mHz is 35 kΩ. This value of R_p is in parallel with R_0, the output resistance of the integrated circuit. The resultant resistance R may now be used to calculate a new $y_{22} = y'_{22}$. This new value of y_{22} is then used to design the matching network so that the load admittance is the complex conjugate of y'_{22}. For $R_p = 35$ kΩ the new y'_{22} is

$$y'_{22} = \tfrac{1}{35} + 0.01 + j.80 \text{ mS}$$

Thus the admittance of y_L is

$$y_L = 0.039 - j0.80 \text{ mS}$$

Using the impedance plane technique, values for L_2, C_3, and C_4 are calculated to be

$$L_2 = 0.26\ \mu\text{H}$$

$$C_3 = 630 \text{ pF}$$

$$C_4 = 27 \text{ pF}$$

The total R and C present across the integrated-circuit output is

$$\frac{1}{R_T} = (2)(0.039)(10^{-3}) \rightarrow R_T = 12.5 \text{ k}\Omega$$

and

$$C_T = C_0 + C(\text{matching network})$$
$$= 28 \text{ pF}$$

Using these values for R and C the calculated bandwidth, Δf, is

$$\Delta f = \frac{1}{(2\pi)(12.5 \times 10^3)(28 \times 10^{-12})} = 0.455 \text{ MHz}$$

The transducer gain, recomputed with y'_{22}, is 30.1 dB. Thus the finite Q of the tuned output reduces G_T by a factor of 6.6 dB. Measured values of transducer gain for the amplifier of Figure (6-2) are shown in Figure 6-10.

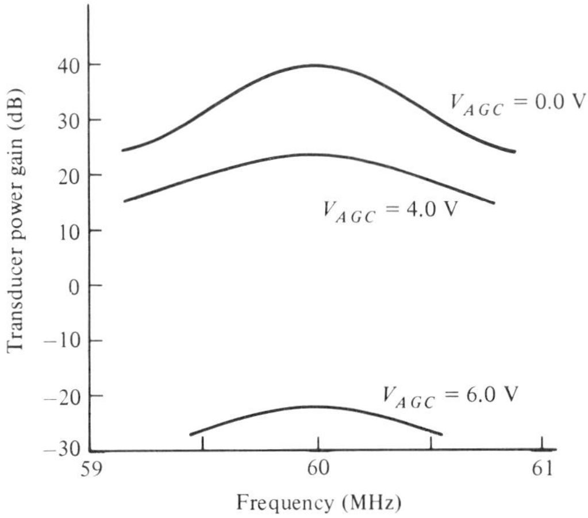

FIGURE 6-10. 60 MHz single-stage amplifier gain.

6-6.3 Wide-band Amplifier Design

The circuit in Figure 6-2 demonstrates the high gain that may be achieved at the expense of bandwidth. The narrow bandwidth is due primarily to the large value of resistance involved (12.5 kΩ). If this value is reduced, the bandwidth will be broadened. At the same time the gain is reduced so that more than one stage may be needed to realize a given gain. The following example illustrates multistage amplifier design techniques.

EXAMPLE 6-2

Design a wide-band tuned amplifier with the following specifications:

- Single tuned
- Center frequency, 45 MHz
- Bandwidth, 15 MHz
- Transducer power gain, 30 dB

SOLUTION

To realize the specified gain and bandwidth using the MC1550, it is necessary to use two stages of amplification with three tuned circuits. This results in an overall bandwidth reduction which may be computed as follows for cascaded states:

$$BW_{overall} = BW_{stage} \frac{1}{\sqrt{2^{1/n} - 1}} \qquad (6\text{-}19)$$

where n is the number of tuned circuits.

For $n = 3$ and an overall bandwidth of 15 MHz, the bandwidth per stage is

$$BW_{stage} = \frac{15 \text{ MHz}}{\dfrac{1}{\sqrt{2^{1/3} - 1}}} = \frac{15}{\sqrt{0.25}} = 30 \text{ MHz}$$

Thus each stage must be designed for 30 MHz. From Figures 6-3 through 6-6, the y parameters at 45 MHz, $V_{CC} = 6$ V dc, and $V_{AGC} = 0.0$ V are as follows:

$$y_{11} = 2.5 + j2.0 \text{ mS}$$

$$y_{12} = \text{(negligible)}$$

$$y_{21} = 16 - j21 \text{ mS}$$

$$y_{22} = 0.01 + j0.55 \text{ mS}$$

The source and load impedance are taken to be 50 Ω, as in Example 6-1.

The simplified circuit for this design is shown in Figure 6-11. By extending the two-port network (as indicated by the dashed lines) to include R_L, the new value of $y_{22} = y'_{22}$ may be found. The total resistance R for each stage is the parallel combination of R_o and R_L. From y_{22} the value of R_o is

$$R_o = \frac{1}{0.01 \times 10^{-3}} = 100 \text{ k}\Omega$$

If we assume that y_s can be accurately transformed into y_{11}^* and y_L into y'_{22}, the value of R_L that will result in a gain of 15 dB per stage can be found from

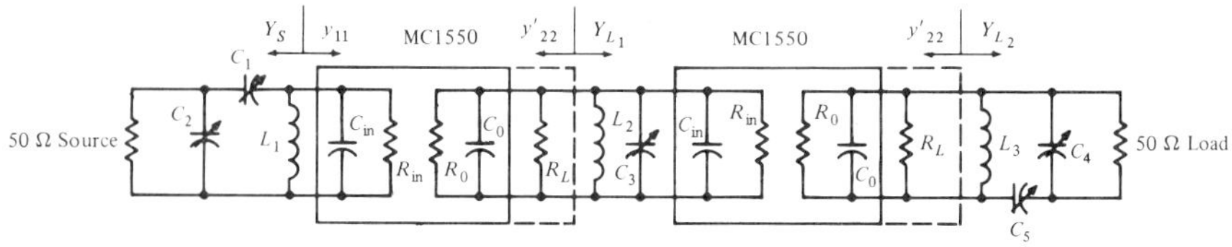

FIGURE 6-11. 45-MHz two-stage amplifier.

the transducer power-gain equation. Thus using equation (6-7),

$$\mathrm{Re}(y'_{22}) = \frac{|y_{21}|^2}{4G_T\,\mathrm{Re}(y_{11})} = \frac{[(16^2 + 21^2)^{1/2} \times 10^{-3}]^2}{4(31.62)(2.5 \times 10^{-3})}$$

$$= 2.88 \text{ mS}$$

The minimum value of R_L may now be computed as follows:

$$\frac{R_o R_L}{R_o + R_L} = \frac{1}{\mathrm{Re}(y'_{22})}$$

or

$$R_{L(\min)} = \frac{R_o}{R_o\,\mathrm{Re}(y'_{22}) - 1} = \frac{10^5}{(10^5)(2.88 \times 10^{-3}) - 1}$$

$$= 350\ \Omega$$

For design purposes choose $R_L = 500\ \Omega$ to ensure the specified gain. Thus $R = R_L \| R_o = 500\ \Omega$.

High-Q inductors ($Q > 30$) must be used to ensure that the effect of their resistance on R_L can be neglected. Using the equations for center frequency and bandwidth of a tuned circuit,

$$f_0 = \frac{1}{2\pi\sqrt{LC}} \qquad \Delta f = \frac{1}{2\pi RC}$$

values for L and C at each interstage are calculated. These values are

$$C_1 = 25 \text{ pF} \qquad L_1 = 0.42\ \mu\text{H} \qquad R_L = 510\ \Omega$$

$$C_2 = 20 \text{ pF} \qquad L_2 = 0.68\ \mu\text{H}$$

$$C_3 = 5 \text{ pF} \qquad L_3 = 0.55\ \mu\text{H}$$

$$C_4 = 20 \text{ pF}$$

$$C_5 = 25 \text{ pF}$$

The complete design for this amplifier is shown in Figure 6-12. Measured values for the gain are shown in Figure 6-13.

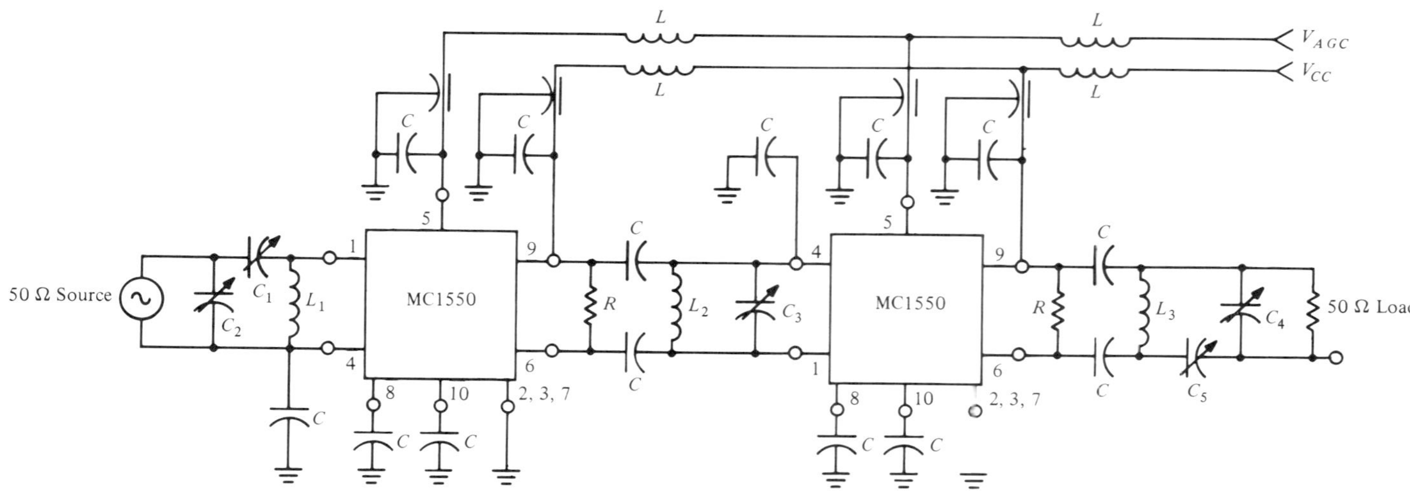

$C = 2000$ pF
$C_1 = C_2 = C_4 = C_5 = 9\text{–}35$ pF
$C_3 = 2\text{–}8$ pF
$L_1 = 0.42\ \mu$H
$L_2 = 0.68\ \mu$H
$L_3 = 0.55\ \mu$H
$L = 1\ \mu$H
$R = 510$ ohms

FIGURE 6-12. Wide-band 45 MHz amplifier.

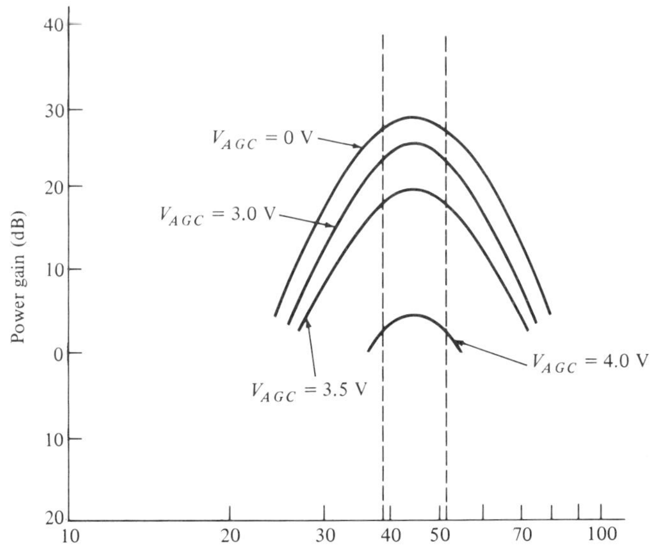

FIGURE 6-13. AGC characteristics of 45-MHz wide-band amplifier.

6-6.4 Stagger-Tuned IF Strip

The terms *stagger-tuning* refers to an amplifier comprising several stages in cascade, in which the stages are not tuned to the same frequency but are ''staggered'' at frequencies above and below the desired center frequency. Generally speaking, the tunings of the individual stages and their bandwidths may be different. The objectives of stagger tuning are twofold:

1. A gain–bandwidth product is achieved which is greater than that of synchronously tuned cascade systems.
2. It is possible to achieve a selectivity curve of prescribed amplitude response.

Thus it is possible to stagger tune n single stages to achieve a flat bandpass, an equal-ripple (Chebyshev) bandpass, or numerous other selectivity curves with very good gain. The following design example illustrates the design procedure.

EXAMPLE 6-3

Design an IF strip for the following specifications:

- Center frequency (f_0), 45 MHz, stagger tuned
- Bandwidth (Δf), 6 MHz, $R_s = 50\ \Omega$
- Power gain (G_T), 70 dB, $R_L = 50\ \Omega$
- AGC control, 50 dB

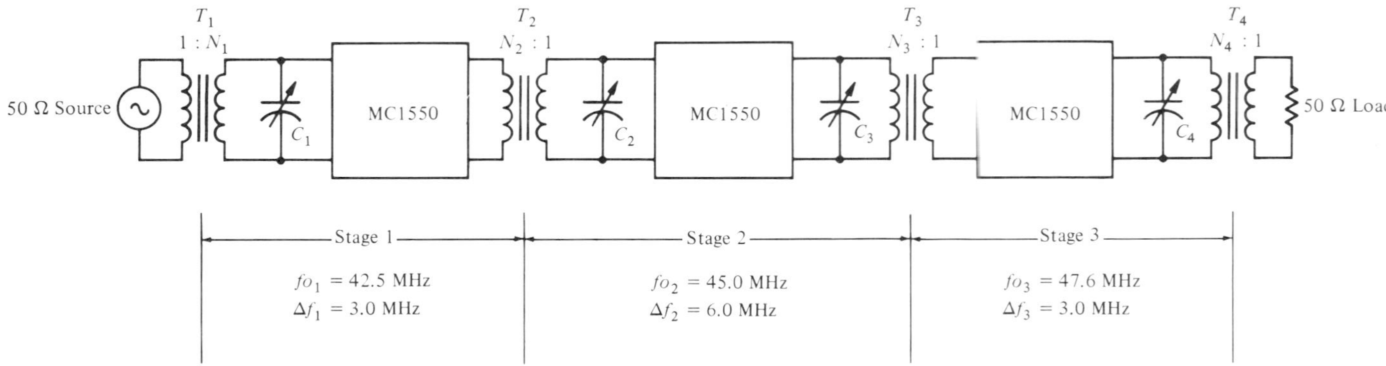

FIGURE 6-14. 45-MHz flat stagger-tuned IF strip.

SOLUTION

A typical circuit for this design is shown in Figure 6-14. The individual stage frequency requirements for this amplifier are as follows:

- *Stage 1:* tuned to f_0 with $\Delta f = 6$ MHz,
- *Stage 2:* tuned to $f_{0\alpha}$ with $Q = 2.0/\delta$,
- *Stage 3:* tuned to $f_{0/\alpha}$ with $Q = 2.0/\delta$,

where $\delta = \Delta f/f_0 = 0.1333$, $\alpha = 1 + 0.433\delta$, and $Q = f_0/\Delta f$. Thus the frequencies for individual stages are:

- *Stage 1:* tuned to 45 MHz with $\Delta f = 6$ MHz.
- *Stage 2:* tuned to 47.6 MHz with $\Delta f = 3$ MHz.
- *Stage 3:* tuned to 42.5 MHz with $\Delta f = 3$ MHz.

The interstage design is easily accomplished if we consider the coupling transformers to be ideal. Referring to Figure 6-15, we see that the load on one side of the transformer can be referred to the other side. The reflected impedance, Z_r, is given by equation (2-66) and is

$$Z_r = \frac{(\omega M)^2}{Z_p + R_s} \approx \frac{(\omega M)^2}{R_s} \tag{6-20}$$

A detailed discussion of networks such as the one shown in Figure 6-15b is contained in Chapter 2. The bandwidth and resonant frequency for the parallel tuned network is given by

$$\Delta f = \frac{1}{2\pi R_T C_T} \qquad f_0 = \frac{1}{2\pi\sqrt{L_T C_T}}$$

where

R_T = total parallel resistance
C_T = total parallel capacitance
L_T = total parallel inductance

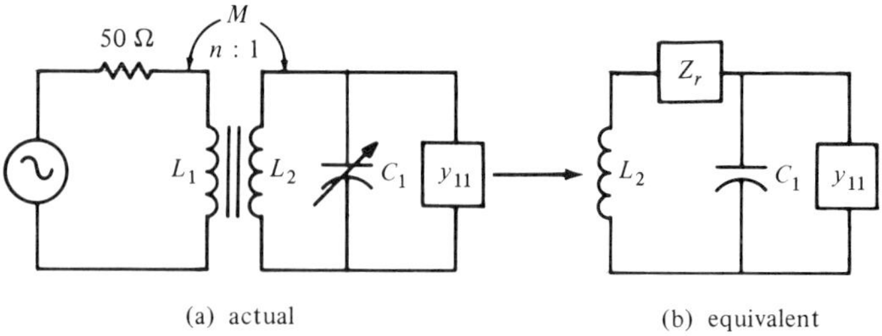

FIGURE 6-15. Interstage network.

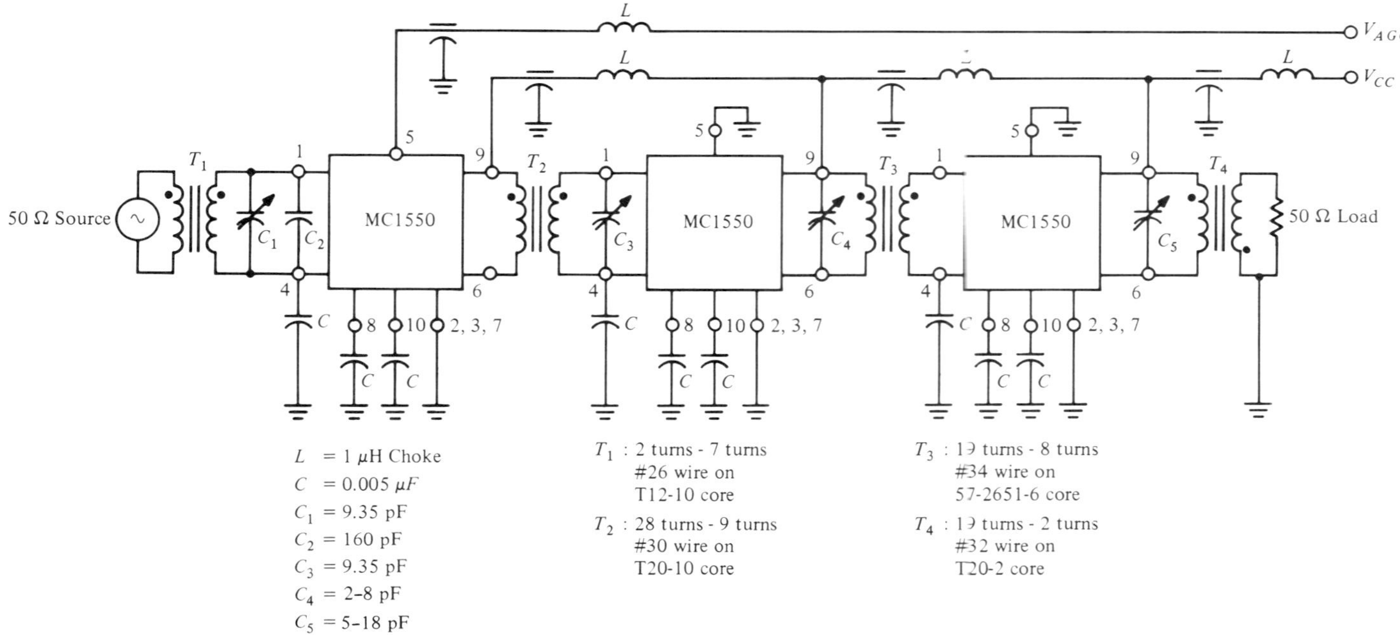

FIGURE 6-16. 45-MHz stagger-tuned amplifier.

A schematic of thc final design is shown in Figure 6-16. Since there are two tuned circuits associated with each stage, there exists an overall bandwidth reduction of each stage. This can be handled by broadbanding the output tuned circuit of stage 1 while achieving the desired selectivity and bandwidth with the input tuned circuit, and vice versa with stage 2. The same procedure could be followed with stage 2. That is, broadband the output tuned circuit and use the input to achieve bandwidth and selectivity. For this particular circuit, the procedure used was to tune the input and output of stage 2 synchronously and take into account the bandwidth reduction factor.

6-6.5 IF Strip Tuning

Since $y_{12} \approx 0$, this facilitates tuning the amplifier. A procedure for tuning is as follows. First disconnect the stages and load each stage with an impedance to simulate the actual impedance. Tune each stage to the desired frequency and bandwidth. Finally, reconnect the stages and fine tune the integrated amplifier. Overall response of the amplifier is shown in Figure 6-17. The choice of which stages to AGC is optimal. In practice, various combinations can be tried to see which is must effective.

6-7 INTEGRATED-CIRCUIT RF AMPLIFIER APPLICATIONS

In the preceding sections we have examined basic design techniques for integrated-circuit RF amplifiers. We see that using admittance parameters is a valuable tool in design since we only have access to the terminals of the device. In addition to their uses as wide or narrowband amplifiers, these amplifiers may be used as mixers, product detectors, frequency generators, and so on. These applications will be examined in later chapters. RF amplifiers are available from a number of manufacturers. An example of these is the RCA CA3004/

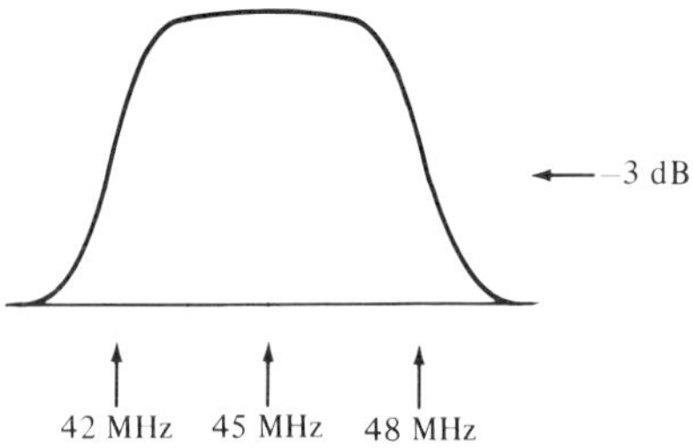

FIGURE 6-17. 45-MHz IF amplifier response.

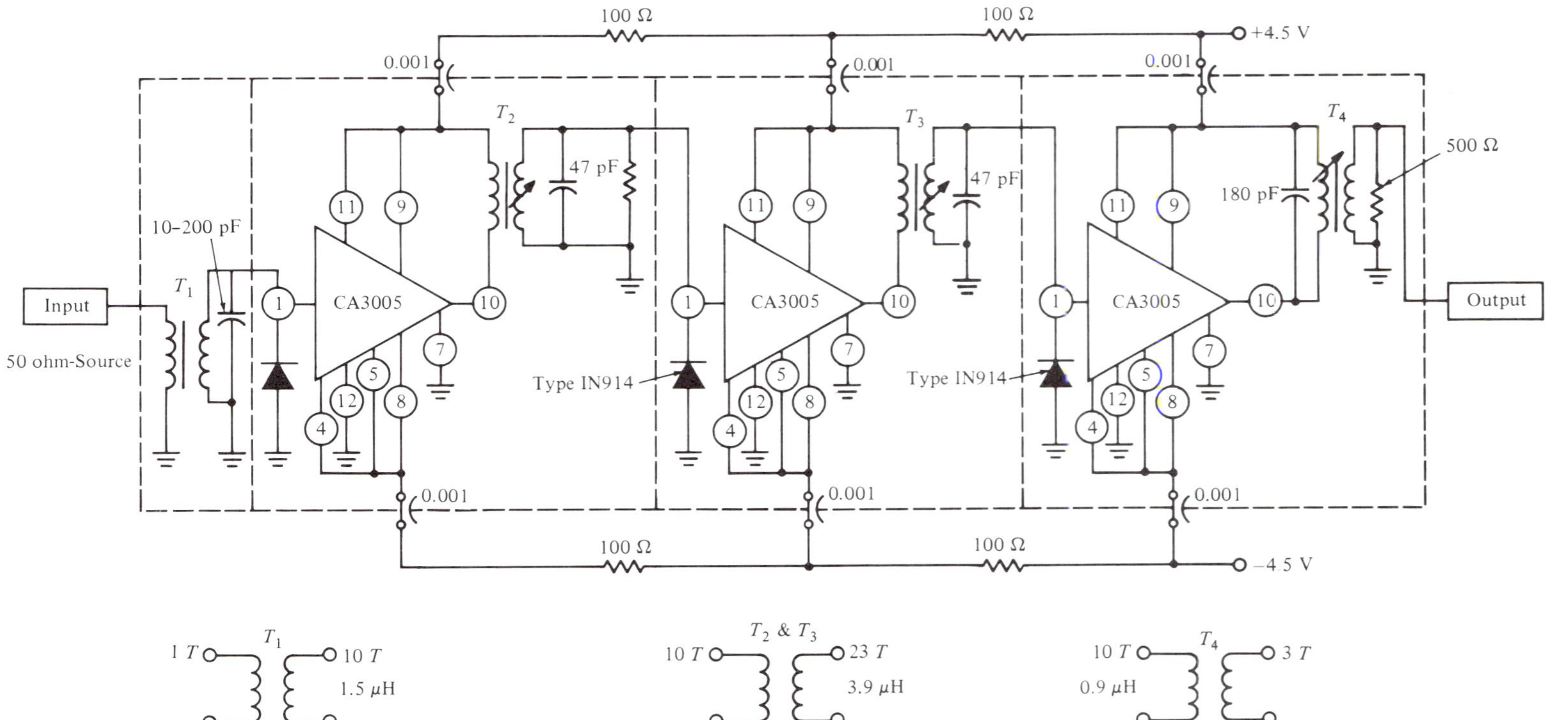

Notes: 1. Transformer T_1 is a Ferramic Q-2 Toroid Type (unloaded Q = 200).
2. Transformers T_2, T_3, and T_4, are slug-tuned with Carbonyl IT-75 material (unloaded Q = 75)

FIGURE 6-18. Three-stage 12-MHz limiting IF amplifier.

CA3005/CA3006 series. A 12-MHz IF amplifier using the CA3005 is shown in Figure 6-18.

PROBLEMS

6-1. Find a reactive coupling network, using two elements, which when terminated in 1 Ω provides approximately the following input impedance.

f (rad/s)	Z_{in} (Ω)
1	$0.5 + j0$
1.25	$0.4 + j0$
1.5	$0.3 + j0$

6-2. Using Figure 6-8, what is the input impedance if the components values are as follows?

$$Z_L = 1.25 - j1\ \Omega$$

$$B_1 = 1.0\ \text{S}$$

$$X_2 = 1.5\ \Omega$$

6-3. For the reactive matching L structure shown, what is the input impedance?

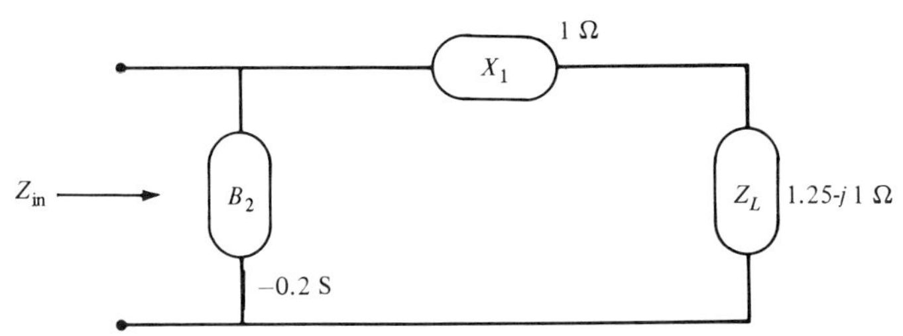

PROBLEM 6-3

6-4. What is the input impedance for the network shown?

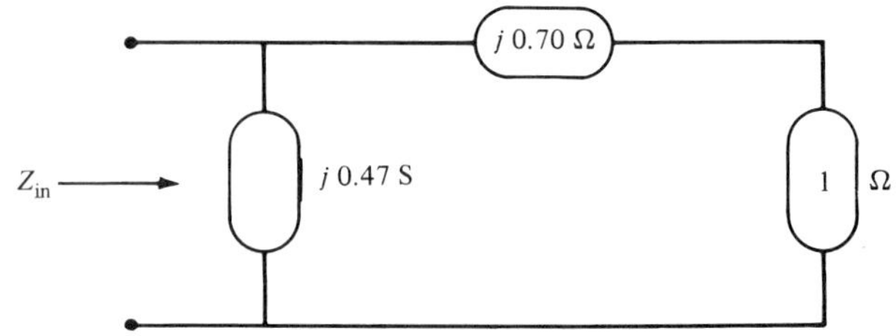

PROBLEM 6-4

6-5. What is the load impedance for the network shown?

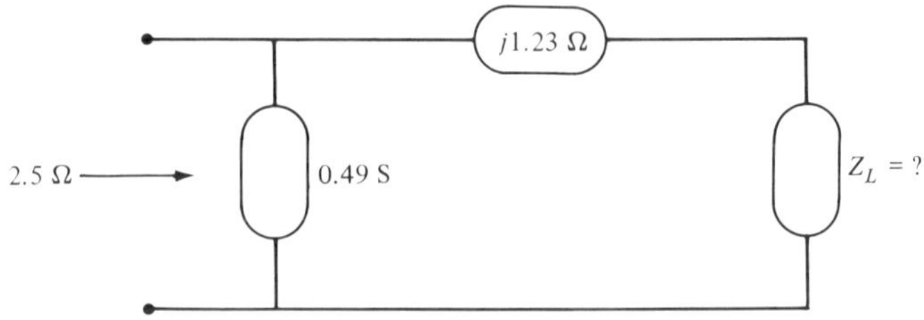

PROBLEM 6-5

6-6. Given the following y parameters, what is the MAG and transducer gain, G_T? $R_L = 1000\ \Omega$.

$$y_{11} = 2.25 + j2.0 \text{ mS}$$

$$y_{12} = 0.0$$

$$y_{21} = 8.0 - j15 \text{ mS}$$

$$y_{22} = 0.02 + j6.0 \text{ mS}$$

6-7. Repeat Example 6-1 for a 45-MHz amplifier.

CHAPTER 7

Oscillators with Phased-Locked-Loop Applications

We indicated in Chapter 1 that the basic function of the receiver was to perform the inverse of modulation: namely, demodulation and detection of the signal. In addition to amplifiers, oscillators are basic to this process. The basic function of oscillators in receivers is to generate sinusoidal and sometimes square-wave voltages of the desired frequency. As an example, every receiver has a local oscillator and it is the frequency of the oscillator that changes as we turn the dial. The output of this oscillator combines with the incoming signal to form a *beat frequency* and this signal is further amplified. We shall see that if two sine waves of frequencies f_1 and f_2 are multiplied ("mixed"), the result is two different sine waves of frequencies $f_1 \pm f_2$.

In modern transceiver design it is often desirable to be able to receive or transmit on a number of different frequency bands. For this purpose we need a local oscillator that changes frequency in discrete frequency steps. A frequency synthesizer consisting of a phase-locked loop is used to perform this operation. Consequently, it is desirable to become familiar with phase-locked loops and their applications. In this chapter we examine oscillator circuits and their relation to receiver design. Following the trend of Chapter 6, the emphasis will be on integrated circuits.

7-1 GENERAL DESCRIPTION OF OSCILLATORS

The process of oscillation is basic to an oscillator. We remember from circuit theory that if positive feedback is applied to an amplifier circuit, the circuit may become unstable and oscillate. In fact, the conditions for instability were examined in Chapter 6. Thus feedback represents a fundamental method for designing an oscillator. In addition, a frequency-sensitive circuit is needed to control the frequency of oscillation. An oscillator, then, is an unstable amplifier and a frequency-selective circuit such as a parallel or series resonant circuit. To make the oscillator practical, a means of controlling the amplitude of the oscillation is necessary. This implies that the instability must be controlled.

A more basic description of an oscillator is in terms of an effective *negative resistance*. It can be shown that the amplifying part of the circuit supplies an effective negative resistance. The negative resistance coupled with the resonant circuit forms the oscillator.

7-1.1 Two-Terminal Oscillators

The simplest type of oscillator is the two-terminal type. This oscillator is constructed by simply connecting a resonant circuit to the negative-resistance circuit. The oscillator is referred to as a two-terminal type since the resonant circuit can be physically separated from the negative-resistance part of the circuit. This is not true for the feedback type of oscillator.

7-1.2 Feedback Oscillators

The feedback oscillator differs from the two-terminal type in that any amplifying device can be made into an oscillator. Hence a single transistor can be used as a feedback oscillator. The two basic types of feedback oscillators are:

- Hartley.
- Colpitts.

Other feedback oscillators are simply variations of these two circuits. These two circuits, without the biasing resistors, are shown in Figure 7-1. Referring to Figure 7-1a, if we consider the transistor as connected in the common-base configuration, the emitter tap couples energy in phase to the emitter. Recall that in the common-base connection there is no phase inversion between emitter and collector. The tank circuit constitutes the collector load and if the Q is high

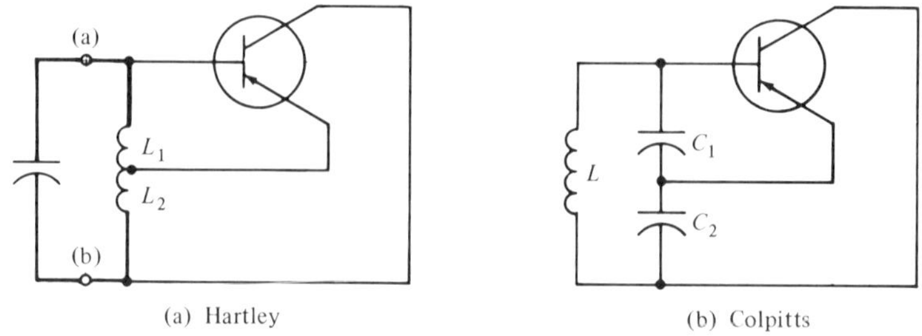

FIGURE 7-1. AC circuit for the Hartley and Colpitts oscillators.

enough, amplification will fall off rapidly on either side of the resonant frequency of the tank circuit. Since the single transistor serves as the amplifier, it follows that the integrated-circuit amplifiers studied in Chapter 6 should be excellent devices for constructing oscillators. An application of the MC1550 amplifier is considered in the next section.

7-2

FUNDAMENTALS OF OSCILLATOR OPERATION

The basic feedback amplifier is modeled in Figure 7-2. The transfer function, e_o/e_i, is arrived at easily. Referring to the figure, the basic equations describing the feedback amplifier are seen to be

$$e_o = A_{\text{ol}}(\omega)(e_i + e_{\text{fb}}) \tag{7-1}$$

$$e_{\text{fb}} = \beta(\omega)e_o \tag{7-2}$$

where A_{ol} is the open-loop voltage gain and β is the fraction of output voltage fed back to the input. Substituting equation (7-2) into equation (7-1) results in the closed-loop gain with feedback, A_{fb}; that is,

$$\frac{e_o}{e_i} = A_{\text{fb}} = \frac{A_{\text{ol}}(\omega)}{1 - A_{\text{ol}}(\omega)\beta(\omega)} \tag{7-3}$$

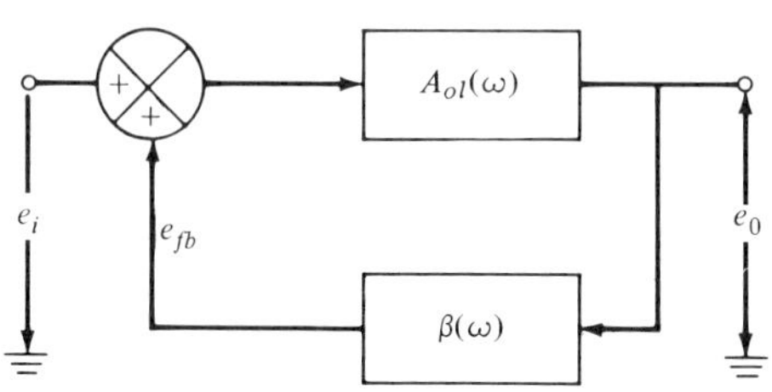

FIGURE 7-2. Basic feedback amplifier.

We recognize this as the standard feedback equation commonly used in control theory. An examination of the full implications of this equation is beyond the scope of this book. Our treatment, then, will be restricted. The reader is referred to a text on control theory for an in-depth analysis.

Referring to equation (7-3), the closed-loop gain increases without bound when

$$A_{ol}(\omega)\beta(\omega) = 1 \tag{7-4}$$

and self-sustained oscillators result. Comparing both sides of equation (7-4), it is evident that the imaginary component of $A_{ol}(\omega)\beta(\omega)$ must equal zero. This provides the means of determining the resonant frequency, ω_0. Once the frequency of oscillation, ω_0, has been determined, $\beta(\omega_0)$ can be found and the threshold, A_T, below which oscillations will not be sustained can be determined from equation (7-5):

$$A_T(\omega_0) = \frac{1}{\beta(\omega_0)} \tag{7-5}$$

The circuit for the MC1550 amplifier is shown in Figure 6-2. If a signal is applied to the base of transistor, Q_1, the MC1550 acts as a common emitter–common base amplifier with high voltage gain and a 180° phase shift at low frequencies. Thus to form an oscillator, the $\beta(\omega)$ network must introduce an additional 180° phase shift from the output to the input to ensure positive feedback.

A number of phase-shift networks are available to produce this. One of the simplest designs is to use an inverting transformer. A simplified diagram for this design is shown in Figure 7-3. If we assume an ideal transformer, the voltage, e_i, at the input is

$$e_i = -\frac{1}{n} e_o \tag{7-6}$$

By definition, $\beta(\omega)$ is the fraction of output voltage fed back to the input. Thus

$$\beta(\omega) = \frac{e_i}{e_o} \tag{7-7}$$

Equating equations (7-6) and (7-7) results in

$$\beta(\omega) = -\frac{1}{n} \tag{7-8}$$

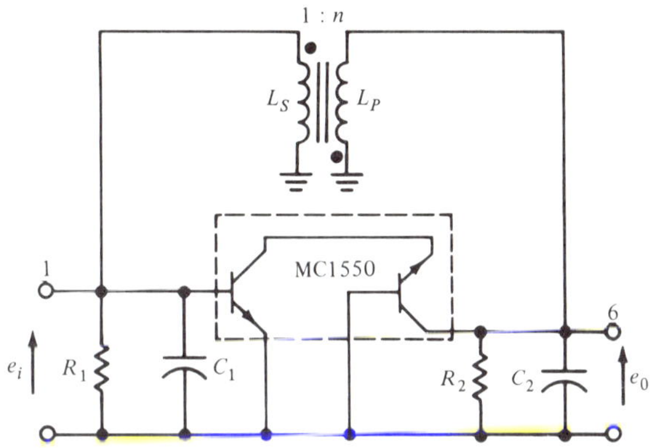

FIGURE 7-3. Transformer-coupled feedback amplifier.

If we now substitute equation (7-8) into (7-4), the following results:

$$A_{\text{ol}}(\omega)\left(-\frac{1}{n}\right) = 1 \qquad (7\text{-}9)$$

or

$$A_{\text{ol}}(\omega) = -n \qquad (7\text{-}10)$$

Thus the open-loop gain of the amplifier must be equal to or greater than the turns ratio of the ideal transformer.

Now referring to Figure 7-3, we see that the voltage gain associated with Q_1 is

$$A_{v1} \approx -\frac{v_{C1}}{v_{be1}} = -\frac{i'_{C1}R_{L1}}{i_b h_{ie}} = -\frac{\beta_{ib}r_{e2}}{\beta_{ib}r_{e1}} \approx -1 \qquad (7\text{-}11)$$

where r_{e1} and r_{e2} are the dynamic impedance of the emitter resistance of Q_1 and Q_2, respectively. The low-frequency voltage gain for Q_2 is

$$A_{v2} \approx \frac{R_{L2}}{r_{e2}} \qquad (7\text{-}12)$$

and the overall amplifier gain is

$$A_{vt} = A_{v1}A_{v2} = -\frac{R_{L2}}{r_{e2}} \approx -\frac{R_{L2}}{r_{e1}} \qquad (7\text{-}13)$$

where R_{L2} is the requivalent load on the amplifier. Now from basic transistor theory r_{e1} is given by

$$r_{e1} = \frac{26(\text{mV})}{I_e(\text{mA})} \tag{7-14}$$

and the equivalent load on the amplifier may be found by rearranging Figure 7-3. For example, using equation (2-55), we may transform R_1 and C_1 to the primary side of the transformer. Thus Figure 7-3 may be redrawn as shown in Figure 7-4.

From Figure 7-4, the ac load on the amplifier is

$$Z_{L2} = R_o \| R_2 \| n^2 R_1 \| Z_p \| X_{C2} \| X_{C1/n^2} \tag{7-15}$$

where

R_o = output resistance of the MC1550
R_2 = external load resistor
$n^2 R_1$ = input resistance of the MC1550 reflected into the primary
$|Z_p|$ = impedance of the transformer primary
$|X_{C2}|$ = reactance of the tuning capacitor
$|X_{C1/n^2}|$ = input capacitance of the MC1550 reflected into the primary

It may be verified from Figure 6-3 that R_o is very large and C_1 is very small (X_{C1} large). Thus equation (7-15) is approximately

$$Z_{L2} \approx R_2 \| Z_p \| X_{C2} \tag{7-16}$$

Substituting equation (7-16) into (7-13) yields an open-loop gain that may be expressed as

$$|A_{01}| = \frac{R_2 \| Z_p \| X_{C2}}{r_{e1}} \tag{7-17}$$

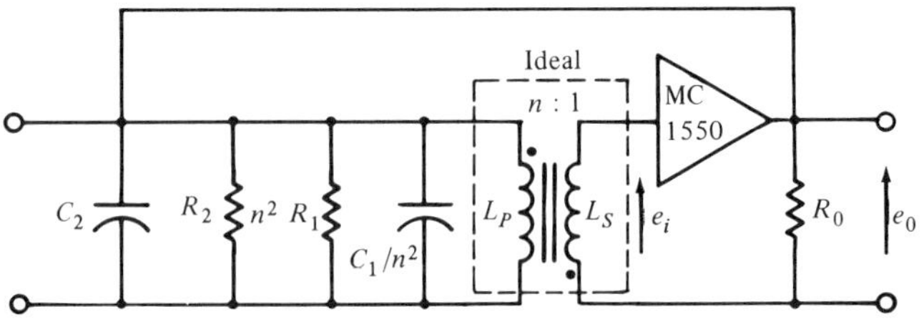

FIGURE 7-4. Transformer coupling referred to primary.

Thus using equation (7-10), the condition for self-starting and sustained oscillation is

$$\frac{R_2\|Z_p\|X_{C2}}{nr_{e1}} \geq 1 \tag{7-18}$$

From Figure 7-4, the frequency of oscillation by the circuit is given by

$$f_0 \approx \frac{1}{2\pi\sqrt{L_p(C_2 + C_1/n^2)}} \tag{7-19}$$

Equations (7-10) and (7-19), then, constitute design equations for the oscillator. If the transformer is less than 100% efficient ($k < 1$), these equations become

$$f_0 = \frac{1}{2\pi\sqrt{(1 - k^2)L_pC_T}} \tag{7-20}$$

$$A_{\text{ol}}(\omega) = -k\sqrt{\frac{L_p}{L_s}} \tag{7-21}$$

EXAMPLE 7-1

Design an oscillator tunable from 5 to 10 MHz. Use the MC1550 amplifier.

SOLUTION

Choose $L_p = 1.3\ \mu$H. Then from equation (7-19),

$$C_2(5\text{ MHz}) \approx \frac{1}{L_p\omega_0^2}$$

$$= \frac{1}{(1.3 \times 10^{-6})[(2\pi)(5 \times 10^6)]^2} = 780\text{ pF}$$

$$C_2(10\text{ MHz}) = \frac{1}{(1.3 \times 10^{-6})[(2\pi)(10 \times 10^6)]^2} = 195\text{ pF}$$

The maximum turns ratio for the transformer is obtained from equation (7-18). The emitter resistance of Q_1, r_{e1}, is obtained from equation (7-14). Referring to Figure 6-1, we have

$$I_e \approx \frac{\dfrac{R_2}{R_1 + R_2}V_{CC} - V_{D1}}{R_2} = \frac{\dfrac{3\text{ k}\Omega}{3\text{ k}\Omega + 3\text{ k}\Omega}(10) - 0.7}{3\text{ k}\Omega}$$

$$= 1.43\text{ mA}$$

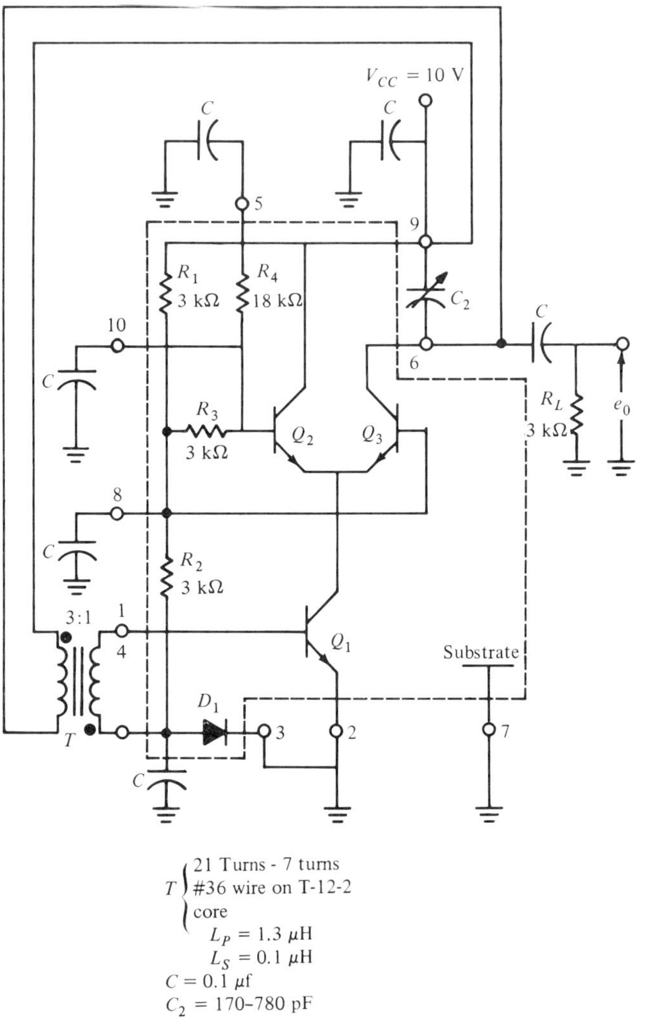

FIGURE 7-5. Five- to 10-MHz oscillator.

Here we are assuming that the diode D_1 is sufficiently matched to the emitter–base diode of Q_1 that the two currents are equal. Then

$$r_{e1} \approx \frac{26}{1.43} = 18.18\ \Omega$$

Now at resonance $X_{C2} = Z_p$ and at the low end

$$X_{C2} = \frac{1}{\omega C_2} = \frac{1}{(2\pi)(5 \times 10^6)(780 \times 10^{-12})}$$

$$= 40.8\ \Omega$$

If we choose $R_2 = 3\ \text{k}\Omega$, the requirements for the turns ratio is

$$\frac{3\ \text{k}\Omega \| 40.8 \| 40.8}{n(18.18)} \geq 1 \Rightarrow n \leq 1.10$$

A final design for the oscillator is shown in Figure 7-5. Observe that in the calculations a 100% efficient transformer ($k = 1$) was assumed. For an air-core transformer ($k = 0.3$) the turns ratio, n, is increased by a factor of approximately 3.

7-3

CRYSTAL-CONTROLLED OSCILLATORS

When extremely stable oscillation is needed, a crystal-controlled oscillator is used. The characteristics of a quartz piezoelectric crystal are shown in Figure 7-6. The mechanical resonance of the crystal makes it look like a very high Q-series resonant circuit, with a capacitor in parallel with it. Referring to Figure 7-6b, the crystal exhibits both a series resonance and a parallel resonance mode. When purchasing a crystal, the mode, series or parallel, must be specified. Since the circuit is high-Q, the crystal is ideal as the resonant feedback mechanism. It is evident from the equivalent circuit that the crystal can be "tuned" by placing a reactance in series or parallel with it.

Many integrated-circuit crystal-controlled oscillators are commercially available, with the user typically supplying the crystal. Typical of these is the MC12561 crystal oscillator shown in block diagram form in Figure 7-7. The crystal for

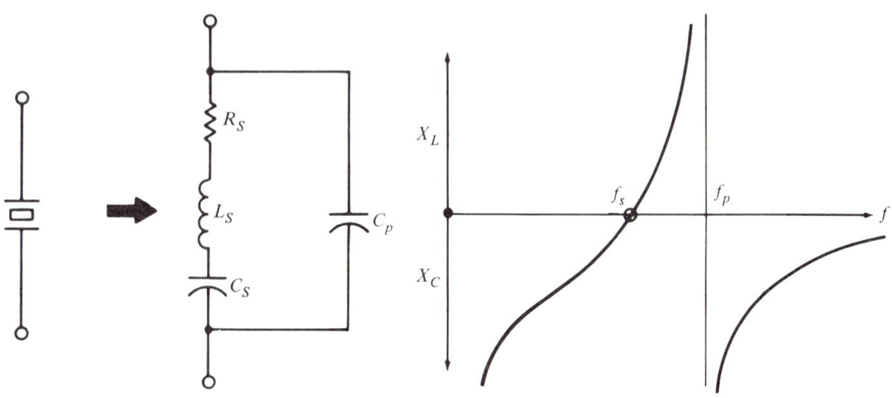

(a) equivalent circuit (b) frequence response

FIGURE 7-6. Quartz piezoelectric crystal.

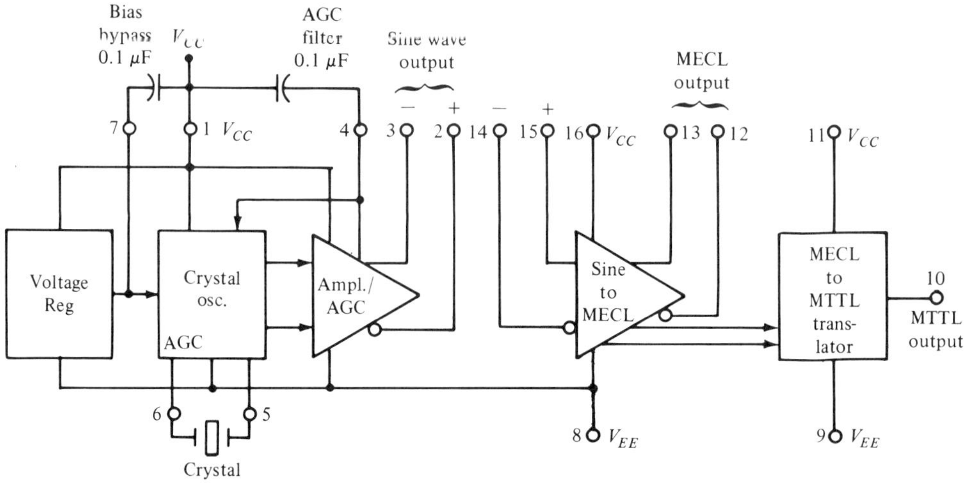

Note: 0.1 μF power supply pin bypass capacitors not shown.

FIGURE 7-7. MC12561 crystal oscillator block diagram.

this oscillator operates in the fundamental series mode. The series resistance, R_s, of the crystal is specified to have a minimum resistance of 155 Ω. For example, a typical specification for a crystal for this oscillator is the following:

- Mode, series
- Frequency, 15 MHz
- Maximum effective resistance at the fundamental, 155 Ω

During startup, stabilization is a function of the crystal series resistance. The lower the resistance, the faster the circuit stabilizes.

7-4 VOLTAGE-CONTROLLED OSCILLATORS

Many oscillator applications require that the oscillator frequency be directly proportional to a dc control voltage. For multivibrator-type oscillators, frequency control is typically controlled by use of a voltage-variable current source that is controlled by varying the resonant frequency of the tank circuit. A typical configuration is shown in Figure 7-8.

The capacitance of the varactor, D, is effectively in parallel with the inductance, L, via the bypass action of C_1 and the 0.1-μF capacitors. The control

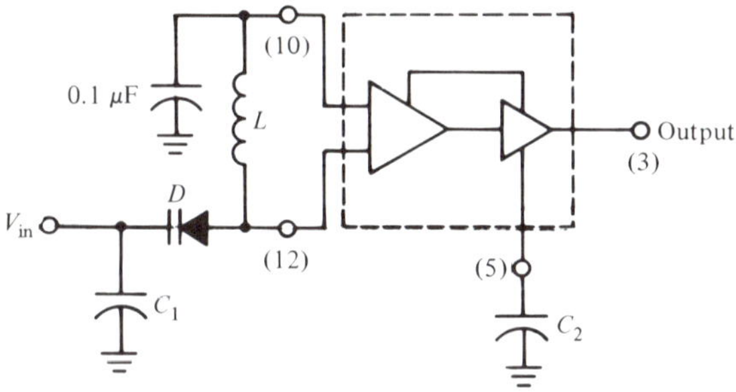

FIGURE 7-8. Feedback oscillator with voltage-controlled feature.

voltage, V_{in}, must correspond to the linear range of the varactor. Voltage-controlled oscillators (VCOs) form an essential feature of phase-locked loops.

7-5 PHASE-LOCKED LOOPS

Electronic phase-locked loops (PLLs) came into use in the 1930s when they were used for radar synchronization and communication applications. This technique for electronic frequency control is used extensively today in satellite communication systems, airborne navigation systems, FM and AM communications systems, computers, and so on.

One of the major applications of the PLL is frequency synthesis in systems that require discrete frequencies or fixed channel spacing. The UHF television band (channels 14 to 83) extends from 470 to 890 MHz in 6-MHz steps, and the aircraft VHF navigation/communication band is 108 to 136 MHz in 25-kHz steps. AM and FM standard broadcast bands are also in channel form. Phase-locked loops are excellent circuit designs for these applications.

7-5.1 Phase-Locked-Loop Principles

The phase-locked loop is a closed-loop electronic servo whose output locks onto and tracks an input reference signal. Refer to Figure 7-9. Phase lock is obtained by comparing the phase of the output signal, f_o, with that of the reference, f_r, and any phase difference is converted into an error correction voltage, V_f, by

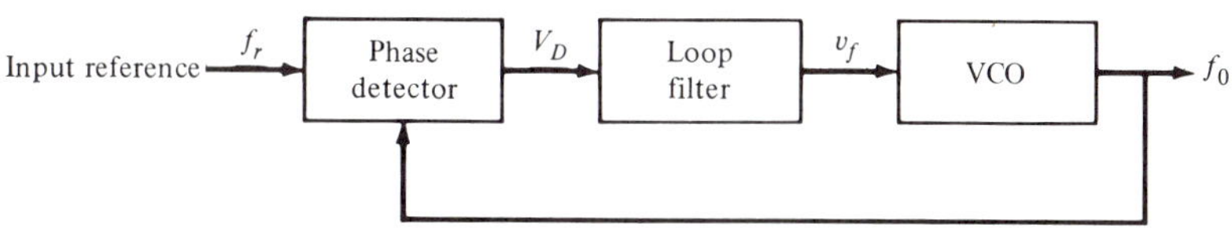

FIGURE 7-9. Basic phase-locked loop.

the loop filter. This error voltage changes the output signal to make it track the input.

Prior to lock, the input may be at any frequency, ω_i. The phase detector in the PLL is such that given an input, ω_i, and a phase detector demodulator reference input from the VCO of ω_o, the phase detector will generate sum and difference frequency components ($\omega_i \pm \omega_o$) and the sum frequency component will be removed by the low-pass filter. If the difference frequency ($\omega_i - \omega_o$) is outside the edge of the filter band, it will also be removed. As the input ω_i, approaches the VCO frequency, the difference frequency decreases and enters the band edge of the low-pass filter. The filter passes some of the dc component of the difference frequency and applies the dc to the VCO. The dc drives the VCO frequency, ω_o, toward the input, ω_i, decreasing the difference frequency and bringing it well within the band edge of the low-pass filter. This allows more of the dc component to be passed by the filter to the VCO, which in turn further decreases the difference component. This positive-feedback action continues until the VCO frequency is equal to that of the input. The difference frequency is now zero and the PLL is in "lock." The VCO will now track the input frequency within the VCO lock range. That range of input frequencies, ω_i, centered about the VCO free-running frequency, ω_c, such that the maximum difference frequency, ($\omega_i - \omega_c$) will allow the initial lock of the VCO to the input is called the *capture range* of the VCO.

Simple trigonometry may be used to illustrate operation of the PLL. Suppose that the input signal is sinusoidal with arbitrary phase. That is,

$$V_i(t) = A_1 \sin (\omega_i t + \theta_i) \tag{7-22}$$

Similarly, the VCO output is

$$V_o(t) = A_2 \cos (\omega_o t + \theta_o) \tag{7-23}$$

where A_1 and A_2 are the amplitudes. The phase detector may be modeled as a four-quadrant multiplier. Thus

$$\begin{aligned} V_D &= V_i(t)(V_o(t) \\ &= K_m A_1 A_2 \sin (\omega_i t + \theta_1) \cos (\omega_o t + \theta_o) \end{aligned} \tag{7-24}$$

and the low-pass filter output, V_f, is

$$V_f = K_m \frac{A_1 A_2}{4} \sin [(\omega_i - \omega_o)t + (\theta_i - \theta_o)] \tag{7-25}$$

where k_m is the gain of the multiplier in volts per radian. The sum frequency

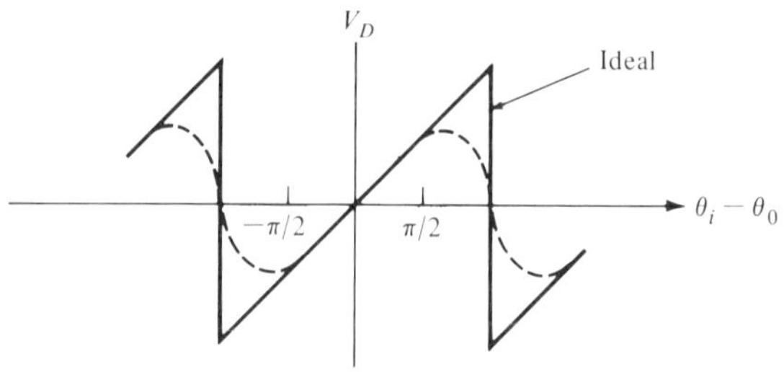

FIGURE 7-10. Typical phase detector characteristic.

terms are filtered by the low-pass filter. If the PLL is frequency locked to the input signal, equation (7-25) reduces to

$$V_f \simeq K_D \sin(\theta_i - \theta_o) \tag{7-26}$$

where $K_D = K_m(A_1A_2/4$ and $\theta_i - \theta_o$ is the phase error.

Equation (7-25) describes the error voltage that is applied to the VCO to maintain lock. The error voltage is indicated in Figure 7-10. If a disturbance enters the PLL and causes it to lose lock $[(\theta_i - \theta_o) > \pm\pi/2]$, the PLL will skip one or more cycles.

7-5.2 Phase-Locked-Loop Characteristics

Fundamental loop characteristics such as capture range, loop bandwidth, capture time, and transient response are controlled primarily by the loop filter. The loop behavior is described by gains in each component block in Figure 7-11.

The error voltage, V_f, is given by equation (7-26) and the frequency of the VCO is controlled by the filtered error voltage, V_f. Deviation of the VCO from its center frequency is given by $\Delta\omega = K_V V_f$. Since frequency is the time rate of change of phase, the VCO operation may be described by

$$\frac{d\theta_o}{dt} = K_V V_f \tag{7-27}$$

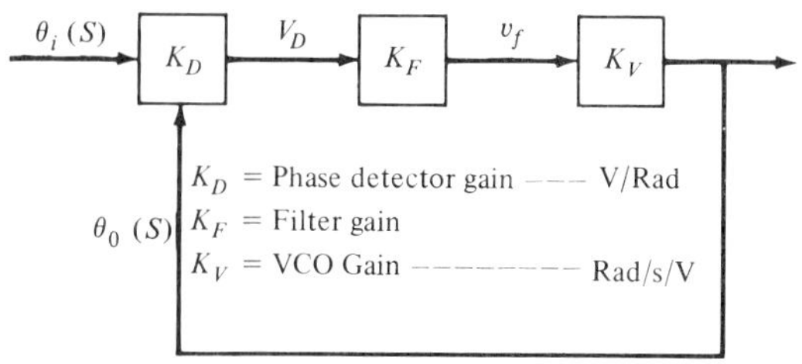

FIGURE 7-11. PLL gain constraints.

Taking the Laplace transform of equation (7-27) yields

$$s\theta_o(s) = K_V V_f(s) \tag{7-28}$$

where s is the Laplace operator introduced in Chapter 3. Solving for $\theta_o(s)$ results in

$$\theta_o(s) = \frac{K_V V_f(s)}{s} \tag{7-29}$$

Equation (7-29) indicates that the VCO output is proportional to the integral of the input control voltage.

The transfer function $\theta_o(s)/\theta_i(s)$ can now be readily determined. If the phase detector is assumed to be linear, equation (7-26) may be approximated as

$$V_D \approx K_D[\theta_i(s) - \theta_o(s)] \tag{7-30}$$

and the loop filter output is

$$V_f(s) = F(S)V_D(s) \tag{7-31}$$

where $F(s)$ is the loop filter transfer function. Combining equations (7-29) through (7-31) results in the basic PLL equation:

$$A(s) = \frac{\theta_o(s)}{\theta_i(s)} = \frac{K_V K_D F(s)}{S + K_V K_D F(s)} \tag{7-32}$$

Loop Filter. We see from equation (7-32) that the PLL transfer function is directly dependent on the filter transfer function $F(s)$. Consider first a loop where $F(s) = 1$. That is, the low-pass filter is removed and the output of the phase detector is tied directly to the VCO input. The loop transfer function reduces to

$$A(s) = \frac{K_V K_D}{s + K_V K_D} = \frac{1}{1 + s/K_V K_D} \tag{7-33}$$

and can be characterized as a first-order loop. Replacing s by $j\omega$, the magnitude of the frequency response is plotted in Figure 7-12.

The *natural frequency* of the loop coincides with the -3-dB frequency and is denoted as ω_n. For a first-order loop we see that $\omega_n = K_V K_D$ and is the same as the bandwidth of the loop. The term $K_V K_D$ is referred to as the *forward loop gain*.

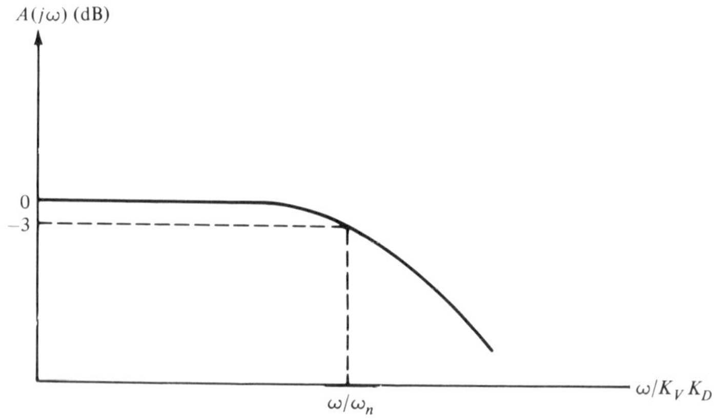

FIGURE 7-12. Loop frequency response.

Next, consider a PLL containing a filter such as the one illustrated in Figure 7-13. Other popular configurations are available. The transfer function for this filter is

$$F(j\omega) = \frac{R_2 + 1/j\omega C}{R_1} = \frac{1 + j\omega R_2 C}{j\omega R_1 C} = \frac{1 + j\omega\tau_2}{j\omega\tau_1}$$

or (7-34)

$$F(s) = \frac{1 + \tau_2 s}{\tau_1 s}$$

where $\tau_1 = R_1 C$ and $\tau_2 = R_2 C$. Therefore, from equation (7-32),

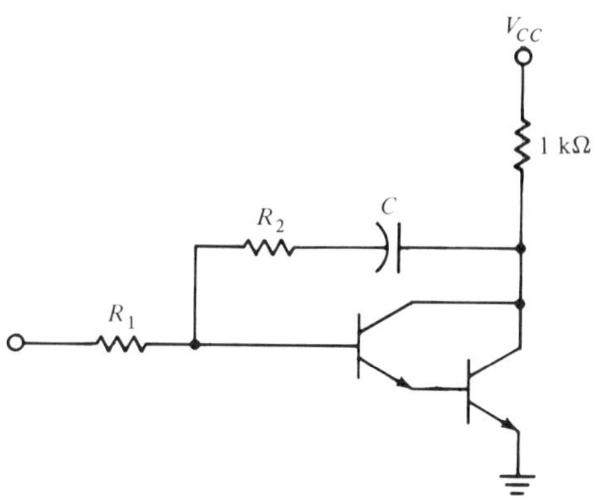

FIGURE 7-13. Active loop filter.

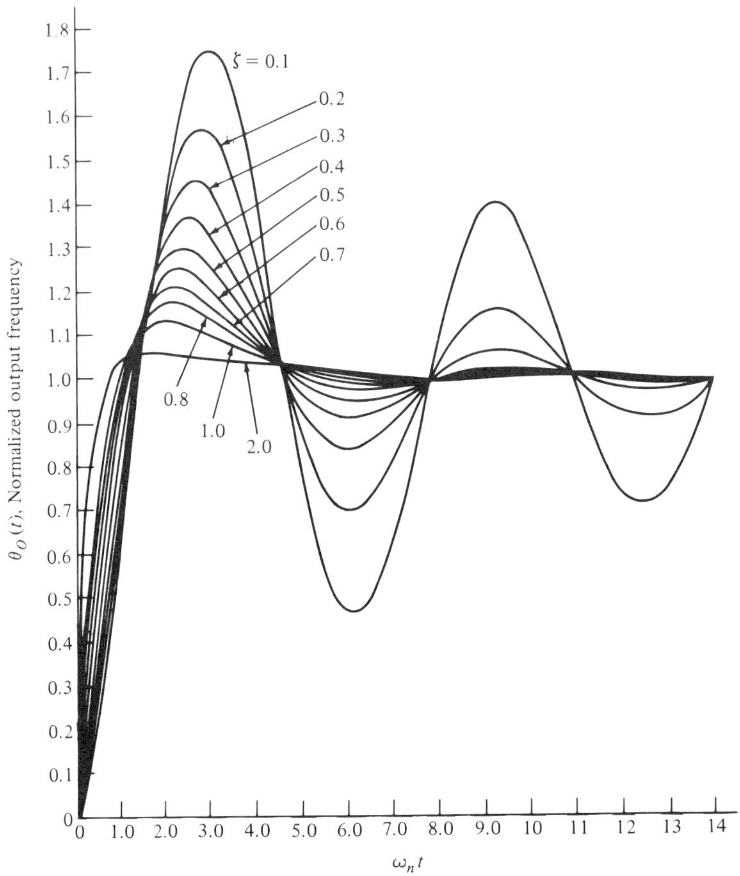

FIGURE 7-14. Second-order step response.

$$A(s) = \frac{K_V K_D \dfrac{1 + \tau_2 s}{\pi_1 s}}{s + K_V K_D \dfrac{1 + \tau_2 s}{\tau_1 s}} = \frac{1 + \tau_2 s}{1 + \tau_2 s + s^2 \tau_1 / K_V K_D} \qquad (7\text{-}35)$$

The *standard form* for a second-order system is

$$A(s) = \frac{1 + \tau_2 s}{1 + (2\delta/\omega_n)s + s^2/\omega_n{}^2} \qquad (7\text{-}36)$$

Comparing equations (7-36) and (7-35), we see that the loop bandwidth or natural frequency, ω_n, and the damping factor, δ, is

$$\omega_n = \sqrt{\frac{K_V K_D}{\tau_1}} \qquad (7\text{-}37)$$

and

$$\delta = \sqrt{\frac{K_V K_D}{\tau_1} \frac{\tau_2}{2}} \tag{7-38}$$

Both ω_n and δ are important to a step input of phase or frequency. The normalized step response for a second-order system is shown in Figure 7-14. This figure represents a particularly important design tool. Values for ω_n and δ may be chosen from a transient response (time domain) or steady-state frequency plot (roll-off point and peaking versus frequency). Once these two values are defined, synthesis of the filter may proceed. Resistor values may be calculated from equations (7-37) and (7-38). That is,

$$R_1 = \frac{K_V K_D}{\omega_n^2 C} \tag{7-39}$$

and

$$R_2 = \frac{2\delta}{\omega_n C} \tag{7-40}$$

EXAMPLE 7-2

Design the loop filter for a phase-locked loop which must lock up in less than 2 ms and will have a peak overshoot of less than 20%. The VCO gain, K_V, is 10×10^6 rad/s/V and the phase detector gain is 0.2 V/rad.

SOLUTION

From Figure 7-14, $\delta = 0.8$ will produce less than 20% overshoot. Also, we see that the loop settles to within 5% of normal at $\omega_n t = 4.5$. Thus

$$\omega_n = \frac{4.5}{t} = \frac{4.5}{2 \times 10^{-3}} = 2.25 \times 10^3 \text{ rad/s}$$

Let $C = 1\ \mu\text{F}$; then

$$R_1 = \frac{K_V K_D}{\omega_n^2 C} = \frac{(10 \times 10^6)(0.2)}{(2.25 \times 10^3)^2(10^{-6})} = 395 \text{ k}\Omega$$

and

$$R_2 = \frac{2\delta}{\omega_n C} = \frac{(2)(18)}{(2.25 \times 10^3)(10^{-6})} = 711\ \Omega$$

FREQUENCY SYNTHESIZERS

Frequency synthesizers are easily produced by dividing the VCO output frequency before feeding it back to the phase detector. To facilitate this, it is convenient to use a VCO whose output frequency is a square wave, thus making it possible to accomplish division using ordinary integrated circuits. Refer to Figure 7-15. Since the phase detector will indicate lock (see Figure 7-10) only for signals of the same frequency, the VCO will drive to the *N*th harmonic of the reference frequency, f_r.

In Chapter 4 we examined the spectrum of a square wave. The phase detector will contain output frequencies which are the sum of the reference frequency and the harmonics of the square wave from the divider network. These frequencies, however, will fall outside the passband of the low-pass filter and the error voltage to the VCO is still described by equation (7-26). If the crystal oscillator output is also a square wave, a digital phase detector may be utilized. The operation of this detector approaches the ideal characteristics illustrated in Figure 7-10. Numerous integrated-circuit phase detectors and VCOs are available commercially. The design problem, then, lies in selecting these components to fit the operating frequency of interest and then tailoring the loop filter and divider network for the specific application.

With a divider network in the loop, equation (7-32) becomes

$$A(S) = \frac{K_V K_D K_n F(S)}{S + K_V K_D K_n F(S)} \tag{7-41}$$

Following the foregoing procedure, equations for R_1 and R_2 are

$$R_1 = \frac{K_V K_D}{N\omega_n^2 C} \tag{7-42}$$

and

$$R_2 = \frac{2\delta}{\omega_n C} \tag{7-43}$$

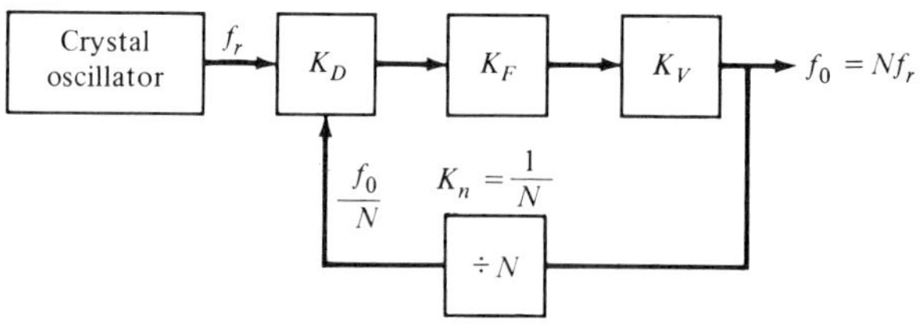

FIGURE 7-15. Frequency synthesizer.

EXAMPLE 7-3

Design a frequency synthesizer to the following specifications.

- f_{out}, 2 to 3 MHz.
- Frequency steps, 100 kHz.
- Lockup time between channels (to 5%), 1.0 ms.
- Overshoot, < 20%.

SOLUTION

From the statement of the problem, first obtain the range of N.

Range:

$$f_r = f_{in} = 100 \text{ kHz}$$

then

$$N_{max} = \frac{f_{0(max)}}{f_r} = \frac{3.0 \text{ MHz}}{0.1 \text{ MHz}} = 30$$

$$N_{min} = \frac{f_{0(min)}}{f_r} = \frac{2.0 \text{ MHz}}{0.1 \text{ MHz}} = 20$$

VCO Range: The VCO output frequency range must extend beyond the maximum and minimum limits specified. This is necessary to accommodate the overshot specification. For 20%, end limits are as follows:

$$f_{0(max)} \geq 3.0 + (0.2)(3.0) = 3.6 \text{ MHz}$$

$$f_{0(min)} \leq 2.0 - (0.2)(2.0) = 1.6 \text{ MHz}$$

The VCO tuning range is then $3.6/1.6 = 2.25{:}1$.

Damping factor and loop natural frequency: From Figure 7-14 a value of $\delta = 0.8$ will produce a peak overshoot of less than 20%. From the same figure we see that the response will settle to less than 5% at $\omega_n t = 4.5$. Thus

$$\omega_n = \frac{4.5}{t} = \frac{4.5}{0.001} = 4.5 \times 10^3 \text{ rad/s}$$

Filter design: To compute C, the gain of the VCO and the phase detector must be known. Choosing for these components, the MC4324/4024 voltage-controlled multivibrator and the MC4344/4044 phase detector, we find from the data sheets that the respective gain constants are

$$K_V = 11 \times 10^6 \text{ rad/s/V}$$

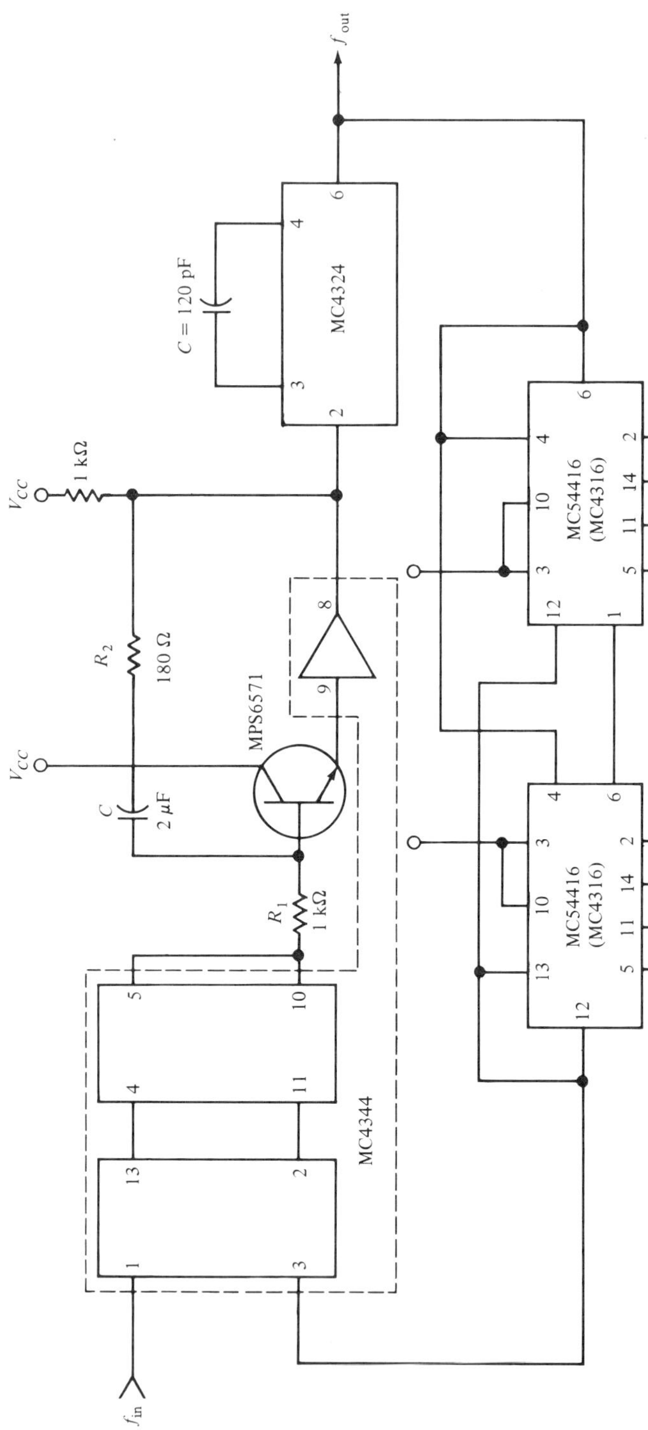

FIGURE 7-16. Frequency synthesizer circuit diagram.

and

$$K_D = 0.1 \text{ V/rad}$$

Choose $R_1 = 1\ \text{k}\Omega$. Then

$$C = \frac{K_D K_V}{N_{\max}\omega_n^2 R_1} = \frac{(0.1)(11 \times 10^6)}{(30)(4.5 \times 10^3)^2(10^3)} = 2\ \mu\text{F}$$

$$R = \frac{2\delta_{\min}}{\omega_n C} = \frac{(2)(0.8)}{(4.5 \times 10^3)(2 \times 10^{-6})} = 180\ \Omega$$

$$\delta_{\max} = \delta_{\min}\sqrt{\frac{N_{\max}}{N_{\min}}} = 0.8\sqrt{\frac{30}{20}} = 0.98$$

From Figure 7-14 we see that a value for δ up to 0.98 will also meet the settling-time requirement. The circuit diagram for this phase-locked loop is shown in Figure 7-16.

7-7 INTEGRATED-CIRCUIT PHASE-LOCKED LOOPS

In many instances design requirements do not dictate a special-purpose design. A large number of monolithic phase-locked loops are available for use at economical prices. Typical of these are the 560 series by Signetics. Generally speaking, these chips may be used with the addition of a few external components; the SE/NE564 provides response up to 50 MHz. Specifications with application information are contained on the following pages. The succeeding chapters on modulation and demodulation will contain additional examples of current uses for phase-locked loops.

SE/NE564-N,I

DESCRIPTION

The NE564 is a versatile, high frequency Phase Locked Loop designed for operation up to 50MHz. As shown in the block diagram, the NE564 consists of a VCO, limiter, phase comparator, and post detection processor.

APPLICATIONS

- **High speed modems**
- **FSK receivers and transmitters**
- **Frequency synthesizers**
- **Signal generators**

FEATURES

- **Operation with single 5V supply**
- **TTL compatible inputs and outputs**
- **Operation to 50MHz**
- **External loop gain control**
- **Reduced carrier feedthrough**
- **No elaborate filtering needed in FSK applications**
- **Can be used as a modulator**
- **Variable loop gain (Externally Controlled)**

PIN CONFIGURATION

I,N PACKAGE

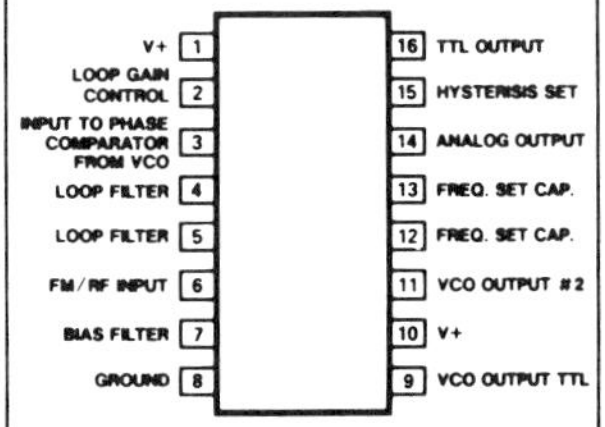

ORDER PART NO.
SE/NE564

ABSOLUTE MAXIMUM RATINGS

PARAMETER			RATING	UNIT
V+	Supply voltage			V
	Pin 1		14	
	Pin 10		6	
P_D	Power dissipation		400	mW
T_A	Operating temperature	NE	0 to 70	°C
	Operating temperature	SE	−55 to +125	
t_{stg}	Storage temperature		−65 to 150	°C

BLOCK DIAGRAM

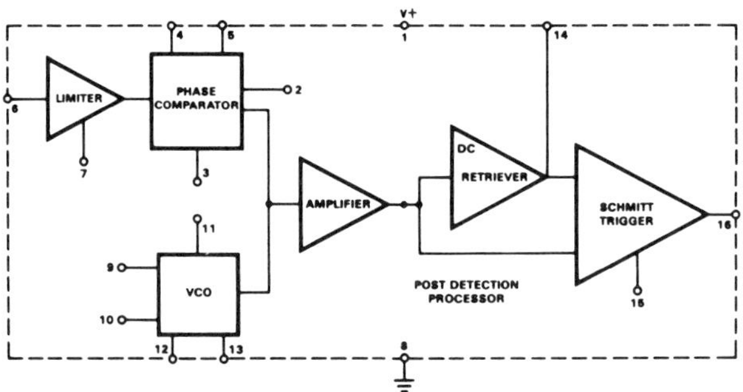

FUNCTIONAL DESCRIPTION

The NE564 is a monolithic phase locked loop with a post detection processor. The use of Schottky clamped transistors and optimized device geometries extends the frequency of operation to greater than 50MHz. In addition to the classical PLL applications, the NE564 can be used as a modulator with a controllable frequency deviation.

The output voltage of the PLL can be written as shown in the following equation:

$$V_O = \frac{(f_{in} - f_o)}{K_{VCO}} \qquad \text{Equation 1}$$

K_{VCO} = conversion gain of the VCO (see figure 7)
f_{in} = frequency of the input signal
f_o = free running frequency of the VCO

The process of recovering FSK signals involves the conversion of the PLL output into logic compatible signals. For high data rates, a considerable amount of carrier will be present at the output of the PLL due to the wideband nature of the loop filter. To avoid the use of complicated filters, a comparator with hysterisis or Schmitt trigger is required. With the conversion gain of the VCO fixed, the output voltage as given by Equation 1 varies according to the frequency deviation of f_{in} from f_o. Since this differs from system to system, it is necessary that the hysterisis of the Schmitt trigger be capable of being changed, so that it can be optimized for a particular system. This is

SE/NE564-N,I

ELECTRICAL CHARACTERISTICS V+ = 5V, T_A = 25°C, fo = 5MHz, I_B = −200μA unless otherwise specified.
Test Circuit: Figure 1

PARAMETER	TEST CONDITIONS	SE564			NE564			UNIT
		Min	Typ	Max	Min	Typ	Max	
Maximum VCO frequency		50	65		45	60		MHz
Lock range	Input ≥ 200mVrms, T_A = 25°C	60	90		60	90		% of fo
	= 125°C	30	50					
	= −55°C	120	150					
	= 0°C				100	120		
	= 70°C				50	70		
Capture range	Input ≥ 200mVrms, R2 = 27Ω	25	35		25	35		% of fo
	= 100Ω	35	50		35	50		
VCO frequency drift with temperature	fo = 5MHz, T_A = −55°C to 125°C		400	1000				PPM/°C
	= 0°C to 70°C					400	1250	
	fo = 500kHz, T_A = −55°C to 125°C		250	500				
	= 0°C to 70°C					400	850	
VCO frequency change with supply voltage	V+ = 4.5V to 5.5V		3	6		3	6	% of fo
Demodulated output voltage	Modulation frequency: 1kHz, fo = 5MHz							
	Input deviation: 10%, T = 25°C	120	140		120	140		mVrms
	1%, T = 25°C	12	14		12	14		mVrms
	T = 0°C				11	13		mVrms
	= −55°C	9	12					mVrms
	= 70°C				13	15		mVrms
	= 125°C	14	16					mVrms
Linearity	Deviation: 1% to 8%		1	3		1	3	%
Signal to noise ratio			40			40		dB
AM rejection			35			35		dB
Supply current	V+ = 5V		35	50		35	50	mA
Leakage current	Pin 9		1	10		1	10	μA
Output current	Pin 9			6			6	mA
Supply voltage	Pin 1	4.5		12	4.5		12	V
	Pin 10	4.5		5.5	4.5		5.5	V

accomplished in the 564 by varying the voltage at pin 15 which results in a change of the hysteris of the Schmitt trigger.

For FSK signals, an important factor to be considered is the drift in the free running frequency of the VCO itself. If this changes due to temperature, according to Equation 1 it will lead to a change in the dc levels of the PLL output, and consequently to errors in the digital output signal. This is especially true for narrow band signals where the deviation in f_{in} itself may be less than the change in f_o due to temperature. This effect can be eliminated if the dc or average value of the signal is retrieved and used as the reference to the comparator. In this manner, variations in the dc levels of the PLL output do not affect the FSK output.

VCO Section

Due to its inherent high frequency performance, an emitter coupled oscillator is used in the VCO. In the circuit, shown in the equivalent schematic, transistors Q_{21} and Q_{23} with current sources $Q_{25}-Q_{26}$ form the basic oscillator. The free running frequency of the oscillator is shown in the following equation:

$$f_o = \frac{1}{16R_CC_1} \qquad \text{Equation 2}$$

$R_C = R_{19} = R_{20}$ = 100Ω (INTERNAL)
C_1 = external frequency setting capacitor

Variation of V_d (phase detector output voltage) changes the frequency of the oscillator. As indicated by Equation 2, the frequency of the oscillator has a negative temperature coefficient due to the positive temperature coefficient of the monolithic resistor. To compensate for this, a current I_R with negative temperature coefficient is introduced to achieve a low frequency drift with temperature.

Phase Comparator Section

The phase comparator consists of a double balanced modulator with a limiter amplifier to improve AM rejection. Schottky clamped vertical PNPs are used to obtain TTL level inputs. The loop gain can be varied by changing the current in Q_4 and Q_{15} which effectively changes the gain of the differential amplifiers. This can be accomplished by introducing a current at pin 2.

Post Detection Processor Section

The post detection processor consists of a unity gain transconductance amplifier and comparator. The amplifier can be used as a dc retriever for demodulation of FSK signals, and as a post detection filter for linear FM demodulation. The comparator has adjustable hysteris so that phase jitter in the output signal can be eliminated.

As shown in the equivalent schematic, the dc retriever is formed by the transductance amplifier $Q_{42}-Q_{43}$ together with an external capacitor which is connected at the amplifier output (pin 14). This forms an integrator whose output voltage is shown in the following equation:

$$V_O = \frac{g_m}{C_2} \int V_{in}dt \qquad \text{Equation 3}$$

g_m = transconductance of the amplifier
C_2 = capacitor at the output (pin 14)
V_{in} = signal voltage at amplifier input

With proper selection of C_2, the integrator time constant can be varied so that the output voltage is the dc or average value of the input signal for use in FSK, or as a post detection filter in linear demodulation.

The comparator with hysterisis is made up of $Q_{49}-Q_{50}$ with positive feedback being provided by $Q_{47}-Q_{48}$. The hysterisis is varied by changing the current in Q_{52} with a resulting variation in the loop gain of the comparator. This method of hysterisis control, which is a dc control, provides symmetric variation around the nominal value.

Design Formula

The free running frequency of the VCO is shown by the following equation:

$$f_o = \frac{1}{16R_CC_1} \text{ in Hz} \qquad \text{Equation 4}$$

R_C = 100Ω
C_1 = external cap in farads

The loop filter diagram shown is explained by the following equation:

$$F(s) = \frac{1}{1 + sRC_3} \qquad \text{Equation 5}$$

$R = R_{12} = R_{13} = 1.3k\Omega$ (INTERNAL)

By adding capacitors to pins 4 and 5, two poles are added to the loop transfer function

at $\omega = \frac{1}{RC_3}$.

FM DEMODULATOR

The NE564 can be used as an FM demodulator. The connections for operation at 5V and 12V are shown in figures 2 and 3 respectively. The input signal is ac coupled with the output signal being extracted at pin 14. Loop filtering is provided by the capacitors at pins 4 and 5 with additional filtering being provided by the capacitor at pin 14. Since the conversion gain of the VCO is not very high, to obtain sufficient demodulated output signal the frequency deviation in the input signal should be fairly high (1% or higher).

MODULATION TECHNIQUES

The NE564 phase locked loop can be modulated at either the loop filter ports (pins 4 and 5) or the input port (pin 6) as shown in figure 4. The approximate modulation frequency can be determined from the frequency conversion gain curve shown in figure 5. This curve will be appropriate for signals injected into pins 4 and 5 as shown in figure 4.

EQUIVALENT SCHEMATIC

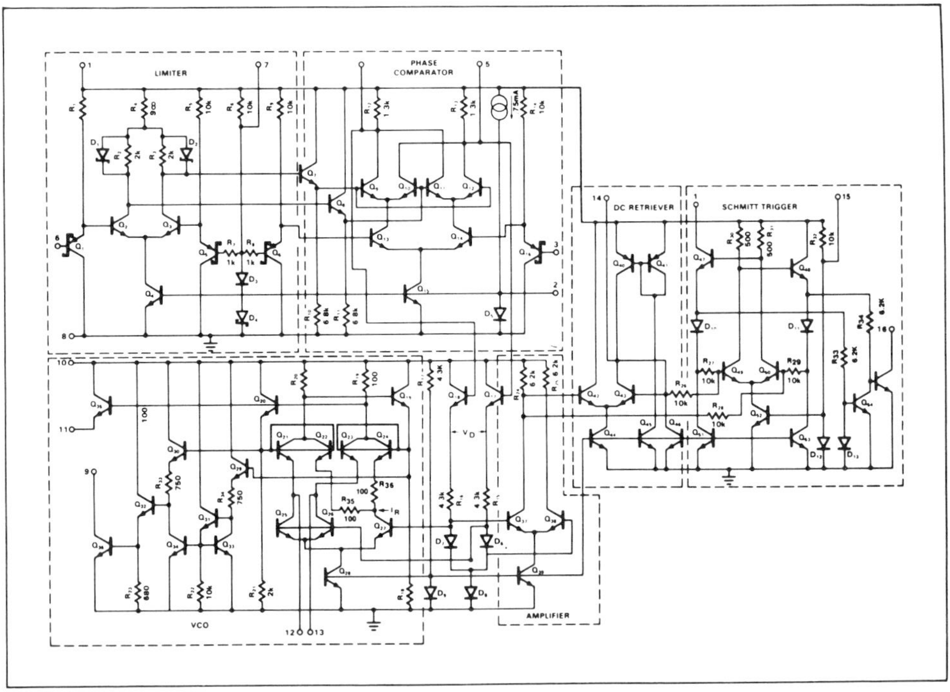

FSK Demodulation

The 564 PLL is particularly attractive for FSK demodulation since it contains an internal voltage comparator and VCO which have TTL compatible inputs and outputs, and it can operate from a single 5 volt power supply. Demodulated dc voltages associated with the mark and space frequencies are recovered with a single external capacitor in a dc retriever without utilizing extensive filtering networks. An internal comparator, acting as a Schmitt trigger with an adjustable hysteresis, shapes the demodulated voltages into compatible TTL output levels. The high frequency design of the 564 enables it to demodulate FSK at high data rates in excess of 1.0M baud.

Figure 6 shows a high-frequency FSK decoder designed for input frequency deviations of ± 1.0MHz centered around a free-running frequency of 10.8MHz. The value of the timing capacitance required was estimated from figure 8 to be approximately 40pF. A trimmer capacitor was added to fine tune f_O' to 10.8MHz.

Figure 9 indicates that the ± 1.0MHz frequency deviations will be within the lock range for input signal levels greater than approximately 50mV with zero pin 2 bias current. While strictly this figure is appropriate only for 5MHz, it can be used as a guide for lock range estimates at other f_O' frequencies.

The hysteresis was adjusted experimentally via the 10kΩ potentiometer and 2kΩ bias arrangement to give the waveshape shown in figure 7 for 20K, 500K, 2M baud rates with square wave FSK modulation. Note the magnitude and phase relationships of the phase comparators output voltages with respect to each other and to the FSK output. The high frequency sum components of the input and VCO frequency also are visible as noise on the phase comparators outputs.

SE/NE 564 TEST CIRCUIT

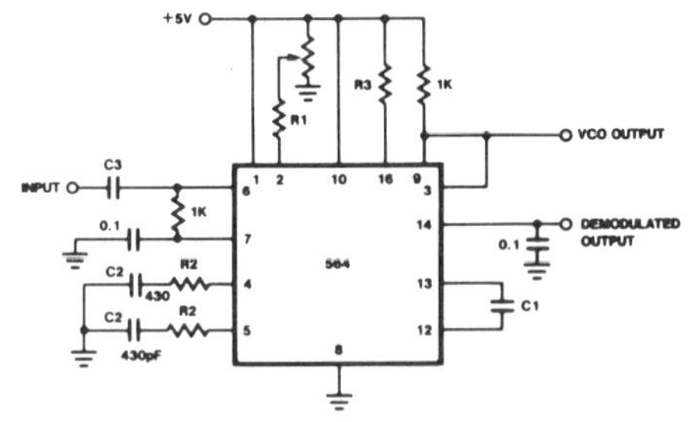

Figure 1

FM DEMODULATOR AT 5V

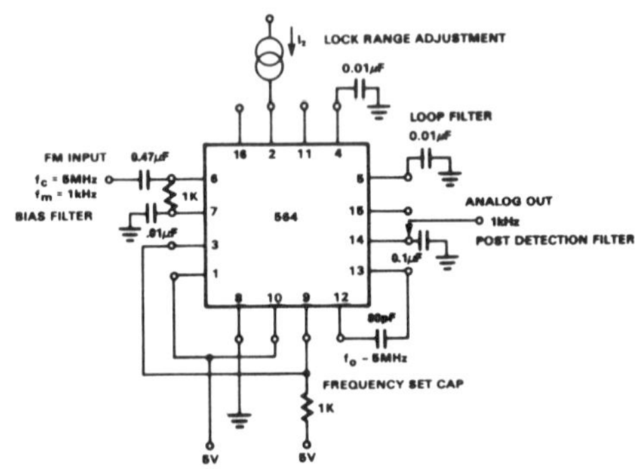

Figure 2

FM DEMODULATOR AT 12V

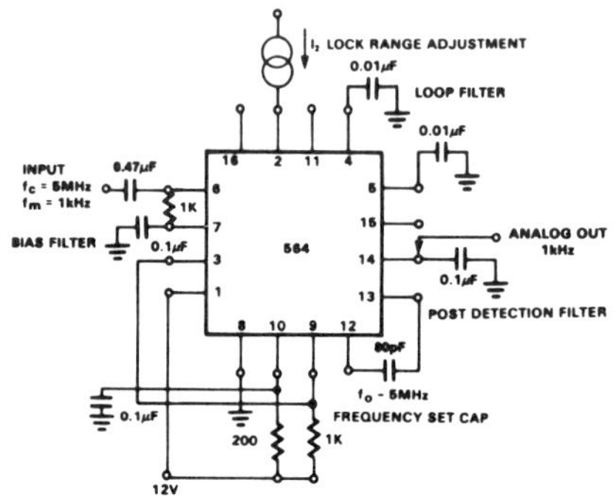

Figure 3

MODULATOR

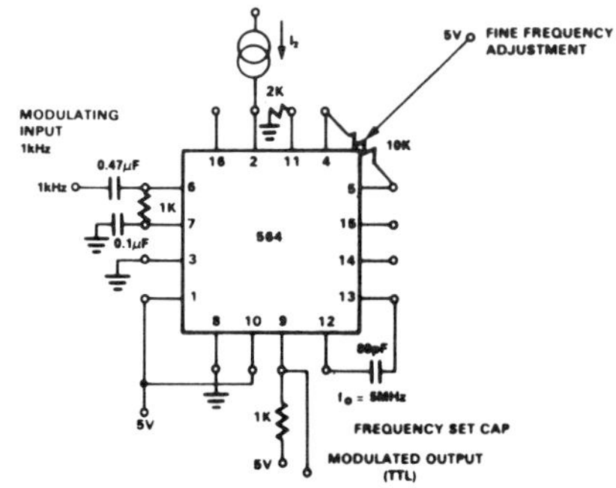

Figure 4

VCO CONVERSION GAIN

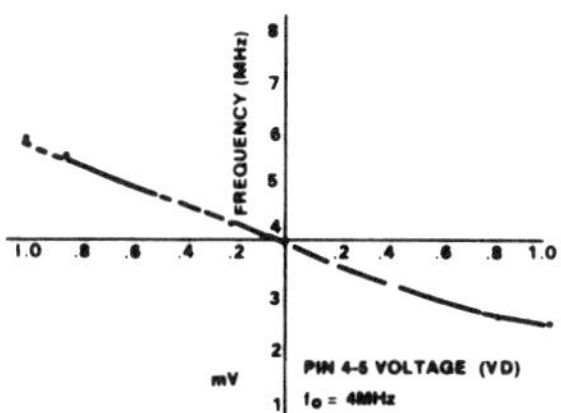

K_{VCO} = CONVERSION GAIN

$K_{VCO} = \frac{\Delta f}{\Delta V_D} \simeq -1.8 \frac{MHz}{V}$

FOR K_O OF PLL MODEL

$K_O = \frac{(\Delta f / fo)}{VOLT} = \frac{-(1.8MHz / 4MHz)}{1VOLT}$

$= \frac{-0.45 \text{ RADIANS}}{VOLT}$

$= -4.5 \times 10^{-4}$ RADIANS/mV

Figure 5

10.8MHz FSK DECODER USING THE 564

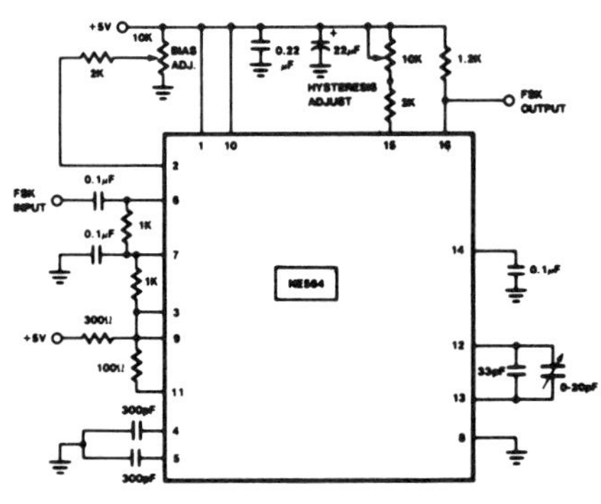

Figure 6

PHASE COMPARATOR (PINS 4 AND 5) AND FSK (PIN 16) OUTPUTS FOR DATA RATES OF

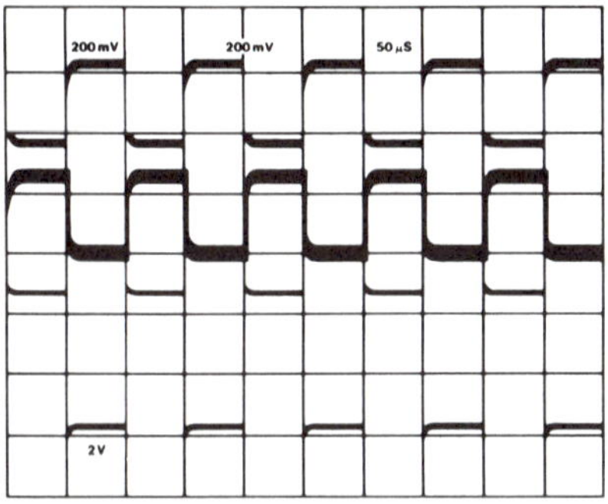

(a) 20K BAUD

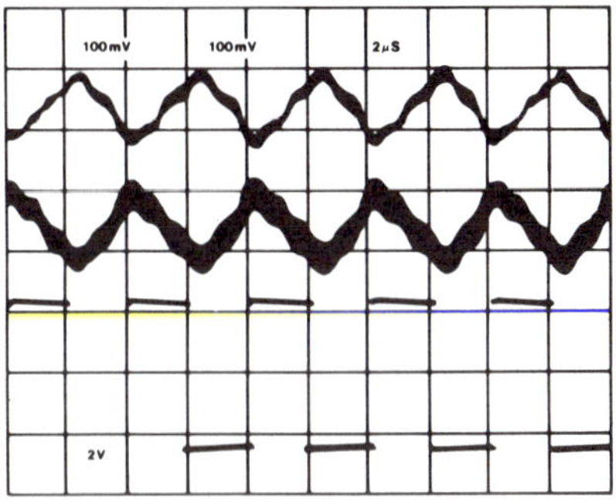

(b) 500K BAUD

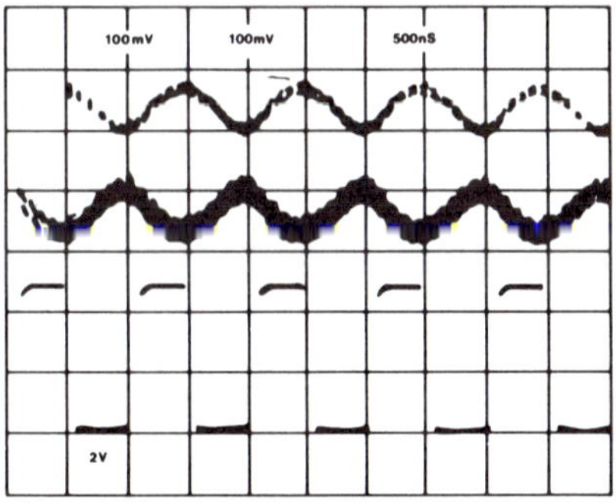

NOTE
Top trace-pin 4
Center trace-pin 5
Bottom trace-pin 16

(c) 2.0M BAUD

Figure 7

TYPICAL PERFORMANCE CHARACTERISTICS

LOCK RANGE vs SIGNAL INPUT

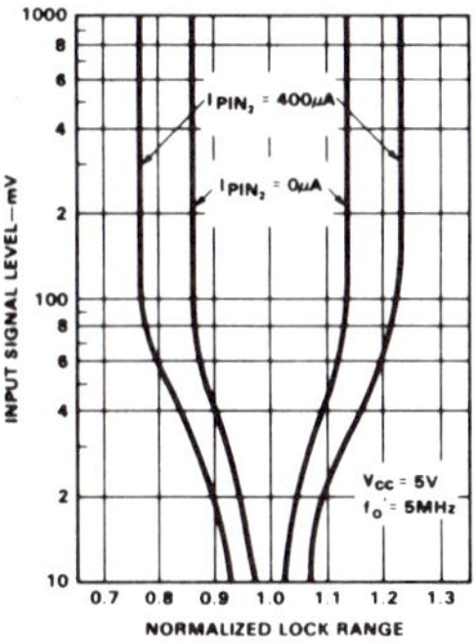

VCO CAPACITOR vs FREQUENCY

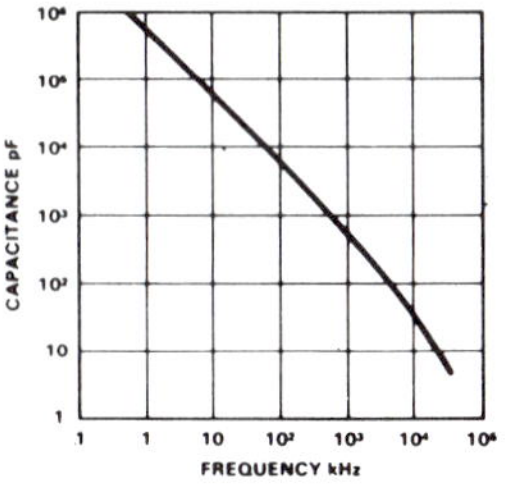

Figure 8

SE/NE564-N,I

TYPICAL NORMALIZED VCO FREQUENCY AS A FUNCTION OF PIN 2 BIAS CURRENT

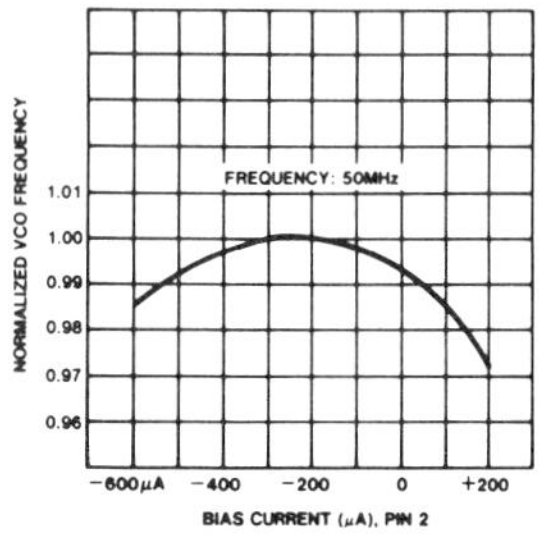

TYPICAL NORMALIZED VCO FREQUENCY AS A FUNCTION OF PIN 2 BIAS CURRENT

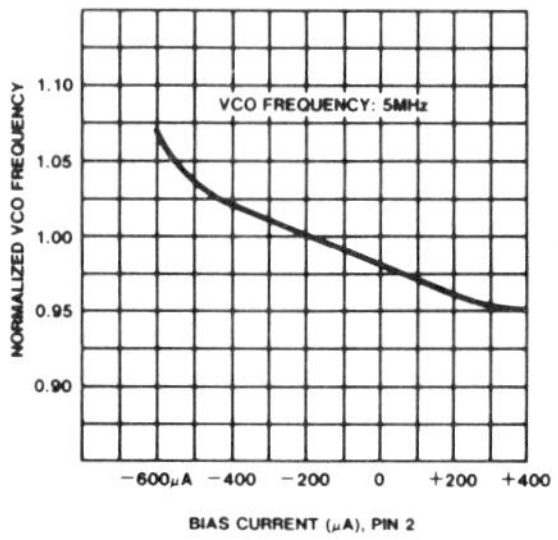

NORMALIZED VCO FREQUENCY AS A FUNCTION OF TEMPERATURE

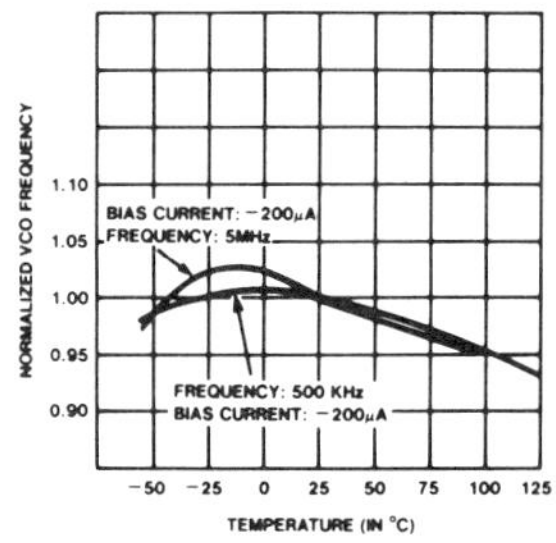

NOTE

Refer to section 10 of Signetics Analog Applications Manual for more detailed information.

CIRCUIT DESCRIPTION OF THE 564

The 564 contains the functional blocks shown in Figure 4.15. In addition to the normal PLL functions of phase comparator, VCO, amplifier and low-pass filter, the 564 has internal circuitry for an input signal limiter, a dc retriever, and a Schmitt trigger. The complete circuit for the 564 is shown in Figure 4.16.

Limiter

The input limiter functions to produce a near constant amplitude output that serves as the input for the phase comparator. Eliminating amplitude variations in the FM input signal improves the AM rejection of the PLL. Additional features of the 564s limiter are that it is capable of accepting TTL signals, operates at high-frequencies up to 50MHz, and remains functional with variable supply voltages between 5 and 12* volts.

BLOCK DIAGRAM OF THE 564

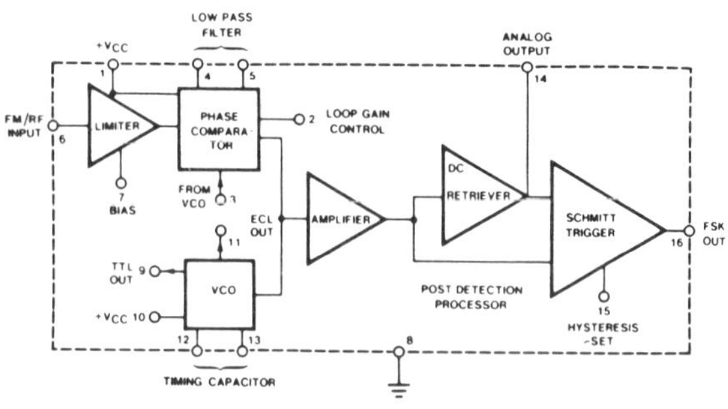

Figure 4.15

SCHEMATIC DIAGRAM OF 564

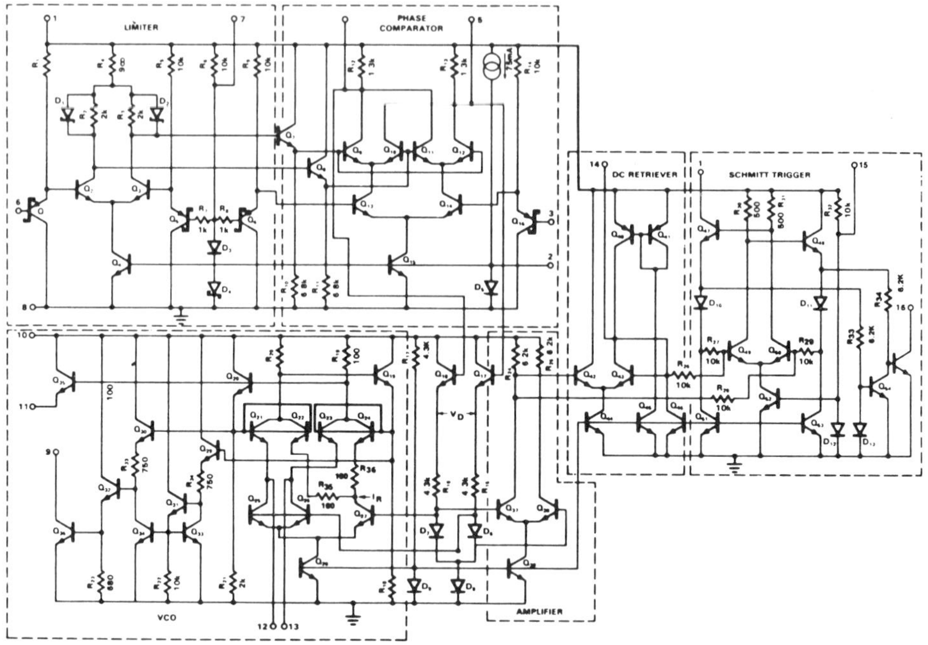

Figure 4.16

Signal limiting is accomplished in the 564 with a differential amplifier whose output voltage is clipped by diodes D_1 and D_2 (see Figure 4.17). Schottky diodes are used because their limiting occurs between 0.3 to 0.4 volts instead of the 0.6 to 0.7 volt for regular IC diodes. This lower limiting level is helpful in biasing, especially for 5 volt operation. When limiting, the dc voltage across R_2 and R_3 remains at the Schottky diode voltage. Good high-frequency performance for Q2 and Q3 is achieved with current levels in the low mA range. Current-source biasing is established via the current mirror of D_5 and Q4 (See Figure 4.16).

Base biasing for Q3 is of concern because of the nature of the input signal which can be either a TTL digital signal of 0 to 5 volts amplitude or a low-level, ac coupled analog signal. Compatibility for either type is achieved by modifying the limiter of Figure 4.17 with the addition of the vertical Schottky PNP transistors Q1 and Q5 shown in Figure 4.18. The input signal voltage appears as a collector-base voltage for Q1 which presents no problems for either high TTL level inputs or low-level analog inputs. Q5 is in turn diode biased by D_3 and D_4 (see Figure 4.16) which places the base voltages of Q1 and Q5 at approximately 1.0 volt. This same biasing network establishes a 1.3 volt bias at the base of Q13 for biasing the phase comparator section. A differential output signal from the input limiter is applied to one input of the phase comparator (Q9 through Q12) after buffering the level shifting though the Q7 - Q8 emitter followers.

*When operating above 5Vdc, a limiting resistor must be used from V_{CC} to pin 10 of the 564

Phase Comparator

The phase comparator section of the 564 is shown in Figure 4.19. It is basically the conventional, double-balanced mixer commonly used in PLL circuits with a few exceptions. The transconductance, g_m, for the Q13 - Q14 differential amplifier is directly proportional to the mirror current in Q15. Thus by externally sinking or sourcing current at pin 2, g_m can be changed to alter the phase comparators conversion gain, K_d. The nominal current injected into this node by the internal current source is 0.75mA for 5 volt operation. If this current is externally removed by gating, the phase comparator can be disabled and the VCO will operate at its free-running frequency.

The variation of K_d with bias current at pin 2 is shown in the experimental results of Figure 4.20. Note the inherent 90° phase error in the loop produces an approximate zero phase comparator output voltage. For any particular bias current, the slope of the line is the K_d conversion gain for the phase comparator. Numerically the data of Figure 4.20 can be expressed as

$$K_d \simeq 0.46 \frac{\text{volts}}{\text{rad}} + 7.3 \times 10^{-4} \frac{\text{volts}}{\text{rad} \times \mu A} \times I_{BIAS} \tag{4.3}$$

where I_{BIAS} is in μA. Equation 4.3 is valid for bias current less than $800\mu A$ where saturation occurs within the phase comparator.

The current level established in Q15 of Figure 4.18 determines all other quiescent currents in the phase comparator (Q9 through Q14). Currents through R_{12} and R_{13} set the common-mode output voltage from the phase comparator (pins 4 and 5). Since this common-mode voltage is applied to the VCO to establish its quiescent currents, the VCO conversion gain (K_o) also depends upon the bias current at pin 2.

VCO

The VCO is of the basic emitter-coupled astable type with several modifications included to achieve the high frequency, TTL compatible operation while maintaining low frequency drift with temperature changes. The basic oscillator in Figure 4.21 consists of Q19, Q20, Q21, and Q23 with current sinks of Q25 and Q26. The master current sink of Q28 keeps the total current constant by altering the ratio of currents in Q25 - Q26 and the dummy current sink of Q27.

The input drive voltage for the VCO is made up of common-mode and difference-mode components from the phase comparator. After buffering the level shifting through Q17 - Q18 and R_{15} - R_{16}, the VCO control voltage is applied differentially to the base of Q27 and to the common bases of Q25 and Q26.

The VCO control voltages from the phase comparator are the pin 4 and pin 5 voltages or

$$V_4 = V_{C9} = V_{B18} = V_{CM} + \frac{1}{2} V_{DM} \tag{4.4}$$

$$V_5 = V_{C12} = V_{B17} = V_{CM} - \frac{1}{2} V_{DM} \tag{4.5}$$

where V_{CM} and V_{DM} are the respective common-mode and the difference-mode voltages.

Emitter followers Q17 and Q18 convert these control voltages into control currents through D_6 and D_7 of the form

BASIC LIMITER STAGE

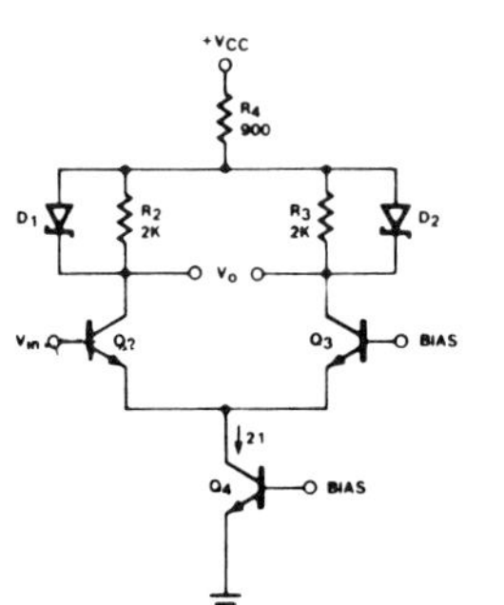

Figure 4.17

LIMITER STAGE WITH INPUT BUFFERING

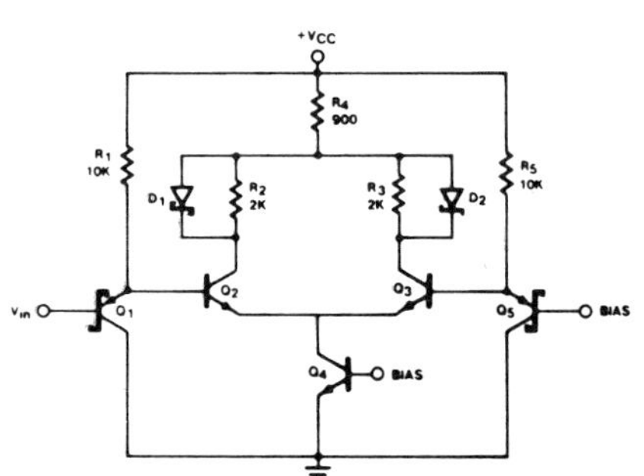

Figure 4.18

$$I_6 = \frac{1}{R_{15}}\left[V_{CM} - \frac{1}{2} V_{DM} - 3\,V_{BE}\right] \quad (4.6)$$

$$I_7 = \frac{1}{R_{16}}\left[V_{CM} + \frac{1}{2} V_{DM} - 3\,V_{BE}\right] \quad (4.7)$$

These individual currents are summed in D_8 and become with $R_{15} = R_{16} = R$.

$$I_8 = I = I_6 + I_7 = \frac{2}{R}(V_{CM} - 3\,V_{BE}) \quad (4.8)$$

Writing I_6 and I_7 as functions of the total I current gives

$$I_6 = \frac{I}{2}\left(1 - \frac{V_{DM}}{RI}\right) \quad (4.9)$$

$$I_7 = \frac{I}{2}\left(1 + \frac{V_{DM}}{RI}\right) \quad (4.10)$$

Now consider variations in I_6 and I_7 while I remains constant.

Let x indicate the current imbalance such that

$$I_6 = (1 - x)\,I = \frac{I}{2}\left(1 - \frac{V_{DM}}{RI}\right) \quad (4.11)$$

$$I_7 = xI = \frac{I}{2}\left(1 + \frac{V_{DM}}{RI}\right) \quad (4.12)$$

where $0 \leq x \leq 1$. Thus x is defined to be

$$x = \frac{1}{2}\left(1 + \frac{V_{DM}}{RI}\right) \quad (4.13)$$

Currents I_6 and I_7 establish proportional currents in Q25, Q26, and Q27 in a manner similar to the analysis above since the current in Q28 is a constant, or

$$I_o = I_{C28} = I_{E25} + I_{E26} + I_{E27A} + I_{E27B}$$

Gilbert(10) has shown that the D_7 - D_8 diode pair will cause identical differential currents to be reflected in both the Q25 - Q26 and the Q27A - Q27B differential amplifier pairs. Consequently the constant current of I_o jointly shared by the differential amplifier pairs will divide in each pair with the same x factor imbalance as in Equation 4.13.

$$I_{E25} + I_{E26} = xI_o \quad (4.14)$$

$$I_{E25} = I_{E26} = \frac{x}{2} I_o \quad (4.15)$$

$$I_{E27A} + I_{E27B} = (1 - x)\,I_o \quad (4.16)$$

$$I_{E27A} = I_{E27B} = \left(\frac{1 - x}{2}\right) I_o \quad (4.17)$$

Now consider placing a capacitor between the collectors of Q25 and Q26 (pins 12 and 13). Oscillation will occur with the capacitor alternately being charged by Q21 and Q23 and constantly discharged by Q25 and Q26. When the Q21 and Q22 pair conducts, Q23 and Q24 will be off causing a negative ramp voltage to appear at pin 13 and a constant voltage at pin 12 as shown in Figure 4.22. During the next half-cycle, the transistor roles and voltages are reversed. Capacitor discharge is via Q25 and Q26 which act as constant-current sinks with current amplitudes as in Equation 4.15.

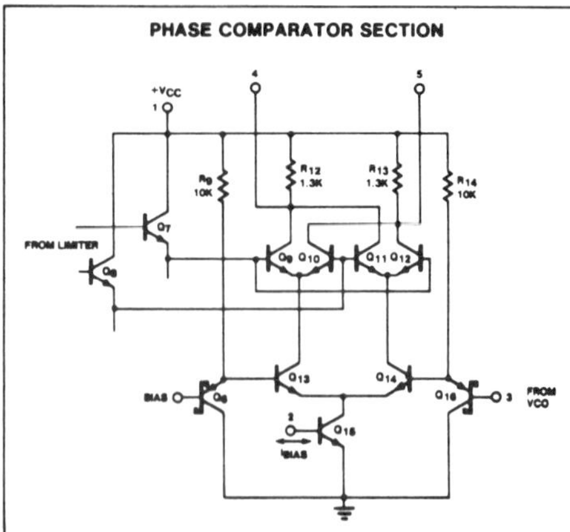

Figure 4.19

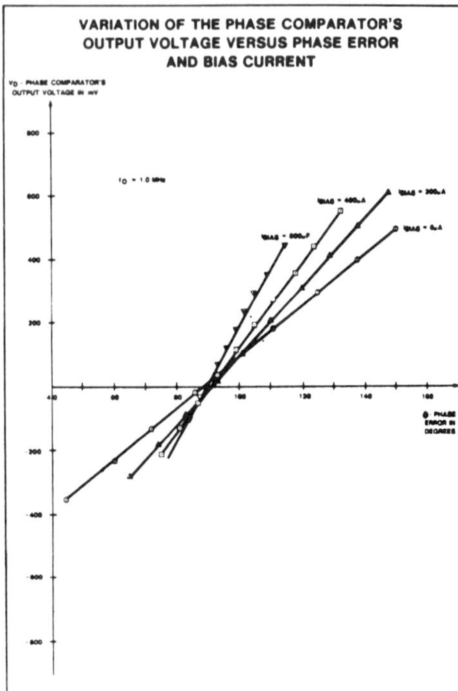

Figure 4.20

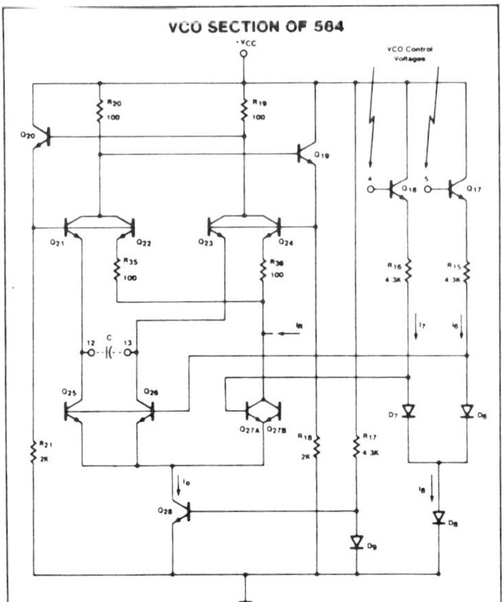

Figure 4.21

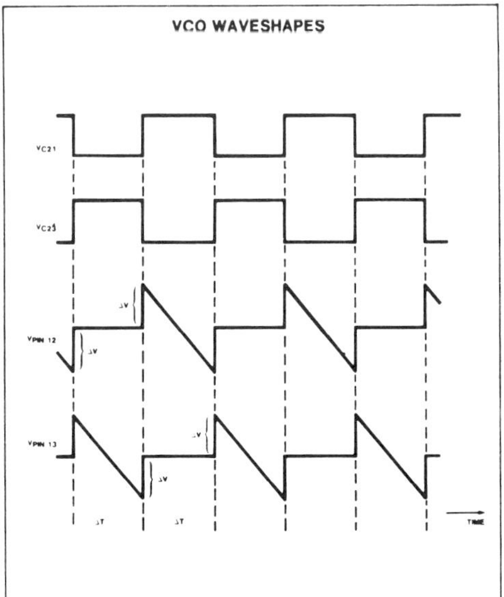

Figure 4.22

During each half-cycle, the capacitor voltage changes linearly by $2\Delta V$ volts in ΔT seconds where

$$\Delta V = 2R_{20} I_o \left(\frac{x}{2} + \frac{1-x}{2}\right) = R_{20} I_o \quad (4.18)$$

and

$$\Delta T = \frac{C2\Delta V}{I_{E25}}. \quad (4.19)$$

Combining these two equations with Equation 4.15 gives a half period of

$$\Delta T = \frac{4C\,R_{20}}{x} \quad (4.20)$$

Utilizing Equation 4.13 with the ΔT expression gives the desired VCO frequency expression of

$$f_o = f_o'\left(1 + \frac{V_{DM}}{RI}\right) = f_o'\left[\frac{V_{DM}}{2(V_{CM} \cdot 3\,V_{BE})}\right] \quad (4.21)$$

where f_o' is the VCOs free-running frequency given by

$$f_o' = \frac{1}{16\,R_{20}\,C} \quad (4.22)$$

Equation 4.21 shows that the oscillator frequency is a linear function of the differential voltage from the phase comparator. Resistors R_{35} and R_{36} function to insure that an initial current imbalance exists between the Q25 - Q26 transistor pair and the dummy Q27. This imbalance insures that the oscillator is self-starting when power is first applied to the circuit.

The VCO conversion gain is determined as

$$K_o = \frac{\partial f_o}{\partial V_{DM}} = \frac{f_o'}{RI} \quad Hz/volt \quad (4.23)$$

which is valid as long as the transistors V_{BE} changes are small with respect to the common-mode voltage. Both f_o and K_o are inversely proportional to R which has a strong positive temperature coefficient. An internal current I_R having an equal and opposite negative temperature coefficient is inserted into the VCO as shown in Figure 4.21.

Experimental determination of K_o can be found from the data of Figure 4.23 where K_o is the slope of either line. Numerically these results are for $I_{BIAS} = 0$,

$$K_o = 0.95\,\frac{MHz}{volt} = 5.9 \times 10^6\,\frac{rad}{volt\text{-}sec} \quad (4.24)$$

and for $I_{BIAS} = 800\mu A$

$$K_o = 1.7\,\frac{MHz}{volt} = 10.45 \times 10^6\,\frac{rad}{volt\text{-}sec} \quad (4.25)$$

It must be noted that the specific values obtained for K_o in the manner above are valid only for the 1.0MHz free-running frequency where the data was taken. However, good estimates for K_o at other free-running frequencies can be obtained by linearly scaling K_o to the desired f_o'. Thus it is sometimes convenient to define a normalized K_o as

$$K_{o(norm)} = \frac{K_o}{f_o'} = 5.9\,\frac{rad}{volt}\,(I_{BIAS} = 0)$$
$$= 10.45\,\frac{rad}{volt}\,(I_{BIAS} = 800\mu A) \quad (4.26)$$

The K_o estimate for any bias then can be obtained by multiplying the normalized conversion gain by the desired free-running frequency, or

$$K_o\,(\text{any } f_o') = K_{o(norm)}\,f_o'. \quad (4.27)$$

The additional VCO circuitry of Q29 through Q36 (Figure 4.16) functions to produce the TTL and ECL compatible outputs at pins 9 and 11.

Amplifier

The difference-mode voltage from the phase comparator is extracted and amplified by the amplifier in Figure 4.16. The single-ended output from this amplifier serves as input signals for both the Schmitt Trigger and a second differential amplifier. Low-pass filtering with a large capacitance at pin 14 produces a stable dc reference level as the second input to the Schmitt Trigger. When the PLL is locked, the voltage at pin 14 is directly proportional to the difference between the input frequency and f_o'. Thus pin 14 provides the demodulated output for a FM input signal.

Schmitt Trigger

In FSK applications the pin 14 voltage will assume two different voltage levels corresponding to the mark and space input frequencies. A voltage comparator could be used to sense and convert these two voltage levels to logic compatible levels. However at high data rates, V_{DM} will contain a considerable amount of carrier signal which can be removed by extensive filtering. Normally this complex filtering requires quite a few components, most all of which are external to the monolithic PLL. Also since the control voltage for the comparator depends upon K_O and the deviations of the mark and space frequencies from f_O', the filtering has to be optimized for each different system utilized. However the necessary dc reference level for the comparator is present in the PLL but buried in carrier frequency feedthrough which appears as noise in the system. A Schmitt trigger with variable hysterisis can be used successfully to decode the FSK data without the need for extensive filtering.

Consider the system shown in Figure 4.24 where the input signal is the single-ended output derived from the amplifier section of the 564. The dc retriever functions to establish a dc reference voltage for the Schmitt trigger. The upper and lower trigger points are adjustable externally around the reference voltage giving the variable hysteresis. For very low data rates, carrier feedthrough will be negligible and the ideal situation depicted in Figure 4.25 results. Increased data rate produces the carrier feedthrough shown in the (b) figure where false FSK outputs result because the feedthrough amplitude exceeds the hysteresis voltage. Having the capability to increase the hysteresis as in (c) produces the desired FSK output in the presence of carrier feedthrough.

Another important factor to be considered is the temperature drift of the f_O' in the VCO. Small changes in f_O' will change the dc level of the input voltage to the Schmitt trigger. This dc voltage shift would produce errors in the FSK output in narrow-band systems where the mark and space deviations in f_{in} are less than the f_O' change with temperature. However this effect can be eliminated if the dc or average value of the amplifier signal is retrieved and used as the reference voltage for the Schmitt trigger. In this manner, variations in the f_O' with temperature do not affect the FSK output.

VCO OUTPUT FREQUENCY AS A FUNCTION OF INPUT VOLTAGE AND BIAS CURRENT

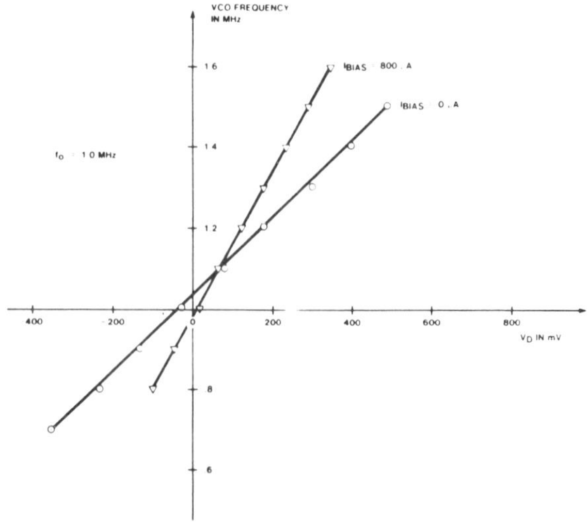

Figure 4.23

POST DETECTION PROCESSOR FOR FSK

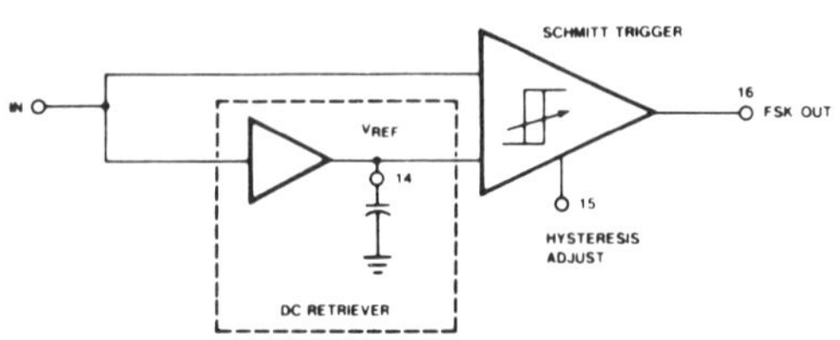

Figure 4.24

WAVESHAPES FOR FSK DECODING IN THE POST DETECTION PROCESSOR

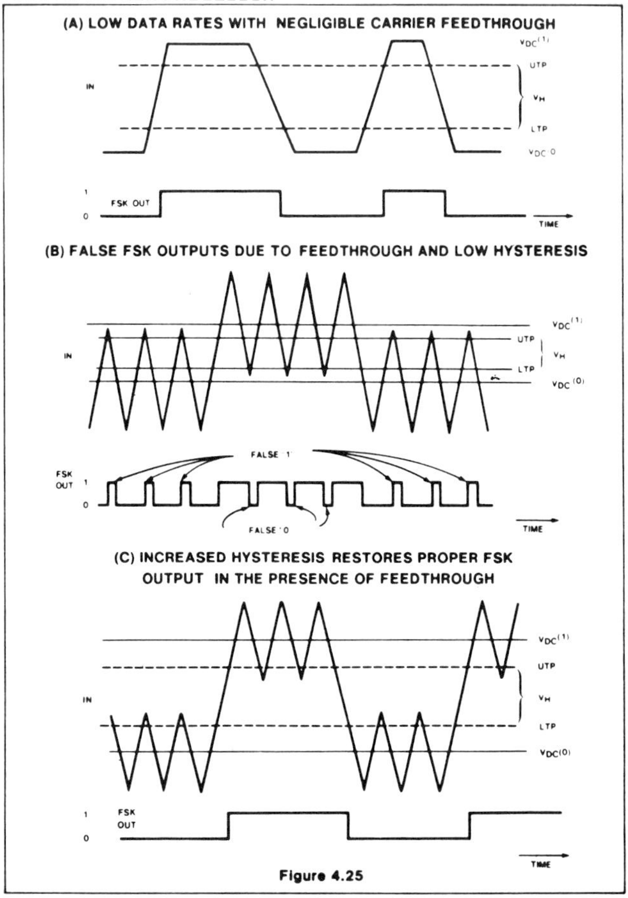

Figure 4.25

PROBLEMS

7-1. Repeat Example 7-1 for a fixed resonant frequency of 20 MHz. Assume a coefficient of coupling, k, for the transformer of 0.5. Choose $R_L = 4\ \text{k}\Omega$.

7-2. Repeat Problem 7-1 for $R_L = 2.5\ \text{k}\Omega$. What is the ac load impedance?

7-3. Calculate the open-loop and closed-loop gain for Problem 7-2. Determine the threshold gain below which oscillations will not be sustained.

7-4. Design a frequency synthesizer to the following specifications:

- f_0, 10 to 15 MHz.
- Frequency steps, 50 kHz.
- Lock-up time between channels (to 5%), 2.0 ms.
- Overshoot, < 15%.

7-5. The following filter is sometimes used as a loop low-pass filter. (a) Derive the transfer function, $H(j\omega)$, (b) find expressions for the break frequencies, and (c) sketch the frequency response.

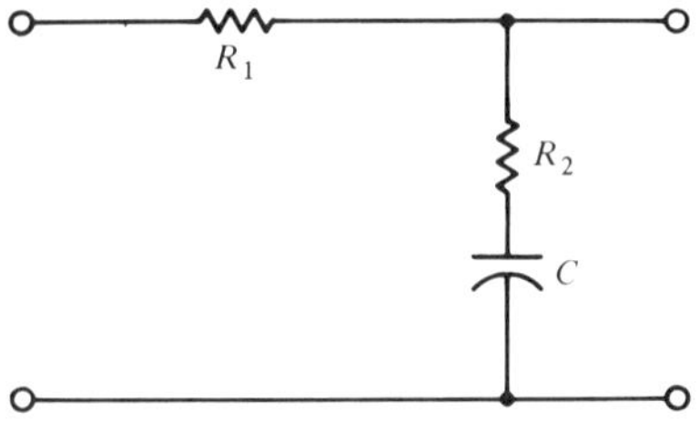

PROBLEM 7-5

7-6. Using the filter configuration of Problem 7-5, find expressions equivalent to ω_n and δ as in equations (7-37) and (7-38).

7-7. Repeat Example 7-2 using the results of Problem 7-6.

7-8. In Figure 7-9 if f_r is a sine wave and f_o is a square wave, find the expression for $V_D(t)$, the phase detector output for (a) $f_r \neq f_o$ and (b) $f_r = f_o$.

PART III

MODULATION/ DEMODULATION TECHNIQUES

CHAPTER 8

Amplitude Modulation and Detection

Many textbooks have been written on the subject of modulation, particularly at the undergraduate level. The emphasis has often concentrated on technique rather than basic theory. Present communication systems are becoming so modular in form that a complete understanding of the system is needed to understand the new developments in construction. Modulation theory provides an opportunity to gain an understanding of the overall communication process which transcends particular techniques.

We indicated in Chapter 1 that modulation of a higher-frequency sinusoidal wave is ordinarily accomplished by varying one of the following parameters: (1) amplitude, (2) frequency, or (3) phase. The latter two are generally classified as *angle modulation*. Modulation of the parameters of a higher-frequency waveform is introduced to make the transmission of information more efficient. The higher-frequency signal is termed the *carrier* and through the process of modulation the information signal is translated to the carrier. Thus the modulated carrier becomes the new information signal.

Modulation of the carrier may be accomplished by continuously varying one of the parameters of the carrier or by interrupting, chopping, and so on, the carrier, dividing it into segments. The latter process is termed *pulse modulation* and is of such importance in transmitting discrete information that a separate chapter (Chapter 11) is devoted to it.

Since modulation is introduced to improve the efficiency of information transmission, the carrier signal must be chosen based on its suitability for the radiation channel. High-frequency carriers are more suitable for radio propagation. Since modulation translates the frequency of the information signal, this same process provides the method of separating information channels (i.e., tuning in the receiver). In addition, it is readily perceived that if several channels of information are translated to different frequencies spaced so that the translated channels do not overlap, we then have a multiple-channel capability.

8-1 AMPLITUDE-MODULATED WAVEFORMS

Amplitude modulation (AM) is historically the oldest technique for transmitting sound on radio signals. Amplitude modulation requires that the amplitude (intensity) of the carrier vary in accordance with the information signal. The carrier is a stable sinusoidal signal and may be represented as

$$e_c(t) = E_c \sin(\omega_c t + \phi) \tag{8-1}$$

where

E_c = peak amplitude of carrier (volts)
ω_c = carrier frequency (radians/second)
ϕ = arbitrary phase angle

Since the phase angle is arbitrary, it may be neglected. Similarly, for modulation by a single audio-frequency signal (tone), the message may be expressed as

$$e_m(t) = E_m \sin \omega_m t \tag{8-2}$$

where

E_m = peak valve of modulating voltage (volts)
ω_m = modulating frequency (radians/second)

Using $e_m(t)$ to modulate E_c requires that equation (8-1) become

$$e_{AM}(t) = (E_c + E_m \sin \omega_m t) \sin \omega_c t \tag{8-3}$$

After factoring E_c and multiplying through by $\sin \omega_c t$, equation (8-3) may be written as

$$e_{AM}(t) = E_c\left(1 + \frac{E_m}{E_c}\sin\omega_m t\right)\sin\omega_c t$$
$$= E_c\sin\omega_c t + m\frac{E_c}{2}\cos(\omega_c - \omega_m)t - m\frac{E_c}{2}\cos(\omega_c + \omega_m)t \quad (8\text{-}4)$$

where $m = E_m/E_c$ is referred to as the *modulation index*. The first term in this equation is the carrier itself. The next two terms are referred to as the *upper and lower sidebands,* possessing frequencies $\omega_c \pm \omega_m$. Thus amplitude modulating with a single-frequency audio tone produces side frequencies above and below the carrier frequency.

An amplitude-modulated wave is shown in Figure 8-1. Because the modulating signal varies slowly compared to the carrier frequency, we may define envelope variation. Note that the positive and negative peaks of the modulating signal causes the carrier amplitude to reach maximum and minimum intensity levels. A symmetrical waveform is formed about the time axis, with the upper and lower part of the waveform having characteristics described by the modulating signal. The slow variations are referred to as the *envelope*. The information displayed by the envelope is readily recovered by diode rectification.

The modulation index, m, is a measure of the intensity of modulation. Note from Figure 8-1 that the minimum value for the modulated carrier is 0 V. The maximum value is $2E_c$. This is illustrated in Figure 8-2. From the figure we see that the maximum value of the modulated waveform is

$$E_{max} = E_{c(max)} + E_{m(max)} \quad (8\text{-}5)$$

and the minimum value is

$$E_{min} = E_{c(max)} - E_{m(max)} \quad (8\text{-}6)$$

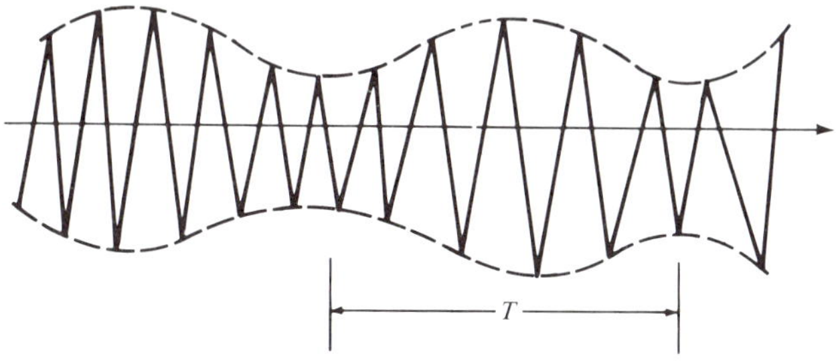

FIGURE 8-1. Amplitude-modulated waveform.

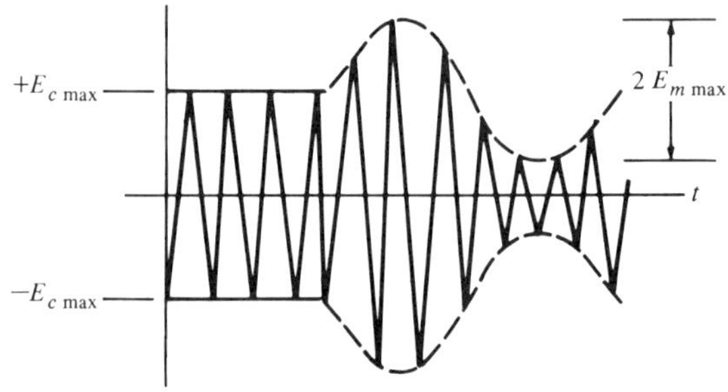

FIGURE 8-2. Sinusoidal modulation.

Taking the ratio of equations (8-5) and (8-6) results in

$$\begin{aligned}\frac{E_{\max}}{E_{\min}} &= \frac{E_{c(\max)} + E_{m(\max)}}{E_{c(\max)} - E_{m(\max)}} \\ &= \frac{1+m}{1-m}\end{aligned} \tag{8-7}$$

Then

$$m = \frac{E_{\max} - E_{\min}}{E_{\max} + E_{\min}} \tag{8-8}$$

Thus for single-tone modulation, the modulation index may be obtained from an oscilloscope display of the modulated waveform. From Figure 8-2 and equation (8-8) it may readily be determined that the maximum value of m is 1.0 for 100% modulation. If the RF peaks more than double on the positive audio peak, a condition known as *overmodulation* will result (see Figure 8-3).

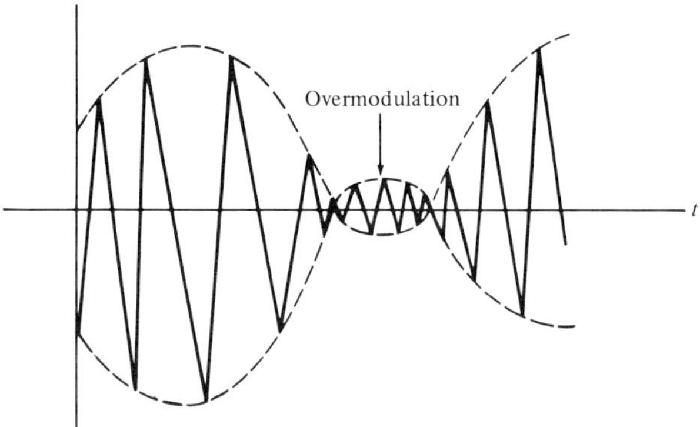

FIGURE 8-3. Overmodulation ($m > 100\%$).

EXAMPLE 8-1

A 5-MHz RF waveform is modulated with a 3-kHz audio tone. The carrier maximum peak-to-peak voltage is 20 V and the minimum peak-to-peak voltage is 10 V. Find (a) the unmodulated RF carrier voltage, (b) the modulation index, and (c) the sideband frequencies.

SOLUTION

(a) $$V_{max(carrier)} = \frac{V_{max(peak)} + V_{min(peak)}}{2}$$

$$= \frac{10 + 5}{2} = 7.5 \text{ V}$$

(b) $$m = \frac{E_{max} - E_{min}}{E_{max} + E_{min}}$$

$$= \frac{10 - 5}{10 + 5} = \frac{5}{15} = 0.33$$

(c) $$f_{L.S.B.} = f_c - f_m = 5 \times 10^6 - 3 \times 10^3 = 4.997 \text{ MHz}$$

$$f_{U.S.B.} = f_c - f_m = 5 \times 10^6 + 3 \times 10^3 = 5.003 \text{ MHz}$$

8-1.1 AM Power Calculations

The power contained in an AM modulated waveform may readily be determined from basic circuit analysis techniques. Assume for the moment that the RF waveform described by equation (8-4) appears across a resistor R. From basic circuit theory the expression for total power is

$$P_T = \frac{1}{T} \int_0^T \frac{e_{AM}^2(t)}{R}\, dt \tag{8-9}$$

where the period T, is defined in Figure 8-1. If we square equation (8-4) and substitute into equation (8-9) nine terms appear. Making use of the orthogonality properties discussed in Chapter 4 all of the cross-product terms are equal to zero. Equation (8-9) may be written as

$$\begin{aligned} P_T = {} & \frac{1}{RT} \int_0^T E_c^2 \sin^2 \omega_c t\, dt + \frac{1}{RT} \int_0^T m^2 \frac{E_c^2}{4} \cos^2 (\omega_c - \omega_m)\, t\, dt \\ & + \frac{1}{RT} \int_0^T m^2 \frac{E_c^2}{4} \cos^2 (\omega_c + \omega_m)\, t\, dt \end{aligned} \tag{8-10}$$

Now the first term obviously represents the power in the carrier and the second and third terms is the power in the lower and upper sidebands. That is,

$$P_T = P_c + P_{\text{L.S.B.}} + P_{\text{U.S.B.}} \tag{8-11}$$

Next, observing that $\int_0^T \sin^2 \omega t \, dt = \int_0^T \cos^2 \omega t \, dt$, equation (8-10) may be written as

$$\begin{aligned} P_T &= P_c + \frac{m^2}{4} P_c + \frac{m^2}{4} P_c \\ &= P_c\left(1 + \frac{m^2}{2}\right) \end{aligned} \tag{8-12}$$

For 100% modulation $m = 1$ and the power in either sideband is

$$P_{\text{S.B.}} = \tfrac{1}{4} P_c \tag{8-13}$$

The total average power then has a maximum value of

$$P_T = 1.5 P_c \tag{8-14}$$

Note from equation (8-4) that either sideband contains all of the information so that only one sideband may be transmitted with no loss of information. The power spectrum for AM with single-tone modulation is shown in Figure 8-4.

EXAMPLE 8-2

For the problem in Example 8-1, determine (a) the carrier power and (b) the total power. Assume a 200-Ω load impedance.

SOLUTION

(a) From equation (8-10),

$$\begin{aligned} P_c &= \frac{1}{RT} \int_0^T E_c^2 \sin^2 \omega_c t \, dt = \frac{E_c^2}{2R} \\ &= \frac{7.5^2}{(2)(200)} = 0.141 \text{ W} \end{aligned}$$

(b) $P_T = P_c\left(1 + \frac{m^2}{2}\right) = 0.141\left(1 + \frac{0.33^2}{2}\right) = 0.149 \text{ W}$

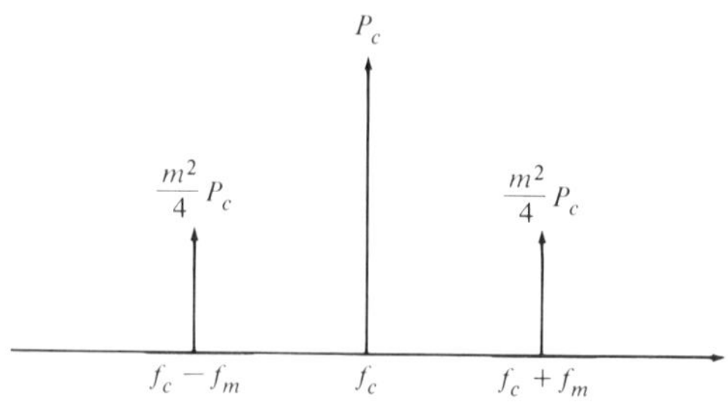

FIGURE 8-4. AM power spectrum.

8-1.2 Effective Voltage and Current

Let the rms value of $e_{AM}(t)$ be E. The total average power is then

$$P_T = \frac{E^2}{R} \tag{8-15}$$

Thus we may equate this result to that expressed by equation (8-12). That is,

$$\begin{aligned} \frac{E^2}{R} &= P_c\left(1 + \frac{m^2}{2}\right) \\ &= \frac{E_c^2}{2R}\left(1 + \frac{m^2}{2}\right) \end{aligned} \tag{8-16}$$

where E_c is the peak value of the unmodulated carrier. Thus the rms value of $e_{AM}(t)$ may be expressed as

$$E = \frac{E_c}{\sqrt{2}}\sqrt{1 + \frac{m^2}{2}} \tag{8-17}$$

Similarly, for the rms value of current, I,

$$I = \frac{I_c}{\sqrt{2}}\sqrt{1 + \frac{m^2}{2}} \tag{8-18}$$

where I_c is the peak value of carrier current.

EXAMPLE 8-3

The rms antenna current of a radio transmitter is 5 A when modulated at the 50% level. Calculate the rms antenna current if the modulation level is increased to 75%.

SOLUTION

$$I_{c(\text{rms})} = \frac{I}{\sqrt{1 + \frac{m^2}{2}}} = \frac{5}{\sqrt{1 + \frac{0.5^2}{2}}}$$

$$= 4.72 \text{ A}$$

At 75% modulation,

$$I = I_{c(\text{rms})}\sqrt{1 + \frac{m^2}{2}} = 4.72\sqrt{1 + \frac{0.75^2}{2}}$$

$$= 5.23 \text{ A}$$

8-1.3 AM Modulation with a Complex Wave

In Chapter 4 we observed that nonsinusoidal waveforms could be described in terms of the Fourier series. In particular, this series consists of a fundamental frequency and higher harmonics. The modulation process described by equation (8-3) is a linear process. Thus we would expect that each additional frequency in the modulating waveform will result in a pair of sideband frequencies. Consider the modulating waveform to be represented by the following series:

$$e_m(t) = E_1 \sin \omega_m t + E_2 \sin 2\omega_m t + E_3 \sin_3 \omega_m t + \cdots \tag{8-19}$$

Equation (8-3) may be written as

$$\begin{aligned} e_{AM}(t) &= E_c\left(1 + \frac{E_1}{E_c} \sin \omega_m t + \frac{E_2}{E_c} \sin 2\omega_m t + \cdots\right) \sin \omega_c t \\ &= E_c\left(1 + m_1 \sin \omega_m t + m_2 \sin 2\omega_m t + \cdots\right) \sin \omega_c t \end{aligned} \tag{8-20}$$

where a modulation index $m_1, m_2, \ldots,$ is defined for each harmonic frequency. Note that continued expansion of this equation yields the sideband frequencies.

An expression for the average power can be derived as in equation (8-12). The general expression for the average power in the modulated wave is

$$\begin{aligned} P_T &= P_c[1 + (m_1^2 + m_2^2 + m_3^2 + \cdots)] \\ &= P_c\left(1 + \frac{m_{\text{eff}}^2}{2}\right) \end{aligned} \tag{8-21}$$

where

$$M_{\text{eff}} = \sqrt{m_1^2 + m_2^2 + m_3^2 + \cdots} \tag{8-22}$$

Similar expressions for the rms voltage and current of the modulated waveform are

$$E = \frac{E_c}{\sqrt{2}}\sqrt{1 + \frac{m_{\text{eff}}^2}{2}} \tag{8-23}$$

and

$$I = \frac{I_c}{\sqrt{2}}\sqrt{1 + \frac{m_{\text{eff}}^2}{2}} \tag{8-24}$$

Note that in setting up systems with complex modulating waveforms it is possible for the effective modulation index to be less than 1 and yet the phase relation between individual components of the modulating wave could result in a voltage that causes the modulating depth to exceed 100%. Typical frequency spectra for AM modulation by complex waveforms are shown in Figure 8-5.

8-1.4 Methods of Obtaining the Amplitude-Modulated Wave

In equations (8-3) and (8-4) we established a mathematical definition of AM modulation together with the frequency spectrum. The next most obvious question is how to produce the amplitude-modulated waveform. We observed that modulation produces a frequency shift of the information signal. That is, new frequencies must be generated. From basic circuit theory we know that linear circuits will not produce new frequencies. We did observe in Fourier analysis that nonlinearities will generate new frequencies. Thus in the modulation process the device or function performed must be nonlinear.

There are many circuits available which can be used to obtain AM modulation. One of the oldest of these is use of the nonlinearity of the characteristic curve of a forward-biased diode, as illustrated in Figure 8-6. The analysis of this circuit is straightforward and is usually considered in basic texts on circuit analysis. Essentially, the forward-biased characteristic curve of the diode may be described as a power series. That is,

$$e_o(t) = k_1 e_{\text{in}}(t) + k_2 e_{\text{in}}^2(t) + k_3 e_{\text{in}}^3(t) + \cdots \tag{8-25}$$

where

$$e_{\text{in}}(t) = E_c \cos \omega_c t + E_m \cos \omega_m t \tag{8-26}$$

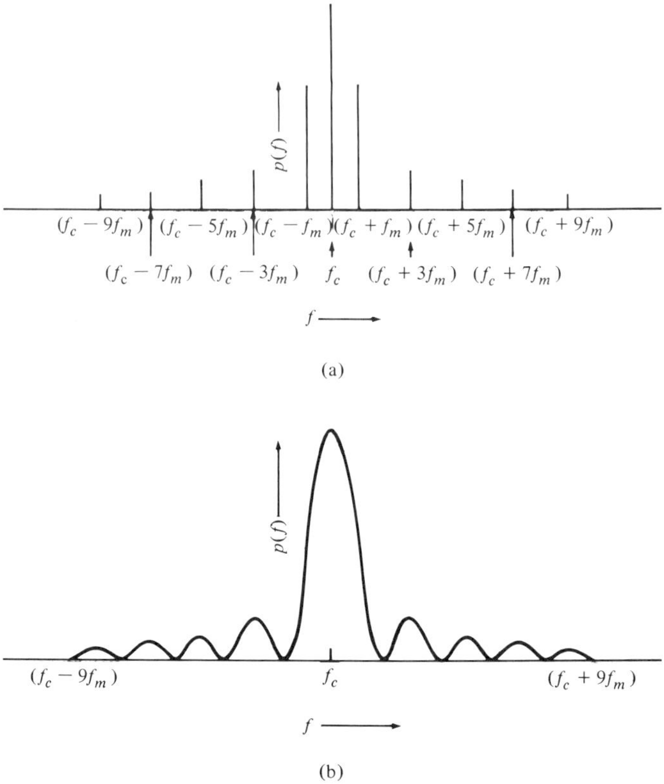

FIGURE 8-5. (a) Square-wave amplitude modulation spectrum; (b) periodic rectangular-wave modulation spectrum.

Considering only the first two terms in the series, $e_o(t)$

$$e_o(t) = k_1(E_c \cos \omega_c t + E_m \cos \omega_m t) + k_2(E_c \cos \omega_c t + E_m \cos \omega_m t)^2 \tag{8-27}$$

Expanding yields

$$e_o = k_1 E_c \cos \omega_c t + k_1 E_m \cos \omega_{mt} + k_2(E_c \cos^2 \omega_c t + 2E_m E_c \cos \omega_c t \times \cos \omega_m t + E_m \cos^2 \omega_m t) \tag{8-28}$$

Regrouping terms, we get

$$e_o = k_1 E_c \cos \omega_c t + 2k_2 E_m E_c \cos \omega_m t \cos \omega_c t + k_1 E_m \cos \omega_m t + k_2 E_c^2 \cos^2 \omega_c t + k_2 E_m^2 \cos^2 \omega_m t \tag{8-29}$$

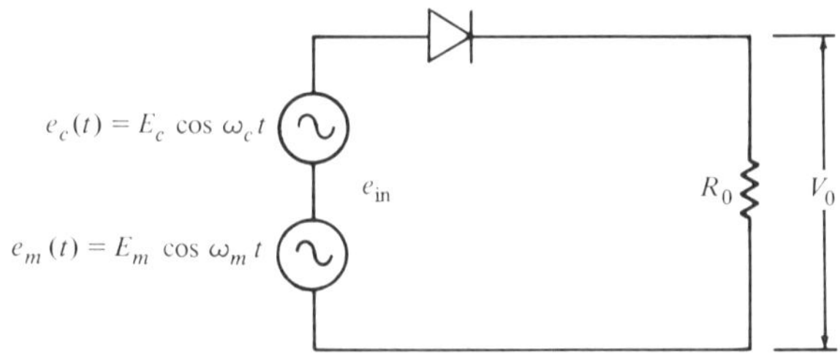

FIGURE 8-6. Diode modulator.

Note that the original frequencies ω_m and ω_c are present plus $\omega_c \pm \omega_m$, in addition to others. Terms corresponding to equation (8-4) may be filtered for AM modulation. In addition, we shall see later that this nonlinear function may be used quite effectively in frequency translation and mixing.

On an integrated-circuit level, a four-quadrant multiplier such as the MC1595 or a balanced modulator/demodulator such as the MC1496 may be used. The MC1596 connected as an AM modulator is illustrated in Figure 8-7. A corresponding oscilloscope display and power spectrum is shown in Figure 8-8. These waveforms indicate the excellent linearity and response that can be obtained with "off-the-shelf" integrated circuits.

The modulation schemes discussed involve very little power. Gain is achieved after modulation with wide-band power amplification. Consequently, this is referred to as low-level modulation. Systems employing modulation of low power levels are low-level modulation transmitters.

In a high-level modulation transmitter, high-gain narrowband amplifiers can be used to achieve the desired carrier power level. However, this also requires

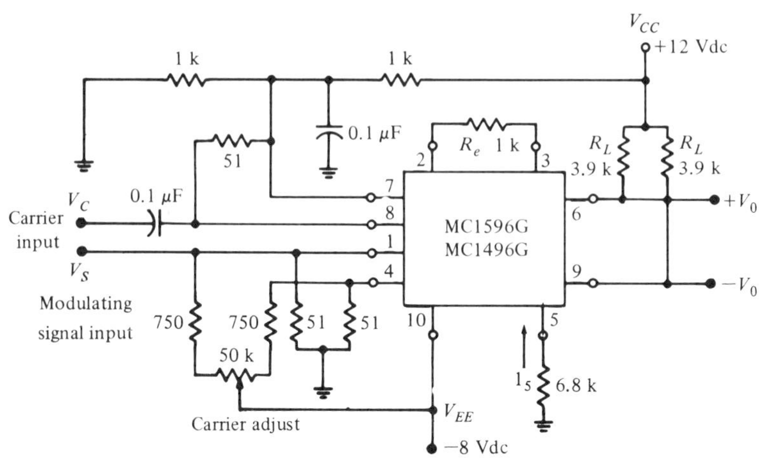

FIGURE 8-7. AM modulator.

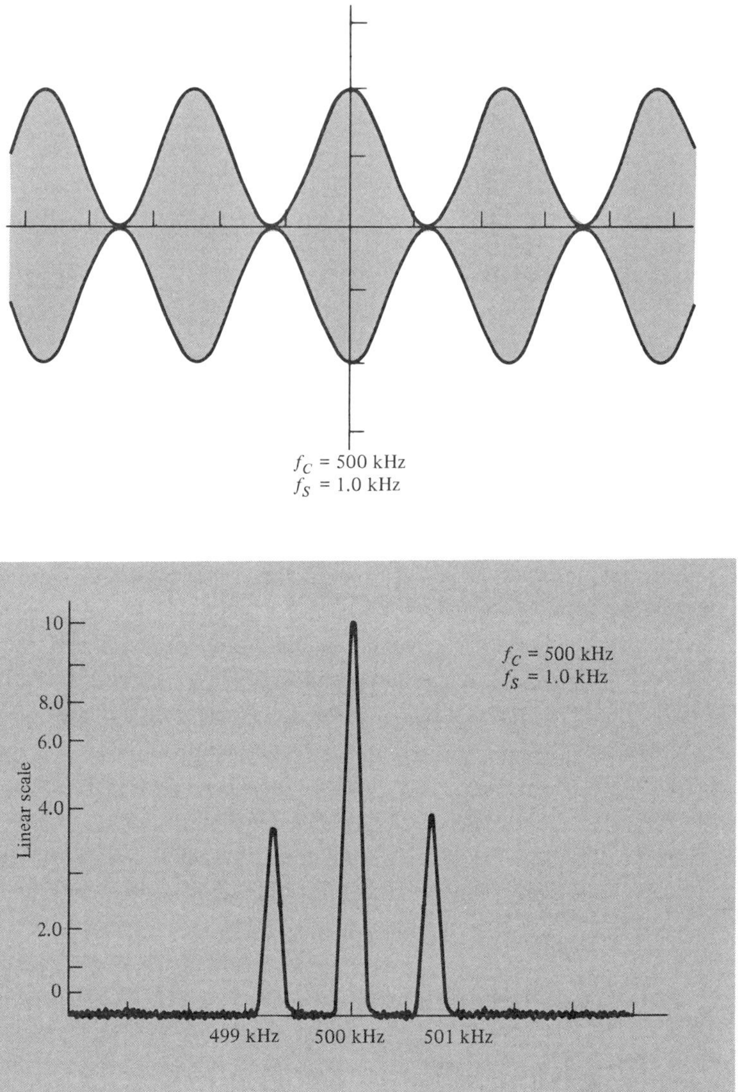

FIGURE 8-8. (a) Amplitude modulation waveform; (b) amplitude modulation spectrum.

the information signal to be of a high power level to achieve modulation. The type of modulation to be used is determined by the designer. Considering a transistor modulator circuit, for example, low-level modulation would be performed in the base circuit. In terms of a very high powered transmitter which uses vacuum tubes, high-level modulation is performed in the plate circuit.

DOUBLE-SIDEBAND SUPPRESSED CARRIER MODULATION

In AM modulation we observed that with no modulation the carrier is still transmitted. This lowers the efficiency of the system. Methods of carrier modulation have been developed to improve the efficiency of transmission. The system considered here involves suppression of the carrier.

Any device that produces a product of signals can be used to produce double-sideband suppressed carrier (DSB-SC) modulation. Expressed mathematically, this is

$$
\begin{aligned}
e_{(t)} &= e_m(t)e_c(t) \\
&= E_m \cos \omega_m t E_c \cos \omega_c t \qquad (8\text{-}30) \\
&= \frac{E_m E_c}{2} \cos(\omega_c + \omega_m)_t + \frac{E_m E_c}{2} \cos(\omega_c - \omega_m)t
\end{aligned}
$$

Note that the sidebands are present but no carrier term appears. The MC1596 circuit discussed earlier provides a means of realizing the function described by equation (8-30). This configuration is shown in Figure 8-9 and the modulation waveform and frequency spectrum are illustrated in Figure 8-10.

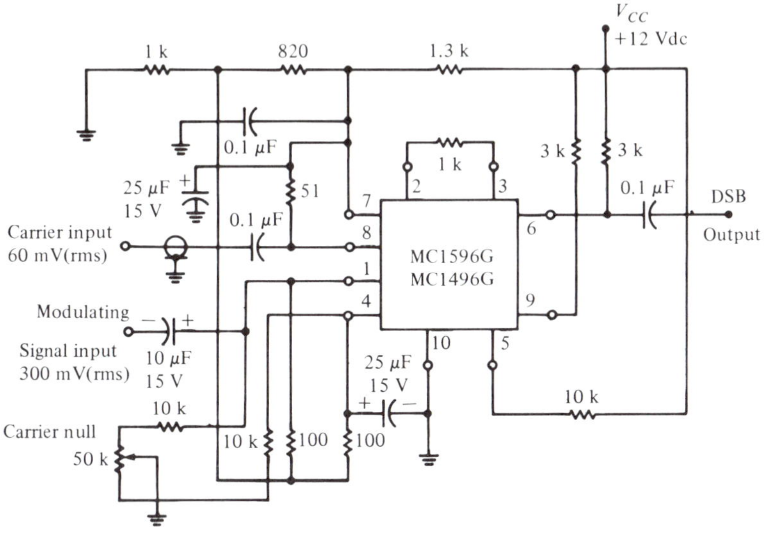

FIGURE 8-9. Balanced modulator circuit.

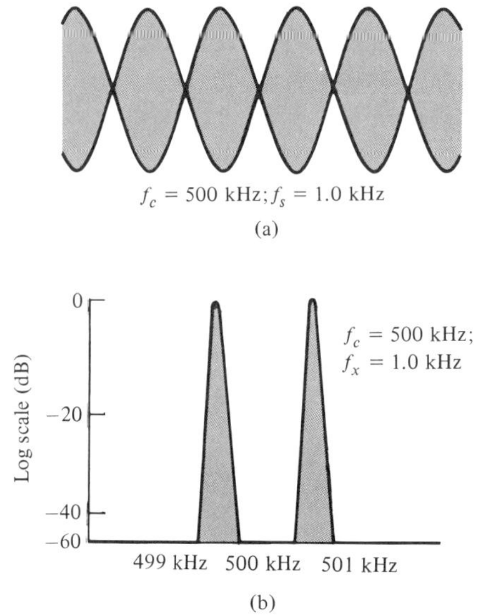

FIGURE 8-10. (a) Suppressed carrier waveform; (b) suppressed carrier spectrum.

The DSB-AM wave is often generated using a *ring modulator*. This circuit is illustrated in Figure 8-11. The name derives from the placement of the diodes. The carrier frequency amplitude is much greater than the information signal and performs the function of a switch. The diodes are biased on and off at the carrier rate. When point *b* is at a higher potential than point *a*, the diodes are all open circuited. Conversely, when point *a* is at a higher potential than point *b*, all the diodes are biased on. Thus the information signal is interrupted at the carrier rate.

Expressed mathematically, this is

$$e_o(t) = E_m \cos \omega_m t \; s(t) \tag{8-31}$$

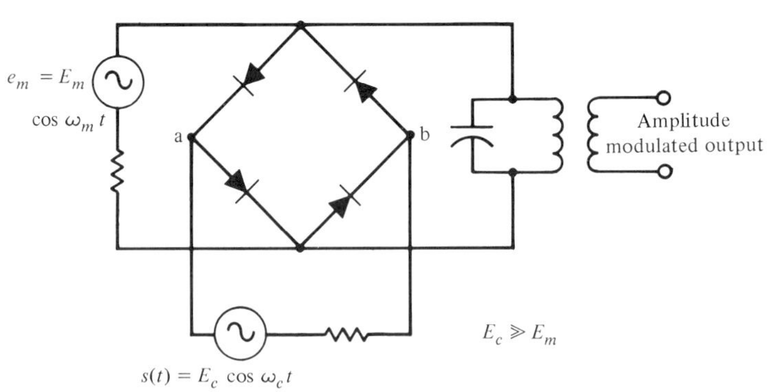

FIGURE 8-11. Ring modulator.

The switching function may be represented by its Fourier series. Thus

$$e_o = E_m \cos \omega_m(t)$$

$$\left(\frac{1}{2} + \frac{2}{\pi} \cos \omega_c t + \frac{2}{3\pi} \cos 3\omega ct + \cdots + \frac{2}{n\pi} \cos n\omega_c t\right) \quad (8\text{-}32)$$

where

$n = 1, 3, 5, \ldots.$

$$e_o = \frac{E_m}{2} \cos \omega_m t + \frac{2}{\pi} E_m \cos \omega_m t \cos \omega_c t + \frac{2}{3\pi} E_m \cos \omega_{mt} \cos \omega_{ct} + \cdots \quad (8\text{-}33)$$

If the transformer is followed by a bandpass filter whose bandwidth is $f_c \pm f_m$, the only term retained in the output is

$$e_o = E_m \cos \omega_m t \cos \omega_c t = \frac{E_m}{\pi} \cos (\omega_c + \omega_m)t + \frac{E_m}{\pi} \cos (\omega_c - \omega_m)t \quad (8\text{-}34)$$

which is the desired amplitude-modulated wave. If the equivalent circuit of the MC1496 is examined, the switching function is noted as a design feature.

We note the complete absence of the carrier in the spectrum. As may be expected, this difference is reflected in the receiver. Double-sideband suppressed carrier is often used in pulse modulation for data transmission. A variation of the PLL discussed in Chapter 7 is used for demodulation. General communication systems employing DSB-AM are not common where continuous modulation is used. The main reason is that another amplitude-modulation scheme offers the additional advantages of greater signal power and a further reduction in bandwidth. This method is single-sideband modulation (SSB).

8-3 SINGLE-SIDEBAND AMPLITUDE MODULATION

Each sideband of the DSB-SC signal carries the information. Obviously, transmission efficiency can be improved if only one sideband is transmitted. Not only is the power reduced by a factor of 2 but the bandwidth is also halved. This reduced bandwidth results in an improvement of the signal-to-noise ratio.

There are several methods for producing the SSB signal. One of the oldest

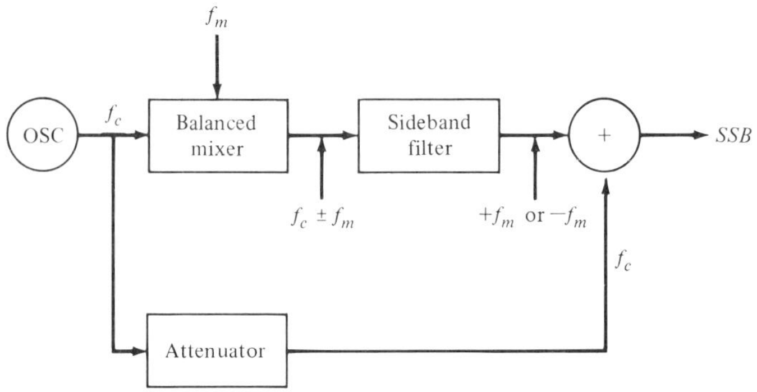

FIGURE 8-12. Block diagram of SSB generator using the filter method.

is the filter method illustrated in Figure 8-12. The balanced mixer generates the DSBSC signal. This signal is passed through a narrowband filter with sharp cutoff characteristics at the filter band edges. Either the upper or lower sideband is selected to produce the SSB. Because of the critical roll-off characteristics of the filter, it is preferable to implement the narrowband filter at low frequencies. Additional frequency translation may be employed to bring the SSB up to the desired transmission frequency.

Frequency translation not involving modulation is referred to as *frequency conversion* or *frequency mixing*. The operation of frequency converters is identical to the operation of modulators. The modulated information signal replaces the information signal and the injected carrier frequency is chosen so that either the sum or difference frequency is the desired translated frequency. Refer to Figure 8-13. The converter output is the product of the inputs. That is,

$$\begin{aligned} e_o &= \frac{E_m}{\pi} \cos (\omega_{c1} + \omega_m)t \cos \omega_{c2}t \\ &= \frac{E_m}{2\pi} [\cos (\omega_{c1} + \omega_{c2} + \omega_m)t + \cos(\omega_{c1} - \omega_{c2} + \omega_m)t] \end{aligned} \tag{8-35}$$

Suppose that $f_{c1} = 1$ mHz and the desired spectrum for transmission is 3.2 mHz + f_m. Then the bandwidth of the information has not been altered. It is simply translated from $f_{c1} + f_m$ to $f_{c1} + f_{c2} + f_m$ or 3.2 mHz + f_m.

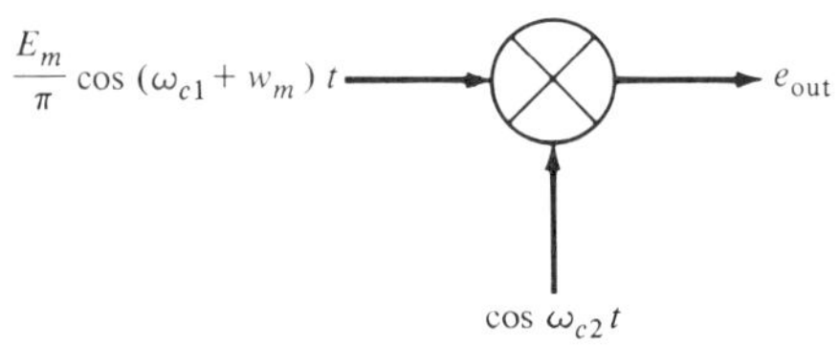

FIGURE 8-13. Frequency converter.

Proper operation of receivers employing synchronous detection requires the presence of the carrier term in the radiated signal. Synchronous detectors utilize a PLL and are discussed in the following section. Referring to Figure 8-12, the carrier is seen to be attenuated and reintroduced before transmission at a greatly reduced level.

A second method for producing SSB is termed a *phase-shift SSB modulator*. Refer to Figure 8-14. This system employs the use of two balanced modulators. From equation (8-30) the output of modulator 1 is

$$e_{o(1)} = \frac{E_c E_m}{2} \cos (\omega_c - \omega_m)t + \frac{E_c E_m}{2} \cos (\omega_c + \omega_m)t \qquad (8\text{-}36)$$

and for modulator 2 it is

$$\begin{aligned} e_{o(2)} &= \frac{E_c E_m}{2} \cos (\omega_m t + 90^\circ) \cos (\omega_c t + 90^\circ) \\ &= \frac{E_c E_m}{2} \cos (\omega_c - \omega_m)t + \frac{E_c E_m}{2} \cos [(\omega_c + \omega_m) + 180^\circ] \qquad (8\text{-}37) \\ &= \frac{E_c E_m}{2} \cos (\omega_c - \omega_m)t - \frac{E_c E_m}{2} \cos (\omega_c + \omega_m)t \end{aligned}$$

The output of the adder is the desired SSB signal. That is,

$$\begin{aligned} e_o &= e_{o(1)} + e_{o(2)} \\ &= E_m E_c \cos (\omega_c - \omega_m)t \end{aligned} \qquad (8\text{-}38)$$

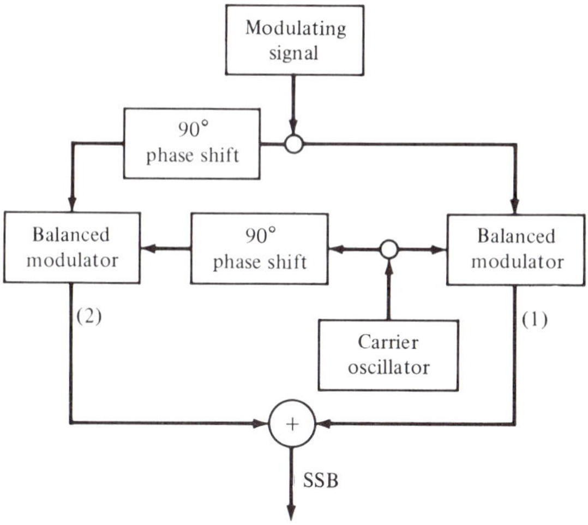

FIGURE 8-14. Block diagram of phase-shift SSB modulator.

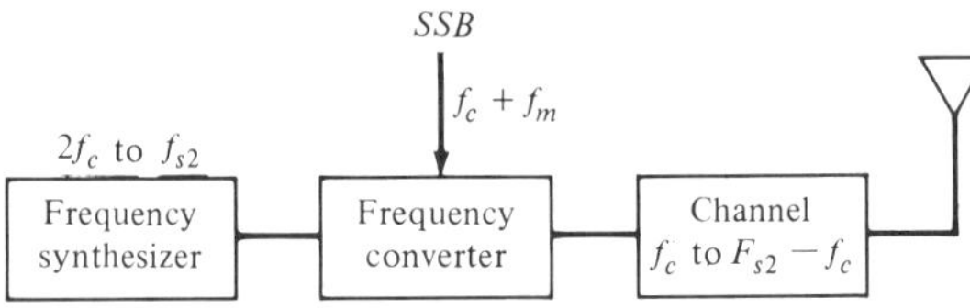

FIGURE 8-15. SSB transmitter.

It is often desired to operate the SSB system over a range of frequencies. Figure 8-15 illustrates a method of accomplishing this.

8-4 DEMODULATION OF AMPLITUDE-MODULATED SIGNALS

Demodulation or detection is the process of recovering the information. The terms *demodulation* and *detection* are generally used interchangeably. We saw that the process of modulation involves frequency translation. Demodulation, then, involves reversing this process. The information signal must be translated back to the original frequency spectrum. In the following we examine several techniques for accomplishing this.

8-4.1 Synchronous Demodulation of Amplitude-Modulated Signals

Consider the demodulation system illustrated in Figure 8-16. The signal into the multiplier is described by equation (8-3). The $\sin \omega_c t$ term is generated locally in the receiver and is synchronized both in frequency and phase to the input carrier $\sin \omega_c t$. Thus the multiplier output is

$$\begin{aligned} e_o &= e_{AM}(t) \sin \omega_c t \\ &= (E_c + E_m \sin \omega_m t) \sin \omega_c t \sin \omega_c t \\ &= [E_c \sin \omega_c t + m \frac{E_c}{2} \cos (\omega_c - \omega_m)t - m\frac{E_c}{2} \cos (\omega_c + \omega_m)t] \sin \omega_c t \\ &= \frac{E_c}{2} (1 - \cos 2\,\omega_c t) + m \frac{E_c}{4} \sin (2\omega_c - \omega_m)t \\ &\quad + m\frac{E_c}{4} \sin \omega_m t - m\frac{E_c}{4} \sin (2\omega_c + \omega_m)t + m\frac{E_c}{4} \sin \omega_m t \end{aligned} \tag{8-39}$$

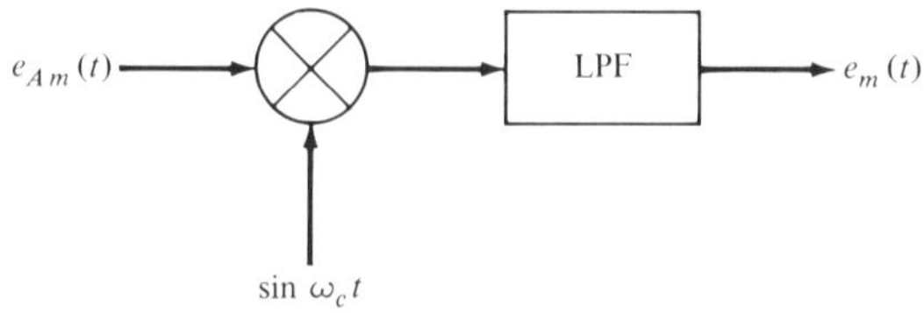

FIGURE 8-16. Synchronous demodulator.

Considering magnitudes only, the frequency spectrum at the multiplier output [equation (8-39)] is indicated in Figure 8-17. Note that the original AM spectrum is translated to twice the carrier frequency. In addition, a dc term is produced together with the original information frequency ω_m. The low-pass filter recovers the information from the spectrum and ac coupling eliminates the dc.

At this point the question arises as to the physical generation of the $\sin \omega_c t$ term. Enter the PLL solution discussed in Chapter 7. A coherent (synchronous) AM detector circuit is shown in Figure 8-18. Referring to Figure 8-4, we see that the AM spectrum contains the carrier, ω_c, which the PLL locks onto. The VCO will track this carrier but with a 90° phase shift. Thus either the VCO output or the AM input must be shifted 90°.

An AM detector using the Signetics NE561 PLL is shown in Figure 8-19. This integrated circuit incorporates a multiplier on the chip and thus it is possible to accomplish synchronous demodulation. By shifting the input 90° the VCO and the input become in phase. To accomplish this phase shift at the input, a simple *RC* lag network may be used. The values for R_1, C_3, R_2, and C_4 are calculated using the following relationship:

$$R_1C_3 = R_2C_4 = \frac{1}{2\pi f_c}$$

Assume that $R_1 = R_2 = 12\ \text{k}\Omega$. Then

$$C_3 = C_4 = \frac{1}{(2\pi)(455 \times 10^3)(12 \times 10^3)} = 29\ \text{pF}$$

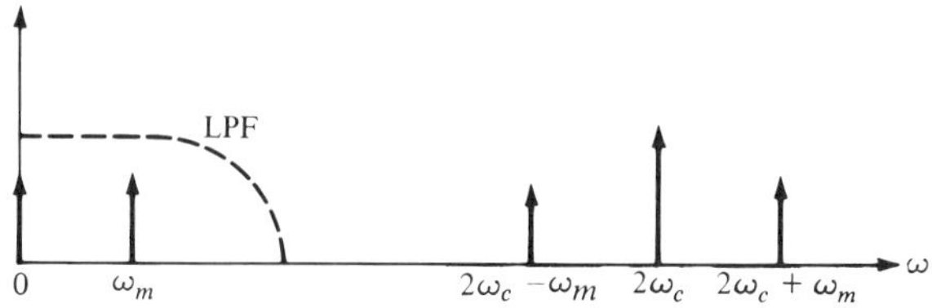

FIGURE 8-17. Signal spectrum at multiplier output.

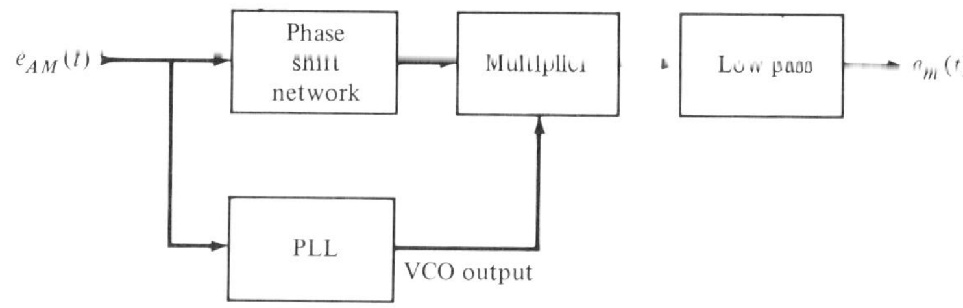

FIGURE 8-18. Coherent AM detector using a phase-locked loop.

The standard value is 27 pF. The timing capacitor for setting the VCO frequency may be calculated as

$$C_6 \approx \frac{300}{f_c} \text{ pF} \qquad (f_c \text{ in MHz})$$

or

$$C_6 \approx \frac{300}{0.455} = 600 \text{ pF}$$

The standard value is 680 pF. By controlling the current into pin 6, the VCO can be fine tuned, so the absolute value of C_6 is not critical. By injecting or loading current into pin 6, the VCO frequency can be adjusted $\pm 20\%$ (with 4 mA of control current). No information is taken from the loop low-pass filter. Thus the filter capacitor C_2 is not critical. A value is selected which will maintain loop stability. A typical value is 0.01 μF. Capacitors C_8, C_9, and resistor R_6 form a post AM filter to eliminate carrier feed-through.

8-4.2 Asynchronous (Envelope) Detection of Amplitude-Modulated Signals

Envelope detection is a popular detection method for AM signals where the envelope of the modulated waveform is identical to the original information signal. A nonlinear element is utilized which still provides frequency translation to transform the information to the audio spectrum. A circuit for accomplishing this is shown in Figure 8-20.

In operation a piecewise linear device is employed to prevent distortion. Any gradual nonlinearity in the diode characteristics curves results in nonlinear detection. The operation of the detector is by direct rectification of the applied modulated signal. From rectifier theory, the output voltage across the load may be expressed as

$$e_o = \frac{R}{\pi(R + r_d)} E_c(1 + m \cos \omega_m t) \tag{8-40}$$

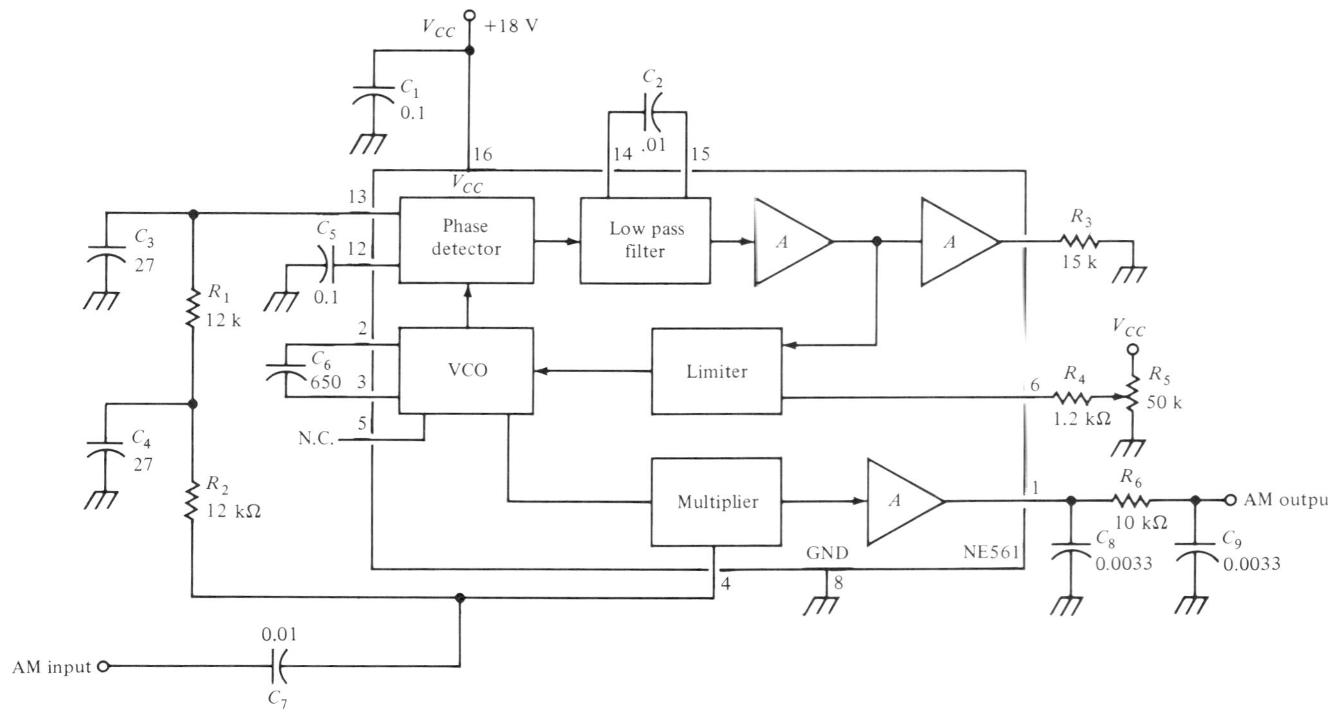

FIGURE 8-19. AM detector for 455-kHz IF using the NE561 PLL.

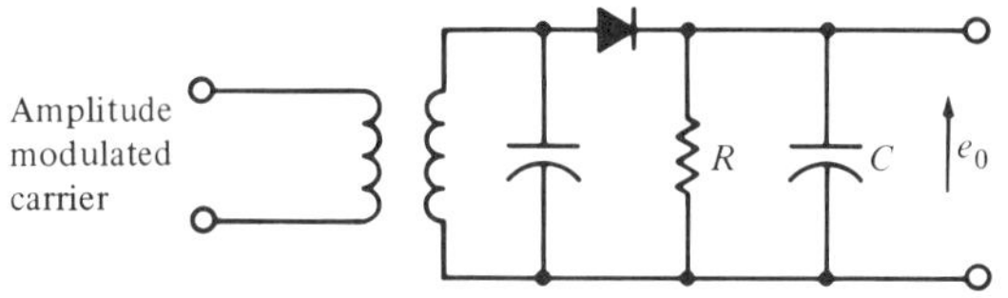

FIGURE 8-20. Envelope detector.

where r_d is the dynamic impedance of the diode. The time constant of the *RC* network is adjusted so that the highest-frequency change in the information signal may be followed. Conversely, the *RC* time constant is such that the network will not respond to the variations of the carrier frequency. Stated mathematically, these criteria are

$$RC < \frac{1}{f_m}$$

and

$$RC >> \frac{1}{f_c}$$

8-5

DEMODULATION OF SINGLE-SIDEBAND SIGNALS

Much of the preceding discussion applies directly to the demodulation of single-sideband (SSB) signals. Consider, for example, the SSB system illustrated in Figure 8-12, where the carrier is reinserted. A PLL detector such as that in Figure 8-19 can then be used. Referring to Figure 8-13, if ω_{c2} is made equal to ω_{c1}, then by equation (8-35) a term exists in the multiplier output equal to $(E_m/2\pi) \cos \omega_m t$. Thus a synchronous detector for AM will also detect SSB where the carrier is reinserted.

The SSB scheme illustrated in Figure 8-14 does not reinsert the carrier. Thus demodulation must be asynchronous. This detection method is implemented by using a nonlinear mixer such as that described by equation (8-29). In general, a beat frequency oscillator (BFO) is employed with a nonlinear mixer (Figure 8-21). As noted previously, the output of the nonlinear element contains the four frequencies shown plus other higher-order frequencies. If the frequency of the BFO oscillator is equal to 450 kHz, for example, and the SSB signal is 450.3 to 450 kHz. Then the information frequency, f_m, may be readily filtered by the low-pass filter. There is no requirement for phase synchronization.

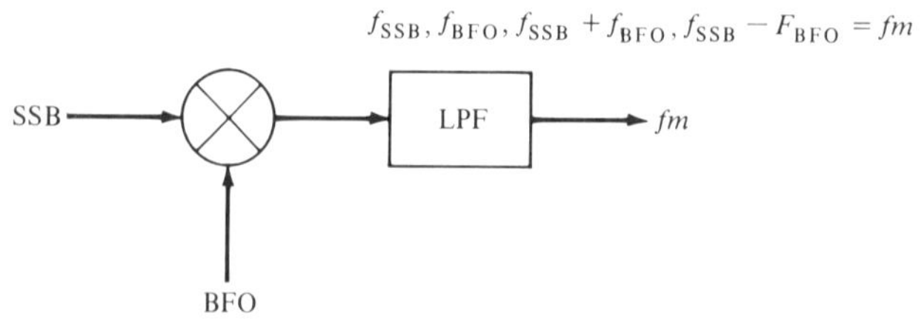

FIGURE 8-21. SSB detection.

8-6

COMPARISON OF AMPLITUDE-MODULATED SYSTEMS

In an amplitude-modulation system with audio bandwidth F_A and a receiver noise density, N_0, the detected noise power is

$$N_{Am} = \int_0^{F_A} 2N_0 d_{fm} = 2N_0F_A \tag{8-41}$$

where the factor of 2 accounts for symmetrically located AM noise components. If the maximum signal detected corresponds to 100% modulation, the signal power detected is P_s, and

$$(S/N)_{AM} = \frac{P_s}{2N_0F_A} \tag{8-42}$$

Now for SSB the noise bandwidth is one-half that of AM. Thus for equal signal power the signal-to-noise ratio is

$$(S/N)_{SSB} = \frac{P_s}{N_0F_A} \tag{8-43}$$

Thus

$$\frac{(S/N)_{SSB}}{(S/N)_{AM}} = \frac{P_s/N_0F_A}{P_s/2N_0F_A} = 2 \Rightarrow 3 \text{ dB} \tag{8-44}$$

In terms of power transmitted, however, the ratio is

$$\frac{P_{TSSB}}{P_{TAM}} = \frac{(m^2/2)P_c}{P_c + (m^2/2)P_c} = \frac{1/2}{1 + 1/2} = \frac{1}{3} \Rightarrow -4.77 \text{ dB}$$

Thus the total power transmitted for SSB is 4.77 dB less than that for AM.

PROBLEMS

8-1. A carrier voltage has a peak amplitude of 10 V and a frequency of 1×10^6 Hz. The amplitude of the RF voltage is modulated by a sinusoidal 1000-Hz signal to make the amplitude of the RF wave vary between 7.5 and 12.5 V. Write an equation for the modulated wave. Determine m.

8-2. The modulated voltage in Problem 8-1 is developed across a 50-Ω load. Determine (a) the total power, (b) the carrier power, and (c) the power in the sidebands.

8-3. An antenna is supplied with 2160 W of RF power, 40% amplitude modulated. Determine the carrier power and the sideband power.

8-4. An audio signal consists of a 10-V 1000-Hz component and a 0.5-V 5000-Hz component. This signal is used to amplitude modulate a 20-V peak carrier with frequency equal to 9 MHz. $R_{\text{ant}} = 50\ \Omega$. Determine (a) the modulation index for each of the audio signal components, (b) the effective modulation index, (c) the power spectrum of the modulated waveform, (d) the total RF power, (e) the power in each sideband frequency, and (f) the bandwidth of the RF spectrum.

8-5. The ac rms antenna current is 10 A when unmodulated. It rises to 11.5 A when modulated. Calculate the modulation index.

8-6. In Figure 8-6, $e_c(t)$ is a 1.5-MHz carrier and $e_m(t)$ is a 5-kHz tone. Determine the frequencies in the waveform appearing across R_o.

8-7. Draw a diagram of the waveform in Problem 8-1. Indicate on the diagram the dimensions from which the modulation index may be obtained.

8-8. An AM waveform equal to 500 W is 75% modulated. Calculate the transmitted power if it is transmitted as a SSB signal.

8-9. For equal transmitted power, compare the signal-to-noise ratio for AM and DSB-SC.

8-10. An AM wave has a carrier of 840 kHz and sidebands extending to 4 kHz on either side. This signal is fed to a mixer with a 1290-kHz local oscillator signal. Find (a) the IF carrier value and (b) the total IF spectrum coverage.

8-11. (a) A DSB-SC signal of the form $f(t) \cos \omega_c t$ is to be demodulated. Show that multiplying this signal by $\cos \omega_c t$ in a product modulator reproduces $f(t)$. (b) The DSB demodulator is phase sensitive. Demonstrate the validity of this statement by multiplying $f(t) \cos \omega_c t$ by $\cos (\omega_c t + \theta)$ and letting θ vary from 0 to π radians.

8-12. (Single-sideband detection). An audio tone cos $\omega_m t$ and a carrier cos $\omega_c t$ are combined to produce an SSB signal given by cos $(\omega_c - \omega_m)t$. (a) Show that cos $\omega_m t$ can be reproduced from the SSB signal by mixing with cos $\omega_c t$ as a local carrier. (b) The local carrier drifts in frequency by $\Delta\omega$ radians to $\omega_c + \Delta\omega$. What is the demodulated signal now? (c) The local carrier shifts in phase to cos $(\omega_c + \theta)$. Find the output signal.

8-13. (a) An amplitude-modulated voltage given by

$$v_c(t) = V(1 + m \cos \omega_m t) \cos \omega_c t \qquad \text{volts}$$

is applied across a resistor R. Calculate the percentage average power in the carrier and each of the two sideband frequencies. (b) Calculate the peak power.

8-14. An SSB radio transmitter radiates 1 kW of average power summed over the entire sideband. What total average power would the transmitter have to radiate if it were operating as a DSB-SC system and the same distance coverage were required?

8-15. An unmodulated carrier is 250 V p:p. Calculate the percent modulation when the p:p value is: (a) 325 V, (b) 400 V, and (c) 500 V.

8-16. A 125 volt RF carrier is modulated by a 2 KHz sinusoidal wave. Find the sideband amplitudes when the modulation index is 0.60.

8-17. The total transmitted power in Problem 16 is 400 watts. Calculate the carrier and the sideband power.

8-18. Calculate the percent modulation if the ac rms current of an antenna is 5.0 amps when unmodulated and rises to 7.2 amps when modulated.

8-19. Explain why a high percentage of modulation is desirable in a communication system.

8-20. What percent of the total transmitted power is in the sidebands when the modulation index is 0.85?

8-21. An AM transmitter operates at 20 MHz. It delivers 10 watts of RF power to a 50 ohm load. A 1.5 KHz sine wave is used to modulate the transmitter alternately at 25 percent and 75 percent modulation. Determine: (a) sideband frequencies; (b) sideband signal voltage and power for both modulation indices; and (c) load current for each modulation index.

CHAPTER 9

Angle Modulation

In Chapter 8 we examined systems in which the amplitude of the carrier was varied in accordance with the intelligence contained in the modulating signal. Another popular method involves varying the frequency of the carrier. An advantage of this system lies in its ability to discriminate against interfering signals. The demodulator is sensitive to frequency variations that contain the intelligence. Noise voltages that vary the amplitude are discriminated against and thus there is little noise effect in the demodulated output.

This system of modulation requires greater bandwidth but because of the superior noise characteristics it has found widespread application. If the frequency of the carrier is varied in accordance with the information signal, the result is termed *frequency modulation* (FM). If the phase of the carrier is varied in accordance with the information, the result is termed *phase modulation* (PM). Both frequency and phase modulation are forms of angle modulation; that is, the instantaneous angle of the carrier is modulated.

FREQUENCY MODULATION

Suggestions for the widespread use of frequency modulation were first made in the early 1920s. The idea at that time was to vary the frequency of the carrier within very narrow limits in order to limit the bandwidth of the modulated signal. This approach was attractive because it promised to ease the problem of crowding of the stations. In addition, narrowband FM was seen as a means of combating static and interference.

A mathematical analysis of the FM spectrum was first performed by John R. Carson, a noted mathematician working at Bell Telephone Laboratories. His discovery of an infinite number of sidebands discouraged efforts at implementing an FM system. Consequently, a successful demonstration of wide-band FM by Major E. H. Armstrong was greeted by widespread disbelief. It is now understood that the enhanced noise performance of the wide-band FM system is achieved by using a bandwidth which is larger than that necessary to transmit the information. It is, in effect, a compromise of spectrum capacity for a better signal-to-noise ratio.

Frequency modulation is easily visualized in terms of an increase or decrease of frequency. Refer to Figure 9-1. Let the signal have the form

$$e_c(t) = A_c \cos \phi(t) \tag{9-1}$$

where A_c is the peak amplitude of the signal and $\phi(t)$ is the phase of $e(t)$. The frequency of $e(t)$ is the time derivative of $\phi(t)$. That is,

$$\omega(t) = \frac{d}{dt} \phi(t) \tag{9-2}$$

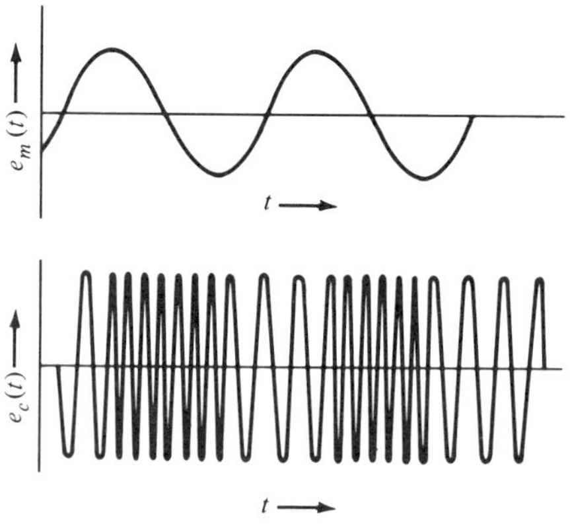

FIGURE 9-1. Frequency modulation by a sinusoidal waveform.

where $\phi(t) = \omega_c t + \theta$. The angle, θ, is the initial phase angle and is the phase of the signal at $t = 0$. The frequency of the unmodulated signal is given by

$$\omega(t) = \frac{d}{dt}(\omega_c t + \theta) = \omega_c \tag{9-3}$$

Now suppose that we have a modulating signal $g(t)$ and it is desired that the instantaneous frequency $\omega(t)$, be

$$\omega(t) = \omega_c[1 + g(t)] \tag{9-4}$$

That is, the frequency varies in proportion to the message $g(t)$. From equation (9-2) we have

$$\begin{aligned}\phi(t) &= \int_0^t \omega(t)\,dt \\ &= \int_0^t \omega_c[1 + g(t)]\,dt = \omega_c t + \omega_c \int_0^t g(t)\,dt\end{aligned} \tag{9-5}$$

An inspection of equation (9-5) reveals that the phase angle varies in proportion to the integral of the message. Substituting equation (9-5) into (9-1) results in the following form for the frequency-modulated wave:

$$e_{\mathrm{FM}}(t) = A_c \cos\,[\omega_c t + \omega_c \int_0^t g(t)\,dt] \tag{9-6}$$

Further investigation of FM requires that we assume specific forms for the message $g(t)$. As with AM, the simplest example to investigate is modulation by a sinusoidal waveform.

9-1.1 Spectra of FM Signals

For convenience the modulating waveform may be written as

$$g(t) = \frac{\Delta\omega}{\omega_c}\cos\omega_m t \tag{9-7}$$

where $\Delta\omega$ is the desired peak frequency change and ω_m is the modulating frequency. From equation (9-5),

$$\begin{aligned}\phi(t) &= \omega_c t + \omega_c \int_0^t \frac{\Delta\omega}{\omega_c}\cos\omega_m t\,dt \\ &= \omega_c t + \frac{\Delta\omega}{\omega_m}\sin\omega_m t + \theta\end{aligned} \tag{9-8}$$

The constant of integration is the phase constant θ. Examining equation (9-8), we see that the peak phase deviation is given by $\Delta\omega/\omega_m$ or $\Delta f/f_m$. This quantity is usually referred to as the *modulation index*. That is,

$$m_f = \frac{\Delta\omega}{\omega_m} \tag{9-9}$$

The phase constant may be eliminated by a suitable choice of time axis and equation (9-6) may be written as

$$e_{\mathrm{FM}}(t) = A_c \cos(\omega_c t + m_f \sin \omega_m t) \tag{9-10}$$

Following the method of Chapter 8 the spectrum of the modulated wave, $e_c(t)$, may be found by expanding equation (9-10) using the familiar trigonometric identity $\cos(A + B)$; that is,

$$e_{\mathrm{FM}}(t) = A_c \cos \omega_c t \cos(m_f \sin \omega_m t) - A_c \sin \omega_c t \sin(m_f \sin \omega_m t) \tag{9-11}$$

At this point in our development it is convenient to consider the case of narrowband FM mentioned previously.

9-1.2 Narrowband FM

Referring to equation (9-11), if the modulation index m_f is small enough such that $m_f \sin \omega_m t$ is always much less than 1, then $e_{\mathrm{FM}}(t)$ may be described by

$$\begin{aligned} e_{\mathrm{FM}}(t) &\approx A_c \cos \omega_c t - A_c \sin \omega_c t \, m_f \sin \omega_m t \\ &= A_c \cos \omega_c t - m_f A_c \sin \omega_m t \sin \omega_c t \end{aligned} \tag{9-12}$$

Comparing the expression with that for amplitude modulation (AM), we readily recognize that the spectrum corresponding to equation (9-12) is a carrier with a pair of sideband frequencies: hence the name *narrowband*.

This similarity of the spectra is the only common thing, however, between the two modulation systems. The differences are apparent from a phasor plot of the two systems. In the AM case for a sinusoidal information signal, we have

$$\begin{aligned} e_{\mathrm{AM}}(t) &= A_c(\cos \omega_c t + \cos \omega_m t \cos \omega_c t) \\ &= \mathrm{Re}[e^{j\omega_c t}(A_c + \tfrac{1}{2}e^{j\omega_m t} + \tfrac{1}{2}e^{-j\omega_m t})] \end{aligned} \tag{9-13}$$

We recognize this as a phasor rotating at an ω_c rate. Assuming a particular instant of time, we need only plot the quantity in brackets (Figure 9-2). We note that the resultant for the phasor is in phase with the carrier term A_c. This is expected since there is no phase variation in AM.

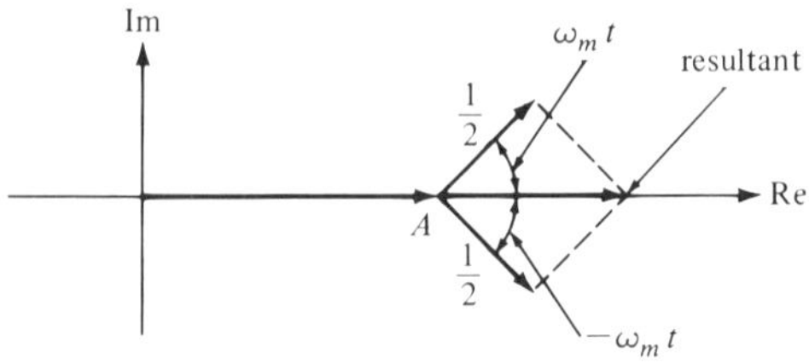

FIGURE 9-2. Phaser plot for AM waveform.

The corresponding expression for narrowband FM [equation (9-12)] may be expressed as

$$e_{FM}(t) = \text{Re}\left[e^{j\omega_c t}\left(A_c - \frac{A_c m_f}{2\omega_m} e^{-j\omega_m t} + \frac{A_c m_f}{2\omega_m} e^{j\omega_m t}\right)\right] \tag{9-14}$$

and the quantity in brackets is plotted in Figure 9-3. Note that as $\omega_m t$ changes the resultant is always the same amplitude. Again, this is to be expected since the amplitude must remain constant as the frequency varies.

At this point we can readily see why early planners envisioned narrowband FM broadcast stations. The bandwidth for narrowband FM is $\approx 2f_M$ and since the detection process is concerned primarily with the frequency variation, the system discriminates against noise.

9-1.3 Wide-band FM

If the modulation index, m_f, is not small enough to make the approximations for the sine and cosine valid, we have what is referred to as *wide-band* FM. Referring to equation (9-11), expressions such as cos (z sin x) and sin (z sin x) are expandable into series form:

$$\cos(z \sin x) = J_0(z) + 2\sum_{n=1}^{\infty} J_{2n}(z) \cos 2nx \tag{9-15a}$$

$$\sin(z \sin x) = 2\sum_{n=0}^{\infty} J_{2n+1}(z) \sin(2n+1)x \tag{9-15b}$$

The $J(z)$ terms are Bessel functions of the first kind. Substituting these expressions into equation (9-11) results in

$$\begin{aligned} e_{FM}(t) = {} & A_c \cos \omega_c t\left[J_0(m_f) + 2\sum_{n=1}^{\infty} J_{2n}(m_f) \cos 2n\omega_m t\right] \\ & - A_c \sin \omega_c t\left[2\sum_{n=0}^{\infty} J_{2n+1}(m_f) \sin(2n+1)\omega_m t\right] \end{aligned} \tag{9-16}$$

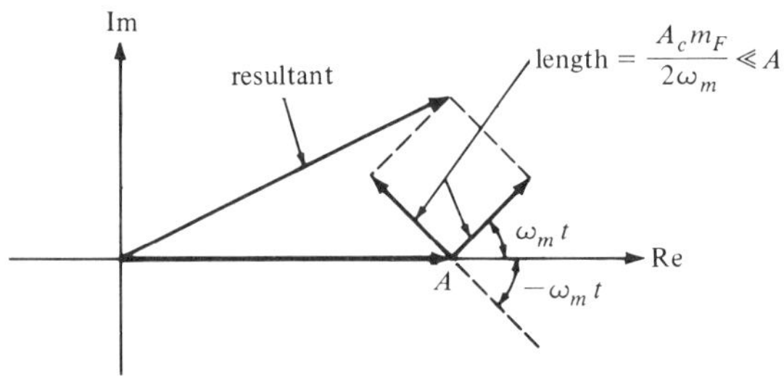

FIGURE 9-3. Phaser plot for narrow-band FM.

This expression in turn may be written as

$$e_{FM}(t) = A_c J_0(m_f) \cos \omega_c t + 2A_c \sum_{n=1}^{\infty} J_{2n}(m_f) \cos \omega_c t \cos 2n\omega_m t$$
$$- 2A_c \sum_{n=0}^{\infty} J_{2n+1}(m_f) \sin \omega_c t \sin (2n + 1)\omega_m t \tag{9-17}$$

Finally by applying the rules for cos A cos B and sin A sin B, we have

$$e_{FM}(t) = A_c J_0(m_f) \cos \omega_c t$$
$$+ A_c J_{2n}(m_f)[\cos (\omega_c t + 2n\omega_m t) + \cos (\omega_c t - 2n\omega_m t)]$$
$$+ A_c J_{2n+1}(m_f) \cos [\omega_c t + (2n + 1)\omega_m t]$$
$$- \cos [\omega_c t - (2n + 1)\omega_m t] \tag{9-18}$$

This expression, although forbidding in appearance, easily reveals the spectrum of the sinusoidally modulated signal. We see that the first term represents a carrier of amplitude $A_c J_0(m_f)$. The remaining terms represent the sidebands. Note that the sidebands are spaced from the carrier and each other by ω_m, and are symmetrical in amplitude. Furthermore, the even-order sidebands are in phase with each other and with the carrier, whereas the odd-order sidebands are 180° out of phase.

The Bessel functions (those of the "first kind") are well tabulated and are plotted in Figure 9-4. The argument of the Bessel function is the modulation index and it is interesting to note that there are values of m_f for which the carrier amplitude $J_0(m_f)$ is zero. The first zero occurs at $m_f = 2.405$. This fact proves useful in the calibration of FM modulators. A narrowband receiver is tuned to the carrier frequency and the deviation of the modulating frequency is increased until the carrier amplitude becomes zero. Knowing that $\Delta f/f_m = 2.405$, the exact frequency deviation may be calculated.

The spectrum of a wide-band FM wave with sinusoidal modulation is shown

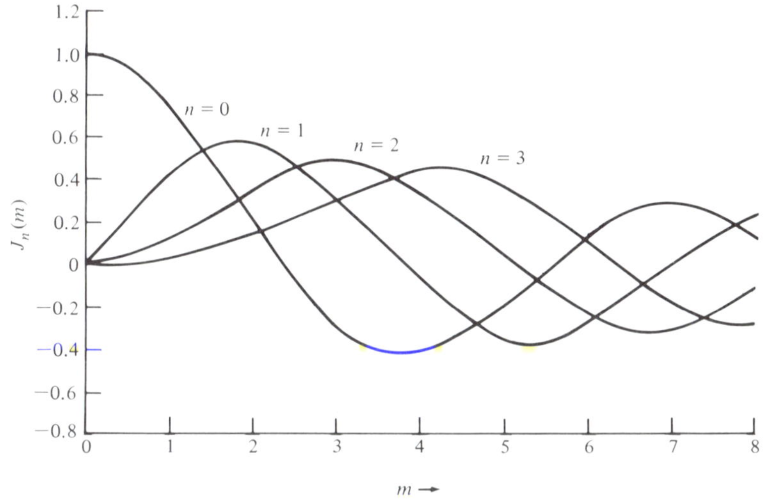

FIGURE 9-4. Bassel functions of the first kind.

in Figure 9-5. Examination of equation (9-18) reveals that the number of sidebands surrounding the carrier are infinite. The number of significant sidebands, however, are limited. A good approximation to the bandwidth covered by the significant sidebands is

$$\text{BW} \approx 2(\Delta f + f_m) \tag{9-19}$$

EXAMPLE 9-1

Find the approximate band of frequencies occupied by an FM wave whose carrier frequency is 10,000 rad/s and $m_f = 5$. The message is 10 cos 20*t*.

SOLUTION

$$\Delta\omega = m_f \omega_m$$

$$= (5)(20) = 100 \text{ rad/s}$$

$$\text{BW} \approx 2(\Delta\omega + \omega_m)$$

$$= (2)(100 + 20) = 240 \text{ rad/s}$$

The band of frequencies occupied is 9880 to 10,120 rad/s.

Equation (9-18) does not reveal much information about the general case when the modulating signal is not sinusoidal. The lengthy analysis required to obtain the spectrum of a single-tone frequency-modulated signal indicates that a much

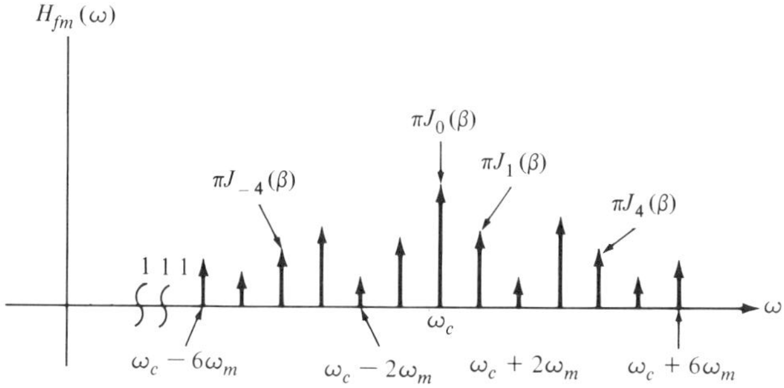

FIGURE 9-5. FM spectrum with sinusoidal modulation.

more difficult problem is encountered when the modulating signal is nonsinusoidal. If the nonsinusoidal signal is decomposed by Fourier analysis into a number of sinusoids, the FM spectrum may be written by assigning to each modulating frequency f_k a modulation index $m_k = \Delta f_k/f_k$. The spectrum is then composed of terms of the form

$$e_{FM}(t) = A_c J_\mu(m_1) J_\gamma(m_2) J_\lambda(m_3) \cdots$$
$$J_\beta(m_k) \cos(\omega_c \pm \mu\omega_1 \pm \gamma\omega_2 \pm \lambda\omega_3 \pm \cdots \pm \beta\omega_k)t \quad (9\text{-}20)$$

where $\mu, \gamma, \lambda, \ldots, \beta$ each takes on all values from zero to infinity. If $\mu, \gamma, \lambda, \ldots, \beta$ are all zero, the amplitude of the carrier component $\cos \omega_c t$ is found to be

$$e(t) = A_c \prod_k J_0(m_k) \quad (9\text{-}21)$$

that is, the product of all the zero-order Bessel functions. If $\mu = 1$ and $\gamma, \lambda, \ldots, \beta$ remain zero, the amplitude of the term $\cos(\omega_c \pm \omega_1)t$ is the amplitude of the first-order sideband of ω_1. This is $J_1(m_1)$ for the single-tone case, but it is now modified by the product of the zero-order Bessel functions corresponding to the remaining modulating frequencies. By allowing $\mu, \gamma, \ldots, \beta$ to assume independently all values from zero to infinity, the spectrum corresponding to each modulating frequency is obtained. Obviously, the spectral approach is almost useless in computing FM problems.

In practical problems the bandwidth is of paramount importance. Filters, amplifier bandwidth, and so on, must be wide enough to process the signal. In practice, the number of significant sidebands are a function of the modulation index m_f and the peak deviation, $\Delta\omega$. As noted previously, for $m_f << 1$ the spectrum is limited to a carrier and one pair of sidebands. In the case of multitone modulation it is evident from equation (9-20) that, as the number of modulating signals increase, the amplitude of the sidebands decreases rapidly since the total energy of the wave is constant and independent of the form of the modulating signal. Since the sidebands extend from zero to infinity, it is impossible to assign a definite bandwidth for the FM wave. An expression that is often used for the bandwidth requirements is

$$BW \approx 2\Delta f\left(1 + \frac{2}{m_f}\right) \tag{9-22}$$

EXAMPLE 9-2

A 50-MHz carrier is frequency modulated by a sinusoidal signal described by 5 cos 5000t. The sensitivity of the transmitter is 1000 Hz/V. Determine the approximate bandwidth.

SOLUTION

$$\Delta f = (1000\text{ Hz/V})(5\text{ V}) = 5000\text{ Hz}$$

$$\Delta\omega = 2\pi\ \Delta f = 10{,}000\pi\text{ rad/s}$$

$$BW = 2(\Delta\omega + \omega_m) = (2)(10{,}000 + 5000)$$

$$= 36{,}400\text{ rad/s}$$

EXAMPLE 9-3

An FM waveform is described by

$$e(t) = 20\cos(3 \times 10^8\pi t + 10\cos 2000\pi t)$$

Find the approximate bandwidth.

SOLUTION

From the equation, $\omega_m = 2000\pi$ rad/s. To compute $\Delta\omega$ it is necessary to compute the instantaneous frequency.

$$\omega_i(t) = \frac{d}{dt}\phi(t)$$

$$= \frac{d}{dt}(3 \times 10^8\pi t + 10\cos 2000\pi t)$$

$$= 3 \times 10^8\pi - 20{,}000\pi \sin 2000\pi t$$

Thus

$$\Delta\omega = 20{,}000\pi$$

$$BW \approx (2)(20{,}000\pi + 2000\pi) = 44{,}000\pi\text{ rad/s}$$

PHASE MODULATION

The difference between FM and PM is subtle. We begin with the unmodulated signal given by equation (9-1). That is,

$$e_c(t) = A_c \cos \phi(t) \tag{9-23}$$

We desire the phase angle $\phi(t)$ to vary in proportion to the modulating signal. Thus

$$\phi(t) = \omega_c t + \theta + g(t) \tag{9-24}$$

Again by a suitable choice of axis we may eliminate the constant phase angle, θ. Let the information signal $g(t)$ be written as

$$g(t) = \Delta\theta \cos \omega_m t \tag{9-25}$$

where $\Delta\theta$ is the peak phase deviation. Substituting equation (9-25) into (9-24) results in

$$\phi(t) = \omega_c t + \Delta\theta \cos \omega_m t \tag{9-26}$$

Using this result, equation (9-23) becomes

$$\begin{aligned} e_c(t) &= A_c \cos [\omega_c t + \Delta\theta g(t)] \\ &= A_c \cos (\omega_c t + \Delta\theta \cos \omega_m t) \end{aligned} \tag{9-27}$$

If we denote $\Delta\theta$ as m_p, this expression becomes

$$e_{\text{PM}}(t) = A_c \cos (\omega_c t + m_p \cos \omega_m t) \tag{9-28}$$

where $\Delta\theta = m_p$ is referred to as the *phase modulation index*.

Now, the instantaneous angular frequency, ω, is, from equation (9-24),

$$\begin{aligned} \omega &= \frac{d}{dt} \phi(t) \\ &= \omega_c + \frac{d}{dt} \phi(t) \end{aligned} \tag{9-29}$$

Comparing this to equation (9-4) for FM, we see that the only difference is that $g(t)$ is differentiated. After performing this differentiation, we have

$$\omega = \omega_c + \frac{d}{dt} m_p \cos \omega_m t$$
$$= \omega_c - \omega_m m_p \sin \omega_m t \tag{9-30}$$

Thus the maximum equivalent frequency deviation is

$$\omega_{\text{eq}} = \omega_m m_p \tag{9-31}$$

The bandwidth of the phase-modulated wave can be obtained by forgetting that it is not an FM wave.

Thus we can use equation (9-19), where Δf is replaced by Δf_{eq}. The maximum phase shift is determined by the receiver characteristics. Recall from Chapter 7 that a PLL will lose lock if the phase shift exceeds $\pm 90°$. In fact, the maximum is often set at 1 rad.

If we compare equations (9-28) and (9-10), we see that expressions for FM and PM have basically the same form. Thus the techniques applied to find the FM spectrum also apply for PM. For a small modulation index, m_p, we have narrowband PM whose spectrum for sinusoidal modulation is a carrier with a pair of sideband frequencies. For a large modulation index, wide-band PM results. The spectrum is obtained by using expansions of basically the same form as those given by equation (9-15). The now familiar infinite sideband structure is obtained.

9-2.1 Phase Modulation Obtained from Frequency Modulation

Phase modulation is often referred to as indirect FM because PM produces a frequency-modulated wave from a fixed frequency. Using equation (9-31), we may write the expression for a phase-modulated wave (9-28) as

$$e_{\text{PM}}(t) = A_c \cos\left(\omega_c t + \frac{\Delta\omega_{\text{eq}}}{\omega_m} \cos \omega_m t\right) \tag{9-32}$$

Next, take the derivative of $g(t) = (\Delta\omega/\omega_m) \cos \omega_m t$ and substitute into equation (9-6). Thus

$$e_{\text{FM}}(t) = A_c \cos\left(\omega_c t + \int_0^t -\frac{\Delta\omega}{\omega_m} \omega_m \sin \omega_m t \, dt\right)$$
$$= A_c \cos\left(\omega_c t + \frac{\Delta\omega}{\omega_m} \cos \omega_m t\right) \tag{9-33}$$

Comparing equations (9-32) and (9-33) for PM and FM, respectively, we see they are identical in form. We see that if the signal used to frequency modulate a carrier were $\frac{d}{dt}\, g(t)$, the waveform is the same as that of a carrier phase modulated with $g(t)$. Recall from Chapter 4 that differentiation in the frequency domain is simply an attenuation function with a +6-dB/octave slope. Conversely, if the signal is integrated before phase modulation, the result is indirect FM.

EXAMPLE 9-4

The following waveform is received:

$$g(t) = 10 \cos (10^6 t + 400 \sin 600t)$$

(a) What is the approximate bandwidth of the signal?
(b) If this is an FM wave, what is the modulation signal and the modulation index?
(c) If this is a PM wave, what is the modulation signal and the modulation index?

SOLUTION

(a) $\text{BW} = 2(\Delta\omega + \omega_m)$

$$\omega_i(t) = \frac{d}{dt}\,\phi(t)$$

$$= \frac{d}{dt}\,(10^6 t + 400 \sin 600t)$$

$$= 10^6 + 24 \times 10^4 \cos 600t$$

$$\text{BW} \approx (2)(24 \times 10^4 + 600) = 2.812 \times 10^5 \text{ rad/s}$$

(b) If this is an FM wave, the instantaneous $\phi(t)$ is

$$\phi(t) = \omega_c t + m_f \int_0^t g(t)\, dt$$

$$g(t) = \cos 600t$$

and

$$m_f = 400$$

(c) If this is a PM wave, the instantaneous phase is

$$\phi(t) = \omega_c t + m_p g(t)$$

Then

$$g(t) = \sin 600t$$

and

$$m_p = 400 \text{ rad}$$

9-3

METHODS OF PRODUCING ANGLE MODULATION

An obvious method for producing frequency modulation is the voltage-controlled oscillator (VCO) discussed in Chapter 7 (Figure 7-8). A varactor is used to alter the reactance of the tuned circuit used as the feedback network. The modulating voltage across the varactor varies the capacitance and thus the resonant frequency. This scheme is often referred to as a *reactance modulator*. The percent of modulation is controlled by the amplitude of the modulating voltage. Linearity is a function of the voltage-to-frequency characteristics of the VCO.

We observed that integrated-circuit PLLs also contain a VCO. Thus frequency modulation may be accomplished by summing the modulating signal with the error control voltage. This changes the dc level to control the VCO at the modulating rate. Figure 8-19 is repeated as Figure 9-6 for convenience. If the current into pin 6 becomes a variable current, the VCO will change frequency at the modulated rate. The VCO output is taken at pin 5.

Another method of FM generation is to insert offset currents across the timing capacitor. A 10.7 MHz FM modulator using an NE560 PLL is shown in Figure 9-7. Note that the input is ac grounded and the VCO output is taken from pin 5.

A common technique for producing phase modulation utilizes a crystal-controlled oscillator. Refer to Figure 9-8. The crystal-controlled oscillator drives the *LC* circuit with a constant current and a constant frequency. Varying the capacitance of the varactor at the modulation rate varies the reactance (admittance) of the *LC* circuit. This variation in admittance shifts the phase of the voltage developed across the tuned *LC* circuit.

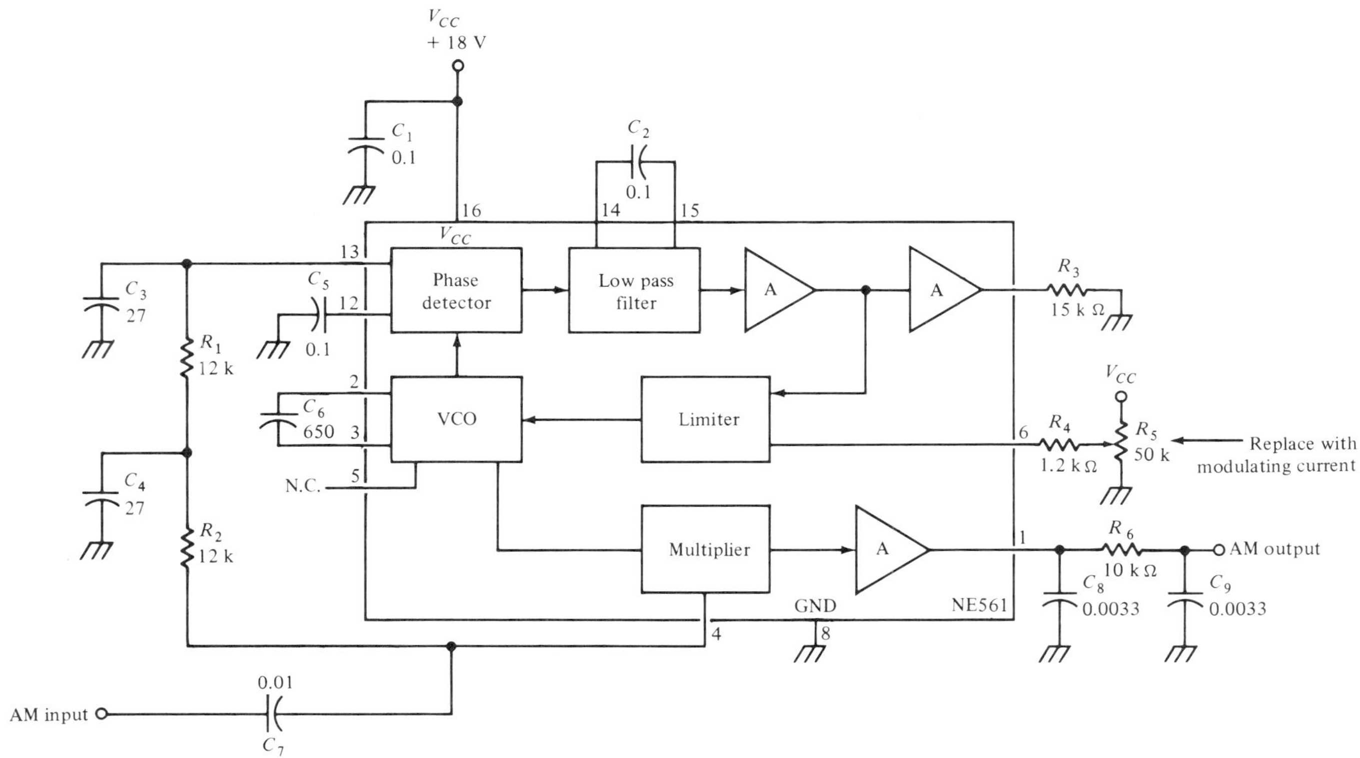

FIGURE 9-6. FM modulation using an AM PLL detector.

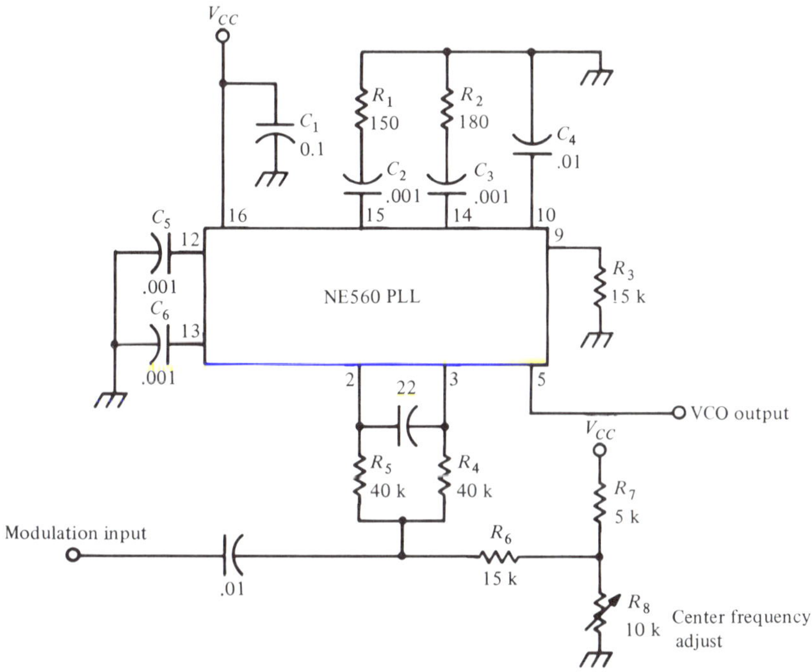

FIGURE 9-7. 10.7-MHz FM modulator using the NE561 PLL.

9-4

DETECTION OF ANGLE-MODULATED SIGNALS

We noted in Chapter 4 that the process of differentiation introduced an amplitude versus frequency response of 6 dB/octave. This, then, perhaps provides a key for demodulating FM signals. Differentiating an FM signal should produce an amplitude variation proportional to the frequency. After differentiating the FM

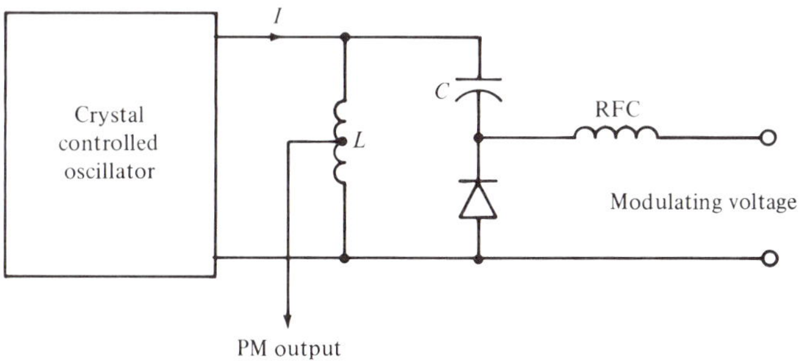

FIGURE 9-8. Phase modulator.

FIGURE 9-9. FM demodulator.

signal described by equation (9-10), we have

$$\frac{d}{dt}\left[A_c \cos\left(\omega_c t + m_f \sin \omega_m t\right)\right]$$

$$= -A_c\omega_c(1 + \Delta\omega \cos \omega_m t) \cos (\omega_c t + m_f \sin \omega_m t) \quad (9\text{-}34)$$

Thus the magnitude of the envelope is defined by

$$|A_c\omega_c(1 + \Delta\omega \cos \omega_m t| \quad (9\text{-}35)$$

and an envelope detector can be used to detect this variation (Figure 9-9). When a differentiator is used in this manner, it is referred to as a *discriminator*.

We see, then, that a system function that converts FM into AM provides a means of demodulating FM. Consider the bandpass filter in Figure 9-10. Observe that if we operate on the ''up'' or ''down'' slope of the bandpass filter, this function effectively performs differentiation. The nonlinearity of the filter function can be essentially eliminated if we take the difference of two bandpass filters whose center frequencies are shifted (Figure 9-11).

A balanced slope detector is shown in Figure 9-12. The tuned circuit consisting of the upper half of the transformer and C_1 is tuned to ω_a and the tuned circuit consisting of the other half of the winding and C_2 is tuned to ω_b. The voltage V_{out} is given by

$$V_{out} = |V_1| - |V_2| \quad (9\text{-}36)$$

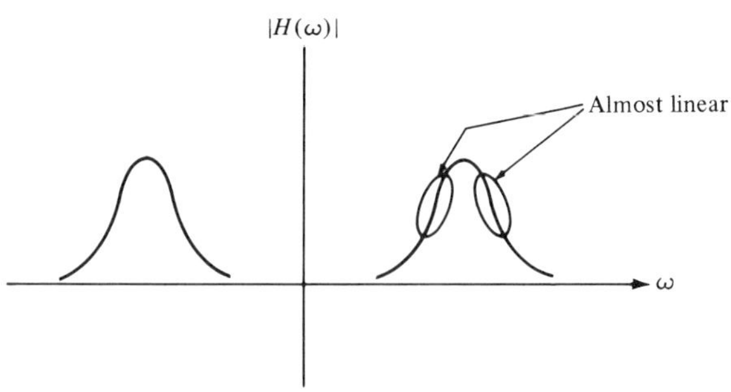

FIGURE 9-10. A bandpass filter as a discriminator.

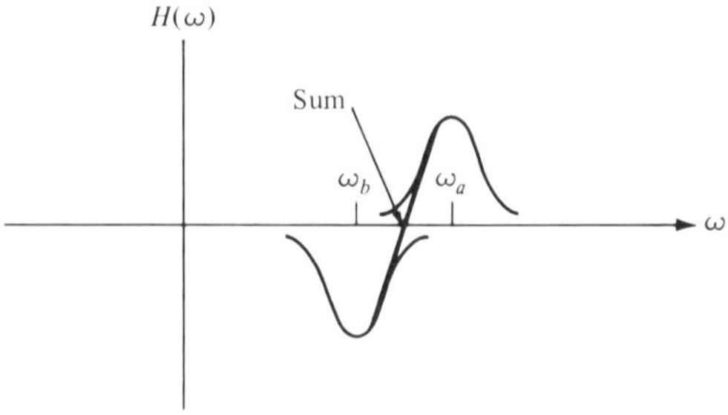

FIGURE 9-11. Improving the linearity of the discriminator.

When the carrier deviates toward ω_a, voltage V_2 is greater than V_1. Conversely, when the carrier deviates toward ω_b, voltage V_2 is greater than V_1. For an unmodulated carrier $V_1 = V_2$ and $V_{out} = 0$.

9-4.1 Foster–Seeley Discriminator

The Foster–Seeley circuit (Figure 9-13) is often used as an FM demodulator. Its operation differs substantially from that of the detuned detector in Figure 9-11. Both the primary and secondary of the Foster–Seeley circuit are tuned to the unmodulated frequency of the carrier. The operation of the circuit depends on the 90° phase shift between the primary and secondary voltages at resonance. The frequency and phase response of the network is used to obtain frequency demodulation. Capacitors C_1, C_2, and C_3 have low reactance at the R_F frequency. Thus the R_f voltage that appears across the primary, $e_{FM}(t)$, also appears across the RF choke, L_1. Thus the voltage across diode D_1 is

$$V_{D1} = K(e_1 + e_2) \tag{9-37}$$

where e_1 is the incoming RF and e_2 is the induced RF in the upper half of the secondary. Also, the voltage across diode, D_2, is

$$V_{D2} = K(e_1 - e_2) \tag{9-38}$$

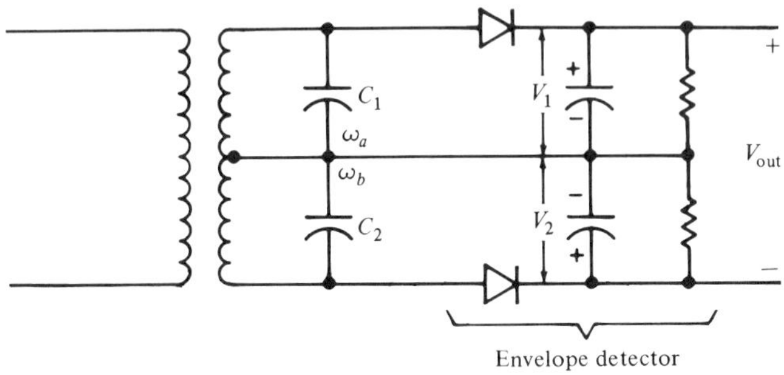

FIGURE 9-12. Slope demodulator.

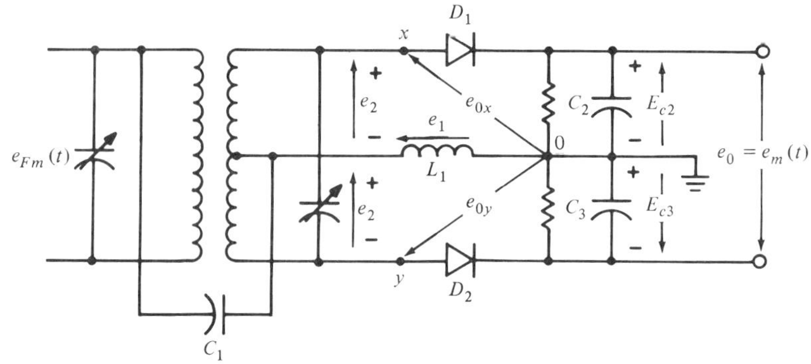

FIGURE 9-13. Foster Seely frequency demodulator.

If the instantaneous frequency applied to the primary is above the resonant frequency, the tuned circuit appears capacitive and the voltage across both secondaries lags the primary voltage. Below resonance, the tuned circuit is inductive and the secondary voltages lead the primary voltage. The output voltage is

$$\begin{aligned} E_o &= E_{c2} - E_{c1} \\ &= K(|V_{D1}| - |V_{D2}|) \end{aligned} \tag{9-39}$$

Phasor diagrams describing the action of the demodulator are shown in Figure 9-14.

It is evident that the output of both the Foster–Seeley network and the slope detector are sensitive to amplitude variations of the RF signal [see equations (9-36) and (9-39)]. Consequently, limiter circuits are applied before detection to eliminate amplitude variations.

9-4.2 Ratio Detector

A detector that is not sensitive to amplitude variations of the incoming signal is the ratio detector (Figure 9-14). The RF voltages that exists in Figure 9-13 are identical to those in Figure 9-15. Note that one of the diodes has been reversed. Thus the output may be taken across either R_1 or R_2 or the algebraic sum of the voltage across R_1 and R_2. The voltages across the combination are essentially constant. That is, if V_{o1} increases, then V_{o2} decreases, and vice versa. Any fluctuations due to amplitude variations may be filtered by a large capacitor connected across the output. A vector diagram showing the result of summing the two output voltages is shown in Figure 9-16. From the figure we may write the following equations:

$$V_o = V_{R1} + V_{o1} \tag{9-40}$$

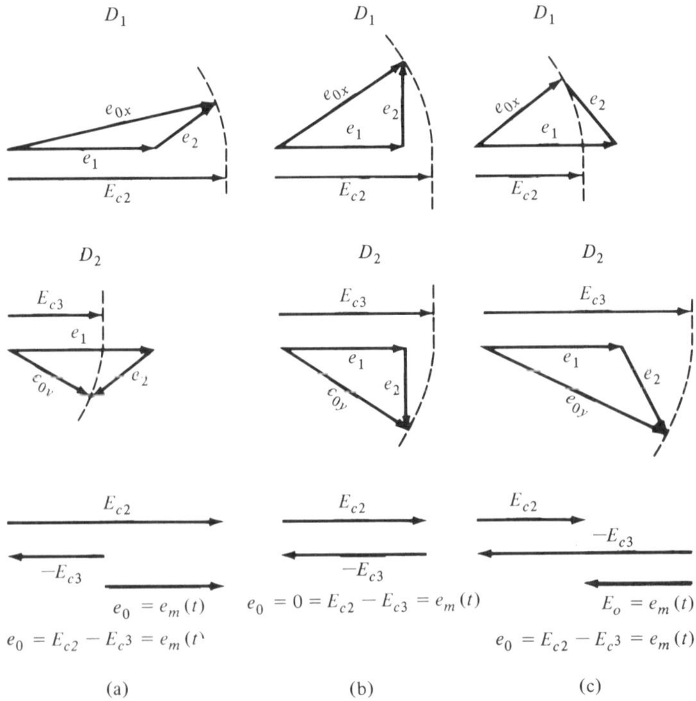

FIGURE 9-14. Phasor diagrams showing operation of angle demodulator: (a) above resonance; (b) at resonance; (c) below resonance.

and

$$-V_o = V_{R2} - V_{o2} \tag{9-41}$$

If $V_{R2} = V_{R1}$, the algebraic sum is

$$V_o = \frac{1}{2}(V_{o2} - V_{o1}) = \frac{1}{2}K(|V_{D2}| - |V_{D1}|) \tag{9-42}$$

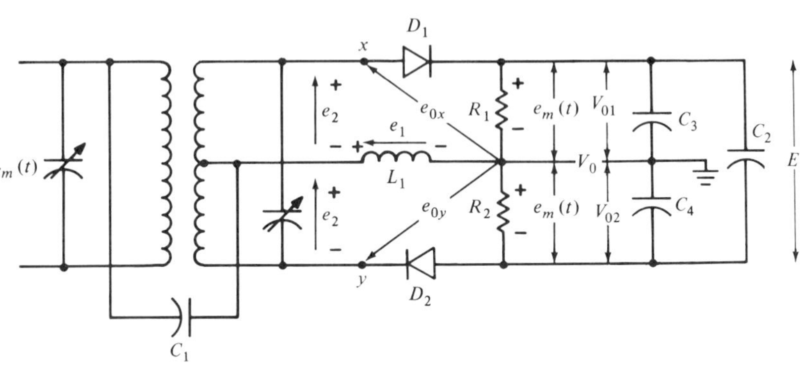

FIGURE 9-15. Ratio detector.

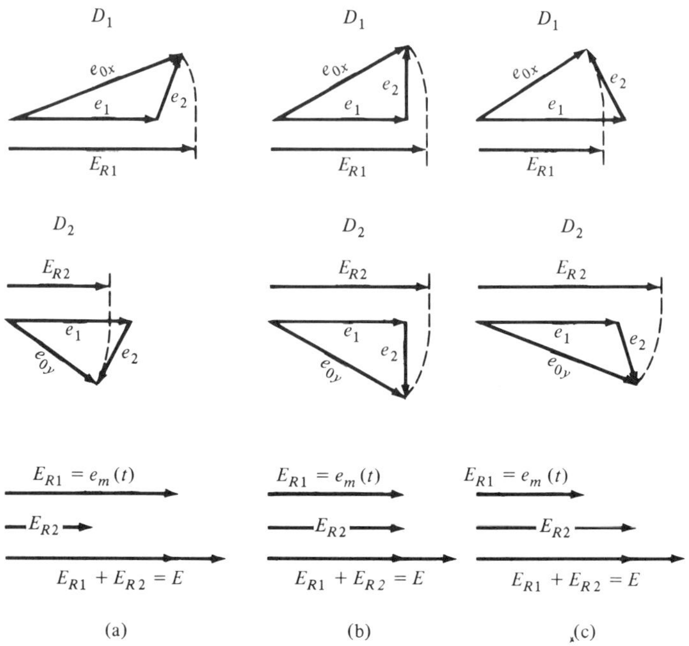

FIGURE 9-16. Phasor diagrams illustrating the operation of a ratio detector: (a) above resonance; (b) at resonance; and (c) below resonance.

The output voltage is one-half that of the Foster–Seeley circuit. The ratio of the two voltages, V_{o1} and V_{o2}, is relatively constant for changes in amplitude of the RF signal.

9-5 PHASE-LOCKED-LOOP DISCRIMINATOR

In this application, the PLL can be used as a complete FM IF strip, limiter, and FM detector. It can be used to detect wide- or narrowband FM signals with greater linearity than the methods previously discussed for frequencies within the range of the VCO. One popular use is where a number of broadcast channels may be monitored by varying the VCO free-running frequency.

The operation of the PLL was discussed in Chapter 7. Without resorting to a mathematical analysis, consider PLL operation with an FM input. When locked, the frequency of the VCO is controlled by the dc voltage from the loop low-pass filter. If the VCO tracks the incoming instantaneous frequency exactly, it is obvious that the varying control voltage from the filter must look exactly like the original message that frequency modulates the RF frequency. Thus the filter output is the demodulated FM.

Consider a practical example. Figure 9-17 is a 10.7-MHz FM demodulator using an NE560 PLL integrated circuit. An approximate value for the VCO timing capacitor is given by

$$C_0 \approx \frac{300}{f_c} = \frac{500}{10.7} = 28 \text{ pF}$$

The design example uses a 22-pF capacitor in parallel with an 8-pF trim capacitor for fine tuning. The low-pass filter is composed of R_1, R_2, C_2, and C_3. Values for R_1 and R_2 are usually between 50 and 200 Ω. Let $R_1 = R_2 = 150\ \Omega$. Assuming that the bandwidth of the demodulated signal is 15 kHZ, values for C_2 and C_3 are

$$C_2 = C_3 = \frac{13.3}{f_m} = \frac{13.3}{15{,}000}\ \mu\text{F}$$
$$= 886 \text{ pF}$$

For this example select standard values of 0.001 μF. Capacitor c_4 constitutes part of a deemphasis circuit. This topic is discussed in the next section on FM noise.

9-6 NOISE IN FREQUENCY MODULATION

Consider the FM noise spectrum in Figure 9-18, where the noise has been bandlimited by a rectangular filter whose bandwidth is $\pm B_{\text{IF}}$. From the previous discussion, the bandwidth must be at least $2(\Delta f_c + f_m)$ where Δf_c is the peak carrier deviation and f_m is the highest modulating frequency. Now, let the noise in Figure 9-18 be divided into many narrowband elements. Each of these noise elements can be treated as a separate entity that interferes with the signal. The linear treatment will be valid only when higher-order terms are negligible. This condition will be satisfied if the rms values of all the modulation indices (m_k) are very small. That is, the noise components are small relative to the signal.

This requirement is assured by setting equation (9-21) (normalized multitone modulation) equal to

$$\prod_k J_0(m_k) \approx 1 \tag{9-43}$$

Here we are assuming that any one component is unaffected by the remaining components. If the noise has a density of N_0 watts/hertz and each noise element

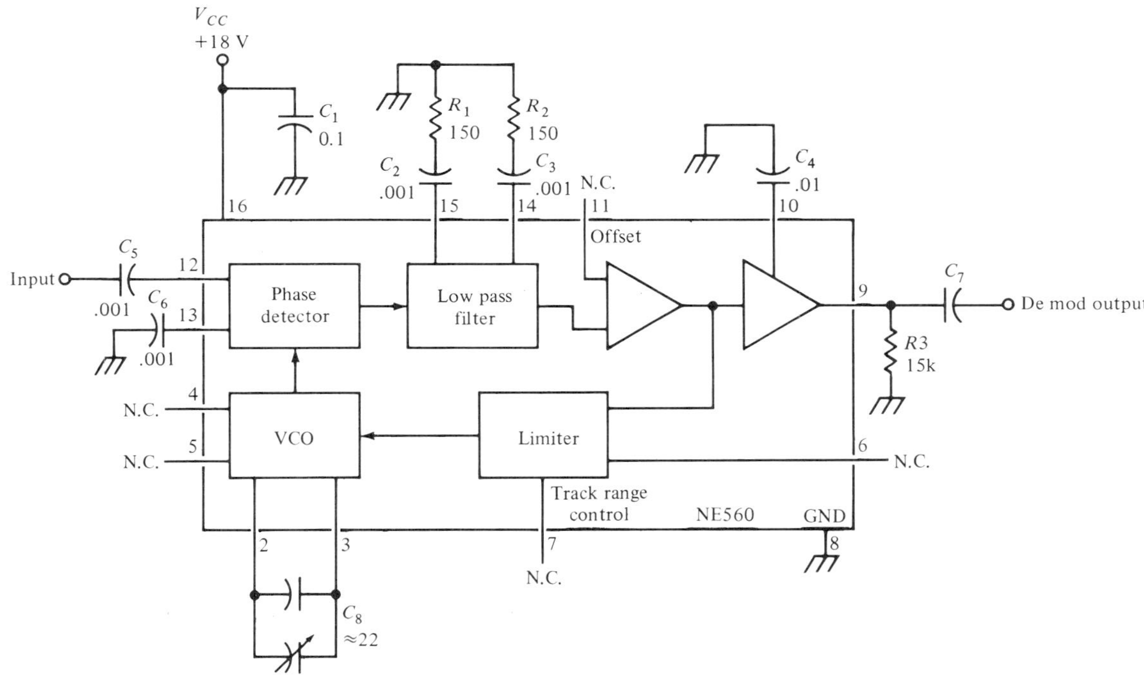

FIGURE 9-17. 10.7 MHz FM demodulator using an NE560 PLL IC.

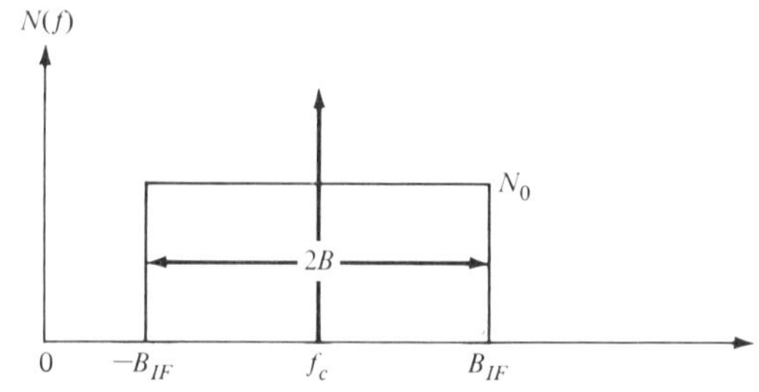

FIGURE 9-18. Two-sided FM noise spectrum input.

has a bandwidth of 1 Hz, the power in each element is N_0 watts. The noise in a 1-Hz bandwidth may be considered a sinusoidal signal whose rms value is $\sqrt{N_0}$ and whose peak value is $\sqrt{2}\,\sqrt{N_0}$. If the carrier power is P_s, its peak value is likewise $\sqrt{2}\,\sqrt{P_s}$.

Refer to Figure 9-19. Two signals are shown in phasor representation. The large signal of normalized amplitude 1 rotates at its angular frequency ω_c. The small signal of normalized amplitude a has a frequency $(\omega_c - \omega_m)$ and is represented as rotating at an angular velocity of ω_m with respect to the larger signal. The resultant R is the vector sum of the two. If we now identify the a component as the noise signal, we see that it changes both the amplitude and the frequency of the overall signal as it rotates.

Assuming that amplitude limiting is employed prior to the discriminator, we are interested only in the frequency variations. As the small phasor rotates with respect to the larger, the resultant oscillates back and forth on both sides of the large phasor. The average frequency of the resultant is the same as that of the large signal. The maximum phase deviation of the resultant from the phase of the large signal occurs when the two signals are in phase quadrature (Figure 9-19b). This maximum phase deviation, $\Delta\phi$, is, from the diagram,

$$\Delta\phi = \tan^{-1} a \tag{9-44}$$

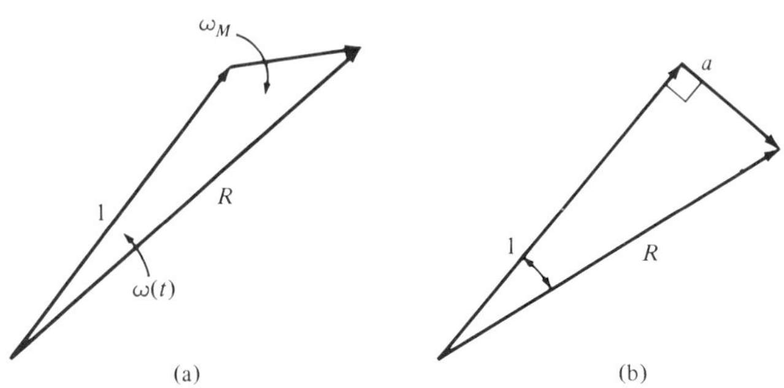

FIGURE 9-19. Phasor diagram of a small interfering signal.

where $a << 1$, $\Delta\phi \approx a$. Thus the resultant signal R, amplitude limited, is

$$\begin{aligned} R &= A \cos (\omega_c t + \Delta\phi \sin \omega_m t) \\ &\approx A \cos (\omega_c t + a \sin \omega_m t) \end{aligned} \tag{9-45}$$

Comparing this to equation (9-10) illustrates that this is a signal of carrier frequency ω_c, frequency modulated at a rate ω_m, with a modulation index $m_f = a$. Thus the maximum frequency deviation is

$$\begin{aligned} \Delta f &= m_f f_m \\ &= a f_m \end{aligned} \tag{9-46}$$

Now, the normalized value of the interfering signal, a, is

$$\begin{aligned} a &= \frac{\sqrt{2}\sqrt{N_0}}{\sqrt{2}\sqrt{P_s}} \\ &= \sqrt{\frac{N_0}{P_s}} \end{aligned} \tag{9-47}$$

and equation (9-46) may be written as

$$\Delta_f = f_m \sqrt{\frac{N_0}{P_s}} \tag{9-48}$$

Equation (9-48) indicates the peak frequency deviation caused by a noise component located f_m hertz from the carrier. The modulation caused by the component occurs at an f_m rate. Referring to Figure 9-17, there is also a symmetrical component located with respect to the carrier which produces the same deviation. Since these are noise components, they are at random phase with respect to each other. The net deviation at a rate f_m, then, is found by an rms addition of the peak individual deviations. That is,

$$\Delta f = \sqrt{2}\, f_m \sqrt{\frac{N_0}{P_s}} = f_m \sqrt{\frac{2N_0}{P_s}} \tag{9-49}$$

This equation describes the usual noise spectrum of FM in which the noise deviation at any frequency is directly proportional to that frequency. Equation (9-49) is plotted in Figure 9-20. The noise is passed through an audio filter so that all components higher than f_m are attenuated.

Generally speaking, the detected signal-to-noise ratio is a more useful expression. Since the signal modulation is variable, it is best to use the maximum

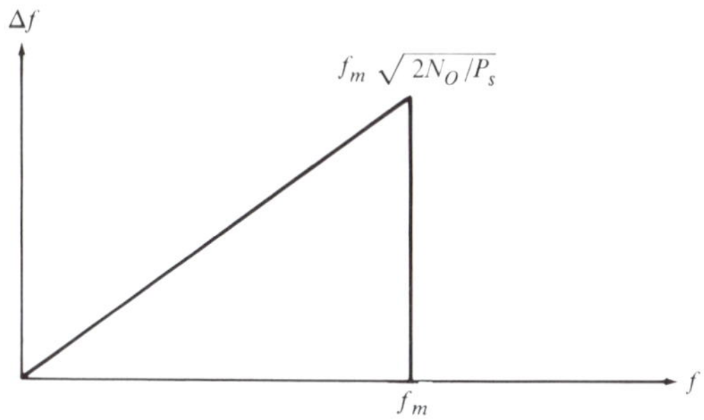

FIGURE 9-20. Detected noise spectrum.

detected signal as a reference. Using equation (9-49), the noise power as a function of audio frequency is

$$n_{\mathrm{FM}} = f_m^2 \frac{2N_0}{P_s} \tag{9-50}$$

If the audio filter cuts off at the audio bandwidth, F_A, the average detected noise power is

$$\begin{aligned} N_{\mathrm{FM}} &= \int_0^{F_A} n_{\mathrm{FM}}\, df_m \\ &= \int_0^{F_A} f_m^2 \frac{2N_0}{P_s}\, df_m = \frac{F_A^3}{3} \frac{2N_0}{P_s} \end{aligned} \tag{9-51}$$

The maximum detected signal voltage is directly proportional to Δf, where the proportionality constant is a function of the discriminator. Since the same proportionality constant applies to the noise, the signal-to-noise power ratio is

$$(S/N)_{\mathrm{FM}} = \frac{\Delta f}{N_{\mathrm{FM}}} = \frac{3\Delta f^2}{F_A^3} \frac{P_s}{2N_0} \tag{9-52}$$

In an amplitude-modulation system with the same audio bandwidth, F_A, and the same receiver noise density, N_0, the noise power from the detector is

$$N_{\mathrm{AM}} = \int_0^{F_A} 2N_0\, df_m = 2N_0 F_A \tag{9-53}$$

The factor of 2 accounts for the symmetrically located AM noise components equivalent to those in Figure 9-17. If the maximum detected signal corresponds to 100% modulation, the detected signal power is P_s and

$$(S/N)_{\text{AM}} = \frac{P_s}{2N_0 F_A} \tag{9-54}$$

Taking the ratio of equations (9-52) and (9-54) results in

$$\frac{(S/N)_{\text{FM}}}{(S/N)_{\text{AM}}} = \frac{3\Delta f^2}{F_A^2} = 3m_f^2 \tag{9-55}$$

where m_f is the modulation index of the highest modulation frequency. The factor $3m_f^2$ is called the *improvement ratio* of FM over AM and is usually expressed in decibels.

The temptation exists to obtain a very high signal-to-noise ratio by using a large value for m_f. If the audio bandwidth F_A is kept constant and Δf is increased, this requires an increase in B_{IF}. From Figure 9-18 we see that the total noise increases and eventually the small rms noise conditions under which the analysis was made becomes invalid. In fact, if the noise becomes larger than the signal, a phenomenon known as *noise capture* results. The noise replaces the signal and is noted as a rushing noise sound in a speaker.

9-6.1 Noise and Preemphasis

We noted in the preceding section that due to phase modulation of the signal by noise the detected noise spectrum increases with frequency (Figure 9-20). Thus, if the modulation index of the transmitted signal is constant with frequency, the signal-to-noise ratio of the higher frequencies is less than of the lower frequencies. In addition, the higher frequencies in a voice signal generally

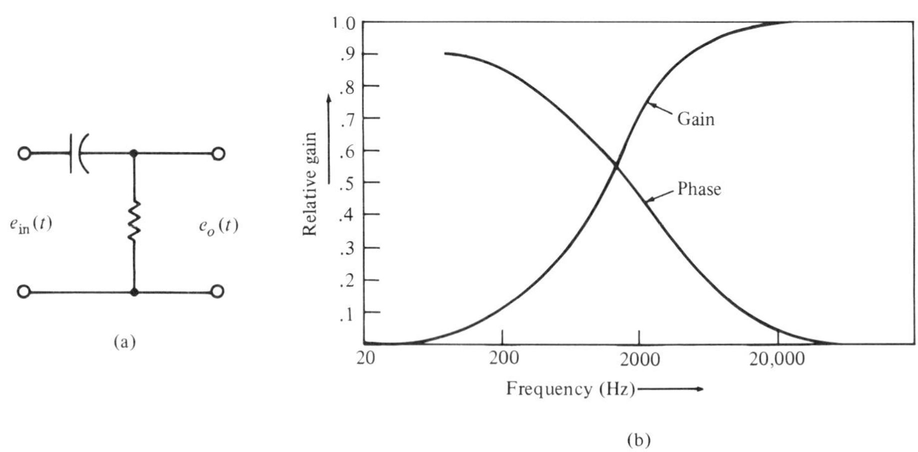

FIGURE 9-21. (a) Pre-emphasis network; (b) gain and phase response.

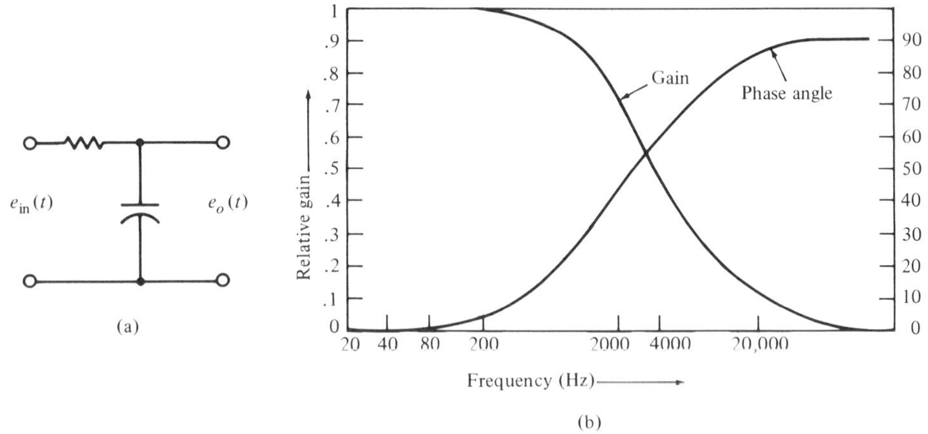

FIGURE 9-22. (a) De-emphasis network; (b) gain and phase response.

have lower amplitude than the lower frequencies. This lowers the modulation index for higher frequencies and thus further degrades the signal-to-noise ratio.

To compensate for this, it is standard practice in FM to preemphasize the higher frequencies before modulating the transmitter. Since this distorts the signal, the receiver must perform the inverse function, that is, deemphasize the signal. Typical networks for performing these functions are shown in Figures 9-21 and 9-22 together with their response curves. For FM stations in the United States the *RC* time constant is set at 75 μs. European stations use a time constant of 50 μs, while several other countries use 25 μs. The deemphasis network of the FM demodulator shown in Figure 9-17 consists of C_4 and the input resistance of pin 10, which is approximately 8 kΩ. The 3-dB roll-off frequency is

$$f_{3dB} = \frac{1}{2\pi R_D C_D}$$

where R_D is 8 kΩ. For a time constant of 75 μs,

$$C_D = \frac{1}{2\pi f R_D} = \frac{75 \times 10^{-6}}{R_D} = 0.0094\ \mu\text{F}$$

In practice a 0.01-μF capacitor was chosen since it is the nearest standard value.

PROBLEMS

9-1. The angular frequencies given below are constantly changing. At the end of 5 s find the angle from a reference time $t = 0$. (a) $\omega = 105t$, (b) $\omega = 4 \cos 20t$, and (c) $\omega = 2 + 3t + t^2$.

9-2. What is the approximate RF bandwidth if a 50-MHz carrier is deviated 30 kHz by a 5-kHz sine wave? What is the modulation index?

9-3. Plot the frequency spectrum for Problem 9-2 if the carrier has a peak amplitude of 200 V and the 5-kHz sine wave has a peak amplitude of 20 V.

9-4. A preemphasis circuit consisting of a resistor and a capacitor is needed to drive a load impedance of 50 kΩ. What value of capacitor is needed to produce the required preemphasis for commercial FM? If the load is reduced to 5 kΩ, what is the value of the capacitance?

9-5. Find the instantaneous frequency of $e_{\text{FM}}(t)$, where

$$e_{\text{FM}}(t) = 20(\cos 10t \cos 30t^2 - \sin 10t \sin 30t^2)$$

9-6. Find the approximate band of frequencies occupied by

$$e_{\text{FM}}(t) = 100(2\pi \times 10^6 t + 20 \cos 1000\pi t)$$

Is this narrowband or wide-band FM?

9-7. Compare the bandwidth of Problem 9-6 with that to transmit the same information by AM.

9-8. If the expression in Problem 9-6 represents a PM wave, what is the expression for the message $e_m(t)$?

9-9. Explain why the receiver output increases in a PM system as the modulating frequency increases even though the amplitude of the modulating signal remains constant.

9-10. A frequency converter is used to downshift a signal of frequency, ω_1, and phase, $\phi = \pi/4$. If the new frequency is $\frac{1}{3}\omega_1$, what is the value of ϕ? Assume that the converting frequency has phase angle $\theta = 0$.

9-11. An FM wave has a carrier of 99.6 MHz and swings ±75 kHz to either side of this center frequency. In a receiver, this signal is mixed with an oscillator frequency of 110.3 MHz. Find the IF center frequency and frequency swing.

9-12. A phase modulation system produces an "indirect" FM of 200 Hz from an audio modulating signal of 10 V at 100 Hz. (a) How much frequency deviation would a 10-V 20-Hz audio signal produce? (b) How much deviation would a signal of 10 V at 2000 Hz produce? (c) How much reduction is needed in the amplitude of the signal in part (b) to make its deviation equal to the original condition? (d) How much reduction is needed to equalize the deviation in part (b) to part (a)?

9-13. An FM transmitter develops an unmodulated carrier power of 10 W. At a certain modulation level, spectrum analysis shows 4 W of power in the sideband components. (a) What is the total power in the modulated wave? (b) What is the carrier power at this modulation level?

9-14. An FM wave has a deviation of 60 kHz. The modulating signal producing this deviation is a 12-V 5-kHz sine wave. Find (a) the modulation index, (b) the number of significant side frequencies, and (c) the total bandwidth required for this signal.

9-15. Find the amount of indirect FM produced when a carrier is shifted 30° by a 10-V 5-kHz sine-wave modulating signal.

9-16. An FM broadcast transmitter has a deviation of ±20 kHz. What will the deviation be in each of the following cases? (a) The audio signal amplitude is tripled; (b) the audio signal power is doubled; and (c) the audio signal frequency is halved.

9-17. Find the modulation index for an FM wave if the frequency swing is ±20 kHz and the modulating signal frequency is 4 kHz.

9-18. An FM transmitter has a resting frequency of 96 MHz. The carrier amplitude is 40 V. When modulated with an 8-kHz sine-wave signal of 10 V amplitude, the frequency deviation is 40 kHz. Find the modulation index.

9-19. Calculate the BW of an FM system given that the maximum deviation, Δf, is 20 KHz and $f_s = 4$ kHz.

9-20. An FM broadcast transmitter is modulated 30 percent by a 4 kHz tone. What is the frequency swing of the transmitter if the percent modulation is doubled?

9-21. A 5 kHz tone is used to 40 percent modulate an FM transmitter. If the frequency of the modulating signal is changed to 6.5 kHz, and the percent modulation is unchanged, what is the transmitter frequency deviation?

CHAPTER 10

Receiver Functions

At this point we have studied reams of material which, as promised, leads to an understanding of communication systems. Happily, at this time an important first step has been achieved. We are in position to understand the standard radio receiver. Continued study past this chapter will lead to an understanding of a larger world, that of communications systems and networks.

A radio receiver must perform the following basic functions:

- Select the one desired signal and reject all others.
- Amplify the desired signals.
- Demodulate (detect) the information, amplify it, and drive the output device (speaker).

In the following we analyze the operation of an AM receiver.

SUPERHETERODYNE RECEIVER

Early receiver designs used the tuned radio-frequency design principle. These receivers were simply a chain of amplifiers tuned to the same frequency and followed by a detector. These receivers suffered from poor channel selectivity. This occurred because the Q of tuned circuits changes with frequency. To improve this situation the superheterodyne receiver was developed.

Channel bandwidth for standard AM broadcast is set at 10 kHz by the Federal Communications Commission (FCC). Thus the highest information frequency is 5 kHz. The frequency spectrum present at the input of an AM receiver is shown in Figure 10-1. We see that to select the input frequency of interest a filter with very sharp cutoff characteristics is needed.

EXAMPLE 10-1

Assuming that a tuned radio-frequency receiver has a bandwidth of 10 kHz at 550 kHz, find the bandwidth at 1600 kHz.

SOLUTION

$$Q = \frac{f_0}{\text{BW}} = \frac{550}{10} = 55$$

If we assume that the Q is unchanging with frequency, then at the high-frequency end,

$$\text{BW} = \frac{1600}{55} = 29.09 \text{ kHz}$$

Obviously, this receiver would be unsuitable for tuning over the AM frequency spectrum shown in Figure 10-1. A block diagram for a superheterodyne

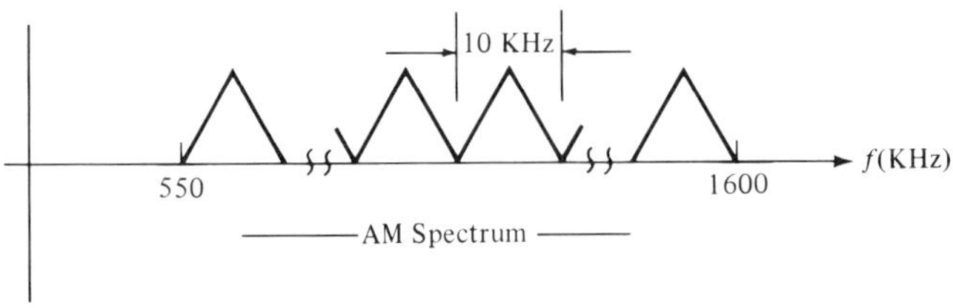

FIGURE 10-1. AM receiver input spectrum.

receiver is shown in Figure 10-2. Starting at the antenna, an AM station of 1200 kHz is being received. This signal is filtered by a tunable RF filter and bandpass RF amplifier. The mixer output consists of sum and difference frequencies, both containing all the information. A series of tuned intermediate-frequency (IF) amplifiers selects the difference frequency 455 kHz. This signal is envelope detected and the recovered audio is amplified.

Now, let us examine the advantage of this receiver over the tuned radio-frequency receiver. By gang tuning the RF filter and the local oscillator, we ensure that the IF frequency is always 455 kHz for any input signal. Thus the tuned IF amplifiers which are centered at 455 kHz have a fixed 10-kHz bandwidth. These amplifiers are referred to as the IF "strip" and are often made up of three tuned circuits which are aligned to generate a Butterworth (maximally flat) response.

Considering receiver operation in conjunction with Figure 10-2, we see that the 455-kHz IF signal is produced with or without the RF filter. Thus its inclusion in the circuit remains to be justified since the signal is selected by the IF amplifiers. The operation of the mixer is referred to as *heterodyning*. This process shifts the incoming frequencies both up and down (sum and difference). Therefore, with more than one signal present at the input, the heterodyning process may "upshift" the desired signal to the IF frequency, while the "downshift" operation shifts an undesired signal to the IF frequency. This undesired signal is referred to as the *image frequency*. The function of the tunable RF filter is to eliminate the image frequency.

If we use the difference frequency, it is obvious from Figure 10-1 that the local oscillator must be tunable over the frequency range of 1005 to 2055 kHz in order to tune in the entire AM spectrum. Choosing the sum frequency would require an oscillator tunable from 95 to 1145 kHz. In practice, it is easier to design an oscillator tunable over this range if it operates at the higher frequencies. This, then, is commercial practice. Now consider the image problem. If the incoming signal is multiplied by 1100 kHz, the station at 645 kHz is shifted

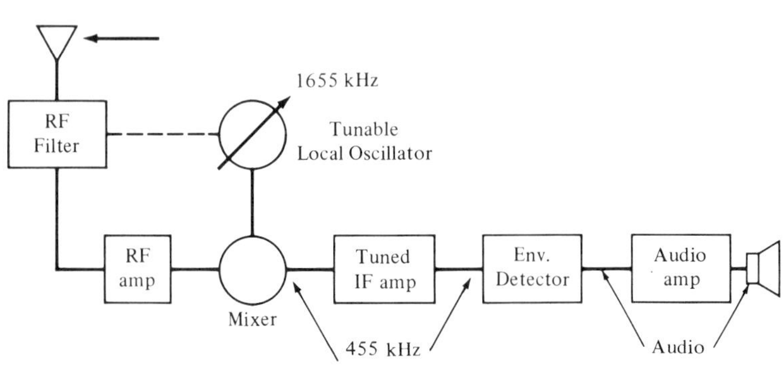

FIGURE 10-2. Block diagram of a superheterodyne AM receiver.

to the IF passband (455 kHz). At the same time, the station at 1555 kHz is also shifted to the IF frequency. This is the image station.

If now the RF filter is centered around the station at 645 kHz but rolls off before the 1555-kHz station frequency, the image signal is rejected. The 910-kHz separation between stations is such that a highly selective filter is not necessary. Refer to Figure 10-3. Note that the frequency separation between the station of interest, f_s, and the image station, f_i, is exactly twice the IF frequency. Thus the RF filter can have a bandwidth up to four times the IF frequency (1820 kHz) and still provide image frequency rejection. In practice, greater attenuation is desired and the filter is not this broadband.

EXAMPLE 10-2

The desired AM station has a frequency of 1400 kHz. Determine the frequency of the local oscillator and the frequency of the image station.

SOLUTION

$$f_{LO} = f_s + f_{IF}$$
$$= 1400 + 455 = 1855 \text{ kHz}$$

The frequency of the image station is

$$f_i = f_s + 2f_{IF}$$
$$= 1400 + (2)(455) = 2310 \text{ kHz}$$

In general, the image frequency is given by

$$f_i = f_s \pm 2f_{IF} \tag{10-1}$$

where the plus sign is used if $f_{LO} > f_s$ and the minus sign when $f_{LO} < f_s$.

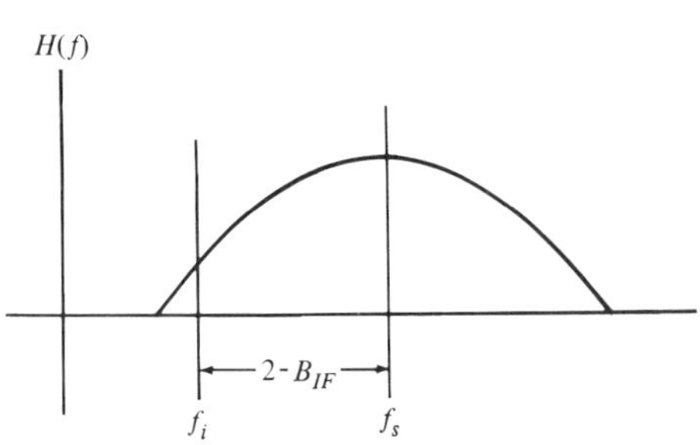

FIGURE 10-3. Relation of image station to station of interest.

EXAMPLE 10-3

An AM receiver has an IF frequency of 455 kHz and is tuned to 600 kHz. What Q is required for the RF filter in order that the power of the image frequency be attenuated by a factor of 10?

SOLUTION

$$
\begin{aligned}
f_i &= f_s + 2I_f \\
&= 600 \text{ kHz} + 2(455 \text{ kHz}) \\
&= 1510 \text{ kHz}
\end{aligned}
$$

Using results from Example 2-2, the Q is found as follows:

$$\left(\frac{\omega}{\omega_0} Q - \frac{\omega_0}{\omega} Q\right)^2 = 9$$

Then

$$
\begin{aligned}
Q &= \frac{3\omega\omega_0}{\omega^2 - \omega_0^2} = \frac{(3)(2\pi)(1.51 \times 10^6)(6 \times 10^5)}{(1.51 \times 10^6)^2 - (6 \times 10^5)^2} \\
&= 8.89
\end{aligned}
$$

10-2

INTEGRATED CIRCUITS FOR AM RADIO APPLICATIONS

Integrated-circuit subsystems for AM radio applications are available from several semiconductor manufacturers. These circuits can assume many forms, depending on the class of service for which it is to be used. Typical of these is the RCA CA3088E, which is designed for high-quality AM superheterodyne receivers. This circuit, shown in Figure 10-4, provides the basic conversion, IF amplification, detection, and audio preamplification sufficient to drive a separate power amplifier, and in addition, auxiliary functions such as power supply regulator, internal AGC for the first IF amplifier, AGC voltage for an optional external RF stage, and an amplified signal to drive a tuning meter output. The device is available in a 16-pin dual-in-line plastic package (DIP). The device is suitable for AM broadcast receivers and may also be used in other AM receiver applications up to 30 MHz.

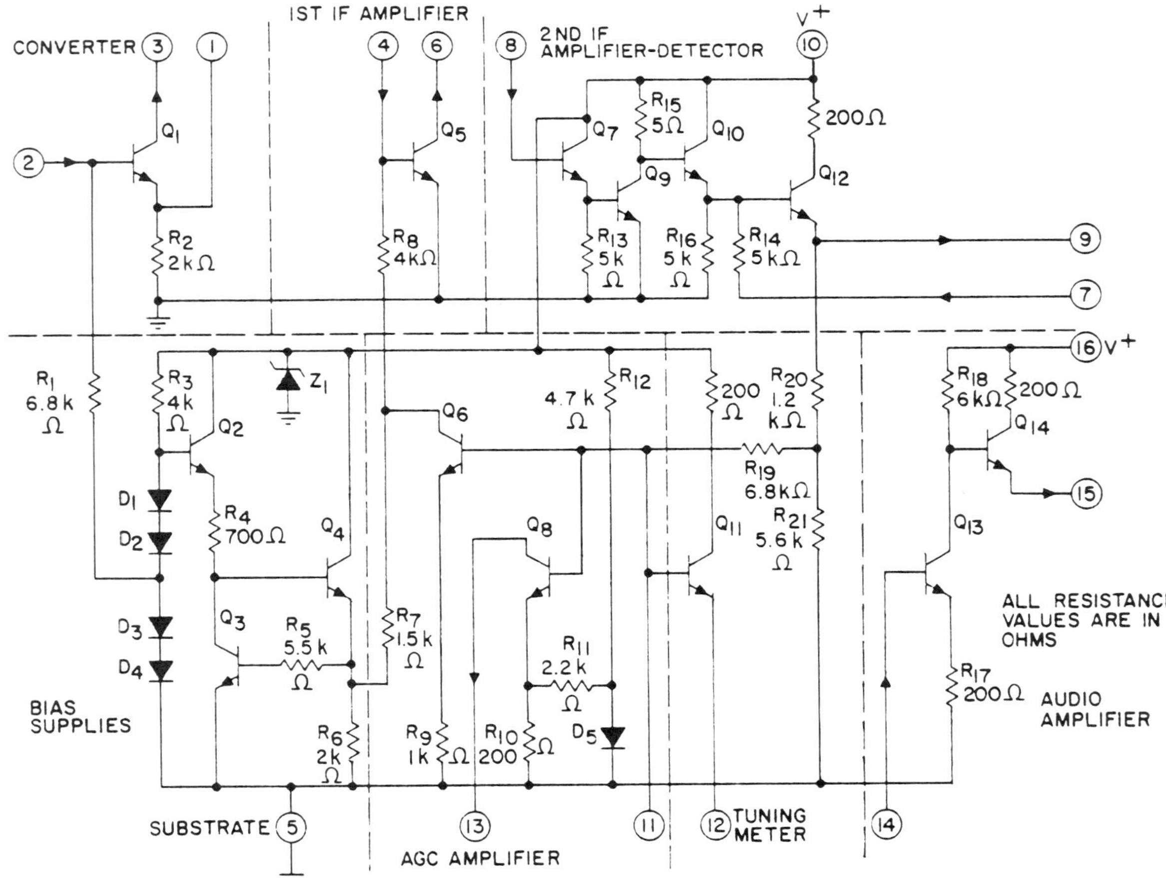

FIGURE 10-4. **Schematic diagram of the RCA CA3088E. All resistance values are in ohms. (© 1972 IEEE)**

10-2.1 Circuit Description

The heart of the circuit is the second IF amplifier–detector combination. The IF amplifier consists of Q_7, Q_9, and Q_{10} with associated components. Transistors Q_7 and Q_{10} are emitter followers which isolate the gain stage, Q_9. The supply voltage at pin 10 is regulated by a zener diode (Z_1). The emitter of Q_{10} is connected to the base of Q_7 through R_{14} and pins 7 and 8 to stabilize the quiescent operation point. Pin 7 should be at ac ground potential to prevent signal feedback. Stage Q_{12} functions as a detector when the proper low-pass filter circuit is connected to pin 9. The rectified voltage at the emitter of Q_{12} also develops a voltage at the junction of R_{20} and R_{19}. This voltage provides the drive for the AGC amplifiers and the tuning meter. The drive voltage across D_5 must be overcome by the drive voltage to the base of Q_8, thus introducing a delay in the gain control to the RF stage.

Internal AGC is applied to the first IF amplifier, Q_5, through transistor Q_6. As signal strength increases and the base drive voltage increases, Q_6 conducts some of the base current being applied to Q_5 and thus reduces the gain of Q_5.

Transistor Q_1 functions as a converter. Diodes D_3 and D_4 in conjunction with emitter resistor R_2 biases the stage to a nominal collector current of 0.35 mA. The emitter of Q_1 is brought out to pin 1 and may be used to apply a signal from the local oscillator. Transistors Q_{13} and Q_{14} function as an audio amplifier.

10-2.2 Applications

A circuit for a typical AM receiver using the CA3088E is illustrated in Figure 10-5. Double-tuned transformer-coupled circuits are shown, but any of the other forms of bandpass filters may be used. A variety of RF integrated circuits are available which may be used for the RF stage.

In addition to its use in a broadcast receiver, this circuit may find applications in other forms of communications equipment. For example, it may be used as a straight IF amplifier for use with a separate tuner, or it can serve as a subsystem in a double-conversion receiver. The space-saving features of the chip make it ideal for small hand-carried receivers. A 10.7-MHz IF amplifier using the CA3088AE and two ceramic filters is shown in Figure 10-6.

10-3

INTEGRATED CIRCUITS FOR FM BROADCAST RECEIVERS

Commercial FM broadcast frequencies occupy the band from 88 to 108 MHz with a BW of 200 kHz. The basic functions of the AM receiver are repeated in the FM receiver. In addition, the FM receiver usually incorporates a limiter amplifier to aid in AM noise rejection. The previous comments regarding image

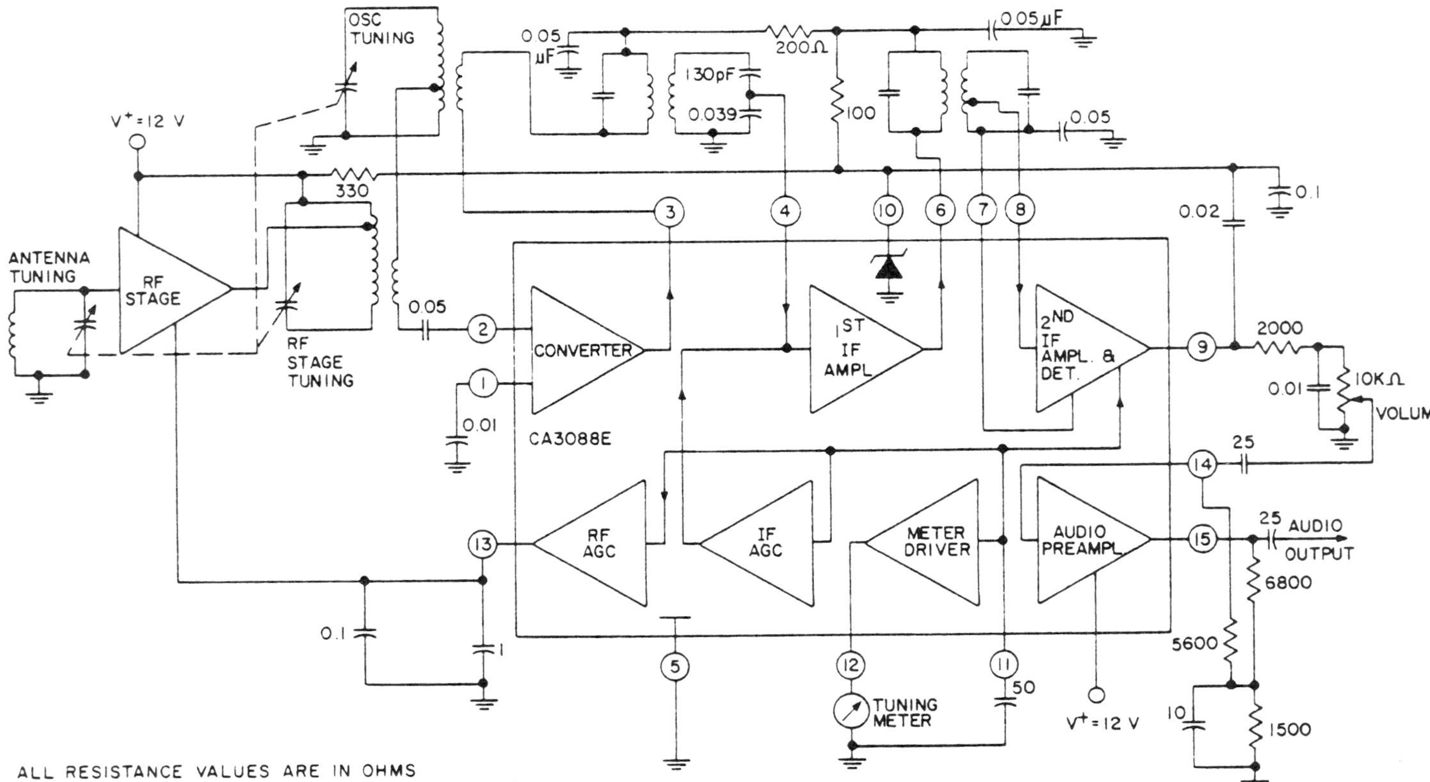

ALL RESISTANCE VALUES ARE IN OHMS
ALL CAPACITANCE VALUES ARE IN MICROFARADS

FIGURE 10-5. Typical AM broadcast receiver using the CA3088E. (© 1972 IEEE)

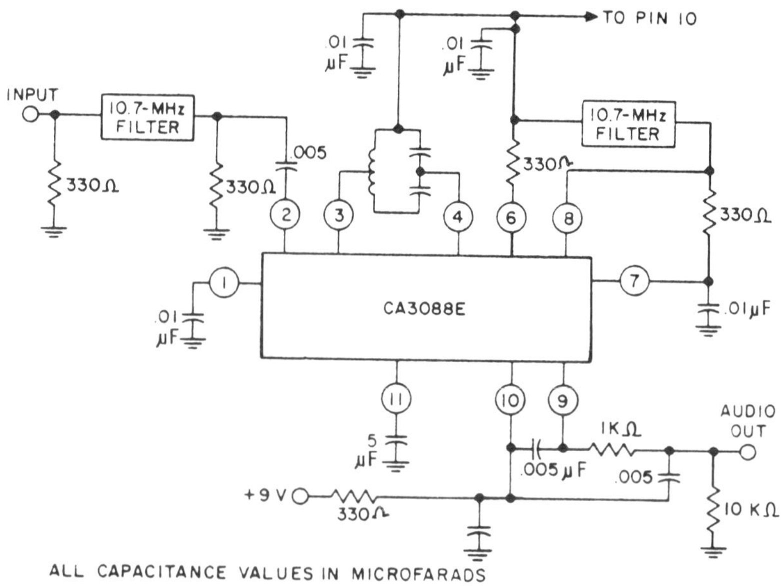

FIGURE 10-6. A 10.7-MHz IF amplifier using the CA30883 and two ceramic filters. (© 1972 IEEE)

frequency rejection apply equally well for FM receivers. As in AM a variety of integrated circuits are available for the design of FM receivers. In the following we examine typical components available as building blocks.

10-3.1 FM Tuner

The use of multifunction integrated circuits are a favorite approach by today's designers. We discussed the application of one of these chips for AM receiver design. The RCA C3005 provides several approaches for FM tuner design. A schematic diagram of the RCA CA3005 is shown in Figure 10-7 and a single-chip front-end design is illustrated in Figure 10-8. This circuit uses one CA3005 and a two-gang tuning capacitor system. The circuit performs the functions of RF amplifier, oscillator, and mixer.

Referring to the figures, we see that transistor Q_3 is a current generator for transistors Q_1 and Q_2, thus establishing common-mode rejection. The oscillator function is established by the positive feedback through the 5-pF capacitor between pins 1 and 10. The frequency of oscillation is determined by tuned circuit L_2, C_2. The RF input signal is injected into the base of transistor Q_3 (pin 3). Transistor Q_3 amplifies the RF and applies it to the emitters of Q_1 and Q_2 to mix with the oscillator frequency. The 10.7-MHz IF signal is obtained from the collector of Q_1 (pin 11).

This approach to the front-end design, although economical, results in only adequate performance. Improved performance is obtained by adding an RF am-

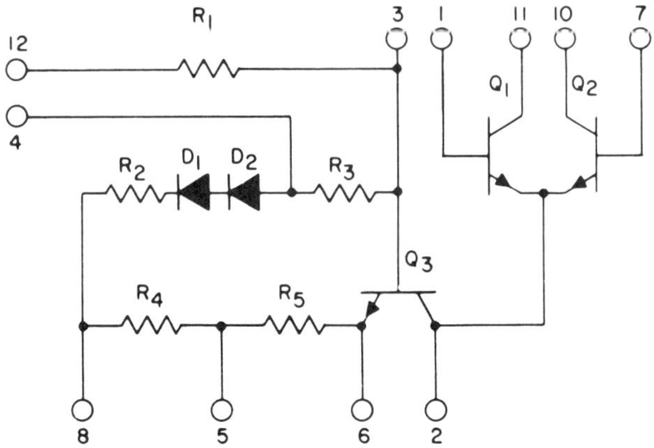

FIGURE 10-7. Schematic diagram of RCA CA 3005 integrated-circuit RF amplifier.

plifier to the mixer–oscillator circuit of Figure 10-8. This improved design is shown in Figure 10-9. The RF amplifier and a three-gang capacitor tuning system adds higher power, lower noise figure, and improved selectivity. Feedback in the circuit is small enough that no neutralization is required.

10-3.2 FM IF Amplifiers, Limiters, and Detectors

A 10.7-MHz FM IF strip and detector are shown in Figure 10-10 with response curves. This circuit uses an RCA CA3005 and an CA3014. Both amplifiers use a differential amplifier which is the best configuration marketed today for symmetrical limiting over a wide input voltage range. The input limiting knee for

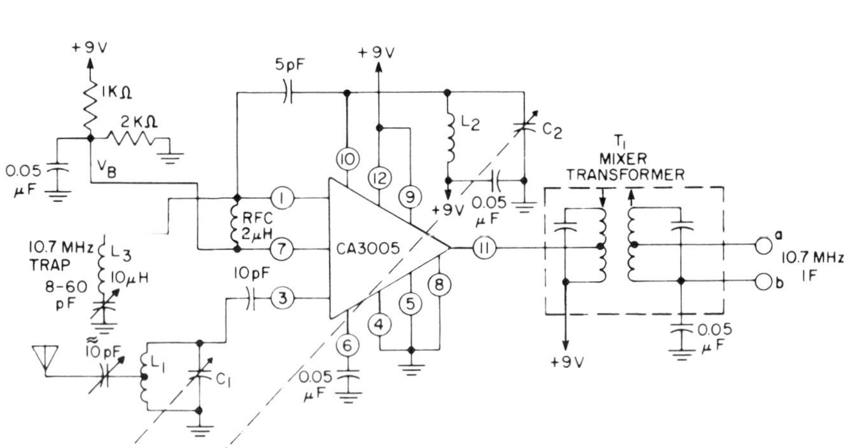

FIGURE 10-8. Single-chip front-end using the CA3005.

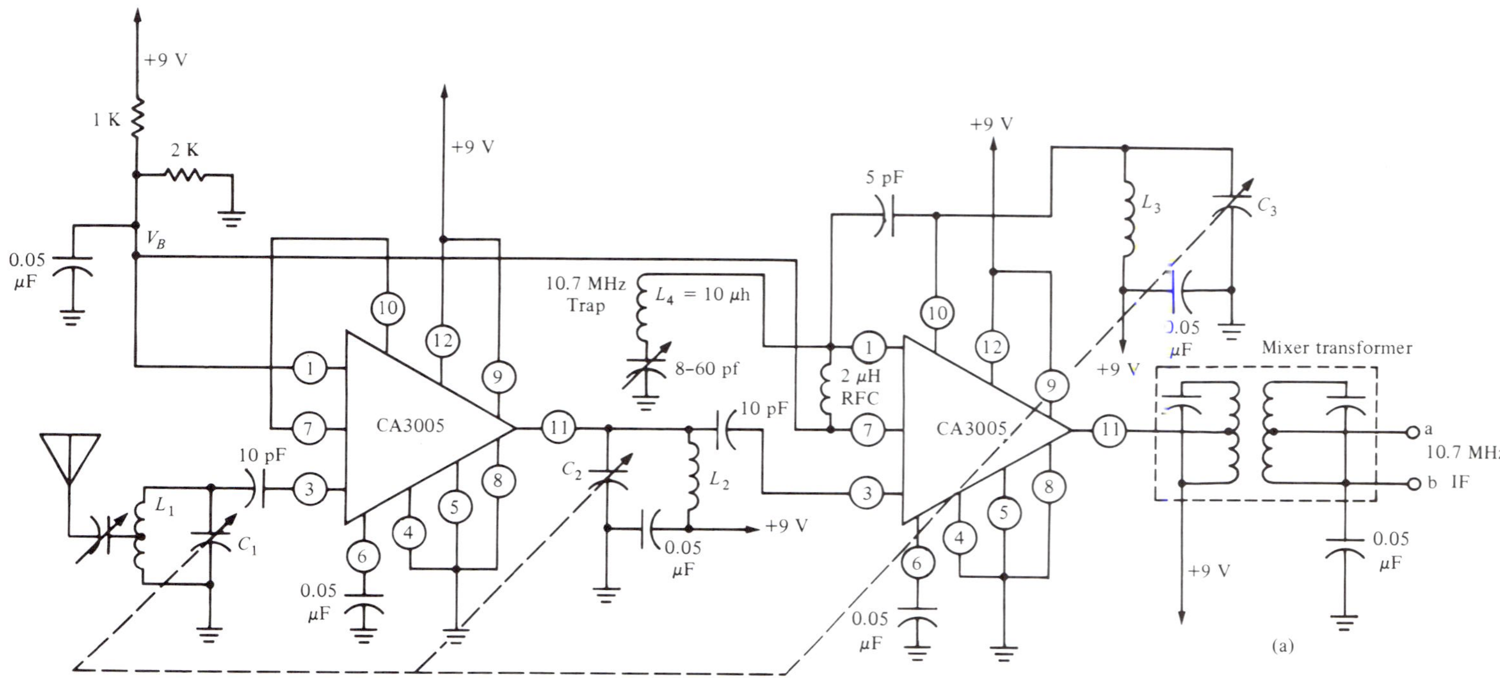

FIGURE 10-9. Two-chip front end.

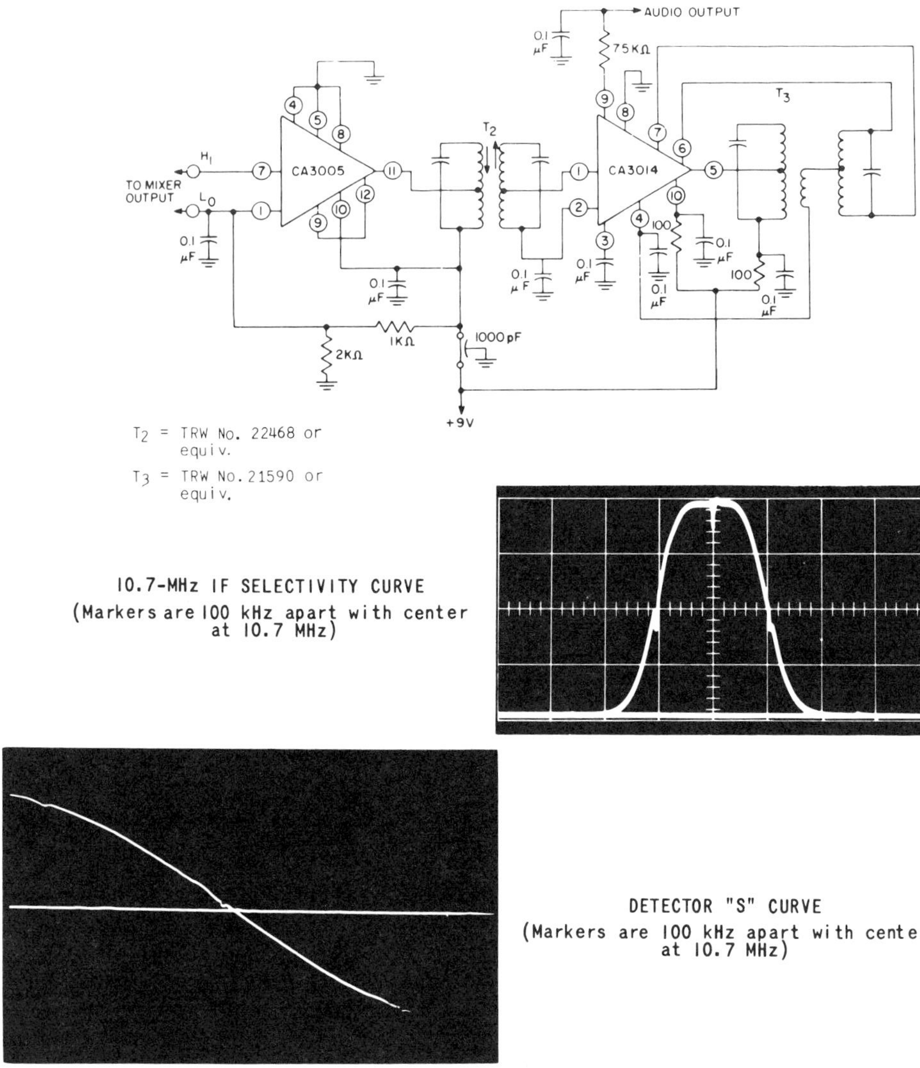

FIGURE 10-10. Selectivity curve and detector "S" curve of two chip 10.7 MHz IF amplifier, limiter, and discriminator using CA3005, CA3014, and interstage transformers.

the IF strip is 30 μv and the recovered audio from terminal 9 of the CA3014 is 220 mV rms. An overall gain of 95 dB is available.

The operation of the circuit is as follows. The 10.7-MHz FM signal from the mixer is applied to pin 7 of the CA3005 and the gain from this point to the input of the CA3014 is 25 dB. The interstage transform T_2 is designed so that the collector output of the CA3005 at terminal 1 does not saturate. This prevents bandpass spreading over large swings in input voltage. From terminal 1 to terminal 5 of the CA3014 there is an additional 70 dB of gain. The FM output at terminal 5 is applied to the primary winding of the phase shift (discriminator) transformer T_3. The secondary winding connected to terminals 6 and 7 has a voltage in phase quadrature with the primary voltage at the 10.7-MHz center frequency. As the FM signal frequency varies, the phase shift of the secondary voltage follows the modulation. The secondary is connected across a resistor–diode network at pins 4, 6, and 7. This network detects the phase shift and consequently demodulates the signal.

Increased selectivity can be obtained in the IF strip by adding double-tuned transformers to the circuit. Figure 10-11 illustrates a circuit in which two transformers are placed immediately after the mixer stage. Placing the transformers immediately after the mixer results in better adjacent-channel rejection.

Most FM front ends come equipped with a double-tuned 10.7-MHz IF transformer in which a secondary high-impedance winding is brought out capacitively. The most effective use of the existing front-end filter is accomplished by the addition of another double-tuned filter. This results in an overall high-insertion-loss filter. Using equation (5-41), the receiver overall noise figure is calculated as

$$F = F_1 + \frac{F_2 - 1}{G_1} + \frac{F_3 - 1}{G_1 G_2} \tag{10-2}$$

where F_1, F_2, and F_3 are the overall noise figures of the first RF, mixer, and IF stages, respectively, and G_1 and G_2 are the gains of the RF amplifier and mixer. If a value of 27 dB is assumed for F_3 (filter plus integrated circuit), 10 dB for the mixer noise figure, and 30 dB power gain for the mixer, the effect of IF noise on mixer noise is given by

$$F_2' = F_2 + \frac{F_{3-1}}{G_2} = 10 + \frac{27 - 1}{30}$$

$$= 10.87 \text{ dB}$$

If the RF stage is assumed to have a power gain of 15 dB and a noise figure of 5 dB, the total receiver noise figure is

$$F = F_1 + \frac{F'_{2-1}}{G_1}$$

$$= 5 + \frac{10.87 - 1}{15} = 5.66 \text{ dB}$$

Thus, as expected, the power gain of the RF amplifier overrides both the IF and mixer noise. A minimum power gain for the RF amplifier is advisable.

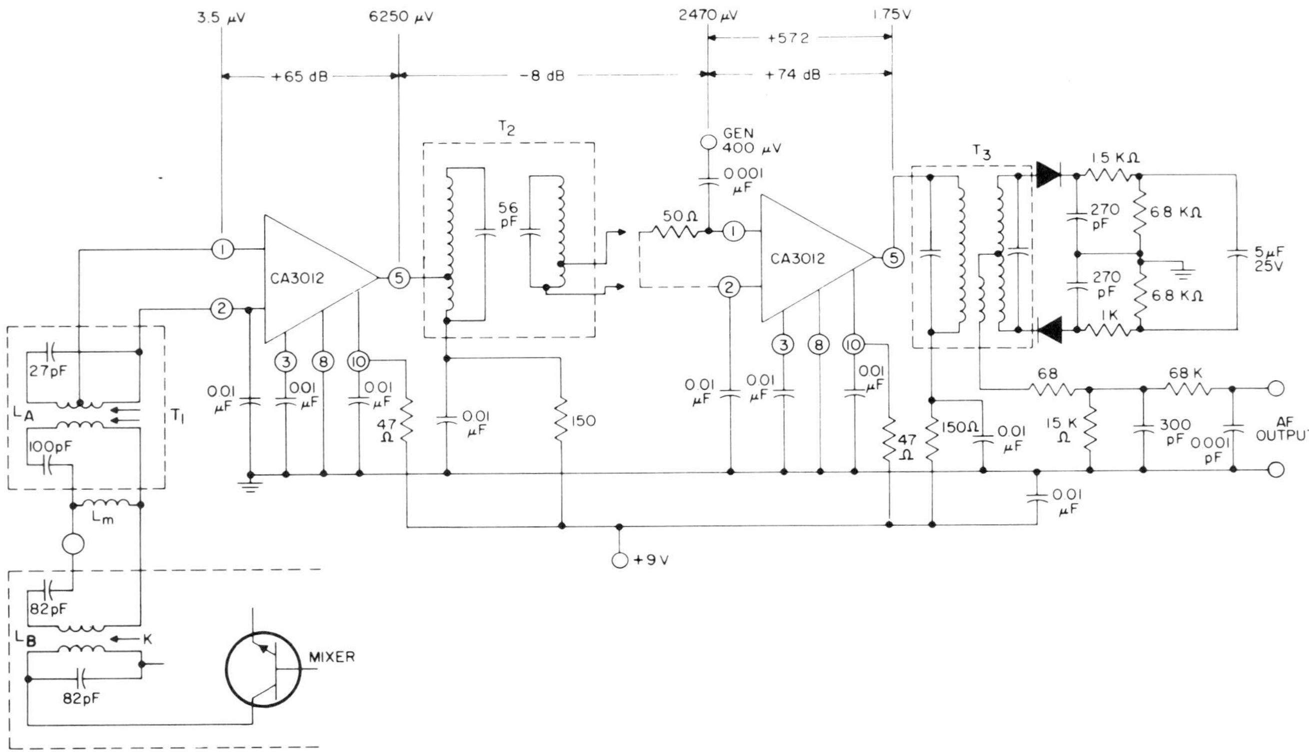

T_2 - TRW No.22960-R2 or equiv.
T_3 - TRW 23148 or equiv.

FIGURE 10-11. 10.7 MHz IF amplifier and ratio detector.

10-3.3 Integrated-Circuit Stereo Decoder

We noted earlier that FM stations have available a bandwidth of 200 kHz in the 100-MHz range. The maximum frequency of the information signal is fixed at 15 kHz. Comparing this with the 5-kHz frequency of AM makes it obvious why high-quality music is transmitted by FM. For narrowband FM, a bandwidth of only 30 kHz is used. Since 200 kHz is available, we see that broadband FM can be used for broadcast. Using the formula for FM bandwidth we see that a frequency deviation of 80 kHz is allowed. In practice, the actual figure used is 75 kHz.

Most good-quality FM receivers include a FM stereo mode of operation. This is generated at the transmitter using subcarriers. A block diagram of a stereo transmitting system is shown in Figure 10-12. The audio is developed by two microphones whose outputs are labeled the "left" and "right" channels. One mixer produces the sum $(L + R)$ of the two channels. The second mixer produces the difference $(L - R)$ of the two signals, where the minus sign indicates phase inversion. A balanced modulator generates sidebands centered around a subcarrier of 38 kHz. The two channels are combined and 19 kHz is summed into the composite spectrum before RF modulation to permit synchronous demodulation of the $L - R$ channel at the receiver. The composite spectrum is shown in Figure 10-13.

The FM receiver is shown in Figure 10-14. Note that the FM is demodulated to recover the stereo composite signal (Figure 10-13). The two signals are then separated using a low-pass filter and the $L - R$ channel is synchronously demodulated. The $(L - R)$ and $L + R)$ signals are passed through a matrix network which separates the L and R signals. Specifically, $L + R + L - R = 2L$ and $L + R - (L - R) = 2R$. Note that a receiver without stereo capability (monoaural) would simply produce the $L + R$ signal. Thus the stereo signal is compatible with monaural receivers.

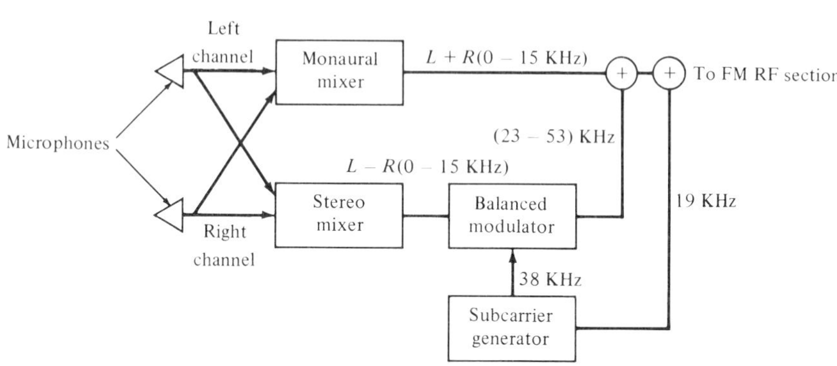

FIGURE 10-12. FM stereo transmitter.

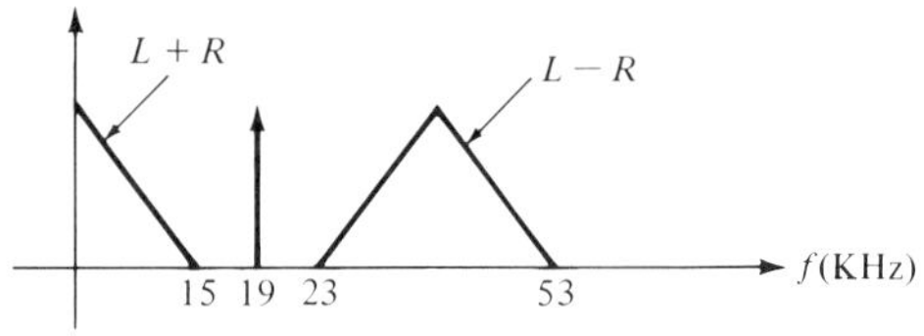

FIGURE 10-13. Composite stereo spectrum.

The unused portion of the allotted band between 53 and 100 kHz is often used for a SCA (subsidiary communications authorization) signal (Figure 10-13). This includes the commercial free music heard in restaurants and stores. It is transmitted on a subcarrier of 67 kHz.

As would be expected, the modulation sequence above produces a natural application of integrated circuits for decoding. Several chips are available commercially to accomplish this. A block diagram of the RCA CA3090AQ system is shown in Figure 10-15.

10-4

CALCULATION OF RECEIVER SENSITIVITY

There are several definitions of receiver sensitivity from which to choose. The choice depends mainly on the type of modulation. These include tangential sensitivity, signal = noise, output signal-to-noise ratio, bit error rate, false-alarm rate, and probability of detection. The latter two are generally character-

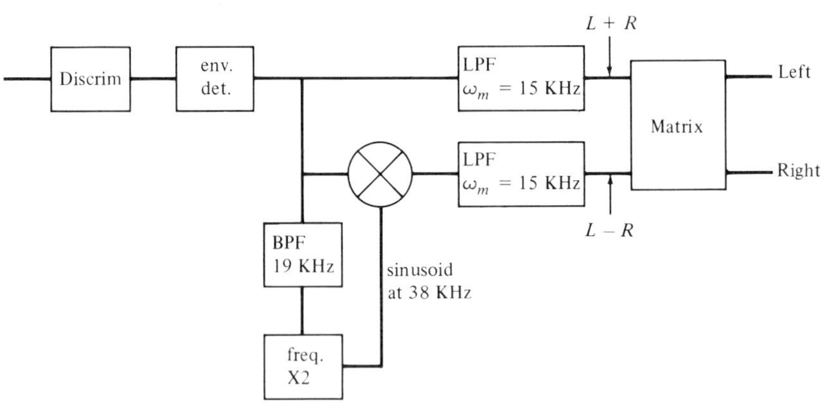

FIGURE 10-14. FM stereo receiver.

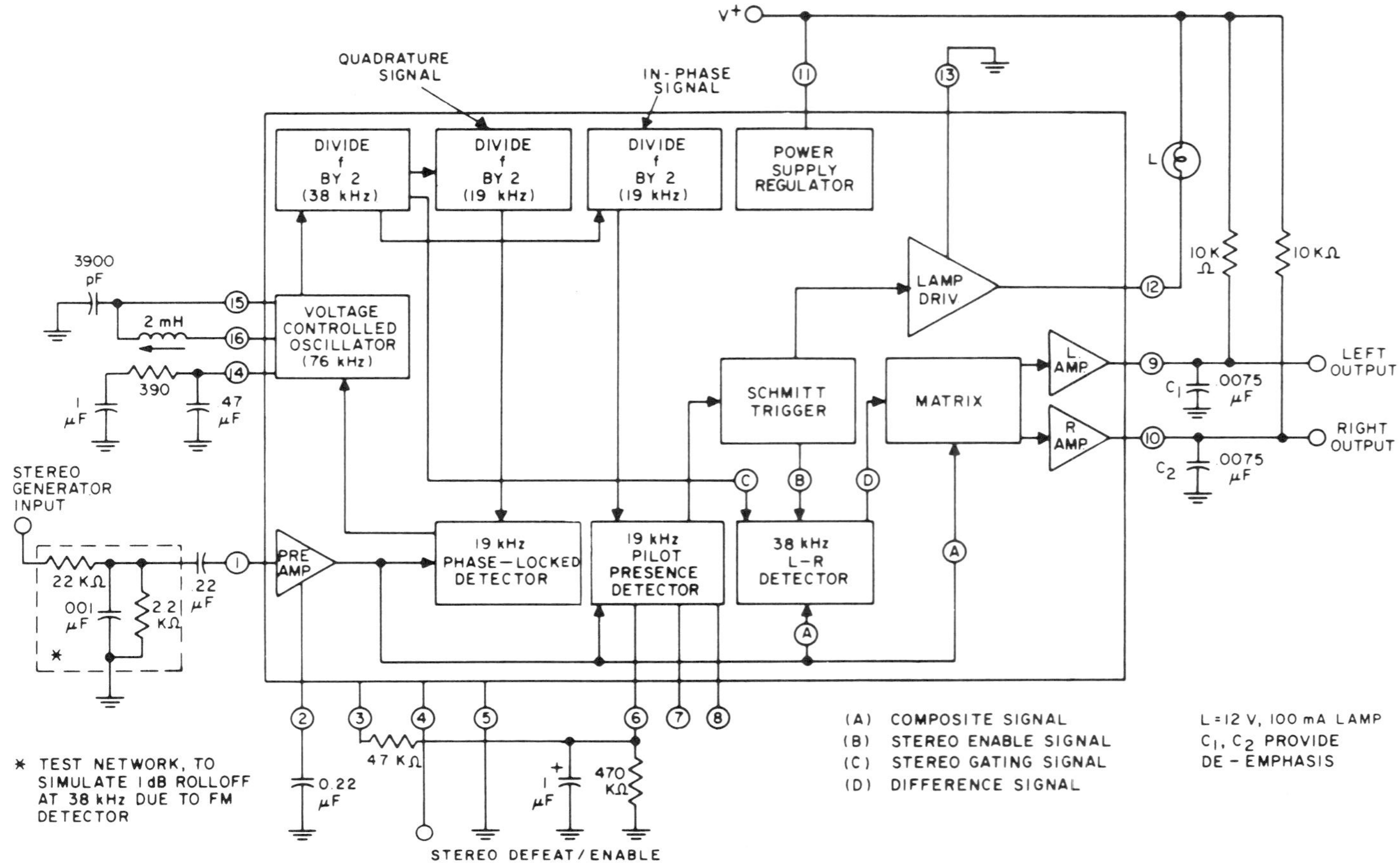

FIGURE 10-15. Block diagram of the RCA CA3090AQ system.

istic of radar receivers. All of these can be converted into terms that are related to other system characteristics, such as required gain, noise figure of individual amplifiers, and insertion loss of components.

System sensitivity is defined as the signal strength that a properly modulated signal must have to provide an acceptable output in a receiver system. It can be predicted as follows:

$$\text{sensitivity} = kTB + F_s + C/N \qquad (10\text{-}3)$$

where

sensitivity = system sensitivity (dBm)
kTB = thermal noise level (dBm)
F_s = system noise figure (decibels)
C/N = carrier-to-noise dB ratio required to produce the acceptable system output

We do not consider antenna gain since sensitivity is conventionally defined at the input to the receiver. Each of the terms in equation (10-3) can be related to specific performance criteria required of the components of a receiver system. Each term will be considered in detail.

We are already familiar with the *kTB* term from Chapter 5. This term represents the rms power level of thermal noise present in a receiver system. If we neglect the effects of the antenna background temperature, it is a function of the noise bandwidth and the system temperature. Note that this term will be the same for all types of modulation. At room temperature a widely used number for *kTB* is -114 dBm per megahertz of bandwidth. In terms of an equation this is

$$kTB = -114 \text{ dBm} + 10 \log \frac{B}{1 \text{ MHz}} \quad \text{dB} \qquad (10\text{-}4)$$

This equation can be verified as follows:

$$kTB = \left(1.38 \times 10^{-23} \frac{\text{W-s}}{°\text{K}}\right)(290°\text{K})\left(\frac{10^6}{\text{s}}\right)$$

$$= 4.0 \times 10^{-15} \text{ W}$$

$$kTB(\text{dBm}) = 10 \log_{10} \frac{kTB}{10^{-3} \text{ W}}$$

$$= 10 \log(4.0, 0 \times 10^{-12}) = -114 \text{ dBm}$$

EXAMPLE 10-4

Calculate kTB in dBm for a system with a 2-MHz bandwidth.

SOLUTION

$$kTB = -114 \text{ dBm} + 10 \log \frac{2}{1}$$

$$= -111 \text{ dBm}$$

Note that kTB is relatively insensitive to changes in temperature. In fact, it requires a 75°C increase in temperature to raise kTB by 1 dB.

The noise figure, F_s, is a measure of the amount of noise, above the basic thermal noise level, that a receiver system adds to incoming signals. It can be derived as follows:

$$F_s = L_1 + F_{\text{pa}} + D \qquad (10\text{-}5)$$

where

F_s = system noise figure (decibels)
L_1 = signal-strength losses (decibels) in circuitry ahead of the first amplifier in a system
F_{pa} = noise figure (decibels) of the first amplifier in the system (normally a preamplifier)
D = a factor (decibels) accounting for degradation of the system noise figure by all circuitry downstream from the first amplifier

Normally, the noise figure of the last predetection amplifier includes the effects of detéctors or discriminators and all postdetection circuitry.

10-4.1 Dependence of the Carrier-to-Noise Ratio of the Receiver

The terms kTB and *noise figure* are common to most receivers and to all definitions of sensitivity. But the carrier-to-noise ratio required to produce an ''acceptable'' output varies widely with both the receiver type and the definition of sensitivity. Here are some examples.

Tangential sensitivity is used with pulse receivers. It is defined as the signal level required to produce a tangential output (Figure 10-16) in which the hard noise on the bottom of the pulse just touches the hard noise on the baseline. At

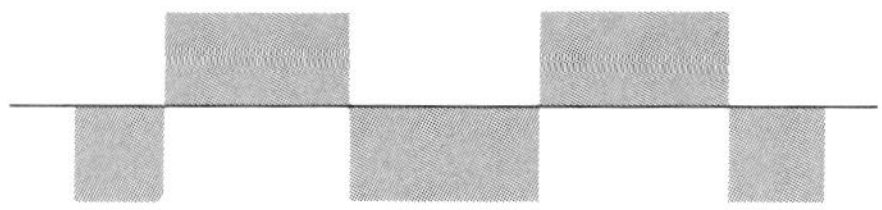

FIGURE 10-16. Tangential output of pulse receiver where hard noise at pulse bottom just touches the hard noise on the baseline.

this point the carrier-to-noise ratio is approximately 8 dB, but it can vary up to 1 dB, depending on scope brightness and pulse waveform.

Output signal-to-noise ratio is used with all types of receivers. The ratio is best determined by comparing the system output readings of a true rms voltmeter both with and without a signal.

In an AM signal, the required carrier-to-noise ratio is equal to the specified output signal-to-noise ratio. However, in FM or digital systems, this specification is more complicated. Some minimum carrier-to-noise ratio will be required for proper operation of the detector or discriminator, but a much higher output signal-to-noise ratio than this minimum C/N is obtained. The difference between the carrier-to-noise ratio and the output signal-to-noise ratio is accounted for by improvement factors.

A special case of the signal-to-noise ratio specification is that for which the ratio is 1 (0 dB), or signal = noise.

Bit error rate ($= 10^{-y}$) is used with digital data receivers. It is defined as the signal level required to produce a digital output in which not more than 1 bit in 10^y will be incorrect.

To convert a bit error rate to a required carrier-to-noise ratio, it is necessary to know the type of RF modulation that is carrying the digital data and the type of threshold used. A graph of carrier-to-noise ratio versus bit error rate is available for various types of modulation. A useful generality is that the bit error rate varies about one order of magnitude for each decibel of signal-to-noise ratio over the normally specified range. The false-alarm rate ($= 10^{-z}$), with probability of detection ($= p$), is used in pulse systems that drive automatic processing circuitry.

At a given carrier-to-noise ratio, the level at which the detection threshold is set establishes the trade-off between the false-alarm rate and the probability of detection. The relationship can be calculated. However, a graph showing the required carrier-to-noise ratio for each value of false-alarm rate and probability of detection is usually available. It will vary with the receiver architecture.

EXAMPLE 10-5

Calculate the system sensitivity for the FM receiver shown in the figure. The signal received has a modulation index (m_f) of 5 and an information bandwidth, f_m, of 20 kHz. Assume room temperature and an output $(S/N)_o = 40$ dB.

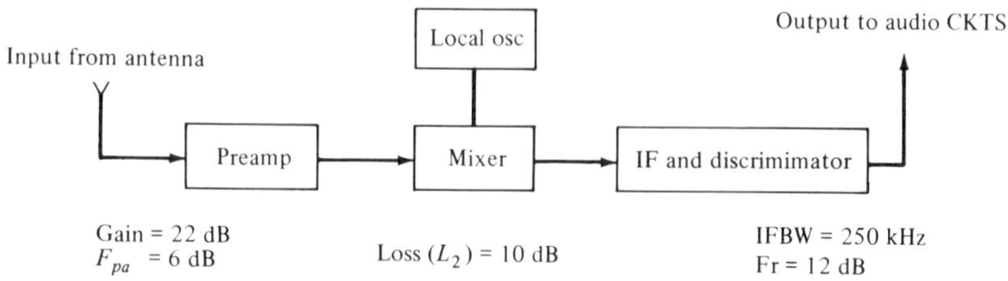

SOLUTION

$$kTB = -114 \text{ dBm} + 10 \log \frac{250 \text{ kHz}}{1 \text{ MHz}} \quad \text{dB}$$

$$= -114 \text{ dBm} - 6 = -119 \text{ dBm}$$

Calculating the noise figure, F_s: mixer loss + IF noise figure = 10 + 12 = 22 dB. The degradation factor, D, is calculated as follows:

$$D = \frac{(L_2 + F_r) - 1}{G_1} = \frac{22 - 1}{22} \approx 1 \text{ dB}$$

$$F_s = L_1 + F_{\text{pa}} + D$$

$$= 0 + 6 + 1 = 7 \text{ dB}$$

Since this is an FM receiver, the relationship between the C/N ratio and $(S/N)_o$ is affected by a SNR improvement factor [see equation (9-52)]. The C/N ratio must be at least 10 dB for proper discriminator operation. The required carrier-to-noise ratio is given by $C/N = (S/N)_o - (S/N)_{\text{improvement}}$ (all terms in decibels), where

$$(S/N)_{\text{improvement}} = 10 \log \left(\frac{3}{2} m_f^2 \frac{\text{IF}_{\text{BW}}}{f_m} \right)$$

$$= 10 \log \left[\frac{3}{2}(25) \left(\frac{250 \text{ kHz}}{20 \text{ kHz}} \right) \right]$$

$$= 27 \text{ dB}$$

Thus

$$C/N = (S/N)_o - 27 \text{ dB} = 40 - 27 = 13 \text{ dB}$$

$$\text{sensitivity} = kTB + F_s + C/N$$

$$= -119 + 7 + 13 = -99 \text{ dBm}$$

EXAMPLE 10-6

Calculate the system sensitivity for the pulse receiver shown. Assume room temperature.

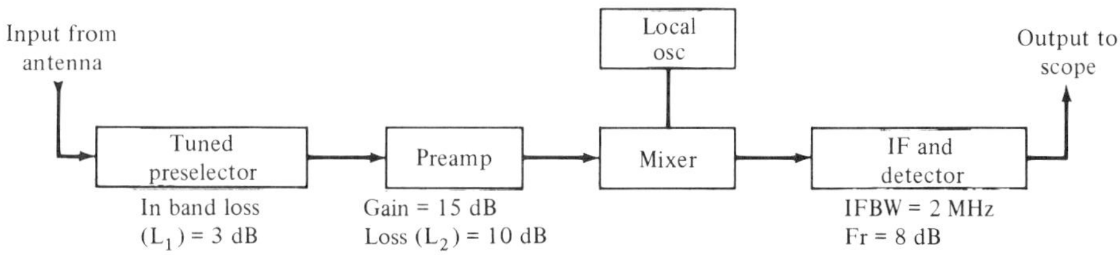

SOLUTION

$$kTB = 114 \text{ dBm} + 10 \log \frac{2 \text{ MHz}}{1 \text{ MHz}}$$

$$= -114 + 3 = -111 \text{ dBm}$$

Calculation of the noise figure, F_s:

$$D = \frac{(10 + 8) - 1}{15} = 1.1$$

Thus

$$F_s = L_1 + F_{pa} + D$$

$$= 3 + 5 + 1.1 + 9.1 \text{ dB}$$

$$C/N = 8 \text{ dB (from definition of tangential sensitivity)}$$

$$\text{Sensitivity} = kTB + F_s + C/N$$

$$= -111 + 9.1 + 8 = -93.9 \text{ dBm}$$

10-5

MEASUREMENT OF FM RECEIVER SENSITIVITY

In the preceding section we developed methods for calculating sensitivity. We wish now to examine techniques for performing measurements. Receiving systems are vulnerable to noise from two main sources: the antenna and the receiver itself.

Noise picked up by the antenna can be either natural or human-made. Natural noise, from cosmic sources, has a fairly constant level, whereas human-made noise, such as vehicle ignition, varies widely, depending on the closeness of the noise source. Receiver noise can be caused by anything from component to line fluctuations. Adding to the signal and noise coming from the antenna, receiver noise degrades the overall signal-to-noise ratio.

At the lower end of the VHF band, around 30 MHz, antenna noise may exceed receiver noise by many decibels. Thus if we have a sensitive antenna, receiver noise can often be neglected. In low-VHF mobile and portable applications, however, the size of the antenna is usually only a fraction of that needed for high efficiency. A low-efficiency antenna picks up less signal and less noise, so the receiver's performance becomes important. At the higher-VHF frequencies, the antenna noise falls off so that the limiting factor on system performance again becomes the receiver noise.

When measuring FM receiver sensitivity, there are three major methods to choose from. The one used depends partially on whether production tests are being made or new equipment is being designed. The three methods are:

- Quieting method.
- SINAD method.
- Noise figure.

The measurement of noise figure was discussed in Chapter 5. Thus our discussion will be limited primarily to the first two.

10-5.1 Quieting Method

The quieting method is simple, but cannot distinguish between IF and audio response. The SINAD method overcomes this problem at the expense of a more elaborate setup. The third technique—noise figure measurement—provides the most information about the front-end stages.

In the quieting method, an unmodulated RF signal is input to the receiver. This produces a specified amount of noise reduction at the output. This method provides a measure of the minimum RF level necessary for limiting action to remove the bulk of the noise in the receiver output. A test setup is shown in Figure 10-17. This test requires a continuous-wave (CW) RF generator with a

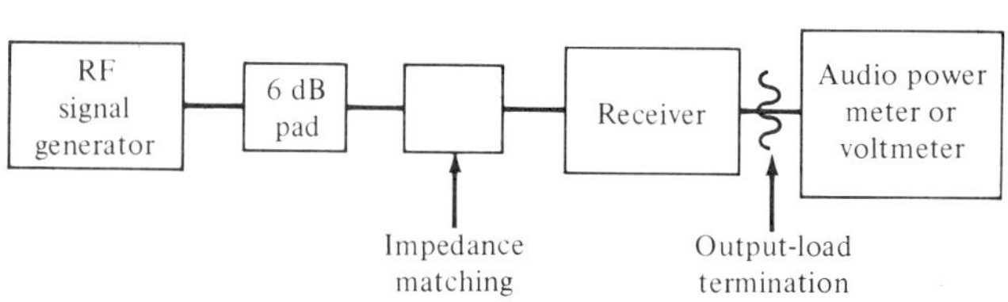

FIGURE 10-17. Receiver quieting measurement setup.

calibrated output control and an audio power meter or voltmeter. The measurement is performed by adjusting the receiver output for a reference audio noise output. Then the RF input level is increased in discrete sets and the corresponding drop in the receiver output noise level is noted.

The data are plotted in Figure 10-18 together with those with a SINAD measurement. Note that a point is eventually reached where further increases in input signal level produce a leveling off of the curve. The decibel difference between this level and the reference or no-signal level represents the maximum quieting of the receiver.

Quieting sensitivity is usually stated as the input necessary for 20- or 30-dB quieting. This specification, then, corresponds to one point on the quieting curve. Another receiver with the same IF bandpass, but different audio roll-off, will produce an entirely different quieting curve. Thus only identical receivers can be compared with the method. The method's main advantage lies in the fact that it is simple and that an FM generator is not needed.

10-5.2 SINAD Sensitivity Measurement

The term *SINAD* stands for "the ratio of measured signal + noise and distortion to noise and distortion," expressed in decibels. As such, this ratio provides a measure of audio output-signal quality at a given audio power level. Generally, the SINAD sensitivity is expressed as the level of frequency-modulated RF necessary to produce a 12-dB SINAD ratio at the output. A test setup is shown in Figure 10-19. An FM signal generator is required with an output control calibrated to 0.1 μV. Also needed is an audio-distortion analyzer with a tunable notch filter and capability for indicating the audio output level. The modulating signal is usually a 1-kHz tone and is adjusted for an FM deviation equal to the maximum expected of the receiver. A standard test deviation may be specified for the receiver.

To make the measurement, begin with the receiver "unsquelched." The "squelch" feature on a receiver disables the audio section on a receiver when

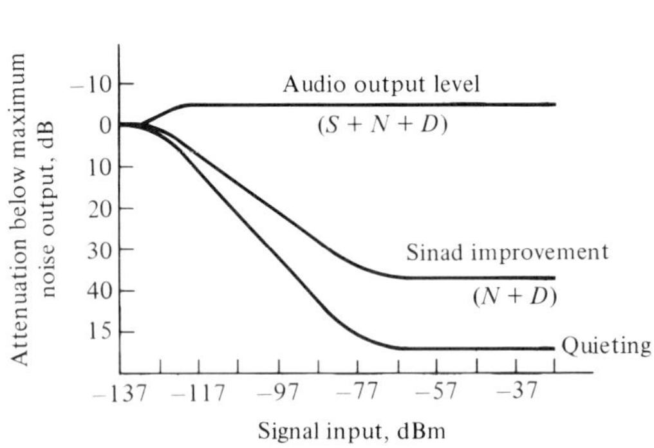

FIGURE 10-18. Typical SINAD and quieting curves.

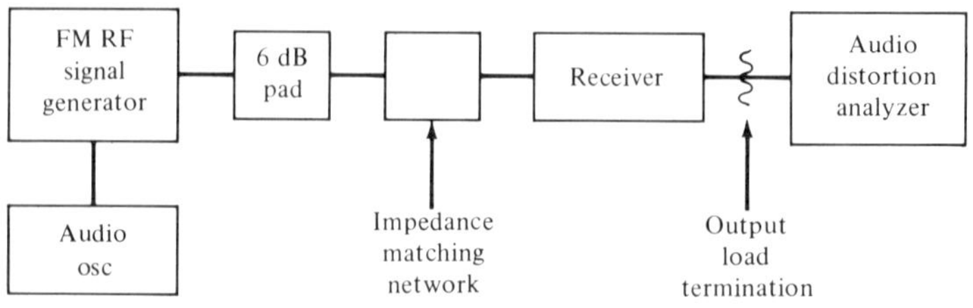

FIGURE 10-19. SINAD measurement setup.

there is no signal present. This eliminates the annoying hiss of amplified noise. Next, set the receiver gain so that the noise produces the rated audio output. Then increase the RF input in discrete steps of approximately 1 dB.

At each input level, switch the distortion analyzer to read first the total audio output + 1 kHz signal + noise + distortion. Next notch out the 1-kHz signal and read the remaining output noise + distortion. The dB difference between the two readings is the SINAD ratio for the RF input level. Continue the measurements until there are no further significant changes in the SINAD ratio. A typical performance curve is shown in Figure 10-18.

This method has the advantage of evaluating the entire receiver, including the audio section. Thus if the receiver produces harmonic distortion of the 1-kHz tone, the distortion products remain when the distortion analyzer rejects the 1-kHz fundamental signal.

10-5.3 Impedance Matching

For both the SINAD and quieting methods a proper impedance match must be maintained between the generator and the receiver. An attenuator pad should be used at the generator output. The design of attenuator pads was discussed in Chapter 2. The pad performs the function of smoothing out reflected impedance variations and also reduces the effect of generator output setting error caused by stray signal leakage at low output levels. Six decibels of attenuation is recommended. Typical pad designs are shown in Figure 10-20.

Input matching becomes more tricky with hand-held receivers or transceivers.

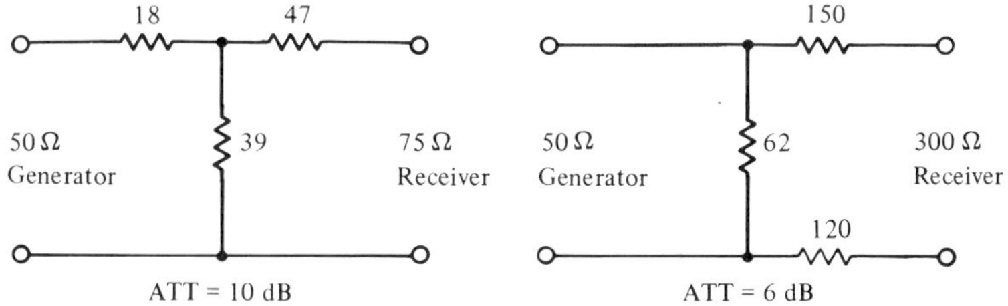

FIGURE 10-20. Impedance matching pad.

In this case the antenna—mounted directly on the set—is the only input connection. To obtain maximum power transfer, we generally make the receiver input impedance equal to the complex-conjugate impedance of the antenna. A dummy antenna—consisting of series *C* or *L* components—must be used to simulate the antenna.

The antenna length usually equals a quarter wavelength or less. If less, the antenna impedance has a small, series-capacitive reactance component in addition to its resistive component: radiation resistance plus ground losses. In the rarer case of an antenna that is longer than a quarter-wavelength, the reactance is inductive. In either case make the *C* or *L* variable and tune for maximum signal transfer.

Hopefully, the manufacturer's data sheet gives the impedance seen looking into the receiver. If it does not, the impedance can be found by trial and error (Figure 10-21). The attenuator pad creates a signal source with an impedance that is close to zero. Set the generator for enough output to give about 10-dB receiver quieting, with no series resistor and with the capacitor, or inductor, tuned for maximum quieting. Then insert trial values of series resistance. Continue until a value is found that requires doubling the generator output to obtain the same receiver quieting.

The value found equals the resistive component of the receiver input impedance. The resistor, together with the *C* or *L* component, forms the dummy antenna. Knowing the receiver input resistance, a matching pad can be designed

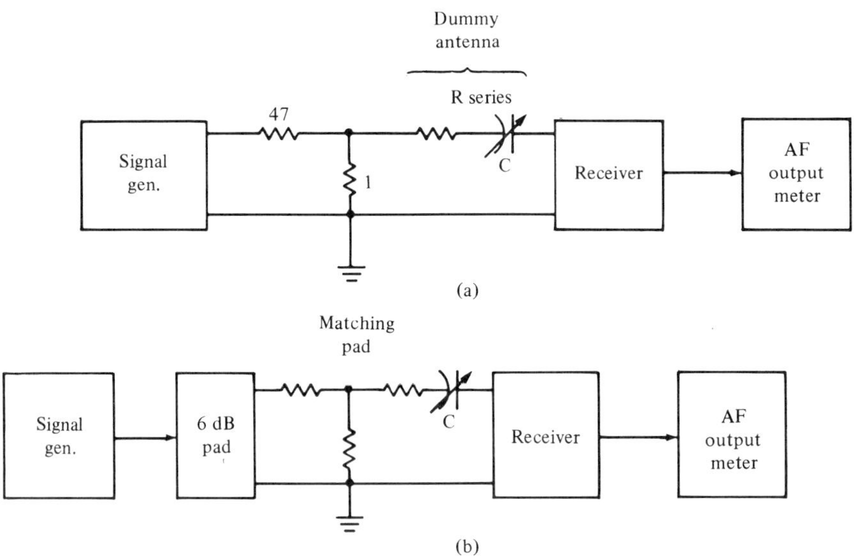

FIGURE 10-21. An unknown receiver input resistance can be found by measurement (a). Value can then be used to build a matching pad that ensures maximum power transer between generator and receiver (b).

(Figure 10-21) to match the generator to the receiver. Remember to calculate and include the decibel loss of the pad when making the sensitivity measurements.

10-6

NOISE FIGURE MEASUREMENT

Noise figure measurement was discussed in Chapter 5. It is one of the most useful to the receiver designer, particularly for the front-end stages. It can be related directly to the active and passive devices that the designer uses. It is not as useful, however, as the SINAD method for production or system testing.

Since a noise generator outputs a wide spectrum of noise, it is necessary to consider the image rejection of the receiver under test. If the image frequency rejection of the receiver is lacking or insufficient, an image frequency trap must be placed at the receiver input.

It is often desirable to be able to calculate the receiver sensitivity if the noise figure is known. From equation (5-34) the noise figure is given by

$$F = \frac{N_o}{N_i G_0} = \frac{N_o}{kT_0 B G_0} \qquad (10\text{-}6)$$

where

N_o = output noise power (watts)
k = Boltzmann's constant
T_0 = standard noise temperature − absolute room temperature, in this case
B = effective noise bandwidth (hertz)
G_0 = receiver power gain

From equation (10-6) we can rewrite the noise figure as

$$F = \frac{N_o}{S_o} \frac{S_i}{kT_0 B} \qquad (10\text{-}7)$$

EXAMPLE 10-7

A receiver has a 10-dB noise figure, a 6-kHz noise bandwidth, and a 50-Ω input impedance. What is the RF input level required for a 10-dB $(S + N)/N$ ratio?

SOLUTION

$$\frac{S_o + N_o}{N_o} \approx \frac{S_o}{N_o} \text{ for a 10-dB ratio}$$

From equation (10-7),

$$S_i = FBkT_0(S_o/N_o) = (10)(6 \times 10^3)(1.38 \times 10^{-23})(290)(10)$$

$$= 2.4 \times 10^{-15} \times 10^{-15} \text{ W}$$

$$= -116 \text{ dBm}$$

For a 50-Ω input the final result is $s_i = 0.35\ \mu V$.

Receiver sensitivity is usually specified in terms of $(S + N)/N$. For values greater than 10 dB, the two ratios are almost equal and can be interchanged with small error. For values less than 10 dB, the graph in Figure 10-22 may be used to convert from one to the other. The conversion of dBm to microvolts for a given receiver input impedance may be calculated or alternatively, the graph in Figure 10-23 may be used.

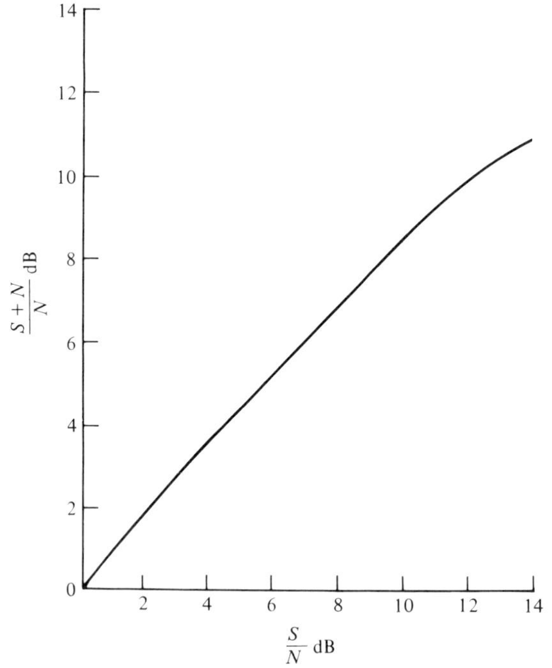

FIGURE 10-22. (S + N)/N to equivalent S/N ratio can be converted with this graph. Above about 10dB, little error results if one is assumed equal to the other.

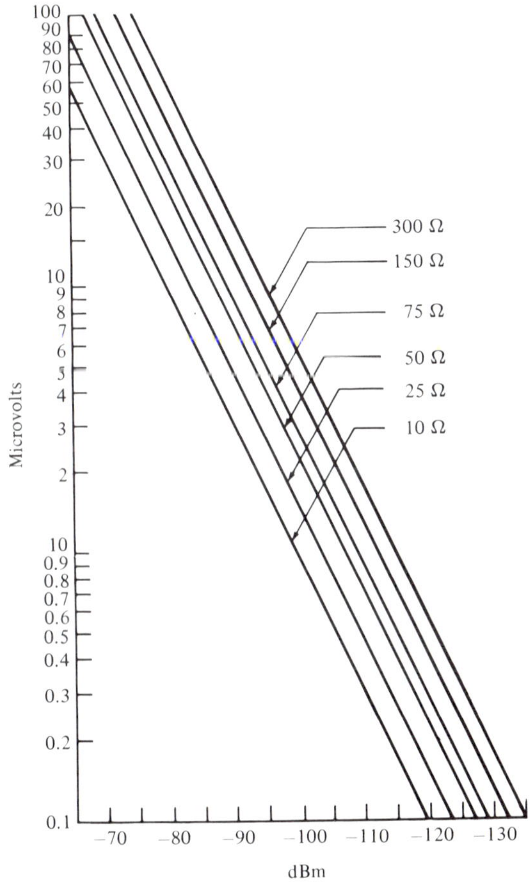

FIGURE 10-23. Conversion of dBm to μV for a given receiver impedance can be performed with the graph.

PROBLEMS

10-1. A superheterodyne receiver is tunable over the range 10 to 15 MHz with an IF of 1.5 MHz. Calculate the image frequency range for the receiver.

10-2. For Problem 10-1, the receiver is tuned to a station with frequency 6 MHz. What should the Q of the RF filter be in order that the image frequency be attenuated by a factor of 20 dB?

10-3. Calculate the system sensitivity for the FM receiver of Example 10-5 if the preamplifier gain is 12 dB and the noise figure for the preamplifier is 9.5 dB.

10-4. Design a T-impedance matching pad if the receiver has an input impedance of 100 Ω and the generator has a 600-Ω output impedance. Calculate the attenuation of the pad.

10-5. A receiver has an 8.2-dB noise figure and a 20-kHz noise bandwidth. The input impedance is 75 Ω. What is the required RF signal level in dBm to produce a 7-dB $(S + N)/N$ ratio? What is the input signal voltage?

10-6. A receiver has an input RF power of −120 dBm. What is the input RF signal voltage if the input impedance is (a) 10 Ω, (b) 50 Ω, and (c) 150 Ω?

10-7. A receiver has an input signal-to-noise ratio of 4 dB. What is the corresponding $(S + N)/N$?

10-8. In each of the following cases, what is the frequency of the image signal? (a) RF at 1200 kHz with an IF of 455 kHz, (b) RF at 650 kHz with an IF of 262 kHz, (c) RF at 106 MHz with an IF of 10.7 MHz, and (d) RF at 3000 MHz with an IF of 60 MHz.

10-9. A broadcast TRF receiver has a bandwidth of 10 kHz at 600 kHz. (a) Assuming that the Q of its tank circuits remain essentially constant, what is the bandwidth of the receiver at 1600 kHz? (b) If the design were altered to produce a bandwidth of 10 kHz at 1600 kHz, what would be the bandwidth at 540 kHz?

10-10. The front end of an AM receiver has a Q of 100 at a received signal of 1000 kHz. The IF frequency is 455 KHz. Calculate the image frequency and its suppression in dB.

CHAPTER 11

Pulse Modulation

The modulation systems considered thus far all used a sinusoidal carrier. Either the amplitude, the frequency, or the phase of the carrier was modulated to impress the modulation on the carrier. This process resulted in a translation of the message spectrum to the carrier frequency. In this chapter we take note of the fact that sinusoidal carriers are not the only possibilities. Rectangular pulses may also be modulated to serve as carriers. A characteristic of pulse modulation is that the information remains at baseband. It is not translated to a higher-frequency carrier.

There are several advantages that accompany the use of pulse modulation. One in particular lies in the fact that if we combine pulse modulation with continuous modulation (AM, FM, or PM), we can realize a multichannel communication system—a desirable feature indeed for data transmission. The transmission of more than one channel of information on one carrier is referred to as *multiplexing*.

In general, we may subdivide pulse modulation into two categories: quantized and unquantized. Considering either category, there are three parameters of a pulse that can be varied: the length, the height, or the position.

In quantized pulse modulation a series or group of rectangular pulses contains the information. For example, a sample of an analog voltage may be digitized or converted to a digital word. This digital word may consist of straight binary or some other digital code. Communication systems using this technique are referred to as *pulse code modulation* (PCM) systems. The advantages of this for a computer are obvious.

In the unquantized category, we may choose to have the width of the pulse vary in proportion to the message. This is referred to as *pulse width modulation* (PWM). With either category, if transmission through a wireless channel is desired, a higher-frequency carrier must be introduced. In the following we consider pulse modulation in terms of the bandwidth, modulation/demodulation schemes, and techniques for producing pulse modulation.

11-1 SAMPLING THEOREM

Analog signals are often sampled and some form of pulse modulation is used to transmit the information. Nyquist has shown that in order that the original signal may be reconstructed from the samples, the signal must be sampled at least twice per highest frequency in the signal. Refer to Figure 11-1. It is obvious

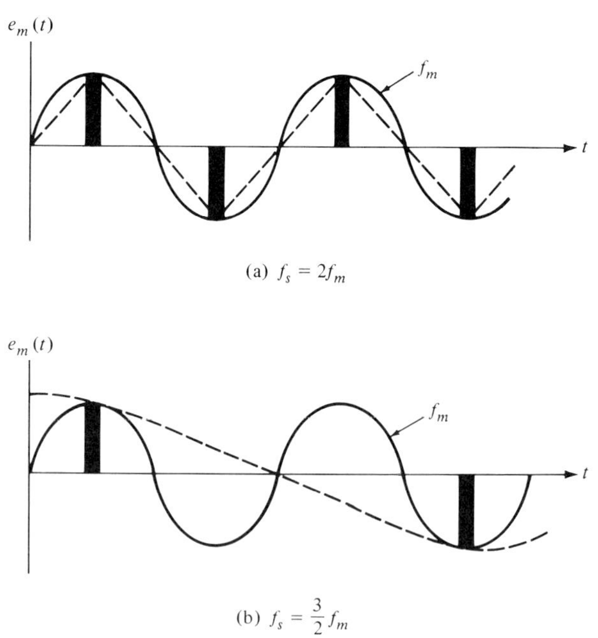

FIGURE 11-1. Illustration of sampling theorem.

in Figure 11-1a that if the maximum values of the samples are connected, we could reasonably expect that with some smoothing the original signal could be reconstructed. In Figure 11-1b, where undersampling is employed, a sinusoidal wave of lower frequency is recovered. We say that the frequency has been "folded" over or that "aliasing" of the signal is occurring as a result of sampling at a rate less than the Nyquist rate. Thus the sampling rate is a primary system consideration where a variable is being sampled.

11-2

PULSE AMPLITUDE MODULATION

A pulse amplitude-modulated system is illustrated in Figure 11-2. Note that a product modulator is used to produce the PAM output. That is, the output is the product of the information signal, $f(t)$, and the switching function, $f_s(t)$:

$$\begin{aligned} e_{\text{PAM}}(t) &= f(t)f_s(t) \\ &= E \cos \omega_m t \; f_s(t) \end{aligned} \tag{11-1}$$

The Fourier series for the switching function, $f_s(t)$, is given in Table 4-1. Making the substitution into equation (11-1) yields

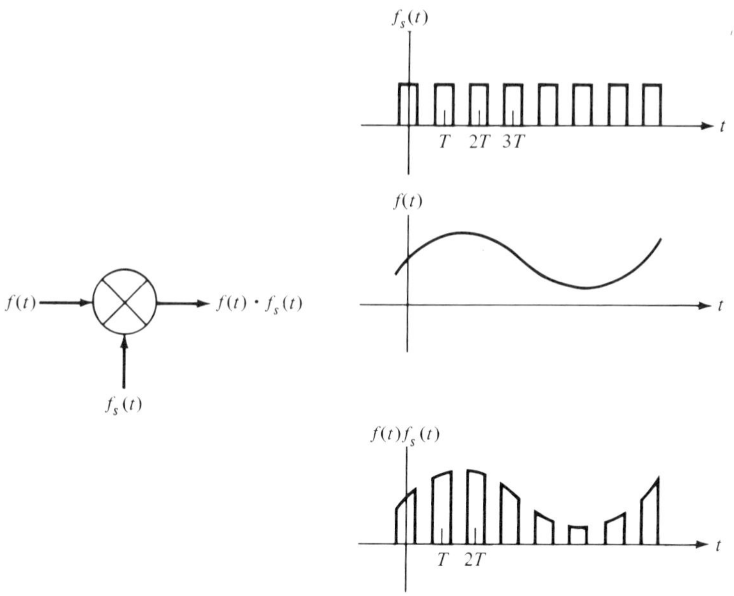

FIGURE 11-2. Pulse-amplitude-modulated system.

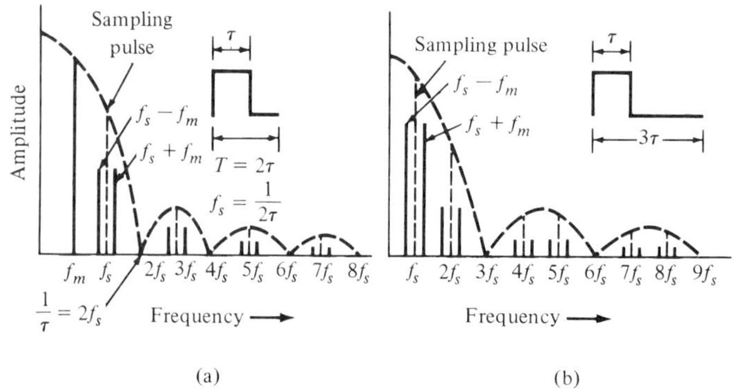

FIGURE 11-3. Frequency spectra of a PAM waveform.

$$
\begin{aligned}
e_{\text{PAM}}(t) &= E\cos\omega_m t\left[\frac{V\tau}{T} + \sum_{n=1}^{\infty}\frac{2V\tau}{T}\frac{\sin(n\pi\tau/T)}{n\pi\tau/T}\cos n\omega_s t\right] \\
&= \frac{VE\tau}{T}\cos\omega_m t + \frac{VE\tau}{T}\sum_{n=1}^{\infty}\frac{\sin n\pi f_s\tau}{n\pi f_s\tau}\cos(n\omega_s + \omega_m)t \qquad (11\text{-}2) \\
&\quad + \frac{VE\tau}{T}\sum_{n=1}^{\infty}\frac{\sin n\pi f_s\tau}{n\pi f_s\tau}\cos(n\omega_s - \omega_m)t
\end{aligned}
$$

where $\omega_s = 2\pi f_s = 2\pi/T$. Frequency spectra for the PAM waveform are plotted in Figure 11-3 for several pulse widths. Note that the time between the samples, that is, the sample rate, controls the fundamental frequency and the spacing of the harmonics (sidebands). The first harmonic to have zero amplitude (zero crossing) is equal to $1/\tau$. The amplitude of the harmonics drops off rapidly past this point. We see also that the information signal frequency, f_m, is contained in the spectrum. This suggests a method of recovering the PAM signal (Figure 11-4). The foregoing discussion does not provide justification for using pulse amplitude modulation. Its usefulness, however, lies in signal multiplexing. Refer to Figure 11-5. A series of signals are sampled and time multiplexed before transmission over a single channel.

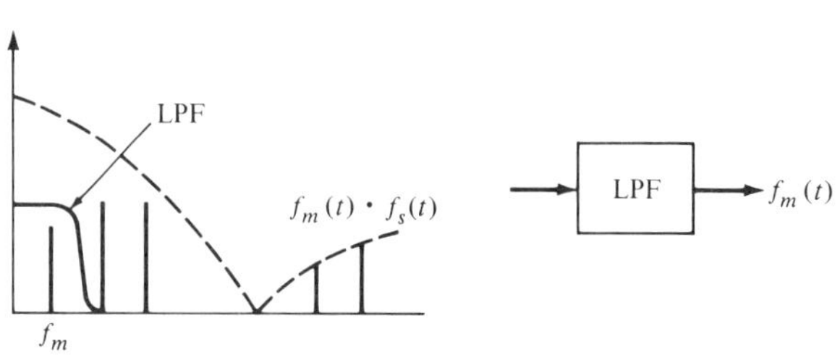

FIGURE 11-4. Recovery of the signal from the PAM waveform.

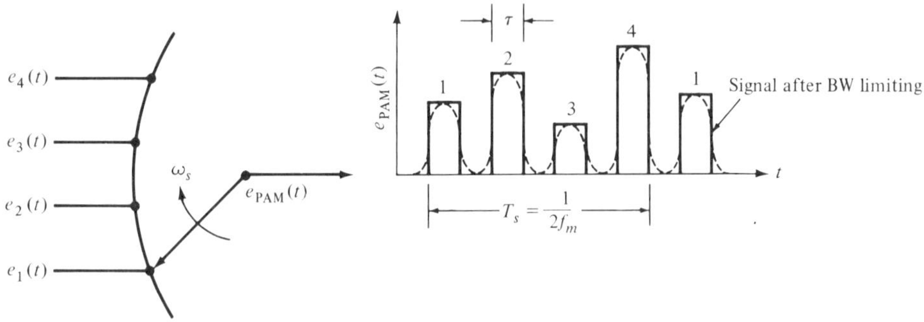

FIGURE 11-5. PAM time multiplexing.

The composite PAM signal will be processed over bandlimited circuits. Due to integration effects, this results in pulse spreading. The "tails" of one pulse will "interfere" with a pulse in the adjacent time slot. This is referred to as *intersymbol interference.* Intersymbol interference may be reduced by widening the transmission bandwidth or by reducing the amount of information contained in a given time interval. This amounts to introducing a "dead time" between samples in the multiplexed PAM wave train. Refer to Figure 11-5. We see that the composite wave train has a fundamental frequency of $1/2f_m$. Within the sample period, T_s, there are variations of one-fourth T_s. Therefore, the bandwidth of the PAM wave train is

$$\text{BW}_{\text{PAM}} \approx 2f_m n \qquad \text{hertz} \tag{11-3}$$

where n is the number of signals sampled. If the dead time between pulses is equal to the pulse width, τ, the sample time is given by

$$\begin{aligned} \tau &= \frac{1}{2}\frac{1}{n}\frac{1}{2f_m} \\ &= \frac{1}{4n}\frac{1}{f_m} \qquad \text{seconds} \end{aligned} \tag{11-4}$$

EXAMPLE 11-1

If $n = 6$ and the highest frequency to be sampled is 6 kHz, determine (a) bandwidth of the PAM wave train and (b) the pulse sample time, τ.

SOLUTION

(a) $\text{BW} \approx 2f_m n$

$= (2)(6000)(6)$

$= 72$ kHz

(b) The sample time, τ, is

$$\tau = \frac{1}{4n} \frac{1}{f_m}$$

$$= \frac{1}{(4)(6)} \frac{1}{6000} = 6.97 \qquad \text{seconds}$$

Note that as more channels of information are added to be sampled in the time T_s, the pulse width, τ, becomes smaller and the bandwidth increases. A scheme for time-division multiplexing and demultiplexing 10 signals is shown in Figure 11-6. To accomplish demultiplexing, the receive switch must rotate in synchronization with the transmit switch. A common method of synchronization is to sacrifice one of the channels and send a known voltage level in its place.

11-3

PULSE DURATION (WIDTH) MODULATION

The sequence of events leading to a pulse width modulation (PWM) wave is illustrated in Figure 11-7. Note that in Figure 11-7c the width of each pulse has been varied in accordance with the instantaneous sample value of $f(t)$. The larger the sample value, the wider the pulse. We see that the primary difference when compared to PAM is that the information has been converted to "time amplitudes."

In a PWM system, either the trailing edge, the leading edge, or both may be altered. A PWM signal is illustrated in Figure 11-8. Mathematically, the pulse

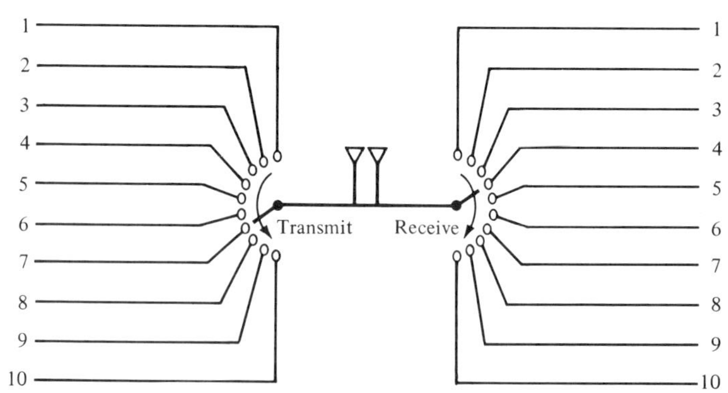

FIGURE 11-6. Time-division multiplexing and demultiplexing of 10 signals.

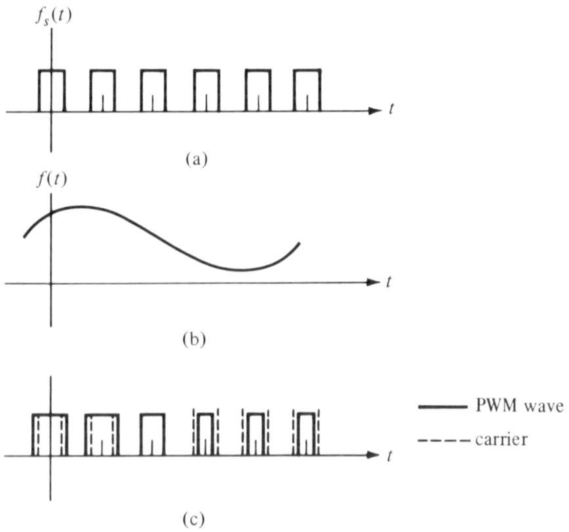

FIGURE 11-7. Pulse width modulation.

width, $\tau(t)$, is described by

$$\tau(t) = \frac{T}{2m} + \frac{KT}{2m} f(t) \tag{11-5}$$

where $T = 1/f_s$ is the sample time and m is the number of signals sampled. K is a system constant. Only one information channel is shown in Figure 11-8 and equation (11-5) for a single channel reduces to

$$\tau(t) = \frac{T}{2} + \frac{KT}{2} f(t) \tag{11-6}$$

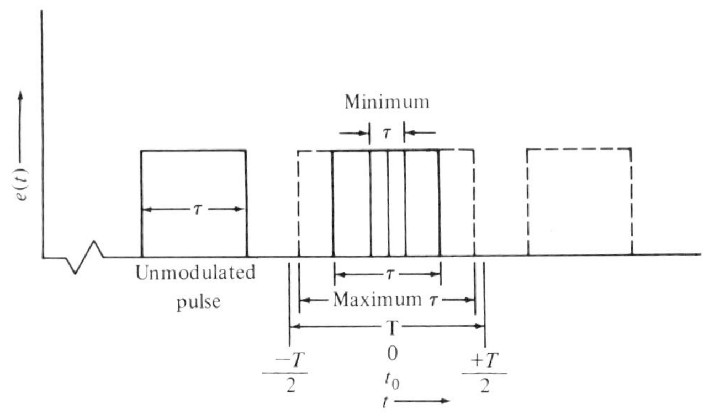

FIGURE 11-8. Modulated pulse.

To find the frequency spectrum of the modulated signal we begin again with the Fourier series for the function $f_s(t)$ (Table 4-1).

$$f_s(t) = \frac{V\tau}{T} + \sum_{n=1}^{\infty} \frac{2V\tau}{T} \frac{\sin(n\pi\tau/T)}{n\pi(\tau/T)} \cos n2\pi \frac{t}{T} \tag{11-7}$$

We substitute equation (11-6) into (11-7) to obtain an expression for the pulse-width-modulated wave. Thus

$$\begin{aligned} e_{\text{PWM}}(t) = \frac{V}{T}\left[\frac{T}{2} + \frac{KT}{2} f(t)\right] + 2\sum_{n=1}^{\infty} \frac{V\left[\frac{T}{2} + \frac{KT}{2} f(t)\right]}{T} \\ \times \sin \frac{n\pi f_s\left[\frac{T}{2} + \frac{KT}{2} f(t)\right]}{n\pi f_s\left[\frac{T}{2} + \frac{KT}{2} f(t)\right]} \cos n\, 2f_s t \end{aligned} \tag{11-8}$$

This expression simplifies to

$$\begin{aligned} e_{\text{PWM}}(t) = V\left[\frac{1}{2} + \frac{Kf(t)}{2}\right] \\ + 2\sum_{n=1}^{\infty} \frac{V}{n\pi} \sin n\pi f_s \left[\frac{T}{2} + \frac{KT}{2} f(t)\right] \cos 2\pi f_s t \end{aligned} \tag{11-9}$$

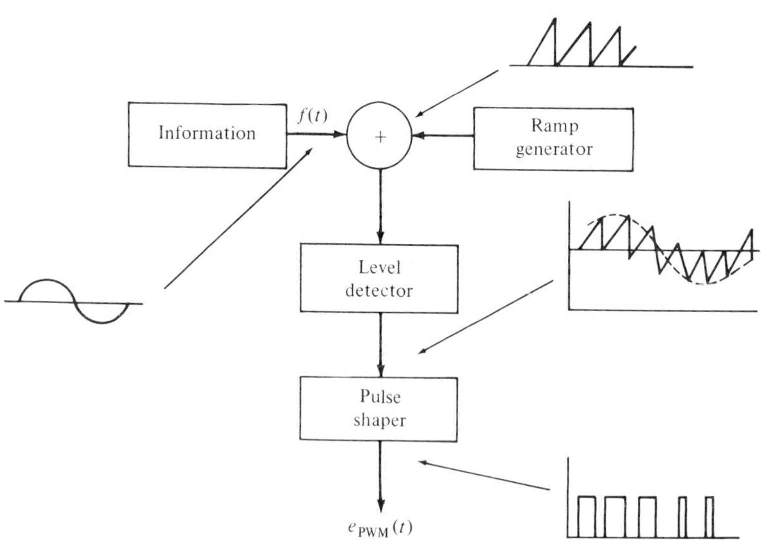

FIGURE 11-9. PWM modulator.

Examining this expression, we see that the first term is the average value of the pulse. The second term is the frequency spectrum of the original information signal, attenuated by a factor of $K/2$. We see immediately that the information, $f(t)$, can be recovered by a low-pass filter, as was the case for PAM. We recognize the third term as having the form of a balanced modulator. Thus the spectra of this term will be sidebands clustered around the harmonics of the sampling frequency. Again, just as in the case of PAM, we must ensure that the sampling frequency is high enough so that this spectra does not overlap the spectra of the information signal, $f(t)$. If this occurs, the information signal cannot be recovered.

One method for generating PWM is shown in Figure 11-9. A sawtooth waveform whose period is equal to the largest pulse width to be produced is added to the information signal. A level detector in conjunction with a pulse shaper is utilized to produce the PWM output.

11-4 PULSE POSITION MODULATION (PPM)

In pulse position modulation (PPM) the position of the pulse varies in accordance with the amplitude of the sampled waveform. The pulse width, τ, is constant. Pulse position modulation may be derived from PWM through the process of differentiating. Refer to Figure 11-10. To produce PPM from the differentiated waveform, we need only rectify the waveform in Figure 11-10b and trigger a device to produce a constant pulse width, τ (Figure 11-11). Note that if we

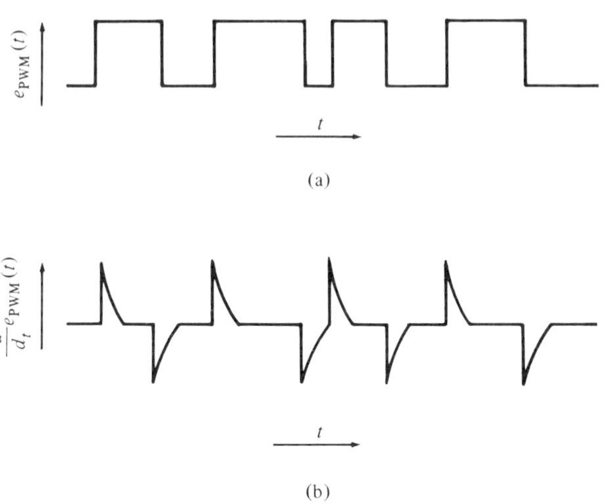

FIGURE 11-10. (a) PWM wave train; (b) differentiated PWM wave train.

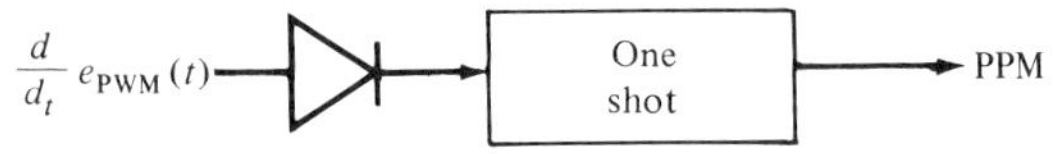

FIGURE 11-11. PPM derived from PWM.

differentiate equation (11-8) the resulting waveform still contains the information frequencies, that is, $d/dT[(KT/2)f(t)]$. Thus the information can be recovered with a low-pass filter. The bandwidth for PPM or PWM is given approximately by

$$\text{BW} \approx \frac{1}{2}\frac{1}{t_r} \tag{11-10}$$

where t_r is the rise time.

11-5 PULSE CODE MODULATION

The preceding discussion centered around characteristics of information that was unquantized. Pulse code modulation (PCM) is the remaining category referred to as quantized. The information signal is separated into discrete packets of pulses (i.e., digital words). This inherently offers an advantage over the preceding pulse techniques, where the information is contained in the pulse amplitude, or in the timing. It is evident that circuit functions such as limited bandwidth, nonlinearities, and so on, may cause these properties to deteriorate.

To implement a PCM system it is necessary to define a code in order to determine how the information packets are to be coded. As before, the information signal is sampled and the sample is replaced by its equivalent digital code. If the original signal is continuous, the process of replacing the continuous signal by discrete signals is called *quantizing*. If the original signal is already discrete, such as numbers, letters, and so on, the replacement of this set of symbols by another set of symbols is called *encoding*.

One of the simplest codes to use with PCM is straight binary. For example, suppose that we have a 5-V video signal which we wish to represent with a 3-bit binary code. Eight combinations (digital words) are possible with 3 binary bits; thus we can quantize the video into seven discrete sample values. That is,

$$\Delta V = \frac{V_{\text{max}}}{2^n - 1} = \frac{5}{2^3 - 1} = 0.7043 \text{ volts}$$

TABLE 11-1 Amplitude Message Codes

Message	Code
0.0	000
0.7043	001
1.4086	010
2.1129	011
2.8172	100
3.5215	101
4.2258	110
4.9301	111

where ΔV is the difference in voltage represented by adjacent digital words. Refer to Table 11-1 for the code corresponding to the amplitude message.

The message, when transmitted, is simply a sequence of 1's and 0's. Detection of the PCM requires a knowledge of the code and the synchronization. In a typical PCM system a group of N code words are transmitted in sequence and are referred to as a *frame*. At the beginning of a frame a "sync" word is transmitted. This digital word consists of a pattern of 1's and 0's which are unlikely to occur in the code group that represents the possible messages. Special circuitry in the receiver detects this code to indicate the start of a frame. Note that the receiver need only decide if a pulse was sent or not. No information is contained in the amplitude, the position, or the duration of the pulse. Distortion of these parameters does not affect the quality of the detected signal. A wrong decision by the receiver will introduce an error. Coding schemes are available for minimizing the effect of a receiver error. The transmission of information by PCM is generally referred to as *digital communications*. This subject is treated in detail in Chapter 12.

PROBLEMS

11-1. In a PAM multiplex system, the highest frequency of the modulating signal is 2.2 kHz. Twenty signals are to be sampled. Determine the optimum sampling time and the sampling frequency. What is the minimum bandwidth for this system?

11-2. Show that if the product modulator in Figure 11-2 is followed by a filter, the output is

$$e_o = K \cos \omega_m t \cos \theta$$

if the input is

$$e_i(t) = E \cos (\omega_s + \omega_m)t + E \cos (\omega_s - \omega_m)t$$

where the switching function $f_s(t)$ has phase error, θ.

11-3. A communication channel is designed to have a bandwidth of 50 kHz. A 200-kHz signal is to be transmitted using PAM. How long would it take to transmit 10 s of the signal?

11-4. Three information signals are to be sent using PAM and time-division multiplexing (TDM). The maximum frequencies of the signals are 30, 10, and 15 kHz. What bandwidth is required for the composite PAM-TDM signal if the three signals are always sampled sequentially? Can a scheme be devised where each is sampled only at its Nyquist rate (two samples per highest frequency)?

11-5. Explain why PPM requires some form of synchronization, whereas PWM and PAM do not.

11-6. Thirty-one shades of gray are desired for a 5-V video signal which is to be transmitted via PCM. How many bits are contained in each PCM word? How many volts does the least significant bit of the PCM word represent?

11-7. What is the sampling rate for the following signals? (a) video at 6 MHz, (b) voice at 5 kHz, and (c) high-fidelity music at 20 kHz.

11-8. Sketch a channel-interleaving scheme for time-division multiplexing the following channels: five 4-kHz telephone channels and one 20-kHz music channel.

11-9. A single, digitized voice channel requires an ideal bandwidth of 32 kHz. What bandwidth is required if 24 such channels are time-division multiplexed?

PART IV

COMMUNICATION SYSTEMS

CHAPTER 12

Digital Communications with Digital Signal Processing

Digital communications using pulse code modulation (PCM) made its debut during World War II. In the subsequent decade it matured and today enjoys supremacy in the digital communications field. The inherent advantages of PCM are quality and security of transmission, system flexibility, and the economy that results from a large amount of common equipment. The advent of cheap integrated circuits resulted in a "marriage" between digital communications and digital signal processing with far-reaching effects.

A code group ($a_0 \cdot \cdot a_n$) of binary digits ($a_i = 1$ or 0) can be used to represent the number

$$N = a_0 + a_1 \cdot 2 + a_2 \cdot 2^2 + \cdot \cdot \cdot + a_n{}^{2n} \tag{12-1}$$

We use this number in PCM to represent the value of a signal voltage at a given instant of time. An instant of time later corresponding to the sample time, a second code group is used to represent this latest sample, and so on. The continuous train of code groups represents the amplitudes of successive samples of the original signal. In a communications setting, these code groups are transmitted as a digital pulse train and at the destination are either processed digitally or are decoded from a number into a voltage proportional to the number.

12-1 FUNDAMENTAL CONCEPTS

Figure 12-1 illustrates a continuous signal sampled at the instants shown and coded into a binary number at the right. This binary (digital) number represents the amplitude of the signal to the nearest half-quantum. Note that if the signal falls anywhere in a given amplitude "zone" it must be coded into the binary number assigned to that zone. If the code groups are transmitted sequentially and then, at the destination, decoded into a voltage proportional to the code, the result is a received signal represented by the dotted "staircase" signal. If, now, the staircase signal is low-pass filtered, a "smoothed" version of the original signal is produced.

The decoded signal can be made a better approximation of the original by decreasing the interval between samples and also by increasing the number of amplitude zones (digits per code). If the sample interval is made small enough, no degradation at all results in the received signal. This is achieved at the expense of an increased number of digits per interval of time that results in an increase in bandwidth. Representation of the signal by the staircase approximation results in the addition of "quantizing" noise to the signal.

12-1.1 Sampling and Reconstitution

The sampling theorem was discussed in Chapter 11. We observed in this discussion that at least two samples per highest frequency must be obtained to

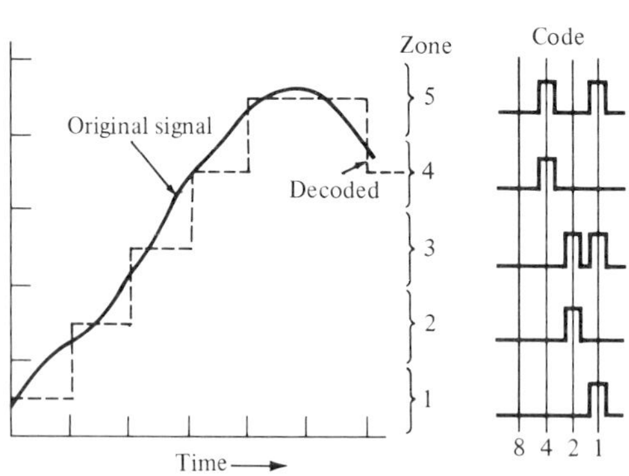

FIGURE 12-1. Pulse code modulation.

prevent "aliasing" or "foldover" of the frequency spectrum. If the original signal, $F(j\omega)$, is bandlimited so as not to contain frequencies greater than ω_m, the spectrum $F*(j\omega)$ of the sampled signal is as shown in Figure 12-2. We see that a sampled signal contains all of the original frequencies and in addition contains modulation products wherein the original signal appears as sidebands about the sampling frequency and all harmonics of the sampling frequency. The spectrum of the sampled signal actually possesses a sin x/x envelope and thus the energy falls off at the higher harmonics.

Note from Figure 12-2 that if $\omega_s > 2\omega_m$, a "guard space" exists between the original signal component and the lower sideband of the sampling frequency, ω_s. If the guard space is preserved, thc original signal may be recovered by a low-pass filter. Thus, to the extent that a "perfect" filter can built, the signal can be recovered with no degradation.

12-1.2 Quantizing Noise

Quantizing noise results from the representation of a continuous signal by a finite number of codes. The extent of this noise may be approximated by examining a typical transfer characteristic for a PCM coder–decoder link (Figure 12-3). Figure 12-3b breaks the characteristics into linear and error components. The amplitude difference between the sampled value and the quantized level is called the *quantization error*. For an n-digit code there are 2^n levels. Then the peak error magnitude is $1/2^n$.

The rms error can be computed for the error curve of Figure 12-3b as

$$\text{rms error} = \left[\frac{1}{\tau}\int_{-\tau/2}^{\tau/2}\left(\frac{1}{2^n}\frac{2}{\tau}t\right)dt\right]^{1/2} = \frac{1}{\sqrt{12}}\frac{1}{2^{n-1}} \qquad (12\text{-}2)$$

That is, the rms error voltage is equal to $1/\sqrt{12}$ times the step size.

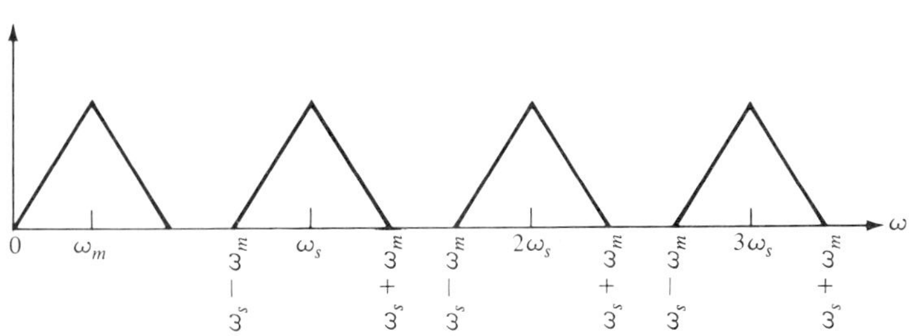

FIGURE 12-2. Spectrum of sampled signal.

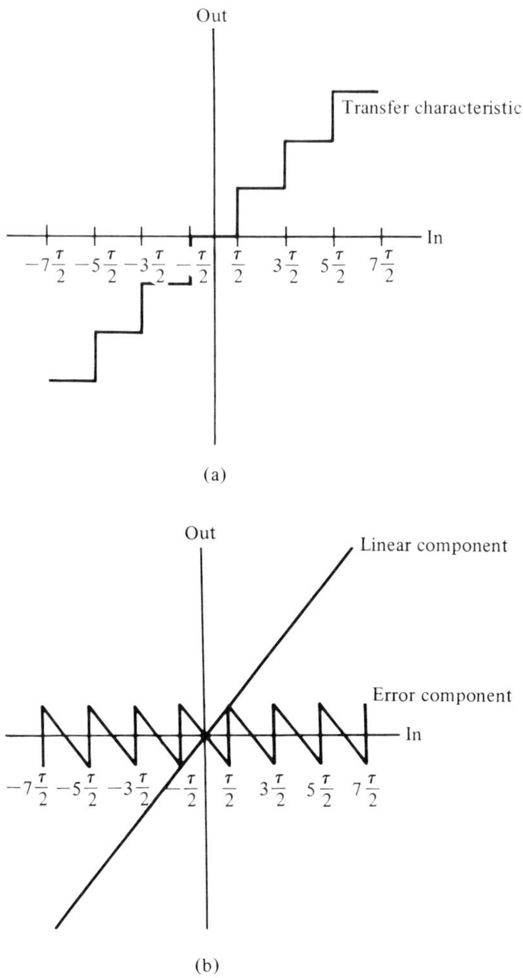

FIGURE 12-3. (a) Transfer characteristic; (b) linear and error components.

EXAMPLE 12-1

A 5-V signal is sampled and digitized to an accuracy of five digits (bits). Determine (a) the quantizing rms error voltage and (b) the signal-to-quantizing noise ratio.

SOLUTION

$$\text{(a) rms error} = (5)\left(\frac{1}{\sqrt{12}}\,\frac{1}{2^5 - 1}\right) = 0.0466 \text{ V}$$

(b) $S/N = \dfrac{5}{0.0466} = 107.30$

$20 \log S/N = 40.61$ dB

Note that the ratio represents rms signal to rms noise, since the rms and peak value of the signal are the same in this example.

The number of bits to which the data are digitized depends on the application and/or nature of the signal. For example, quantization results in contouring for black-and-white television. The effect for one digit is no shade of gray. Two bits results in a picture with elements that are either black or white. Five or more bits yields an acceptable picture. For a voice signal, 1 bit is understandable; 2 bits is "mushy" and less understandable than 1 bit. For 3 or more bits the reconstructed signal steadily improves. Only a slight coarseness remains at 6 bits.

12-2 DATA ACQUISITION

At this time let us consider briefly the problem of data acquisition. Unless the data already exist as discrete values, the digital aspects of a data acquisition system begins at the analog-to-digital (A/D) converter. The conversion of analog signals can be accomplished in several ways, including integration, counting, successive approximation, and parallel conversion. The operation of A/Ds is discussed in texts on digital circuits and will not be discussed here.

In terms of digital communication, the A/D converter can be considered as a black box. The specifications we are interested in are the highest frequency that can be sampled by the data acquisition system, the accuracy, and the conversion time. The block diagram for a typical data acquisition system is shown in Figure 12-4. Systems such as this are readily available in integrated-circuit form. Note that the output is time multiplexed as either parallel or serial digital output.

Degradation in accuracy can occur whenever the input to the system varies appreciably because the input may be changing significantly while the system is still preoccupied with an earlier sample of data. If a system is to convert to within a specified accuracy, the rate at which the input signal may change must somehow be attuned to the time taken by the A/D converter to examine a given sample. It is the sample-and-hold circuit that acts as the intermediary.

To establish this speed/accuracy relationship, the systems designer starts with

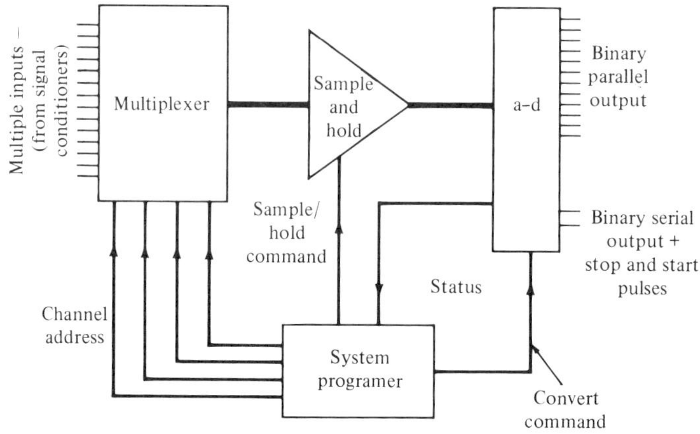

FIGURE 12-4. Data acquisition system.

the A/D conversion time, which is expressed as the frequency of a sinusoid. That is,

$$e(t) = E \sin \omega t \tag{12-3}$$

where E is the maximum amplitude. The rate of change of $e(t)$ with respect to time is

$$\frac{d}{dt} e(t) = \omega E \cos \omega t \tag{12-4}$$

This rate will be maximum when $\cos \omega t = 1$. Thus the maximum rate of change is

$$\frac{\Delta e}{\Delta t} \approx \omega E = 2\pi f E \tag{12-5}$$

Now, for an n-bit converter, the magnitude of the least significant bit (LSB) is 1 LSB $= E/2^n$. To obtain n bits of accuracy, the change in the amplitude of the input, Δe, must not exceed the value of the least significant bit, $E/2^n$. Thus

$$\Delta e = \frac{E}{2^n} = 2\pi f E \, \Delta T \tag{12-6}$$

Finally, solving for f_{max}, we have

$$f_{max} = \frac{1}{2\pi \, \Delta T \, 2^n} \tag{12-7}$$

where ΔT represents the aperture time.

This then is the highest frequency that can be sampled by a data acquisition system if it is to maintain n bits of accuracy and if the signal must be observed ΔT time while the system converts. Equation (12-7) shows the trade-off between the highest frequency a data acquisition system can handle, the aperture time, and the desired accuracy.

EXAMPLE 12-2

A 12-bit A/D converter has a conversion time of 24 μs. Find the maximum frequency that it can sample.

SOLUTION

$$f_{\text{max}} = \frac{1}{2\pi \, \Delta T \, 2^n} = \frac{1}{(2\pi)(24 \times 10^{-6})(2^{12})}$$

$$= 1.6 \text{ Hz}$$

This is an extremely low data input. However, if a sample-and-hold circuit is placed in front of the A/D, the signal could change at a rate many orders of magnitude higher than that the A/D is capable of handling. Very fast sample-and-hold circuits can respond in about 1 ns. This results in an f_{max} of about 39 kHz. At the expense of added circuitry, parallel converters are available which produce a digital word in a time limited only by the propagation delay of the logic, thus resulting in extremely fast conversion time.

The serial output of the multiplexer effectively time multiplexes the output of the coded samples. Each channel is sampled at a rate of at least $2f_m$ per second. If there are M channels and n bits per sample, the bit rate out of the coder is

$$f = 2Mnf_m \qquad \text{bits/s} \tag{12-8}$$

Since there is no guard space between code groups, time-division multiplexing puts M channels into a bit rate that is exactly M times the rate required for one channel.

Equation (12-8) ignores the need for "framing." Framing may be accomplished by transmitting a special code group of p pulses at the beginning of each line of PCM data. The framing pulses are detected by the demultiplexer so that each received code group can be identified with its respective channel. The bit rate for one line (frame) of data is then

$$f = 2(Mn + p)f_m \qquad \text{bits/s} \tag{12-9}$$

EXAMPLE 12-3

Compute the line rate for a 24-channel system with 8-bit coding, 4-kHz bandwidth, and one digit for framing.

SOLUTION

$$f = 2[(24)(8) + 1]4 \times 10^3 = 1.544 \text{ megabits}$$

12-3 PULSE REGENERATION

PCM reduces any signal to a random binary pulse train. Thus all signal types look alike, once coded, and a transmission link with a given bit rate capability may be used for video, voice, telemetry, or any combination. After transmission through a bandlimited, noisy medium, the digital pulse train may be regenerated into its original sequence of rectangular pulses. A bit rate synchronizer typically accomplishes this task. For example, Figure 12-5 is typical of the possibilities for transmission of digital signals over cable pairs. Clean binary pulses applied to the cable pair are filtered by the low-pass characteristics of the cable. Frequency-shaped amplification produces equalized and amplified pulses that are pretty well confined to their correct time slots. A regenerator then looks at each time slot. If it finds a voltage greater than half-peak pulse height, a "new" output pulse is generated. This technique eliminates noise if it does not exceed half-peak pulse height.

Now for random noise, there exists a finite probability that the noise will exceed half-peak pulse height, and the regenerator will introduce errors into the pulse train. The error rate is a rather sharp function of signal-to-noise ratio. For a peak signal-to-rms noise ratio of 20 dB, it is on the order of one error per 10^6 time slots. This is typically characterized as the bit error rate (BER). A typical bit error rate curve is shown in Figure 12-6. In speech errors become noticeable at about one error in 10^5 time slots (BER $= 10^{-5}$).

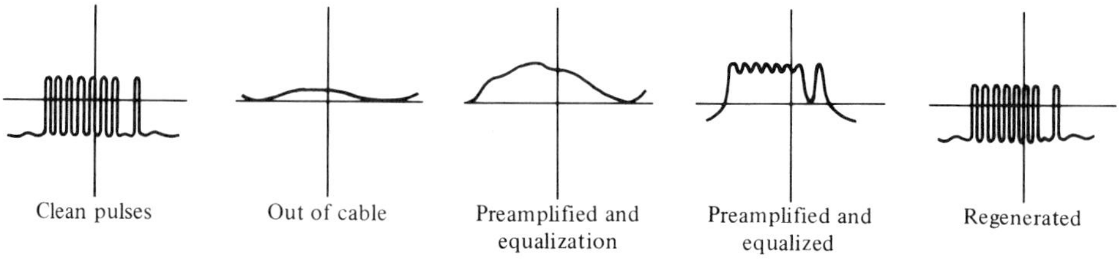

FIGURE 12-5. PCM regeneration principle.

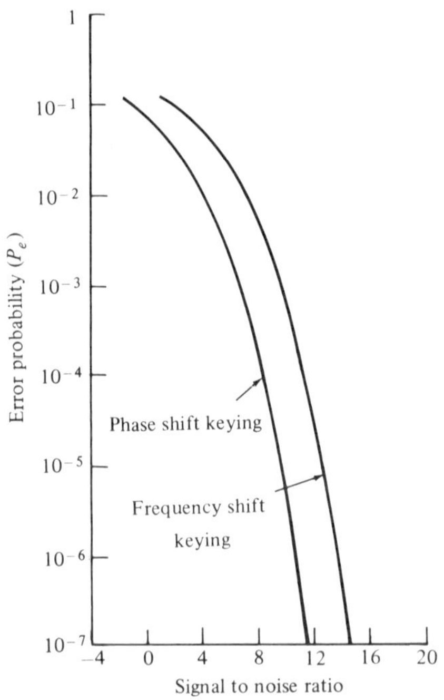

FIGURE 12-6. Typical bit-error-rate curves.

12-4

DATA COMPRESSION AND CODING

It was shown [equation (12-2)] that the rms quantizing error is on the order of $1/\sqrt{12}$ times the step size. A weak signal that does not utilize the full range of codes, then, has a poorer signal-to-quantitizing noise ratio than the stronger signals. In terms of speech, for example, a weak talker activates so few codes that the S/N ratio may be intolerable. As shown in Figure 12-7, compression can be used to give preferential gain to the weak talker in order to keep the S/N ratio roughly independent of talker level. Compression can be obtained by inserting a nonlinear network ahead of a linear encoder.

One of the most common compression characteristics has the form

$$E_o = \frac{\ln (1 + \mu E_i)}{\ln (1 + \mu)} \tag{12-10}$$

where E_i is the normalized sample amplitude at the compressor input and E_o is the normalized compressed output voltage. μ is a parameter that may be used to improve the signal-to-quantizing noise ratio for low-level signals at the expense of performance by high-level signals. Values for μ may vary from 0 to

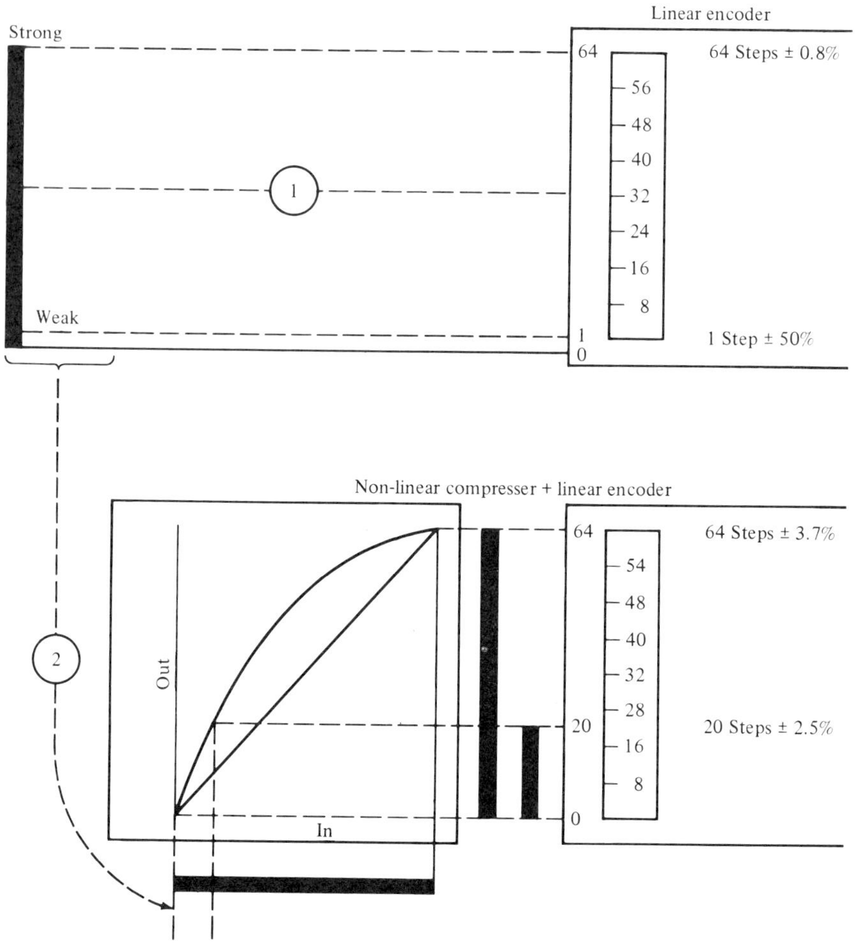

FIGURE 12-7. PCM signal compressor.

200. At the decoder output an ''expanding'' network must remove the effects of compression. The system is often referred to as *companding*.

12-4.1 Differential PCM

Differential PCM, often termed *delta modulation,* has received considerable attention. The differences between successive samples may be coded, transmitted, decoded, and summed to form the original signal. The advantage is a reduction in bandwidth. Common information in successive samples need not be transmitted again. The disadvantage is that an error has considerable effect on samples following the sample in error. Differential schemes have particular applications for signals where ''slow changes'' must be coded precisely but rapid changes may be roughly approximate. An example is black-and-white television.

The actual code used depends on design requirements, the most straightforward being binary. In some instances a reflected binary or Gray code may be used. In this code, only one digit changes for a quantum increase in the signal to be coded. The advantage of this is that no severe code misinterpretations are made due to small deviations from ideal performance. Using straight binary, as one crosses from level 7 to level 8, 4 bits must change (0111 = 7, 1111 = 15, 1110 = 8). Using the Gray code, level 7 progresses directly to level 8. There are fewer circuits that must change state and thus less possibility of error. Other codes, such as the Hamming codes, can be used for error detection and error correction.

12-5

FRAMING

There are two pieces of information needed by the demultiplexer in addition to the sample codes. These are:

- Channel identification.
- Code groups (words) identification.

For single point-to-point systems the two identification problems merge. That is, if the first digit of the first channel can be identified, a counting procedure can be used that groups the codes and the channels. A PCM network consisting of many PCM systems requires a more elaborate procedure.

Framing circuits are responsible for synchronizing all scanning and distribution gates so that the channels are properly routed. These circuits must monitor the system at all times and initiate reframing if the receiving system appears out of synchronization. In addition, the framing circuitry must not be fooled by a reasonable number of digital errors made in transmission. This may consist of a scheme as simple as demanding that the detection of frame "sync" fail twice in a row.

There are many variations possible in frame formats. Successive bits or words of different channels may be interleaved. Sample rates for different channels may be related harmonically. One possible format is illustrated in Figure 12-8. Note that a special framing pulse is transmitted at the end of the frame. This pulse is assigned an "on–off" statistic that is not likely to be the same as a data pulse. If we assign an "on" pulse every other frame, at least two frames must be searched to establish synchronization. This is termed a *backward-acting* system.

A *forward-acting* system results if a readily identifiable frame digit or digits is added. A unique code group is often used, but this increases the bit rate.

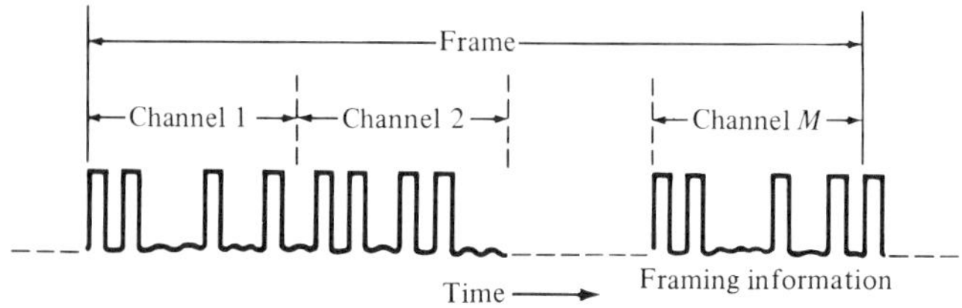

FIGURE 12-8. Typical frame format.

Special comparator circuitry continuously monitors the pulse train and compares it with the stored sync word. More complex systems may group the frames into *master frames*. Thus every 10 or 100 frames, a master-frame sync word is inserted.

12-6

BASEBAND PULSE TRANSMISSION

The manner in which a pulse train is formed for transmission is determined by the characteristics of the medium. For example, if dc transmission or dc restoration is difficult, it is preferable to transmit a wave with no dc component. If high-frequency interchannel crosstalk is a problem, a wave without excessive energy at the upper portion of the spectrum is preferable.

A number of pulse shapes are available for baseband transmission. A number of variations with their spectra are illustrated in Figure 12-9. Note that all of these pulse trains represent the digital word "1011." The first two are unipolar with 50% and 100% duty cycle, respectively. Both have a continuous and a dc component in the frequency spectrum. The spectrum of the 100% duty cycle pulse is zero at the pulse repetition frequency, f_r. Neither of the bipolar pulses has a dc component. The spectra of the pulses are important in designing pulse regeneration equipment. For example, a timing component may be added at the spectra nulls in Figure 12-9b and c. This component may be extracted by the regenerator to serve as a clock.

The effect of intersymbol influence due to transmission over a line or through a filter may be evaluated using an "eye pattern." An eye diagram for a random pulse train is sketched in Figure 12-10. This pattern may be obtained from an oscilloscope if the pulse output from the line is fed to the vertical input of the scope and the sampling clock $C(t - nT_m)$ is fed to the external trigger of the scope. The horizontal time base is set approximately to the symbol duration. The sampling clock occurs at the center of the pulse and thus the pulse should reach its maximum height at this time. Whether or not the distorted pulse from the line reaches its maximum height at the sampling time depends on the value

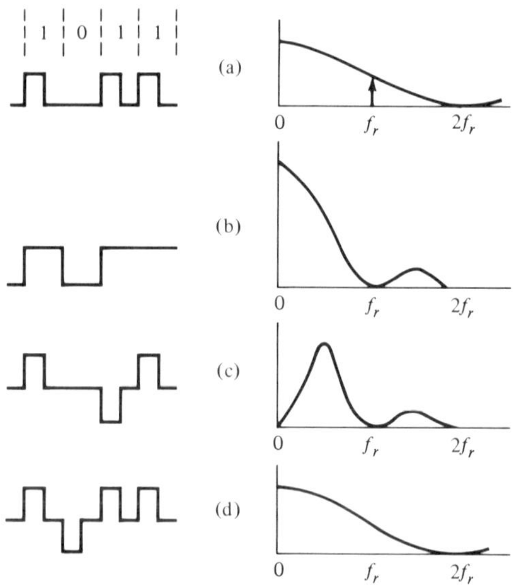

FIGURE 12-9. Spectra of various random pulse trains: (a) unipolar 50%; (b) unipolar 100%; (c) bipolar 50%; (d) polar 50%.

of the previous bits. An imperfect filter on the line will introduce intersymbol interference (ISI). If, for example, the pulse reaches only 90% of its height, the 90% eye opening corresponds to 20 log (90/100) = 0.9 dB ISI degradation.

12-7

TRANSMISSION TECHNIQUES FOR DIGITAL SIGNALS

We have seen in previous sections that pulses have wide spectrums. Consequently, if air is the transmission medium, the pulse train cannot be efficiently

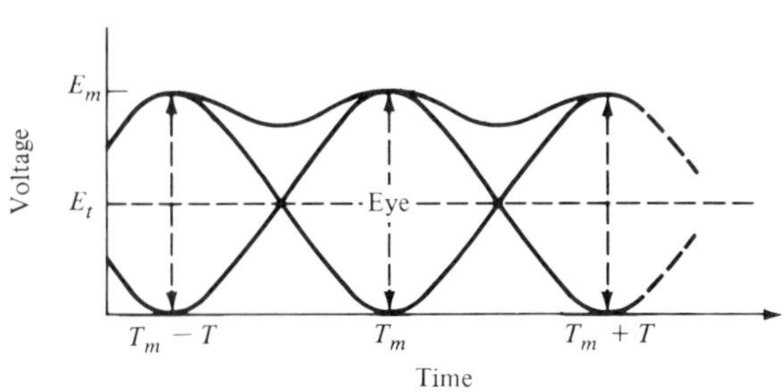

FIGURE 12-10. "Eye" diagram for a random pulse train.

transmitted through the air. The most commonly used systems for transmitting binary information through the air are *frequency shift keying* (FSK) and *phase shift keying* (PSK).

In the case of FSK, two different frequencies are employed to transmit a binary "0." We transmit a burst of cos $\omega_0 t$ to transmit a binary "1"; we transmit a burst of cos $\omega_1 t$ to transmit a binary "0." We see that this is a form of FM modulation where the RF signal is always one of two possible frequencies. The transmitter can be as simple as driving a VCO with the logic "0" or "1" states of the binary signal. This is commonly referred to as a *mark* or *space*. A PLL provides an excellent means of decoding FSK. A circuit using the NE565 is shown in Figure 12-11. In this circuit a mark is transmitted as 1070 Hz and a space at 1270 Hz.

The second form of binary transmission, PSK, is a form of phase modulation. In this case the phase of the transmitted signal represents the binary information. A balanced modulator (multiplier) is used to produce PSK (Figure 12-12). A logic "1" would be transmitted as sin $(\omega_0 t + \pi/2)$ and a logic "0" is represented by sin $(\omega_0 t - \pi/2)$. Thus there is 180° phase shift.

A synchronous demodulator is required to demodulate PSK. A Costas loop (Figure 12-13) is often used for the demodulation of double sideband signals. Referring to the figure, the data are assumed to be random switching simultaneously between $\pm\pi/2$ radians each data transition. At the Costas loop input, the signal is of the form

$$f(t) = \sqrt{2}A \sin [\omega_0 t + \pi/2\ d(t)] \qquad (12\text{-}11)$$

where $d(t)$ takes on values of ± 1 and A is the rms value. Then the input signal is equivalent to

$$f(t) = \sqrt{2}A\ d(t) \cos \omega_0 t \qquad (12\text{-}12)$$

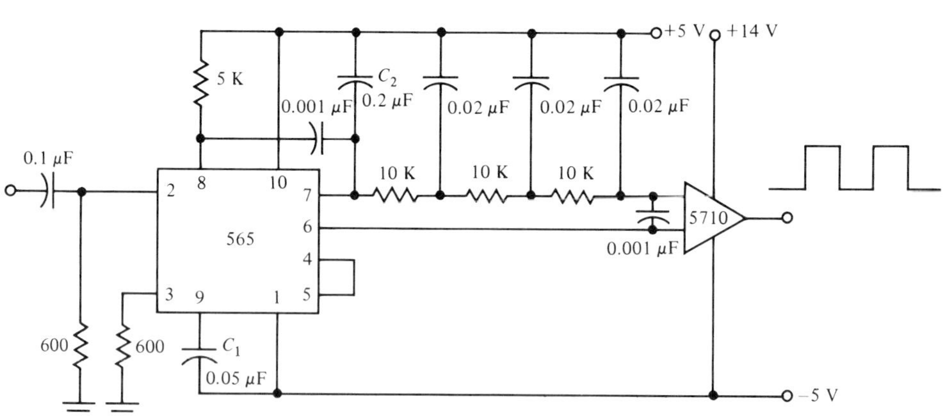

FIGURE 12-11. FSK decoder.

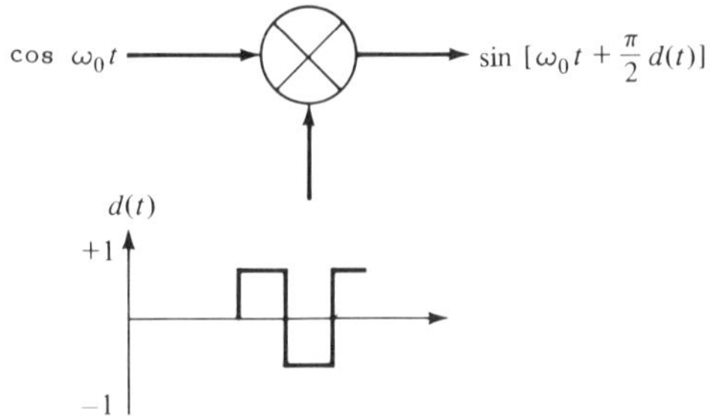

FIGURE 12-12. PSK modulator.

Now assuming low loss in the "arm" filters, the LPF outputs are approximately

$$Y_1(t) = \frac{\sqrt{2}A\ d(t)}{2} \sin \theta \tag{12-13}$$

and

$$Y_2(t) = \frac{\sqrt{2}A\ d(t)}{2} \cos \theta \tag{12-14}$$

where a "locked" condition is assumed. In this condition the VCO is running at the frequency of the unmodulated carrier and is also phase locked. The input to the loop low-pass filter is the product of $Y_1(t)$ and $y_2(t)$. Thus the error voltage

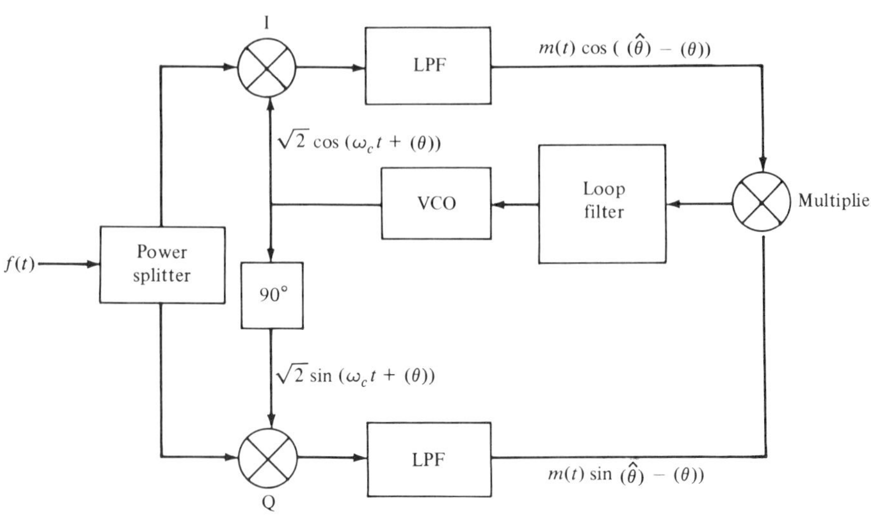

FIGURE 12-13. Costas carrier recovery loop.

to the VCO is

$$e(t) = \frac{K_m}{4} A^2 d^2(t) \sin 2\theta$$
$$= \frac{K_m}{4} A^2 \sin 2\theta = K \sin 2\theta \qquad (12\text{-}15)$$

when K_m is the multiplier gain and $d^2(t) = (\pm 1)^2 = 1$.

Note that the error voltage is a function of sin 2θ instead of sin θ as in the conventional PLL. Thus this loop will lock at 0 or 180°. Since the data are recovered at the output of the low-pass "arm" filters, this means that the arm at which data are recovered may switch. To remove this uncertainty, the data are usually coded in a differential scheme before modulating.

Bit synchronization is an important consideration in digital transmission. The receiver must be able to partition the received signal into time segments corresponding to the sample intervals. To aid in synchronization, three techniques are commonly used as formats in transmitting digital data. These are known as "return to zero" (RZ), "non-return to zero," (NRZ) and "Manchester coding." These pulse transmission schemes are illustrated in Figure 12-14. Note that the first two correspond to parts (a) and (b) in Figure 12-9. We will return to a discussion of these pulse format and coding techniques in Chapter 14.

12-8

DIGITAL SIGNAL PROCESSING

The advent of medium-scale and large-scale integrated circuits has provided a forum in which it is advantageous in many instances to accomplish postdetection digital signal processing in the receiver. Some of the most significant developments are centered around radar receivers. An in-depth treatment of digital signal

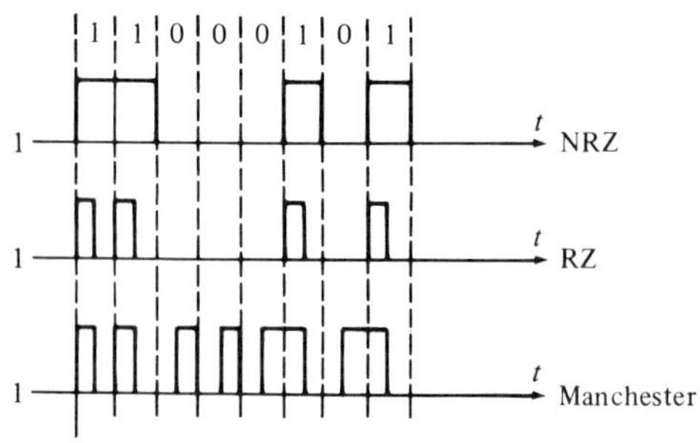

FIGURE 12-14. Three pulse transmission schemes.

processing is beyond the scope of this text, but the concepts are easily demonstrated with the theory and techniques developed thus far.

12-8.1 Phasors

To begin our discussion, let us examine the concept of phasors applied to a sampled signal. Consider the digital sinusoid shown in Figure 12-15. The kth sample value of the sinusoid can be written as

$$F(k) = A \cos(\omega k + \phi) \tag{12-16}$$

when k is an integer corresponding to the sample number. The signal is described by the amplitude A, the frequency ω, and the phase ϕ. The phase ϕ represents the starting point of the cycle relative to the origin, $k = 0$. It is of interest to consider the range of values that ω can take on. Equation (12-16) can be rewritten as the real part of a complex exponential.

$$F(k) = \text{Re}[Ae^{j(\omega k + \phi)}] \tag{12-17}$$

We refer to the function $Ae^{j(\omega k + \phi)}$ as a *phasor* and it can be interpreted as a point that moves around a circle of radius A in the complex plane (Figure 12-16). The signal $F(k)$ is the real part of the phasor and is the projection of the moving point on the real axis. When ω is small, the point moves slowly. When $\omega = \pi$, the point moves directly from $+A$ to $-A$ and back (Figure 12-17). Note that this corresponds to two samples per cycle (i.e., the sampling theorem). Thus the highest frequency that can be represented by a digital signal is π radians per sample. In fact, if ω is greater than π radians, the phasor in Figure 12-1b will reverse itself and rotate clockwise.

Next, let us relate the actual frequency of the signal being sampled and the time, T, between samples. The period of the phasor $e^{j\omega k}$ is obtained by letting $\omega K = 2\pi$. Then

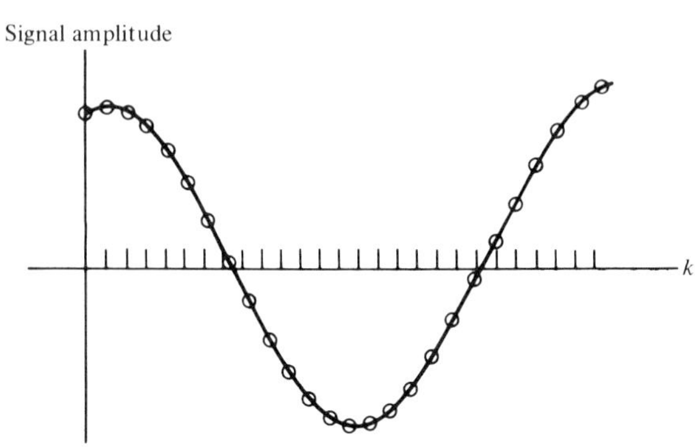

FIGURE 12-15. A digital sinusoid.

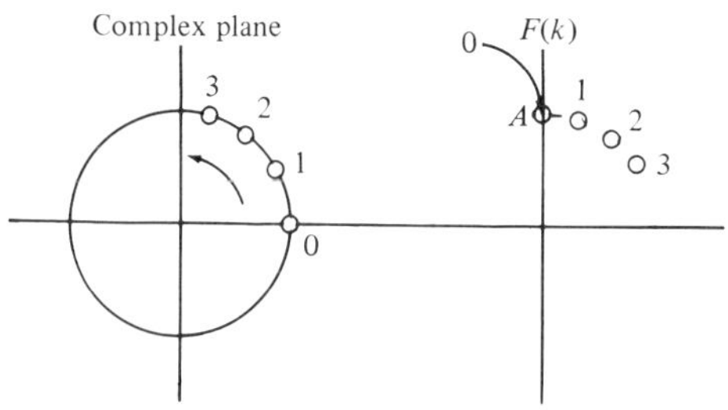

FIGURE 12-16. A Complex digital phasor, together with its real part, a digital sinusoid.

$$K = \frac{2\pi}{\omega} \quad \text{samples/cycle} \tag{12-18}$$

The period of the signal is then

$$KT = \frac{2\pi T}{\omega} \tag{12-19}$$

and the actual frequency of the signal is

$$f = \frac{1}{KT} = \frac{\omega}{2\pi T} \quad \text{herz} \tag{12-20}$$

where ω is the radians between samples and T is the time between samples.

12-8.2 Simple Digital Filter

We have all added a quantity of numbers and then divided the sum by the number to obtain the average. How many realized, however, that this operation constitutes a low-pass digital filter operation? Consider the averaging process

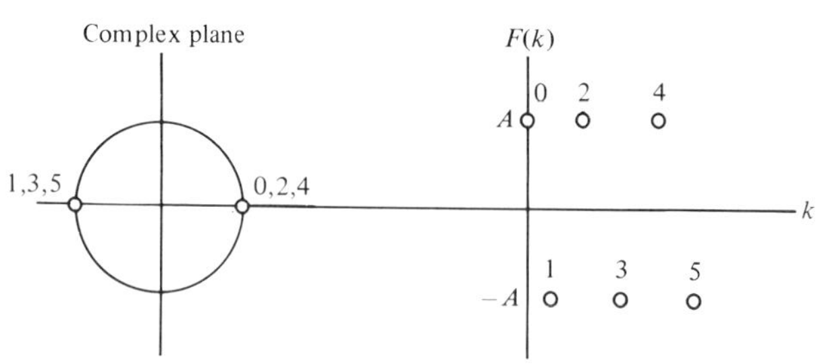

FIGURE 12-17. A Complex digital phasor of frequency π radians per sample interval.

expressed by

$$Y(k) = \frac{X(k) + X(k - 1)}{2} \tag{12-21}$$

where $X(k)$ represents samples of the original waveform and $Y(k)$ is the average or "smoothed" value. Equation (12-21) represents the present value of Y as the average of the present value of X and the preceding value of X.

To obtain the frequency response of the filter represented by equation (12-21), let X be a phasor of frequency ω:

$$X(k) = e^{j\omega k} \tag{12-22}$$

As before, we consider only the real part when we wish to relate to physical signals. Substituting equation (12-22) into (12-21) results in

$$\begin{aligned} Y(k) &= \frac{1}{2}\left[e^{j\omega k} + e^{j\omega(k-1)}\right] \\ &= \frac{1}{2}(1 + e^{-j\omega})e^{j\omega k} \\ &= \frac{1}{2}(1 + e^{-j\omega})X(k) \end{aligned} \tag{12-23}$$

or

$$\frac{Y(k)}{X(k)} = H(j\omega) = \frac{1}{2}(1 + e^{-j\omega}) \tag{12-24}$$

We interpret $H(j\omega)$ as the frequency transfer function, as was done for analog filters in Chapter 3. The function is a complex number with both magnitude and phase.

The magnitude of the transfer function is

$$\begin{aligned} |H(j\omega)| &= \left|\frac{1}{2}(1 + e^{j\omega})\right| \\ &= \left|\frac{1}{2}e^{-j\frac{\omega}{2}}(e^{j\frac{\omega}{2}} + e^{-j\frac{\omega}{2}}\right| \\ &= \left|e^{-j\frac{\omega}{2}}\right| \left|\cos\frac{\omega}{2}\right| = \left|\cos\frac{\omega}{2}\right| \end{aligned} \tag{12-25}$$

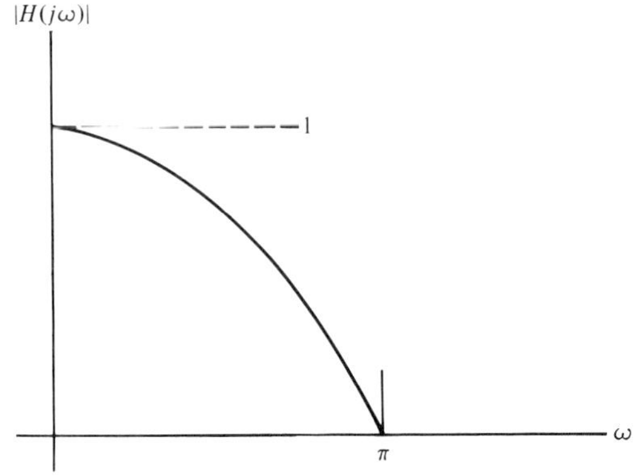

FIGURE 12-18. Amplitude of the low-pass filter *H*.

This function is plotted in Figure 12-18. Note that the function is low pass with cutoff at $\omega = \pi$ radians, the largest frequency that can be present in a digital signal. This response can be easily verified from the phasor input, $X(k)$. At $\omega = 0$, the average of the input is 1. At $\omega = \pi$, the input is alternately $+1$ and -1, so that Y is always 0.

Next consider the phase of $H(j\omega)$.

$$\begin{aligned}\text{Arg } H(j\omega) &= \text{Arg}\left[e^{-j\frac{\omega}{2}} \cos \frac{\omega}{2}\right] \\ &= \frac{\omega}{2}\end{aligned} \tag{12-26}$$

This function is plotted in Figure 12-19. Note that the phase angle is one-half the sample frequency. That is, the phase represents an output delay of one-half the sample interval. This type of filter is known as a *nonrecursive filter*.

Now consider the same filter function but with the sign changed. That is

$$Y(k) = \frac{X(k) - X(k - 1)}{2} \tag{12-27}$$

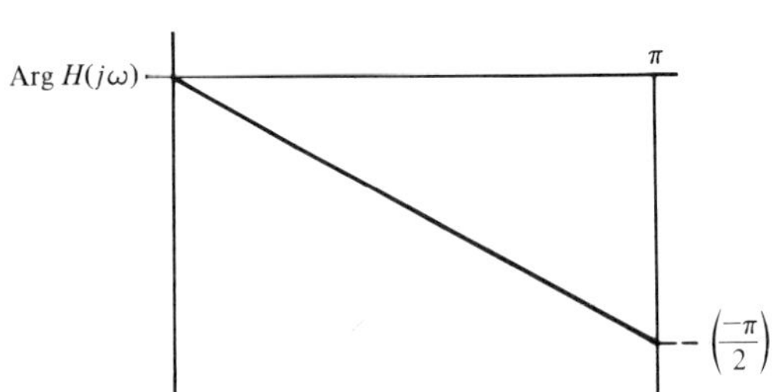

FIGURE 12-19. Phase of the low-pass filter *H*.

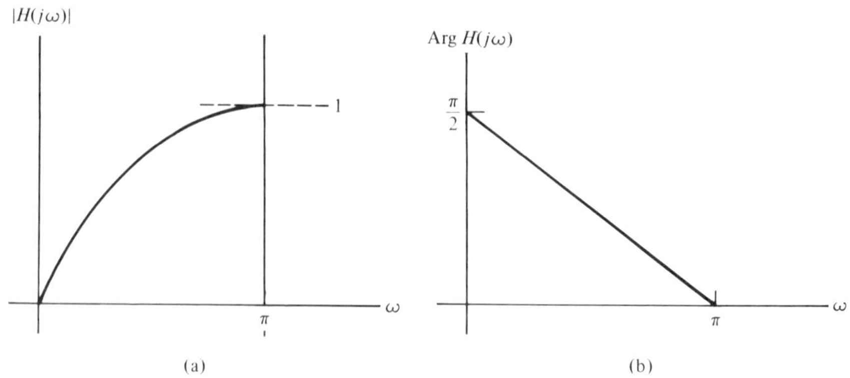

FIGURE 12-20. (a) Amplitude; (b) phase.

Then proceeding as before, the transfer function, $H(j\omega)$, is

$$H(j\omega) = \frac{1}{2}(1 - e^{-j\omega}) \tag{12-28}$$

and the magnitude and phase are

$$|H(j\omega)| = \left|\frac{\sin \omega}{2}\right| \tag{12-29}$$

and

$$\begin{aligned} \text{Arg } H(j\omega) &= \text{Arg}\left(e^{-j\frac{\omega}{2}} j \sin \frac{\omega}{2}\right) \\ &= \frac{\pi}{2} - \frac{\omega}{2} \end{aligned} \tag{12-30}$$

These functions are plotted in Figure 12-20. Note that this represents a high-pass filter.

At this point, we see how to construct a low-pass and a high-pass digital filter. The next logical thought may well be how to vary the filter response. One way is to add more samples and to let the samples have different "weights." For example, we may wish the filter output, Y, to be influenced more by recent samples than by samples in the more distant past (time).

EXAMPLE 12-4

Calculate the (a) magnitude and (b) phase angle for the filter defined by

$$Y(k) = 0.6X(k) + 0.4X(k - 1)$$

SOLUTION

(a) For a sinusoid input:

$$Y(k) = (0.6 + 0.4e^{-j\omega})e^{j\omega k}$$

$$= (0.6 + 0.4e^{j\omega})X(k)$$

$$H(j\omega) = \frac{Y(k)}{X(k)} = 0.6 + 0.4e^{j\omega}$$

$$\text{Re}[H(j\omega)] = 0.6 + 0.4 \cos \omega$$

$$\text{Im}[H(j\omega)] = -0.4 \sin \omega$$

$$|H(j\omega)| = \sqrt{(\text{Real})^2 + (\text{Imag})^2}$$

$$= \sqrt{(0.6 + 0.4 \cos \omega)^2 + (-0.4 \sin \omega)^2}$$

$$= \sqrt{0.52 + 0.48 \cos \omega}$$

(b) The phase angle is given by

$$\text{Arg } H(j\omega) = \arctan \frac{\text{Im}[H(j\omega)]}{\text{Re}[H(j\omega)]}$$

$$= \arctan \frac{-0.4 \sin \omega}{0.6 + 0.4 \cos \omega}$$

See the figure.

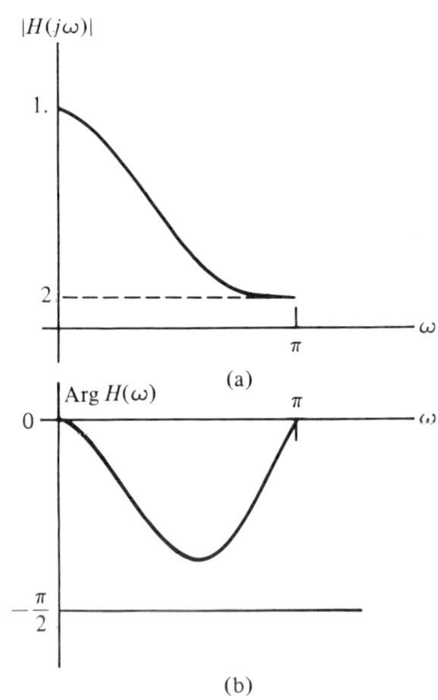

EXAMPLE 12-4

Generalized Nonrecursive Filter. The filters we have described thus far can be generalized where averaging takes place over m samples. That is,

$$Y(k) = C(1)X(k) + C(2)X(k - 1) + \cdots + C(m)X(k - m + 1) \quad (12\text{-}31)$$

where the C's are the weighting coefficients. Now, if we let $X(k) = e^{j\omega k}$ in equation (12-31), this equation may be written as

$$Y(k) = [C(1) + C(2)e^{-j\omega} + \cdots + C(m)e^{-j(m-1)}\omega]X(k) \quad (12\text{-}32)$$

or

$$H(j\omega) = C(1) + C(2)e^{-j\omega} + \cdots + C(m)e^{-j(m-1)\omega} \quad (12\text{-}33)$$

For the sake of simplicity we may let $z = e^{j\omega}$. Equation (12-33) may be written as

$$H(z) = C(1) + C(2)z^{-1} + C(3)z^{-2} + \cdots + C(m)z^{-(m-1)} \quad (12.34)$$

We can think of z as a sort of operator which shifts the sample back in time. Thus a term such as Cz^{-n} implies $(C)X(k - n)$, which is the input delayed by n sample times and multiplied by a "weight" $C(k)$.

12-8.3 Recursive Digital Filter

The preceding digital filter ouputs are a function of only the present and past inputs. Now, let the filter output consist also of past outputs. In terms of analog filters this is analogous to charge stored on capacitors in the filter. Thus past output conditions together with the input affects the present output. Consider the following filter:

$$Y(k) = X(k) + 0.5Y(k - 1) \quad (12\text{-}35)$$

In words, the current filter output consists of the present input plus one-half the immediate filter output. Specifically, a filter that uses previous values of the output as well as present and previous values of the input is called a *recursive filter*.

The operation of the filter described by equation (12-35) can be demonstrated by assuming an input. Suppose that the input is a unit step function. The initial condition for the filter is that past outputs, $Y(-1)$, are zero. Thus the following sequence takes place. For $k = 0$ and $X(k) = 1$, the $Y(k)$ are

$$Y(0) = X(0) + 0.5Y(-1) = 1 + 0.5(0) = 1$$

$$Y(1) = X(1) + 0.50Y(0) = 1 + 0.5 = 1.5$$

$$Y(2) = X(2) + 0.5Y(1) = 1 + 0.5(1.5) = 1.75$$

$$\vdots$$

$$Y(4) = 1.9375$$

The filter output is shown in Figure 12-21.

Generalized Recursive Filters. The general form of a recursive filter uses present and past values of the input as well as past values of the output. We may express this mathematically as

$$Y(k) = C(1)X(k) + \cdots + C(m)X(k - m + 1) - D(1)Y(k - 1) - \cdots - D(L)Y(k - L) \quad (12\text{-}36)$$

In this equation we have m arbitrary coefficients multiplying input values, and L arbitrary coefficients multiplying output values. Further analysis of the filter defined by equation (12-36) dictates that we become familiar with the z-transform. The z-transform applied to a sampled signal transforms from the time domain to the frequency domain. This topic, although not complicated, is beyond the scope of this text. Further information may be obtained from numerous texts on sampled data systems.

12-8.4 Discrete Fourier Transform

In Chapter 4 we introduced the Fourier transform [equation (4-33)] and applied it to a rectangular pulse to find its frequency spectrum. For digital computation

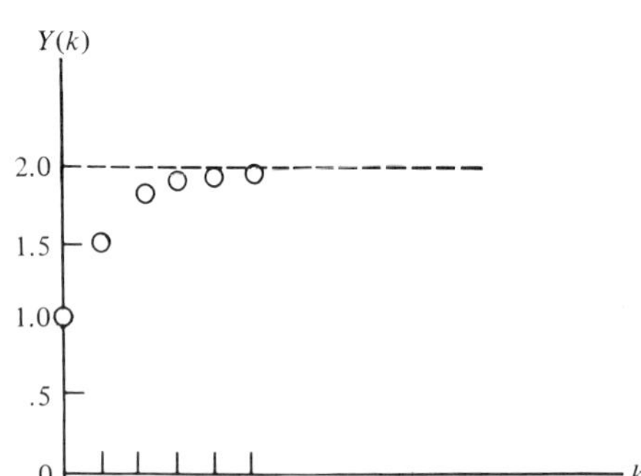

FIGURE 12-21. Output signal *Y(k)*.

there is an analogous transform. This discrete Fourier transform (DFT) is defined as

$$F_n = \sum_{k=0}^{n-1} f_k e^{-j}(2\pi/N)nk \qquad (12\text{-}37)$$

In this equation n is the frequency spectral component, k is the sample number, and N is the number of samples. The computation is best illustrated by an example.

Consider the operation shown in Figure 12-22. To illustrate the computation, let us evaluate F_3 for the four-point signal. Using equation (12-37), we have

$$\begin{aligned} F_3 &= \sum_{k=0}^{3} f_k e^{-j(2\pi/N)nk} \\ &= f_0 e^{-j0} + f_1 e^{-j(3/2)\pi} + f_2 e^{-j3\pi} + f_3 e^{-j(9/2)\pi} \\ &= (3)(1) + 2(0 + j1) + 5(0 - 1) + 0(0 - j1) \\ &= 2 + j2 \end{aligned}$$

The remaining three spectral components can be found in the identical fashion.

The resulting array of complex numbers can be transformed into polar coordinates as shown in Figure 12-23. This frequency-domain representation enables a signal to be valued in terms of sinusoids of various magnitudes and phases at multiples of some fundamental frequency, ω. That is, the time-varying signal can be written as the finite series.

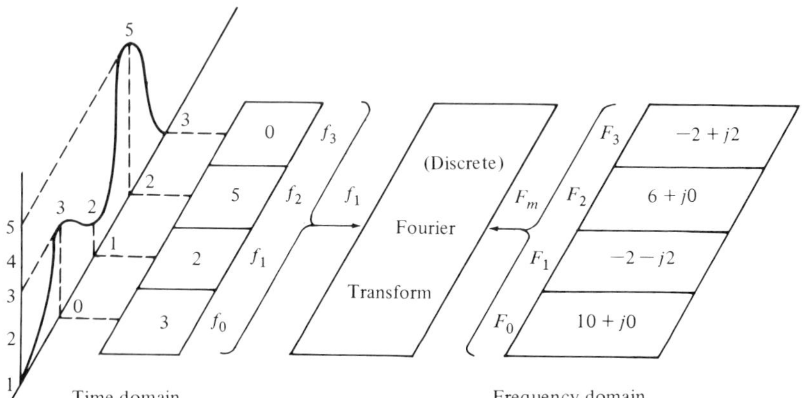

FIGURE 12-22. Fourier transformation.

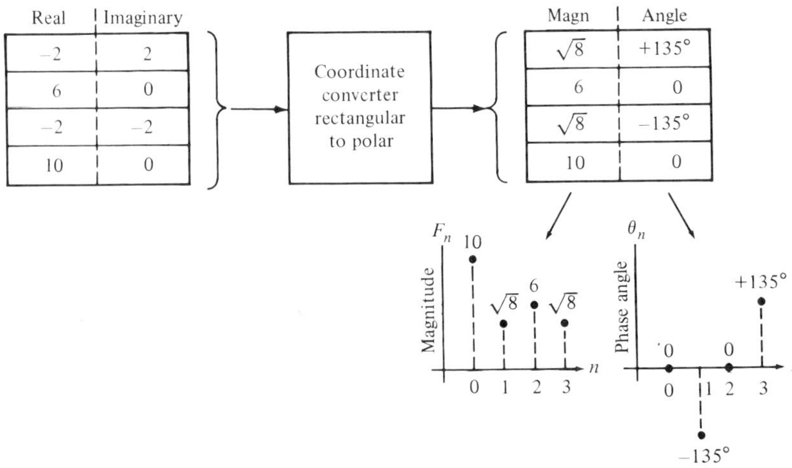

FIGURE 12-23. Rectangular-to-polar conversion.

$$y(t) = \frac{1}{4}[10 + \sqrt{8}\cos(\omega_1 t - 135°) + 6\cos(2\omega_1 t - 0°) + \sqrt{8}\cos(3\omega_1 t + 135°)] \qquad (12\text{-}38)$$

where $\omega_1 = 2\pi/N = 2\pi/4 = 90°$.

Now, to convert the spectrum indicated in Figure 12-23 to a physical frequency, we can use equation (12-20). For example, if the time between samples is 1 ms, the fundamental frequency, f_1, is

$$f_1 = \frac{\omega}{2\pi T} = \frac{\pi/2}{2\pi(10^{-3})} = 250 \text{ Hz}$$

Applications. We see at this time the mechanics of computing a discrete Fourier transform. At this point most readers have read reports of the use of Fourier transforms in voice synthesis and speech recognition: for example, talking cash registers. In a communications setting, however, an interesting example is its use in radar systems. A radar transmits N pulses and the radar receiver looks for N reflected returns. Reflections, of course, indicate a target and constitute N samples of a continuous signal. Thus we can compute an N-point DFT. In practice, a bank of filters are employed, as shown in Figure 12-24.

The N detected target returns, $r(t)$, are stored and then applied to a bank of filters. Each filter then computes an N-point DFT. A threshold device contains a digital word against which the output magnitude of each filter is compared. Target detection is declared if one or more adjacent filter outputs exceed the threshold. The Ku-band rendezvous radar on the U.S. Space Shuttle employs this approach for target detection.

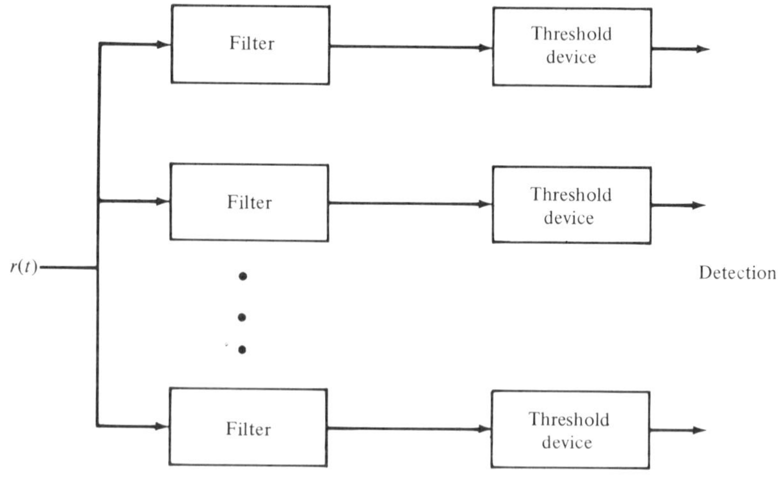

FIGURE 12-24. Doppler filter bank.

12-8.5 Digital Signal Processing Circuit Implementation

With a reasonably good knowledge of digital design, we see the ease with which the previous concepts may be implemented. At the time of this writing, however, even greater possibilities are beginning to exist in digital signal processing. A generation of single-chip processors are appearing on the market. One of these, the Intel 2920, contains programmable elements that permit real-time signal

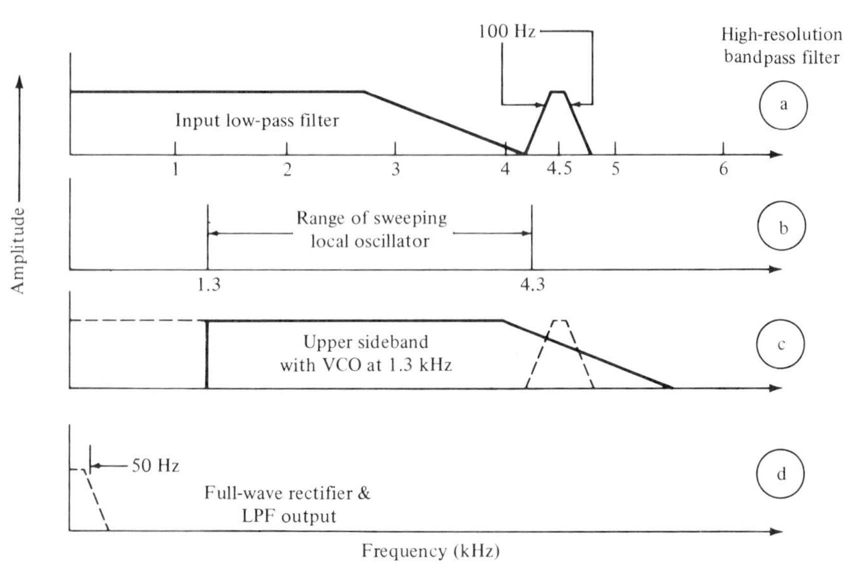

FIGURE 12-25. Intel 2920 programmed as a spectrum analyzer.

processing. Figure 12-25 illustrates its operation as a spectrum analyzer. The operation indicated in Figure 12-25 uses 107 program instructions and 22 RAM locations. Other applications include programming as a recursive digital filter or applications as diverse as approximating a Butterworth filter.

PROBLEMS

12-1. The following voltages are input to a five-bit A/D converter. Find the binary output of the ADC. Assume that full scale is 5 V. (a) 3.24 V, (b) 1.02 V, (c) 0.325 V, (d) 4.32 V.

12-2. An analog signal $[f(t) = \sin 2t]$ is to be transmitted via a 5-bit PCM. Find the PCM signal for the first 6 seconds and sketch the result.

12-3. For Problem 12-2, find the rms error voltage and the signal/quantizing ratio.

12-4. Two signals with an upper frequency of 10 kHz are to be transmitted with 32 levels of quantization. What is the bit rate and the minimum bandwidth of the channel?

12-5. What is the line rate in Problem 12-4 if 5 bits is allowed for framing?

12-6. In equation (12-10), assume that $\mu = 100E_i$ is normalized for a range $-1 \leq E_i \leq 1$ and that E_o is expressed as a 5-bit binary word. Plot this equation using E_o as the y axis and E_i as the x axis.

12-7. Repeat Problem 12-6 with $\mu = 255$.

12-8. The input to a delta modulator is $f(t) = 5t + 1$. The sampler operates at 10 samples/s, and the step size is 1 V. Sketch the output of the delta modulator.

12-9. A color TV channel is to be sampled using 8 bits/sample and a 12-MHz sampling rate. The signal has frequencies from 0 to 5.5 MHz.
(a) In terms of the Nyquist rate, is this sampling rate sufficient?
(b) If the signal varies between $\pm\frac{1}{2}$ V, what is the rms quantizing error?

12-10. Calculate and plot the amplitude and phase of the following digital filters. (a) $H(z) = 2 + z^{-2}$, (b) $H(z) = z^{-1}$, and (c) $H(z) = (z + z^{-1})/2$.

12-11. Calculate the four-point Fourier transform for the signal shown. If the time between samples is 1 s, what is the fundamental frequency, f_1? Write a series for $f(t)$ as in equation (12-38).

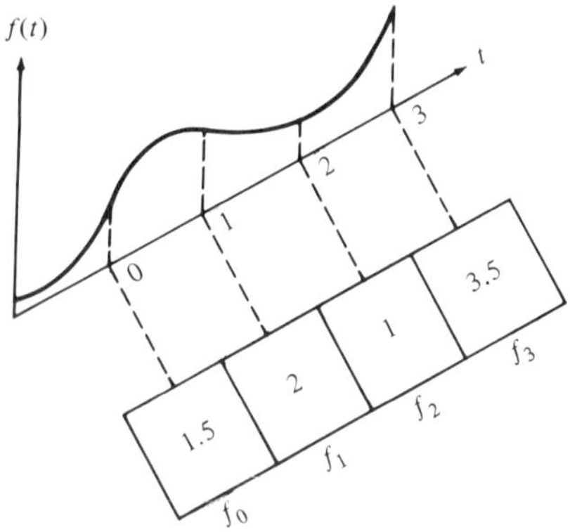

PROBLEM 12-11

12-12. What is the peak signal (rms) or quantization-to-noise ratio for (a) a 6-bit binary code, (b) a 9-bit binary code, and (c) a 16-bit binary code.

CHAPTER 13

Data Communications

With the advent of personal computers, low-cost terminals, standard interfaces, and networks, data communications is becoming as commonplace as telephone communication. In this chapter we attempt to establish the basis for design of a data communications link and perhaps more important, define some of the terminology in current use regarding data communications. Unfortunately, baseline definitions of terminology do not exist today and people reading the same material exhibit large variances in their interpretation of the material.

At this point, we have accumulated the technical knowledge to understand the theory utilized in data communication system design. What is needed now is to define the problem. In an earlier time, data communication system design was the domain of the electronic engineer. The intervening years has seen the introduction of a variety of systems or building blocks whose use requires primarily that interface standards be met.

In this chapter we examine the problem involved in the transmission of digital data over one or more lines. The order of discussion will be as follows:

- Transmission channel.
- Modems.

- Serial communication interfaces.
- Data-link protocols.
- Network architecture and protocol.

13-1 DATA COMMUNICATIONS TRANSMISSION CHANNEL

To design an effective data communication system we must first make sure that we have an effective communication channel. The selection of the proper data communication channel involves first considering a set of data transmission requirements. For example:

- Is the channel required to accommodate transmission in both directions, and if so, simultaneously?
- Will two or more channels be connected? If so, will the connections be switchable or fixed?
- Which station will initiate calls, and how far must the channel transmit baseband signals?
- What data rates and error rates must the channel accommodate?
- Will the data be analog or digital? What type of modulation scheme should be selected?

In designing a data communications system these questions must be answered. A starting point lies in examining the three basic transmission schemes available for data communications systems (Figure 13-1). These are

- *Simplex:* transmission from *A* to *B* but never from *B* to *A*.
- *Half-duplex:* transmission from *A* to *B* during one interval, then from *B* to *A* during another interval.
- *Full-duplex:* simultaneous transmission from *A* to *B* and *B* to *A*. Generally, this is accomplished by four wire circuits, however, two wire circuits can be adapted for full-duplex operation by using two different frequencies for transmit and receive. This technique is used in 1200- and 2400-bit/s full-duplex two-wire modulator/demodulators (modems).

Note from Figure 13-1 that the channels come in just three basic configurations: point-to-point, multipoint, and circuit switched. The point-to-point channel connects just two parties. As indicated in Figure 13-1a and b, these channels can operate in either half- or full-duplex mode. In multipoint systems, several secondary stations share a common line (Figure 13-1c). In operation, the primary

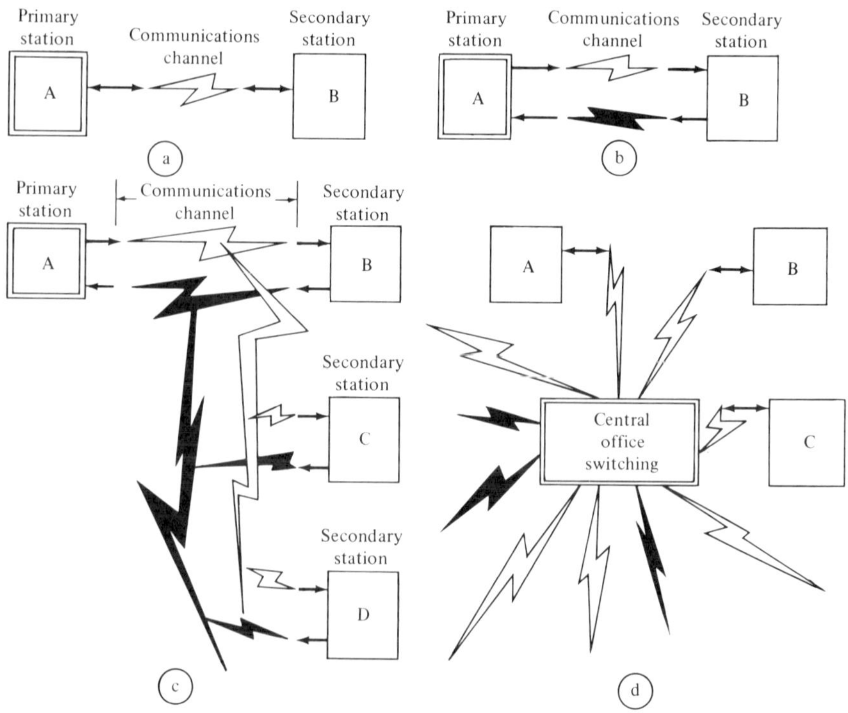

FIGURE 13-1. (a) Half-duplex transmission connects A to B over a two-wire channel. Not only can one end send while the other receives, but they can reverse roles as well. (b) Duplex transmission over nonswitched four wire circuits allows simultaneous transmission and reception. (c) In nonswitched multipoint transmission, several secondary stations share one duplex channel. The primary station polls its tributaries. (d) Point-to-point switched systems establish a new transmitter-to-receiver path for each call, thus volatility is the price of versatility.

station solicits transmission from the secondaries by polling them. It also picks which secondary is to receive a data transmission. It is possible for the primary to transmit to one secondary while receiving from another. Usually, however, nonswitched multipoint systems are restricted to half-duplex transmission. A circuit-switched arrangement (Figure 13-1d) establishes a transmitter-to-receiver path for every call. Calls are initiated by public or dual-tone multifrequency dialers (DTMF) either manually from rotary or keypad devices, or automatically.

13-1.1 Channel Bandwidth

In addition to transmission direction, a channel is also characterized by bandwidth. We recognize that greater bandwidths generally allow higher transmission rates. This is usually specified as the number of line-signal elements per second, or the *baud rate*. For binary signal elements, the baud rate equals the bit rate. When a signal element represents more than two states, as in multilevel modulation, the bit rate exceeds the baud rate.

We have seen via the sampling theorem that a signal whose highest frequency is x hertz need not be sampled more than $2x$ per second. Thus an x-hertz bandwidth can carry $2x$ samples per second. Binary-coded data or bits require two voltages. Thus transmitting 2 bits simultaneously requires four voltage levels. Then $2x$ samples per second are coded into $4x$ bits/s. Generalizing, in one time slot using 2^n signal levels, we can encode n bits. Thus $2nx$ bits/s can be transmitted over an x-hertz bandwidth channel. Let the number of signal levels, L, be represented by

$$L = 2^n \qquad (13\text{-}1)$$

where n is the number of bits in each binary word. Then $n = \log_2 L$. The maximum data rate for the channel is

$$C = 2nx = 2x \log_2 L \quad \text{bits/s} \qquad (13\text{-}2)$$

where x is the channel bandwidth.

EXAMPLE 13-1

A noiseless telephone channel has a bandwidth of 2.7 kHz (300 to 3000 Hz). Find the channel capacity (maximum data rate) for signal levels of four and eight.

SOLUTION

The signal level $L = 2^n = 4 \Rightarrow n = 2$ bits per sample. Then

$$\begin{aligned} \text{channel capacity, } C &= 2x \log_2 L \\ &= (2)(2700)\log_2 L \\ &= (2)(2700)(2) = 10.8 \text{ kbits/s} \end{aligned}$$

For eight levels,

$$\begin{aligned} L &= 2^n = 8 \Rightarrow n = 3 \text{ bits/sample} \\ C &= (2)(2700) \log_2 8 \\ &= (2)(2700)(3) = 16.2 \text{ kbits/s} \end{aligned}$$

13-1.2 Basic Data Communications Channel

A typical data communications channel is shown in Figure 13-2. Note that there are three main hardware sections. Each holds a pair of the following:

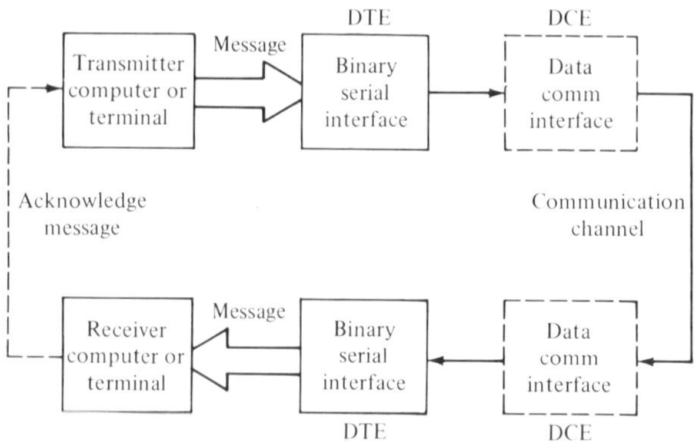

FIGURE 13-2. Typical data communications system.

- Computers or terminals, one transmitting, the other receiving.
- Binary serial interfaces, also called data-circuit terminating equipment (DCE).

The transmitting computer or terminal sends messages in a sequence of parallel characters to a DTE (data terminal equipment). The DTE then serializes each character in the message for a 1-bit-at-a-time transmission. Serial transmission is necessary for a signal pair of wires and it is also easy to synchronize the received data.

To facilitate asynchronous transmission, the DTE appends start and stop bits to each serial character. Also, a parity bit may be added for vertical redundancy checking. Actual techniques for accomplishing this are discussed in Section 13-2. This serial stream of pulses corresponds to one of the electrical standards, such as EIA's RS232C, RS422, or RS423.

The DCE converts these pulses into communications-channel-compatible signals—thus the data communications interfaces that are appropriate depend largely on the channel to be driven. Dial-up or leased lines usually require compatible models. For in-house or short distances, limited-distance modems often suffice.

AT&T's Dataphone Digital Service (DDS) requires data service units. DTEs containing today's RS422 and RS423 line drivers can reliably send dc baseband (unmodulated) signals up to 4000 ft to their counterpart line receivers.

At the receiving end of the channel, each action of the transmission section is reversed. Here a DCE demodulates (decodes) transmitted signals and extracts the information. The interface produces a serial pulse stream that is voltage-level converted and then assembled into a sequence of parallel characters. These characters, in turn, are read by the receiving computer or terminal, which then acknowledges reception of one or more messages from the sender. Messages received incorrectly prompt a request for retransmission of the faulty data.

Although the figure shows it as a separate dashed line, message acknowledgment takes the same path as the data. The difference is that the receiver becomes the sender and the former sender becomes the receiver.

The advent of distributed processing and a growing number of intelligent terminals has placed a growing demand on data communications equipment, especially higher data rates and longer lines. To accommodate these demands, interface specs RS422 and RS423 emerged (see Table 13-1). Note that compared to the old specification, RS232, line lengths up to 4000 ft are allowed. Longer line lengths may require the insertion of a modem.

TABLE 13-1 Comparison of the Old and New Interface Standards

Driver Configuration	**RS232C Single-Ended (Unbalanced)**	**RS423 Single-Ended (Unbalanced)**	**RS422 Differential (Balanced)**
Max. recommended cable length	50 ft	4000 ft (1200 m)	4000 ft (1200 m)
Max. signaling (data) rate	20 kbits/s	100 kbits/s at 40 ft	10 Mbits/s at 40 ft
Driver output voltage, open circuit (V_o)	± 3 to ± 25 V	± 4 to ± 6 V	≤ 6 V between A and B terminals
Driver output voltage, loaded (V_t)	± 5 to ± 15 V 3- to 7-kΩ load	$\lvert V_t \rvert \geq 0.9 \lvert V_0 \rvert$ 450-Ω load	2 V between A and B 100-Ω balanced load (2- to 50-Ω loads)
Driver output short-circuit current (I_s)	$\lvert I_s \rvert \leq \pm 500$ mA	$\lvert I_s \rvert \leq \pm 150$ mA	$\lvert I_s \rvert \leq \pm 150$ mA
Driver output slew rate (dV/dt)	30 V/μs max.	Based on cable length and signaling rate	No control necessary
Receiver mark (off = 1)	≤ -3 V	A = negative with respect to B'	$A < B$
Receiver space (on = 0)	$\geq +3$ V	A = positive with respect to B'	$A > B$
Receiver input threshold	± 3 V	200 mV differential	200 mV differential
Receiver input impedance	3 to 7 kΩ, 2500 pF	$\geq$4 kΩ	$\geq$4 kΩ
Receiver input voltage range	-25 to $+25$ V	-12 to $+12$ V	-12 to $+12$ V
Power off measurement (I_x) (V_o applied to unpowered device)	2 V $\leq V_o \leq$ 25 V	$\lvert I_x \rvert \leq \pm 100$ μA -6 V $\leq V_0 \leq$ 6 V	$\lvert I_x \rvert \leq 100$ μA 0.25 V $< V_o \leq$ 6 V

13-1.3 Data Choice: Analog or Digital

Once the characteristics of the channel are determined, it is necessary to generate the appropriate signals for transmission, either digital or analog. In digital transmission, regardless of its original form, information travels over the channel as a stream of pulses. Before noise and distortion degrade them and cause errors, however, the pulses must be detected and regenerated by repeater circuits.

Pulses get distorted by channel capacitance, inductance, and leakage, which cause reception errors to increase with line length and pulse rate. Thus regenerative repeaters must be spaced closely—approximately 4000 ft with T1 carriers (Figure 13-3). A T1 carrier consists of 24 voice grade channels time multiplexed together.

Pulses can be transmitted up to 10 miles over a cable or common-carrier wire pair with no modulation other than simply turning dc on and off (baseband transmission). Usually, either some Electronic Industry Association (EIA)-type line drivers and receivers or limited-distance modems are needed. For longer-distance digital communications, access is needed to either AT&T's digital-transmission facilities or to microwave circuits from specialized terrestrial or satellite carriers.

Analog transmission, on the other hand, sends a continuous signal range (amplitude, frequency, or phase) over a channel; filters and linear amplifiers maintain signal quality. Analog voice channels are interfaced to the digital output of terminals and computers by a modem, within which digital pulses modulate a phone-channel-compatible carrier. This carrier then passes through the telephone network as though it were a voice signal. The receiving end demodulates the signal back to digital form.

Digital Carrier System. The T1 carrier system (Figure 13-3) digitally transmits 24 voice-grade channels over a single line by time-division multiplexing (TDM) 1.544-Mbit/s PCM data. Each voice channel is sampled at the 8-kHz international rate, and the amplitude of each 8-kHz sample is encoded into 8-bit PCM under the μ-255 law (see "Companding," below).

The total rate of 1.544 Mbits/s results from the T1 framing process, which

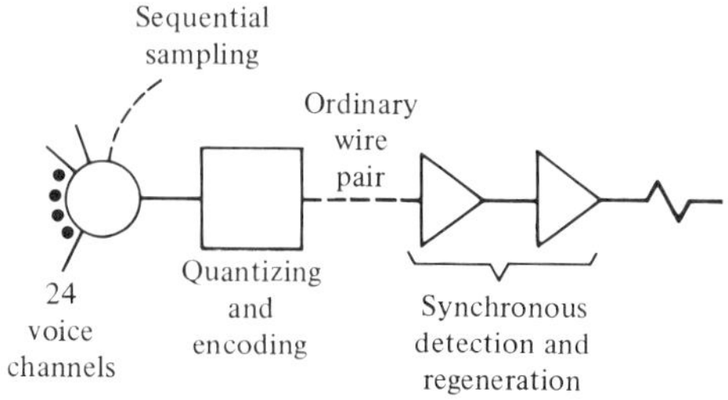

FIGURE 13-3. Digital carrier system.

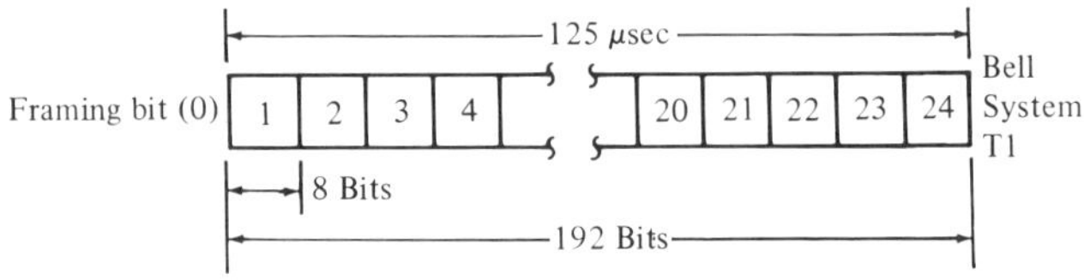

FIGURE 13-4. North American PCM frame.

includes "sync"-bit stuffing. Each T1 frame is made of 8-bit samples from each of the 24 channels plus one "sync" bit, for a total of 193 bits per frame. Since the channel is sampled at 8 kHz, there are 193 × 8000, or 1.544 Mbits/s (see Figure 13-4).

The data are transmitted as pulses over wire pairs containing repeaters. Placed approximately every mile, the repeaters regenerate and retime the pulse waveforms with very low error probability. At the end of the line, the pulse train is decoded and demultiplexed into accurate reproductions of the 24 voice signals.

When it was introduced in 1962, the T1 carrier was meant to carry only digitized voice signals for a 50-mile maximum between central offices. However, AT&T now uses T1 carriers for data as well.

The Bell System started using a family of special-purpose T1 data terminals in 1965. The most popular, the T1WB3, is used in short-haul sections of wideband channels. It employs transition, or sliding-index, encoding, and transmits 50-kbit/s asynchronous data over three 64-kbit/s digital voice channels. The T1WB3 transmits 50-kbit/s asynchronous data over three 64-kbit/s digital voice channels—it would take 12 voice-grade channels to do the same transmission over analog facilities. Not only that, but it is possible to transmit voice or data over the T1WB3's unused T1 line capacity.

In Europe, where PCM framing conforms to specifications of the International Telephone and Telegraph Consultative Committee (CCITT) of the International Telecommunications Union, each 125-μs frame accommodates 32 channels—also 8 bits each. This increases to 256 bits per each 125-μs (CCITT) frame, or a data rate of 2.048 Mbits/s. The lower framing diagram shows why there are no added bits in the CCITT frame. Two full 8-bit channels, sampled in each frame, are reserved for the framing and signaling data (see Figure 13-5).

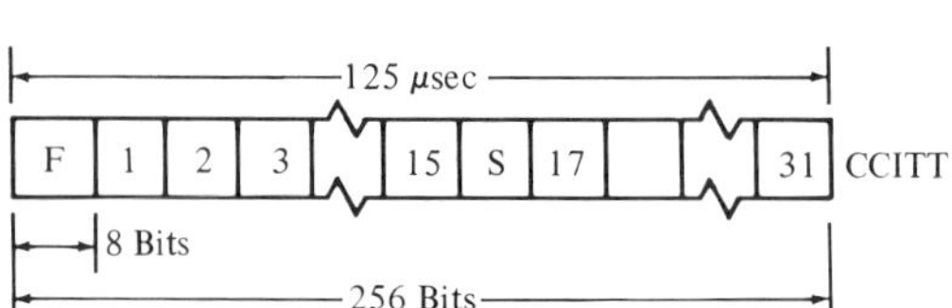

FIGURE 13-5. European PCM frame.

Companding. Compressing and expanding (companding) signals were discussed in Chapter 12. A typical exchange using companding is shown in Figure 13-6. Simply stated, companding is a deliberately nonlinear amplitude modulation that strengthens weak signals and reduces strong signals for transmission. After reception, a companion demodulation restores the original speech. In this way, analog telephone channels maintain acceptable signal-to-distortion levels over wide dynamic ranges of input level.

Like framing, the 8-bit coding of PCM words is subject to geography. Although telephone signals are companded throughout the world, the algorithms vary with the telecomm authority. The Bell System compands with the μ-255 law, while much of the rest of the world uses the A87.6 law. Fortunately, both methods increase the dynamic range (see Figure 13-7). The μ-law, when normalized for a coding range of ±1, is

$$Y(s) = \text{sgn}(x)\,\frac{\ln(1 + \mu|x|)}{\ln(1 + \mu)} \qquad (13\text{-}3)$$

for x falling between ±1. In this expression 255 is used for μ—thus the Bell System's μ-255 law.

The CCITT's standard for companding, on the other hand, is the A-law, and it takes two equations to define it. For the absolute value of x, $|x|$, falling past $1/A$,

$$Y(x) = \text{sgn}(x)\,\frac{1 + \log A|x|}{1 + \log A} \qquad (13\text{-}4)$$

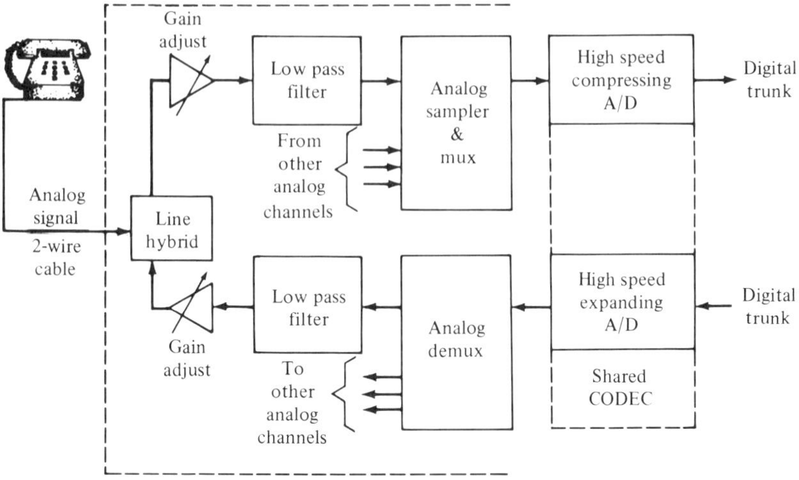

FIGURE 13-6. Local exchange.

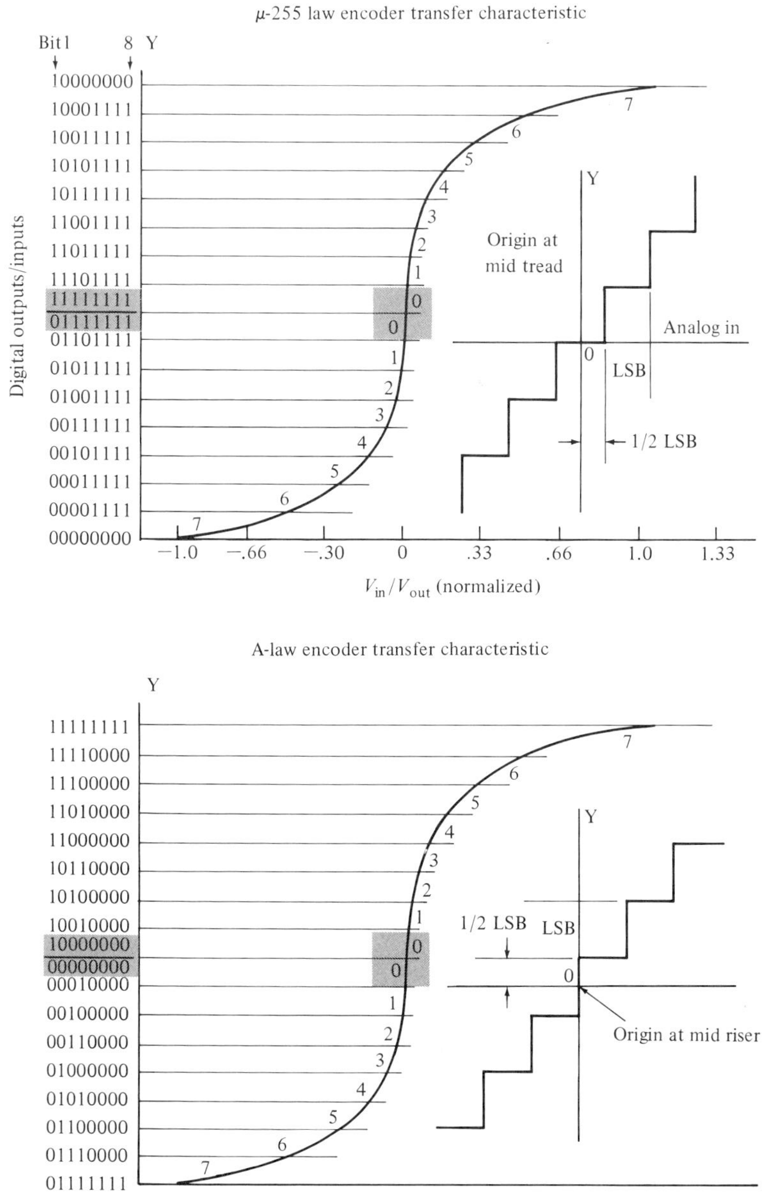

FIGURE 13-7. Companding curves.

For $|x|$ between 0 and $1/A$,

$$Y(x) = \operatorname{sgn}(x) \frac{A|x|}{1 + \log A} \tag{13-5}$$

The value used for the variable A is 87.6.

The expressions for both μ- and A-law companding give the smooth encoding curves shown. Obviously, the decoding characteristics must complement the encoding so that a round trip results in the input as it was before companding.

Codecs approximate the smooth analog functions, defined by the μ- and A-law equations, with straight-line segments. In the most popular approximations for both laws, the first of each word's eight bits defines its segment's polarity (sign bit). With the sign, the three bits following then determine one of 15 major segments (chords). A word's last 4 bits divide each major segment into 32 steps. The steps are uniform within any chord, but their sizes increase as chords go farther from zero. In the longest two of the μ-law chords, steps are 128 times those in chord 1.

The differences between μ and A-law coding are most apparent for weak signals. The transfer characteristics show that the μ-law specifies two codes for 0 V, thereby giving the completely quiet state a midtread position on the amplitude staircase waveform. On the other hand, the A-law characteristic shows no quiet codes, so the quiet state interpolates to a midriser position. This first A-law chord breaks into 64 separate divisions.

The low-level resolution under 8-bit μ-law coding corresponds to that with 13 bits of linear coding. The low-level resolution under 8-bit A-law corresponds to 11-bit binary. Thus, for 8-bit coding, the dynamic range is approximately 78 dB for the μ-law and 66 dB for the A-law. The signal-to-distortion ratio stays nearly constant over a 40-dB range with μ coding, and over the range 30 to 35 dB with A coding.

In telephony, the primary measurement unit is the decibel. This is the basis for various measurements. With decibels we can specify absolute and relative voltage and power. Decibels can be used to calibrate systems whose impedances do not match. The basic decibel unit is the ratio of two power levels:

$$\text{dB} = 10 \log \frac{P_1}{P_2}$$

The same unit can also represent the ratio of two current or voltage levels, at constant impedances. Thus when the impedances Z_1 and Z_2 are equal,

$$\text{dB} = 20 \log \frac{V_1}{V_2}$$

For power relative to a reference level, we use the terms *dBr* or *transmission level point* (limited to the United States). Zero decibels or 0 TLP describes a point in a system where the relative transmission level is zero.

For power referenced to an absolute level, the most commonly used unit is dBm. Here the absolute level referenced is 1 mW. Thus setting P_2 to 1 mW yields

$$\text{dBm} = 10 \log \frac{P_1(\text{watts})}{10^{-3}}$$

When the impedance is 600 Ω,

$$\text{dBm} = 10 \log \frac{V^2/600}{10^{-3}}$$

$$= 10 \log \frac{V^2}{0.6}$$

As a result, the 0-dBm voltage for a 600-Ω impedance is 0.775 V rms.

To relate an absolute power level in dBm to a transmission level of 0 dBr, we first remember that decibels are additive. Therefore, it is necessary, to subtract the reference power in dBr from the absolute power in dBm units to get power in dBm units.

Because telephony deals with human hearing, the dBmp unit has been developed for measuring absolute power of a signal that is filtered to compensate for the ear's characteristics. Figure 13-8 shows that the weighing curves for the two popular filter types, phosphometric and C-message, are closely related. Normally, phosphometric filters are used in *A*-law companded systems, while C-message filtering is used in North America.

By suffixing rnC to decibels we get dBrnC, a noise-referenced unit tailored

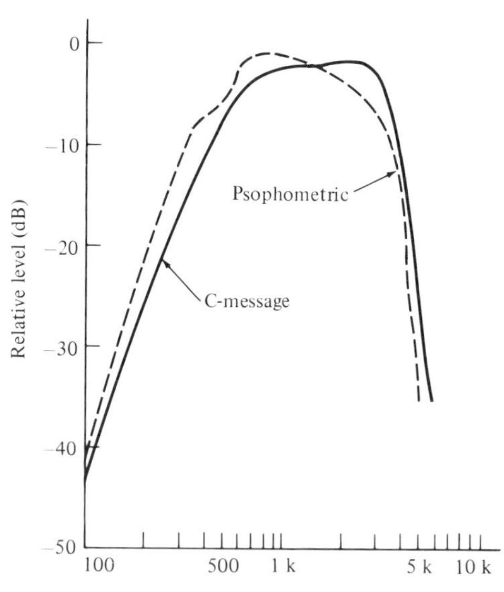

FIGURE 13-8. C-Message and psophometric message filter response curves.

to the ear. The rn stands for reference noise and the C signifies that the signal has passed through a C-message filter. By definition,

$$0 \text{ dBrn} = -90 \text{ dBm}$$

Obviously, then,

$$0 \text{ dBrnC} = -90 \text{ dBmC}.$$

We can reference any of the absolute-power-level units to 0 dBr. To indicate 0-dBr referencing, simply suffix a 0 to a unit's symbol (e.g., dBrnC0) and, of course, adjust the numerical value.

EXAMPLE 13-2

Express 2 pW of power in terms of dBrn.

SOLUTION

$$\text{dBrn} = 10 \log_{10} \frac{2 \times 10^{-12}}{10^{-12}} = +3$$

13-1.4 Choosing a Line: Dial-up Versus Leased

Dial-up lines are the two-wire pairs the common carriers supply on the public switched network. The usual transmitting mode is half-duplex, but full duplex can be obtained using frequency-band-splitting modems. At present, two dial-up services exist:

- Direct distance dialing (DDD).
- Wide-area telecommunications service (WATS).

Dial-up service provides two major advantages: (1) it can reach any point on the worldwide telephone network, and (2) expenses incurred in using the line apply only for the time the line is in use. Balanced against this, however, there are four switched network problems to contend with: noise, delay distortion, limited data throughput, and difficulty in accessing.

Although switched-network noise usually is not much of a problem in conversations, it is dangerous to data. The human brain can interpret speech in the face of line noise that causes errors in computers. Delay distortion stems from the nonuniform transmission speeds of the various frequency components. Data throughput goes down because the switched network's connect, disconnect, and turnaround functions all add to dead time. And accessing is difficult because it is sometimes hard to get a connection or, for that matter, even a dial tone.

For light traffic, leasing a private line costs more than using a dial-up line, and such a line connects to only one location (either point-to-point or multipoint). But leased lines give fewer problems and several advantages, including ready availability—without busy signals—fixed cost, and high quality (via optional conditioning). These advantages lead to higher bit rates and greater throughput. In addition, leased lines are generally four-wire circuits, meant for half- or full-duplex operation. As a result, we can transmit and receive simultaneously without having to turn the lines around. All in all, for point-to-point or multipoint data communications requiring high volume, quality, and speed, a leased line is the better choice.

13-1.5 Specialized Services

There are specialized alternatives to switched, leased, or WATS lines. In the United States, these common carrier services fall into four categories:

- Land-line voice/data.
- Satellite.
- Packet-switched.
- Facsimile and electronic-mail message.

Land-line carriers operate microwave-radio or broadband-cable links for their backbone (long-haul) network and use voice-grade lines (leased or dialed) to reach local customers. Land-line specialized common carriers provide two major classes of voice/data transmission service: interstate, shared, private line; or common-control switching.

Interstate, shared, private lines bring the economy of shared private lines to small and medium-volume users with multiple interstate locations. The primary data communication use for this branch of land-line service is in interactive data processing and the associated data-base management. The characteristically short random transmissions between remote interacting terminals and a centralized data base are served effectively by a pay-as-you-go switched network. Interstate, shared, private lines are offered by MCI Communications Corp. as EXECUNET, SPL, and Quickline services; by Southern Pacific Communications (SPC) as SPRINT; and by ITT as SPNS.

The two most extensive networks, EXECUNET and SPRINT, use different means of accessing customers. EXECUNET uses the local switched DDD network to reach its own long-distance transmission channels. In the destination city (one of 18), the DDD again connects the customer to the channel. With SPRINT, calls are routed through shared trunk lines between SPC switching centers. Every call must either originate or terminate on a dedicated-access line connecting the customer's telephone and the SPC switch. Any telephone in a local city can reach a remote office by dialing the local telephone number connected to the SPC switch. As of June 1978, the company had 25 switching centers in the United States with nine more slated for the near future.

Common-control switching-arrangement service is primarily for the high-volume user requiring special services. MCI's CCSA, ITT's CCSE, and AT&T's EPSCS typify this kind of service. Common features include four-wire switching, a seven-digit numbering plan, abbreviated dialing, alternate call routing, adaptive routing, hot lines, conference calling, customer-management center, and call-record logging.

SPC also offers DATADIAL, an interstate, switched, WATS-like service via the former DATRAN's digital-microwave network. Here, distance does not govern transmission cost. Instead, you pay for network time in 6-s increments. Synchronous full-duplex transmissions of 2.4, 4.8, and 9.6 kbits/s are available.

Satellite carriers lease domestic-satellite circuits, whose bandwidth is then sold piecemeal to users. A transmitting earth station beams signals to a geostationary domestic satellite. The bird's transponder returns the transmission to a receiving earth station where terrestrial links carry the signals to subscribers. The satellite is, in effect, a two-way antenna.

To date, private-line services from satellite carriers have not attracted a substantial number of commercial users for either data or voice communications. One big reason is the throughput degradation caused by long propagation delays—for example, 0.25 s for a coast-to-coast call. Another drawback is that the only service usually available is point to point. Furthermore, there are only a limited number of terminal cities with earth stations. At these satellite "tails," the local loops connecting customers to the earth station have low bandwidth. Satellites have a further difficulty with voice transmissions: echoes. As if all these were not enough, satellite transmission becomes more cost-effective than terrestrial only for links longer than 800 miles.

The primary customers of the satellite operators are the government, radio, the Public Broadcasting Service, cable-TV stations, newspaper wire services and publishers, and other specialty users. When they do go commercial, satellite services will require small, inexpensive two-way antennas that can stand on rooftops or in parking lots. Most current two-way dishes are expensive and so large that they usually must stand outside cities. Their received signals are piped back to subscribers via microwave or telephone lines.

13-2 MODEMS

Modems (modulator/demodulator) units bring together digital signals and analog lines. Refer to Figure 13-9. The modulator accepts digital inputs from computers or terminals and converts the baseband pulses into analog audio-frequency sig-

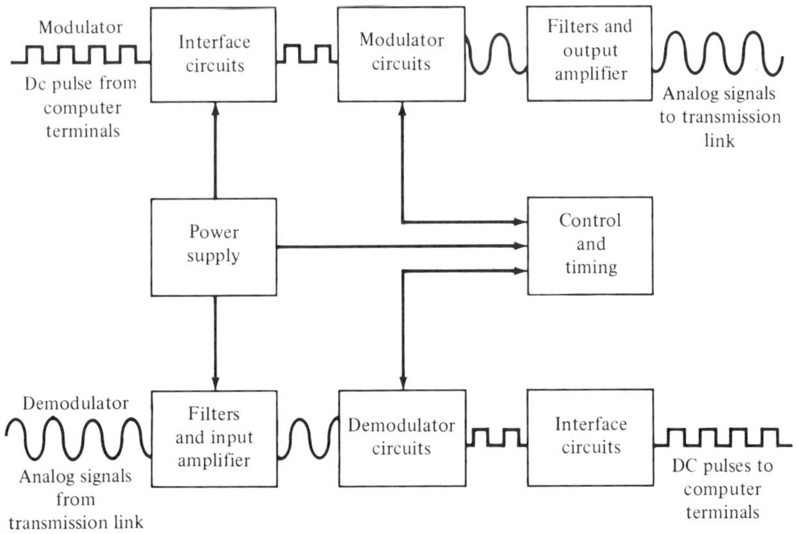

FIGURE 13-9. Modem interfacing block diagram.

nals for transmission. A demodulator in the second modem at the other link end reconverts the audio-frequency analog signals to baseband digital.

Modems are referred to by different names. Some of these are: data sets, data communications equipment, and data-circuit terminating equipment (DCE). Binary serial DCEs fall into four general categories:

- Voice-grade modems and telephone couplers.
- Wide-band modems.
- Short-haul modems.
- Modem eliminators or line driver/receivers.

The characteristics of these modems are summarized in Table 13-2. The most commonly used of the data-transmission mediums are voice-grade telephone lines with 2.7-kHz effective bandwidth. Generally, only Bell System modems or their equivalents from independents work together. This makes Bell compatibility an important modem feature for system integration.

Voice-grade modems transmit either aynchronously or synchronously. Asynchronous units (with start and stop bits) operate at maximum data rates of 1.8 kbits/s over dial-up facilities, and at 2 kbits/s on conditioned leased lines. Synchronous units' maximum data rate is 4.8 kbits/s over dial-up lines and 9.6 kbits/s on private leased lines.

13-2.1 Voice-Grade Modem Applications

Typical applications for low-speed (up to 2 kbits/s) asynchronous voice-grade modems include data acquisition, data logging, time sharing, and terminal-to-

TABLE 13-2 Modem Characteristics

Modem Type	Communications Channel	Data Rate (bits/s)	Use
Voice-grade			
High-speed synchronous	Leased line (3002) Dial-up	4.8 k, 7.2 k, 9.6 k 4.8 k	High-volume machine-to-machine communications; 600 to 1200 bits/s
Medium-speed Synchronous Asynchronous	Leased line/dial-up Leased line Dial-up[a]	2.4 k, 3.6 k 1.8 k, 2 k 1.2 k	Interactive or low-speed remote batch operations; 150 to 300 bits/s
Low-speed asynchronous	Dial-up[a]	300, 600	Interactive teleprinters and glass teletypewriters, data acquisition and collection; 30 to 60 bits/s
Wide-band			
Super group (60 voice-grade) Group (12 voice-grade)	5700, 5800 (TELPAK)[b] 8801	230.4 k 40.8 k, 50 k, 56 k	Large-volume telephone-line multiplexing, dedicated computer-to-computer links
Half-group (6 voice-grade) Lineplexer (2 voice-grade)	8803 2 leased lines	19.2 k 19.2 k	
Short-haul			
Limited-distance (<10 miles)	Private-wire/cable Nonloaded, nonconditioned, noncarrier line[b]	2k to 1 M 2 to 19.2 k	Data communications in plant (private wire) or off premises where distance is 10 miles or less (leased line)
Medium-distance (<50 miles)	Leased line	2 to 9.6 k	Intermediate-distance (10 to 50 miles)
Modem eliminators and line drivers/receivers[c]	Private wire/cable	2 k to 1.544 M	On-premises data communications; typical distances are 500 ft to 2 miles

[a]If several low- or medium-speed asynchronous devices are located in the same area, a frequency- or time-division multiplexer operating on a leased line is an effective alternative to dial-up operation. An acoustic coupler is another alternative.

[b]See Bell Technical Reference 43401 for line operating specifications.

[c]These devices are not really modems since they do not perform the modulation/demodulation functions. The required EIA interface signals are amplified or regenerated and sent down the line as a pulse stream.

computer interactive or data-entry systems. Medium-speed voice-grade modems are usually 1.2-kbit/s asynchronous or 2.4-kbit/s synchronous units. Obviously, the distinction between low- and medium-speed units is blurred by an overlap. Full-duplex modems for dial-up lines like the Bell 103 and 113 used to operate at only 300 bits/s. The need for faster transmission resulted in developing modem technology for full-duplex 1.2-kbit/ units and on to 2.4-kbit/s sets that operate on the switched telephone network. An operator reading 120 characters/s needs 1.2 kbits/s if the characters contain 10 bits—1 start bit, 7 information bits, 1 parity bit, and 1 stop bit. New modems attain this data rate on the switched telephone network or on a two-wire leased line.

While 1.2-kbit/s modems are used with terminals on dial-up lines, you also find 2.4-kbit/s modems on either dial-up lines or leased lines that are connected in a multipoint configuration. Bell System's 212A (1.2 kbits/s) and Codex's LSI 24/24 (2.4 kbits/s) operate asynchronously or synchronously with respect to a terminal. Vadic's Series 34XX modems appear asynchronous to a terminal. All three transmit and receive synchronously.

Synchronous modems that operate at 4.8 to 9.6 kbits/s on leased circuits are considered high speed, and special attention must be paid to conditioning and equalizing at these speeds. Particularly suited for multiplexing, high-speed terminal-to-computer applications and in airline-reservation systems, many of the high-speed modems sold by independent suppliers are not Bell System compatible.

13-2.2 Transmission Synchronizing

Asynchronous data usually come from low-speed terminals with data rates below 2 kbits/s. In asynchronous systems, idle transmission lines are in a mark (or 1) state. A start bit, or transition from mark to space (or 0), precedes each transmitted character and indicates to the receiving terminal that a character is being transmitted. The receiver first detects the start bit, then the data bits that make up the character. At the end of the transmitted character, one (or more) stop bits return the line to mark, where it gets ready for the next character. The process is reported for each character until the entire message is transmitted. The start and stop bits permit the receiving terminal to synchronize its circuitry to the transmitter on a character-by-character basis. EIA RS404 specification governs the procedures for start/stop operation.

13-2.3 Transmission Codes

Asynchronous character lengths depend on the code used: 5 bits for Baudot code, 7 with an additional optional parity bit for the American Society for Communications Information Interchange (ASCII) code, and 8 for EBCDIC (see Figure 13-10).

With synchronous transmission, an internal clock synchronizes the channel's transmitter and receiver. Once the receiving terminal senses one or more syn-

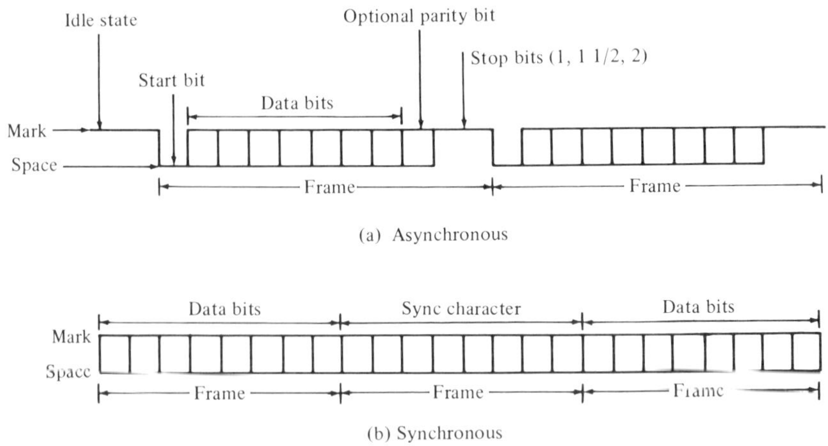

FIGURE 13-10. Asynchronous and synchronous codes.

chronization characters, data transmission proceeds character by character without any intervening start or stop bits. An incoming modulated carrier containing data blocks or frames is interpreted by a modem's receiving section with the help of a phase-locked loop. The PLL produces a receiving clock by locking on transitions in the received data. The receiving terminal, then, samples each bit in the receiving data stream using the modem-supplied receiving clock.

Clock derivation (or bit synchronization) depends on transitions in the serial-data input. To prevent a string of 1's or 0's from causing a synchronous modem's PLL to lose bit synchronization, modems include scrambler/descrambler circuitry. Before transmission, a scrambler changes the data pattern to ensure there are enough transitions. After reception, a companion descrambler restores the original data. Modem scrambling algorithms often differ.

13-2.4 Message Termination

The receiving terminal accepts data from its modem until it detects a special character or character-ending count that indicates the message is over. The particulars of the termination character or count depend on the data-link control procedure (or protocol).

The simple modem circuitry required for asynchronous transmission makes it advantageous for irregular transmission below 2 kbits/s. On the other hand, synchronous transmission uses the transmission facility more efficiently by eliminating the start and stop bits on each character. Furthermore, with multilevel modulation, synchronous data rates can be higher than asynchronous. The multilevel technique encodes 2 or 3 bits in each signal sample through four or eight distinct levels. The result is serial-data rates of 2.4 and 4.8 kbits/s with a 1.2-kbaud signaling rate. The hardware price of synchronous modems is higher, however, because of the precise sync required between data and clock, multilevel modulation, and adaptive-line equalization.

13-2.5 Modulation in Modems

Whether the best design is the dial-up network or leased line depends largely on the modem's modulation scheme for transmitting data on a phone line. Some modulation techniques permit higher transmission rates than others, and all directly affect the maximum data and error rates.

Three basic modulation schemes are used in modems:

- Frequency shift keying (FSK).
- Amplitude modulation (AM).
- Phase modulation (FM).

Figure 13-11 shows representative waveforms for each of the common modem-modulation schemes.

FSK, the most popular technique, is a form of frequency modulation. An FSK system shifts a carrier frequency (say, 1700 Hz) 500 Hz higher to represent a 1, and 500 Hz lower to represent a 0. Thus 1200 Hz represents a 0, while 2200 Hz represents a 1. Generally suitable for low-speed devices like teleprinters, FSK allows operation up to 1800 bits/s.

With AM, a modem can transmit and receive carrier amplitudes that represent binary 1's and 0's. Several AM levels are possible, allowing more data to be sent in the same period. Although AM uses bandwidth more efficiently than FSK, FSK has a signal-to-noise advantage over AM. AM is frequently combined with phase modulation in multilevel modulation schemes, particulary at 9.6 kbits/s. Some specialized AM forms are single-sideband, vestigial-sideband, and quadrature AM (or four-level AM).

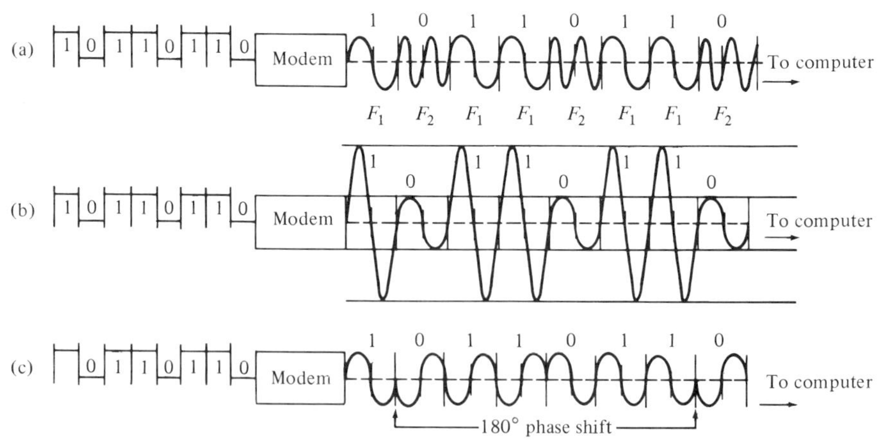

FIGURE 13-11. Modulation schemes: (a) FSK; (b) AM; (c) PM.

PM modems are generally described in terms of the number of phase shifts they use. The most popular are four and eight, which support bit rates up to two and three times the baud rate, respectively. PM shifts a transmitted signal's phase by a set number of degrees in response to the incoming bit pattern. For example, in a two-phase modem a 180° shift indicates a change of state—to a 1 (or 0). Thus no shift represents a series of 1's (or 0's), which eventually ends with a 180° shift. Most 4.8-kbit/s modems use PM, while the 9.6-kbit/s units often use quadrature amplitude modulation (QAM). The QAM method uses two amplitude levels and four distinct phases.

13-2.6 Multiport Modems

When a need exists for several input–output devices at one end of a voice-grade communication line, a multiplexer (*mux*) or a modem-sharing unit can be used. There are many to choose from and this will lower communications cost by sharing one communication line (see Figure 13-12).

Multiplexers take relatively low-speed inputs from a number of terminals and combine them into one higher-speed synchronous data stream for simultaneous transmission on a single channel, usually a type 3002 four-wire full-duplex leased telephone line. At the receiving end, a *demux* reconverts the high-speed data stream into a series of low-speed inputs for the host computer. The channel is split into preassigned time slots when time-division multiplexing (TDM) is used, or frequency bands with frequency-division multiplexing (FDM). A modem is inherent in FDM systems but not in TDM systems, which use added 4.8- or 9.6-kbit/s modems.

The increasingly popular intelligent or statistical muxes (IMs) allocate time

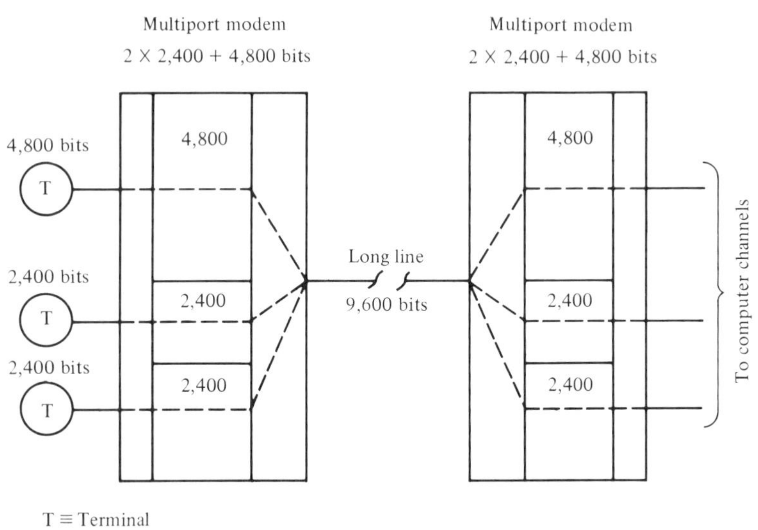

FIGURE 13-12. Multiport modems.

slots among channels through a line-activity-based algorithm. The microprocessor-controlled muxes usually have buffering and retransmission capability as well as network diagnostics. Up to 12 data sources or sinks can share one FDM line; up to 16 can share an intelligent-muxed line; and up to 24 can share a time-division-multiplexed line.

Multiport or split-stream modems partition a high-speed channel into various lower-speed subchannels. For example, a 9.6-kbit/s channel can be subdivided into a set of four 2.4-kbit/s paths or a combination of two 2.4-kbit/s and one 4.8-kbit/s paths. With channel splitting, two, three, or four medium-speed terminals can share one 9.6-kbit/ terminal.

13-3 LSI-BASED TELECOMMUNICATION SYSTEM DESIGN

One writer has written that designing systems that communicate is "like standing on the shoulders of giants." Already in place are systems such as the switched public-telephone network. Along with this, complex communications subsystems come loaded on LSI chips. Thus data and telecommunications equipment designs use these small, reliable, and yet inexpensive building blocks.

An obvious starting point is the interface between the communications equipment and the transmission medium. By converting baseband-digital signals to ones compatible with analog local distribution lines, modems connect both worlds. The modem, once a complex system of products, shrinks to the chip level in chips such as the TMS99532. The TMS99532 can be acoustically coupled to a handset as shown in Figure 13-13. It is housed in an 18-pin DIP (dual-in-line plastic package). Except for the crystal and a few other parts, the 99532 contains everything needed for a duplex modem. For directly connecting to the phone lines, the phone company (and FCC) insist that a specially approved direct-access arrangement (DAA) be used if acoustic coupling is not employed (see Figure 13-14).

The 99532 employs asynchronous, frequency-shift-keyed (FSK) modulation at speeds up to 300 bits/s. It is compatible with the Bell 103J data-set specifications (see Table 13-3). The FSK modulator circuit operates in a continuous-phase mode, during mark/space frequency transitions. The chip's answer/originate (A/O) input selects between the pairs of frequencies and the transmit-serial-data (XMTD) input selects between mark and space carrier frequencies. Bringing the answer-tone-enable (ATE) pin low selects the answer tone and disables the A/O XMTD pins. The answer tone normally is used in European applications. In the United States, Bell-103-compatible modems employ the high-band mark-carrier signal of 2225 Hz as an answer tone. The

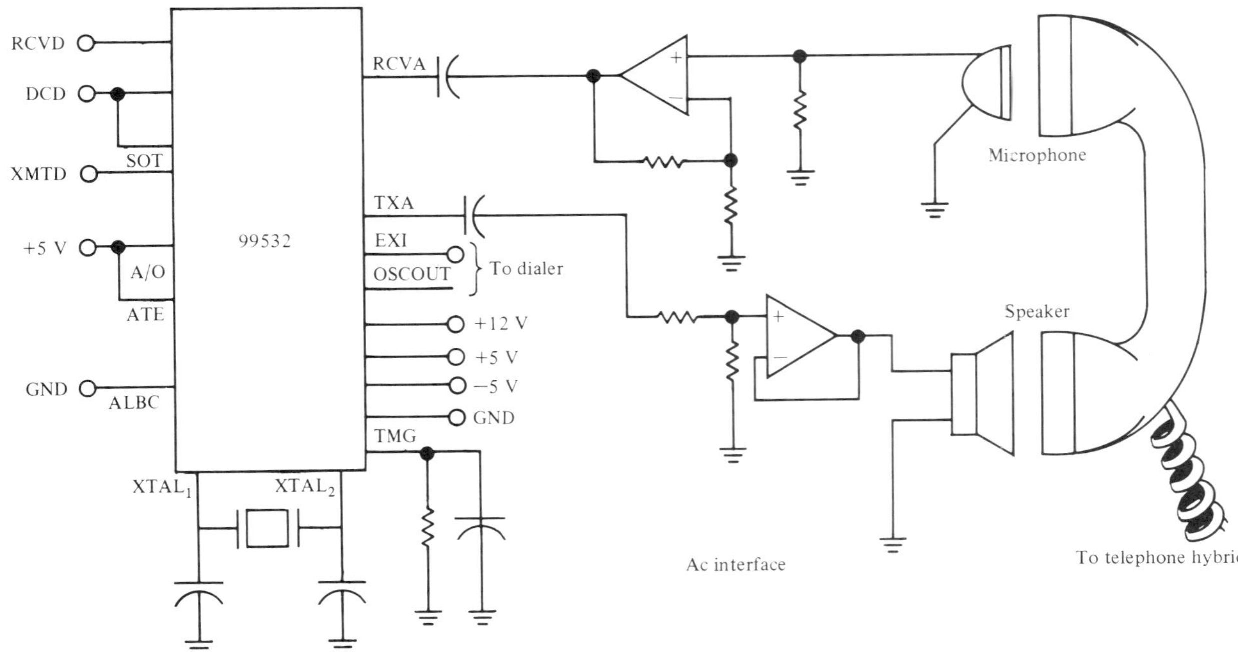

FIGURE 13-13. Modem handset interface.

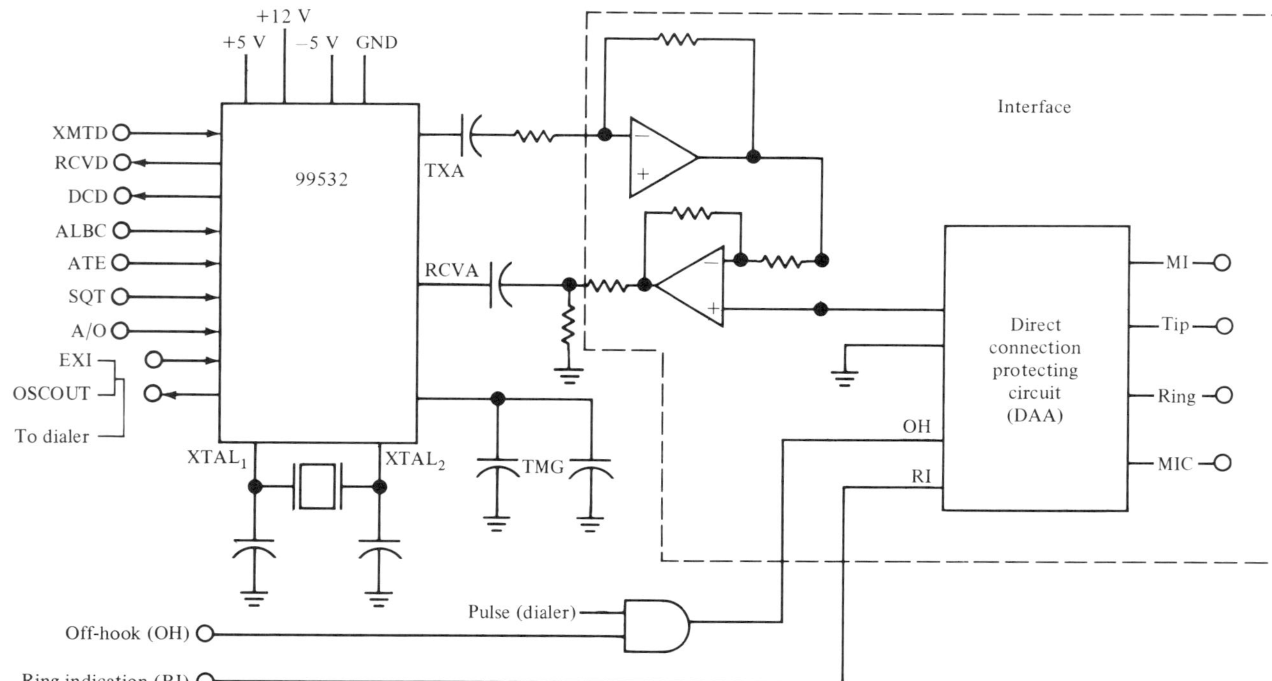

FIGURE 13-14. TMS modem with protection circuitry.

TABLE 13-3 Bell 103J Data-Set Specifications

Function	Specification
Data rate	0 to 300 bits/s
Modulation type	Phase-coherent FSK
Carrier frequencies	
Answer mode	
Mark	2225 Hz } high
Space	2025 Hz } band
Originate mode	
Mark	1270 Hz } low
Space	1070 Hz } band
Diagnostics	Local analog loopback Local digital loopback
Line configuration	Switched
Data-terminal equipment interface	RS232C
Transmission level	−4 to −12 dBm
Reception level	−12 to −50 dBm
Carrier-detection level	
On	−43 dBm
Off	−48 dBm

electrical and functional line specifications that must be met are contained in Table 13-1.

13-4 DATA-COMMUNICATIONS PROTOCOL STRUCTURE

Anyone applying data to lines for remote reception must become familiar with the standard software languages, codes, and protocols available for data-link control (DLC). Only then can we design the most efficient teleprocessing software and hardware.

A DLC is simply a set of rules for orderly information interchange between physically linked DTEs. It is the second level of a layered protocol architecture that has been worked out by the various standard-making bodies. The DLC should be able to support switched point-to-point, leased point-to-point, and leased multipoint channels. There are three categories of link protocols: character controlled, character count, and bit oriented (Table 13-4). Standards makers such as the American National Standards Institute (ANSI), International Standards Organization (ISO), Consultative Committee for International Telephone and Telegraphy (CCITT), and the U.S. government have adopted bit-oriented DLC procedures as a Level 2 protocol.

TABLE 13-4 Level 2 Protocols Defined

Character-Controlled	*Description/Name*
ANSI X3.28	Character-controlled ASCII
ISO 1745	Basic mode for IA5 (ASCII)
ECMA STD 16	Same as 1745
IBM BISYNC[a]	Binary synchronous communications
Many others	
Character Count	
DEC DDCMP	Digital data communications message protocol
Bit-Oriented	*Publication*
ANSI ADCCP	BSR X3.66
ISO HDLC	ISO 3309, 4335
	DIS 6159, 6256
CCITT LAP/LAPB	X.25 Recommendation
ECMA HDLC	ECMA - 40
U.S. govt HDLC	FED STD 1003
+ optional 32-bit FCS	
IBM SDLC[a]	GA27-3093-1

[a]One manufacturer's version of a U.S. or international standard.

13-4.1 Protocol Rules

Rules must be obeyed for orderly information exchange. A protocol is a set of laws that encompasses electrical, mechanical, and functional characteristics of a data communications link, as well as the control procedures for transferring data across the interface and into the hands of user application programs. Standards-making organizations, such as the the Electronic Industries Association (EIA), American National Standards Institute (ANSI), International Consultive Committee for Telegraphy and Telephony (CCITT), International Standards Organization (ISO), and U.S. government (FED) have divided protocols into layered architectures. Each distinct level clearly defines various data communications functions and logical operations. Each level is functionally independent of the others, but builds on these preceding it. To function, higher levels depend on correct operation of the lower levels. Data are transparent from higher to lower levels.

ANSI task groups X3S3 and DISY have defined a five-level and a six-level architecture respectively. ISO's SC-16 group and ECMA have come up with seven levels. Each of these layered protocols is summarized in Table 13-5 and the levels performing similar functions appear next to one another in the architecture control-level comparison table. The blocks in the diagram depict the layers of the SC-16 reference model. Here, Levels 1 through 4 deal with data-communications transport service, while Levels 5, 6, and 7 must be implemented

TABLE 13-5 Seven-Level Protocol Architecture

7	Process control	⟷	Process control	7
6	Presentation control	⟷	Presentation control	6
5	Session control	⟷	Session control	5
	//////////	⟵ Transport service access ⟶	//////////	
4	Transport end-to-end control	⟷	Transport end-to-end control	4
3	Network control	⟷	Network control	3
2	Link control	⟷	Link control	2
1	Physical control	⟷	Physical control	1

in the transport-service-user's software. The seven levels of ISO's hierarchy are as follows:

Level 1: Physical, electrical, and functional interchange that establishes, maintains, and disconnects the physical link between the data terminal equipment (DTE) and the data-circuit terminating equipment (DCE), or between two DTEs. Examples are RS232C and RS449/RS422.

Level 2: Contains the functions that reliably transfer data over one link. This level is the data-link control (DLC); as such it controls the two network nodes that Level 1 physically connects. Examples of DLCs are given in Table 13-4.

Level 3: Provides the functions for intranetwork operation, such as addressing and routing over and above the Level 2 functions for individual links. This is the network control level. While level 3 facilities direct control of switching points, they do not provide for data transfer between the switching points. On a public data network, under the CCITT X.21 synchronous protocol, Level 3 establishes circuit-switched calls through defined control blocks between a DTE and the network. In a packet-switched network, under the CCITT X.25 Recommendation, Level 3 controls the DTE-to-network logical connections that span multiple links. ANSI X3-281 is an attempt to define code-independent message handlings.

Level 4: Provides transport service users with network-independent interfacing for end-to-end (source-to-destination node) control and information exchange across either simple or complex networks. In addition to data-flow control, Level 4 ensures a reliable connection and verifies data.

Level 5: Supports dialogue through session control, if the program is free and if services can be allocated.

Level 6: Provides transformation of information transferred, on request. The presentation controls in Level 6 include compaction, encryption, peripheral-device coding, and formatting.

Level 7: Performs application and system activities in support of the enterprise's total information-processing function.

Whatever category a DLC operates under, it must perform the following functions:

- Initializing.
- Framing.
- Link management.
- Error control.
- Sequence control.
- Flow control.
- Transparency.
- Abnormal-condition recovery.

Initialization, or startup control, is the process of getting transmissions started in a communication system that has been idle. A primary station may be responsible for bootstrapping one or more secondary stations on an unbalanced point-to-point or multipoint link. The initialization process sets the initial logical states and control modes of all such secondaries.

Framing locates transmission-block beginnings and endings via message-delimiting bit groups. Each message constitutes one frame or block. Framing techniques for the three types of DLCs differ radically. Synchronization at the bit level is left up to the modem (DCE) or other hardware-interface components (UARTs or synchronous self-clocked DTEs).

Link management controls transmission and reception and supervises data-transfer sequences. This includes logically establishing and terminating a connection, choosing transmission direction, selecting sources, and identifying senders and receivers on multidrop or dial-up lines.

Error control involves detection of errors by means of vertical (VRC), longitudinal (LRC), or cyclic-redundancy checks (CRC). It also acknowledges correctly received messages and requests retransmission of faulty messages via a negative acknowledgment or reject command. A CRC can detect multiple-bit-error bursts, common on telecommunications channels.

Sequence control is responsible for numbering messages to eliminate duplicates, identifying lost messages, and properly requesting retransmission.

Flow control regulates the passing of messages across the link. The maximum number of frames that can be sent prior to an acknowledgment depends on the retransmission-request scheme of the DLC in use. The most popular of these are the frame-by-frame STOP and WAIT and the continuous GO BACK N types. These require, respectively, either a positive and negative acknowledgment (ACK/NAK) after each message is received, or that the frames remain unacknowledged until the maximum number is reached.

In continuous systems, transmitters and receivers keep a separate send and receive frame count, with the receiver counting only errorless frames. The transmitter compares its send and receive counts when it receives a request for retransmission (Reject command). The difference, N, is the number of frames that must be retransmitted. ADCCP/HDLC's Selective Reject Supervisory command requests retransmission of one specific frame, thereby implicitly acknowledging all others sent and received.

It is clear that STOP and WAIT restrict transmission to two-way-alternate (half-duplex) modes, while GO BACK N permits two-way-simultaneous (full-duplex) operation. However, unacknowledged frames must be stored temporarily in a buffer memory to provide for possible retransmission.

Control messages can also regulate the flow of information. A BOP (bit-oriented-protocol) station that is temporarily unable to receive a new frame can issue a Receive Not Ready command (RNR). (The station's buffers may be full or it may be busy processing the last frame.) When the station is able to receive, it sends a Receive Ready command (RR).

13-4.2 Bit-Oriented-Protocol Transparency

Transparency allows a link to treat all transmitted and received data, including normally restricted control characters, as just a pure data steam. Thus a transparent DLC can accept diverse binary coding, such as floating-point and packed BCD, as well as unique codes and machine-language programs. However, a higher-level protocol (Level 5 or above) must distinguish between the information and control characters of the code set in use: Hollerith, ASCII, or EBCDIC. Transparency is inherent in BOPs and in count protocols. In character-control schemes, a special transparent operating mode is required.

Abnormal-condition recovery encompasses the actions that must be taken when a DTE detects an illegal condition or receives a protocol-defined exception message. Examples are sequencing errors, time-outs when the message flow ceases, and illegal commands, addresses, or responses. No response or acknowledgment after a time-out means that either the station or the link is down.

Synchronous transmission normally maximizes channel-information content and so is used for transmission in most DLCs (exceptions are asynchronous

ANSI X3.28 and DDCMP). All DLCs transmit information as blocks (or frames), with the general message format shown in Figure 13-15a. The frame is subdivided into Header, Information, and Error Checking fields, any of which can be optional, depending on the DLC.

One strength of BOPs: all messages and link configurations use the same frame format (Figure 13-15b). Thus BOP systems are independent of codes, line configurations, and peripheral devices. One distinguishing feature of a BOP with respect to character-controlled protocols is that it defines fields within a frame by bit position rather than control characters. Figure 13-15 shows that a BOP's frame always includes FLAG, ADDRESS, and CONTROL fields. The FLAG field is fixed in both length and pattern: 01111110. After recognizing a FLAG, each secondary station on a link looks for its address to come up. The ADDRESS field in ADCCP/HDLC can be extended recursively to accommodate any number of stations.

The CONTROL field is the hub of a BOP message and can extend for one or two bytes in ADCCP/HDLC. This field determines such vital processing parameters as message type and send and receive frame counts. Through its CONTROL field, a BOP must recognize only the three message types shown in Figure 13-16.

- Information transfer (bit 1 of byte 1 = 0).
- Supervisory control (bit 1 = 1, bit 2 = 0).
- Nonsequenced commands and their responses (bit 1 = 1, bit 2 = 1).

The Information field can be any length—even zero bits—but in IBM SDLC it must be a multiple of 8 bits. Outside the defined field, in addition to the FLAG, a BOP must deal with only two reserved bit patterns:

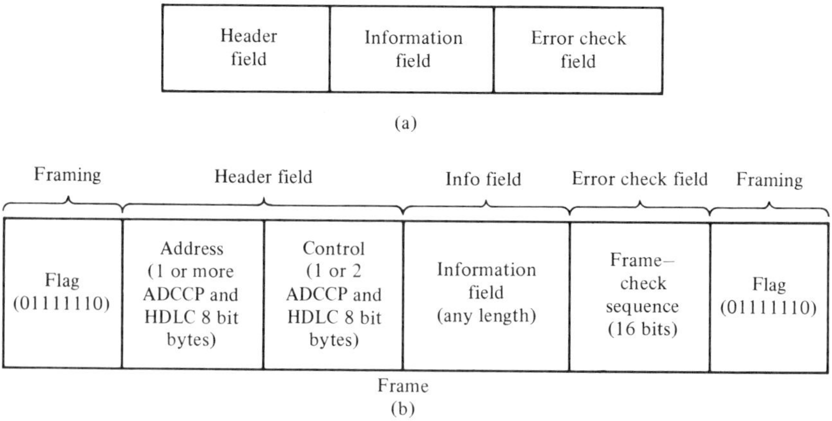

FIGURE 13-15. Message format for synchronously transmitted data.

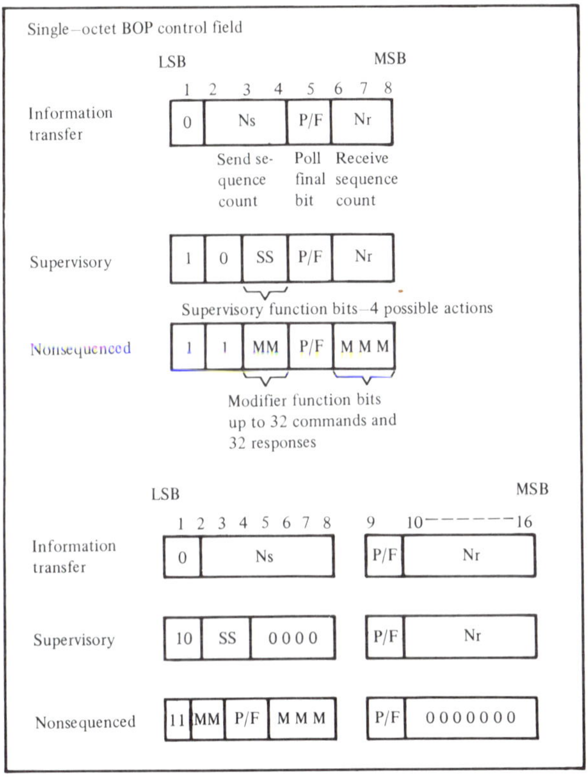

FIGURE 13-16. Control fields for BOPs can span one or two bytes. Whether a single octet (a) or two 8-bit bytes (b), the first 2 bits of the first control byte establish information-transfer, supervisory, or nonsequenced operation. The least significant bit is transmitted and received first.

ABORT = 01111111... (7 to 14 1's)

IDLE = 111111111111111... (15 or more 1's)

An ABORT terminates a frame prematurely. The system can use this unique code when its transmitting station has a problem, such as an underrun. FLAGs following an ABORT keep the link active, so transmission can continue.

The IDLE code identifies an inactive link state. When a half-duplex station detects the IDLE pattern, it can reverse the direction of transmission (turn the line around) and then transmit. Also, an asynchronous response mode can be activated after IDLE detection.

The representative character-oriented protocols in Figure 13-17 rely on an optional header field to specify control parameters, while the information field holds communications-control characters. The result: DLCs operating under these protocols are sensitive to the information code set. In addition, character-controlled systems are heavy with overhead software required to recognize numer-

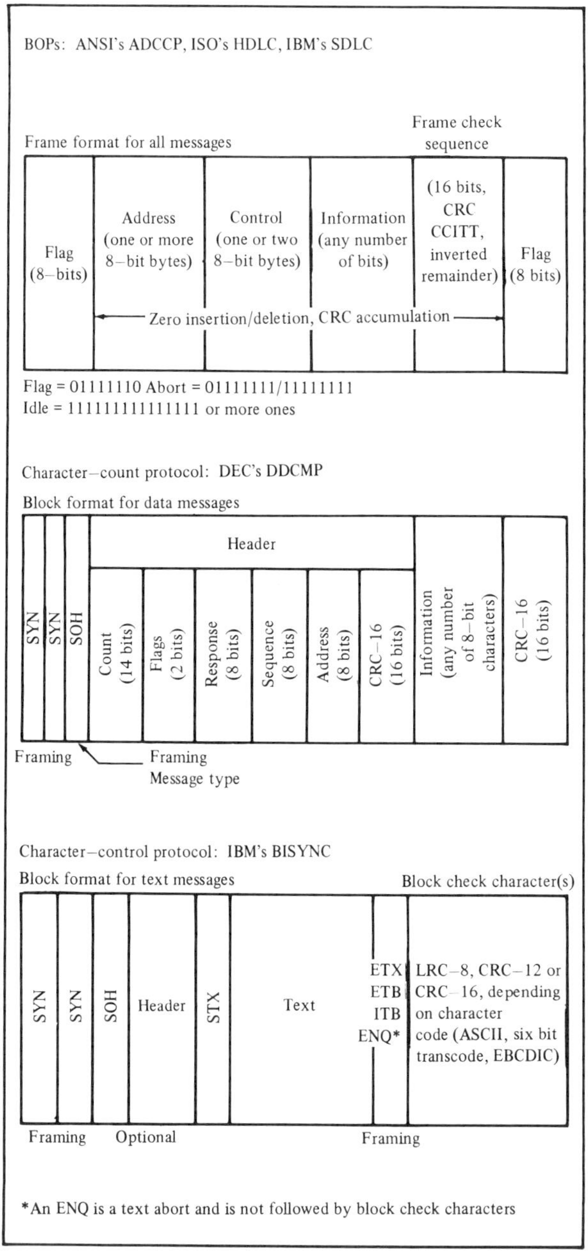

FIGURE 13-17. **In BOPs just about everything is arranged in 8-bit groups. The only special characters that must be searched out are FLAG, ABORT and IDLE. All the rest fall into their predetermined positions. Character-count systems are more complex and in character-control protocols the system must decode many separate characters.**

ous control characters and character sequences. Various message formats handle text (data), control, and acknowledgments. Station selection depends on the line configuration, with a special polling sequence for a multipoint line.

13-5

NETWORK ARCHITECTURE

The key to efficient distributed processing is sharing—of communication lines; of computer hardware and software; of terminals, peripherals, and data bases; and of control over network resources. In fact, it is the sharing of resources within a system that has made distributed processing and networking so popular as a strategy for linking remote computers, terminals, and storage devices. But since the sharing must accommodate all potential users in a highly demand-oriented environment, it is essential to select the right networking scheme. That requires knowing the various network architectures and protocols.

Most computer manufacturers already offer architectures for distributed processing, while emerging private and public data networks promise to slash the cost of the teleprocessing inherent in remote computing. Mainframe and minicomputer-based network architectures share common objectives. These are:

- Accommodate functionally different terminals and peripherals.
- Accommodate multiple-host computers and other intelligent equipment, such as front-end processors, concentrators, and multiplexers.
- Make the network transparent to end users and application programs via standard interfaces and protocols.
- Support communications channels such as leased point-to-point, leased-multipoint, circuit-switched, and packet-switched.
- Manage network-element changes for orderly system reconfiguration.

These objectives require the clear-cut functional separation of layered protocols. Layering insulates functions from one another by placing standard interfaces between layers. Thus a high-level function, like an application program, is not affected by lower-level peculiarities in, say, the Level 2 data-link control or the Level 1 physical-link control. In fact, a change at one level should not affect operation at any other level. For example, a Level 2 switch from BISYNC to HDLC or a Level 1 change from RS232C to RS449/RS422 should not be felt at any higher level.

The protocol layers dealing with network architecture begin with Level 3 of ANSI's five-level X.3S3 Communications Network Control. Because Level 3 defines information formatting and control procedures for establishing a logical path between a network's origination and destination nodes, it is often called the *path level* by computer vendors.

13-5.1 Choice of Direction

Physically, the path may be fixed (leased or circuit-switched) or variable (packet-switched) during a communications operation. In either case, destination nodes are logically identified within the network and translated into physical addresses.

Once a path is established, the network is responsible for maintaining error-free source-to-destination communications, and for managing the channels used. When intermediate intelligent nodes are involved, each path covers more than one data link. Figure 13-18 shows a scheme in which an intelligent concentrator and a front-end processor (FEP) are connected via a path that joins three data links. Private wire pairs connect the FEP and host.

The FEP and/or host contain Levels 4 and 5—the system user and control layers—of a total system protocol that takes the form of a telecommunications operating system. Pairs of end users accessing the network as sources and sinks are logically connected through Levels 4 and 5. These users, ranging in diversity from application programs to I/O devices to data files, gain network access through nodes linked via Level 3.

Once the network is accessed, Levels 4 and 5 ensure communications between these paired users, which might include a remote-site CRT operator interacting with an application program or data file at the host-computer site. Also a nodal processor's application program could be talking to an application program in the host. Specifically, the host could send the nodal processor a program to execute—a procedure called down-line loading.

Although all network architectures generally adhere to the same connection logic and layered structure of Levels 3, 4, and 5, each vendor allocates different functions to different levels. Thus there is no direct mapping between the same-numbered levels of the architectures for different vendors. Moreover, comparable devices from different manufacturers cannot access one another in any one network, which consequently can support only its vendor's products. In addition, there is usually no provision for linking different vendors' networks, although some do attempt a connection to IBM's SNA.

There are other drawbacks to any computer vendor's network architecture. First, high startup costs inhibit small-scale or pilot operations and make incremental expansion expensive. Before making a major commitment, a prudent user should insist on a limited tryout.

Second, network management is the user's responsibility. This includes monitoring performance, reporting status and usage, and isolating faults. Essentially, then, the network user is forced into the data-communications business. Thus networks are hard pressed to accommodate new applications and are unable to add advanced equipment not supported by the vendor.

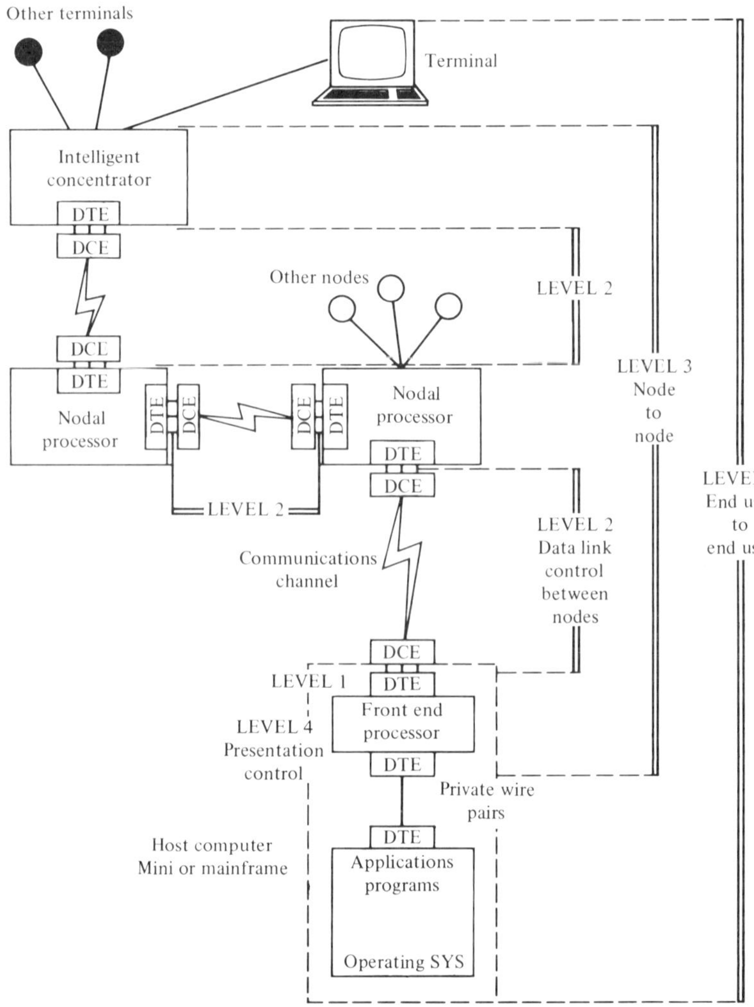

FIGURE 13-18. Distributed processing under the ANSI X.3S3 five-level protocol architecture joins this system's intelligent concentrator and front-end processor via three data links. Private-wire pairs connect the front-end processor and host. ISO has standardized on a seven-level architecture.

13-5.2 Public Data Networks

Solutions to these private-network problems are offered by public data networks (PDNs), which comply with international standards. PDNs exist or are being installed in Canada, Europe, Japan, and the United States. In addition, packet switching, which has spearheaded recent PDN activity, appears to be well on its way toward international acceptance. The protocol for operation with networks providing virtual-circuit package services is CCITT Recommendation X.25.

X.25-based packet-switched networks use leased lines as transmission links and minicomputers for store-and-forward message switching and network control. The access-node minicomputer is called the DCE (data circuit-terminating equipment). The actual user (the DTE, for data terminal equipment), formats data into fixed-length blocks called *packets*. Several different packet types are sent between DTE and DCE.

The X.25 protocol presents a *virtual circuit* interface to a DTE, which means that the DTE gets as many as 4095 point-to-point links to other DTEs (hosts, concentrators, or terminals). Each virtual circuit is set up through special Call Request/Incoming Call and Call Accepted/Call Connected packets with a specific logical channel number for the source-to-destination path. An incoming call coincident with an outgoing call on the same logical channel is a Call Collision—the incoming call is then dropped while the outgoing call is routed through the network. Once the virtual call is set up, data packets can be exchanged. A Call Clear/Clear Confirmation packet performs the logical disconnect function.

The public packet-switched data networks typically operate central offices in major cities to which users can connect their computers and data terminals by means of telephone lines. These central offices contain computer-controlled interfacing and switching equipment and are connected to one another by multiple high-capacity communications channels.

A DTE gains access to the data network through a communications line that connects it to a DCE. If the DTE uses start/stop transmission, the "terminal handler" software must follow the procedures specified by CCITT X.3, X.28, and X.29 to interface that DTE to the packet network. A synchronous DTE must have the required Level 1, Level 2, and Level 3 X.25 interfaces to access the network directly.

The packets of data are transmitted from node to node over optimal long-haul paths, which can be assigned on a per packet basis when used with adaptive routing. (Intranetwork routing is not covered under X.25.) At each intermediate node, the packet is passed to the receiver in a standard format.

Many of the packet-switched network's desirable characteristics result from the adaptive routing. Here, the network path between any two points is not chosen in advance, but is a dynamic function of network condition at any time. This way, a packet network can overcome adverse effects of temporary congestion and failed links or switches (Figure 13-19).

This dymanic allocation of bandwidth particularly benefits interactive data traffic since subscriber billing is usually on a per packet basis. Batch or high-volume communications is better served by conventional circuit-switched and leased-line facilities.

All the sophisticated network routing, monitoring, and error-correction techniques are invisible to the host-computer system. With the X.25 network-access protocol, the network appears to the host system as a remote concentrator at-

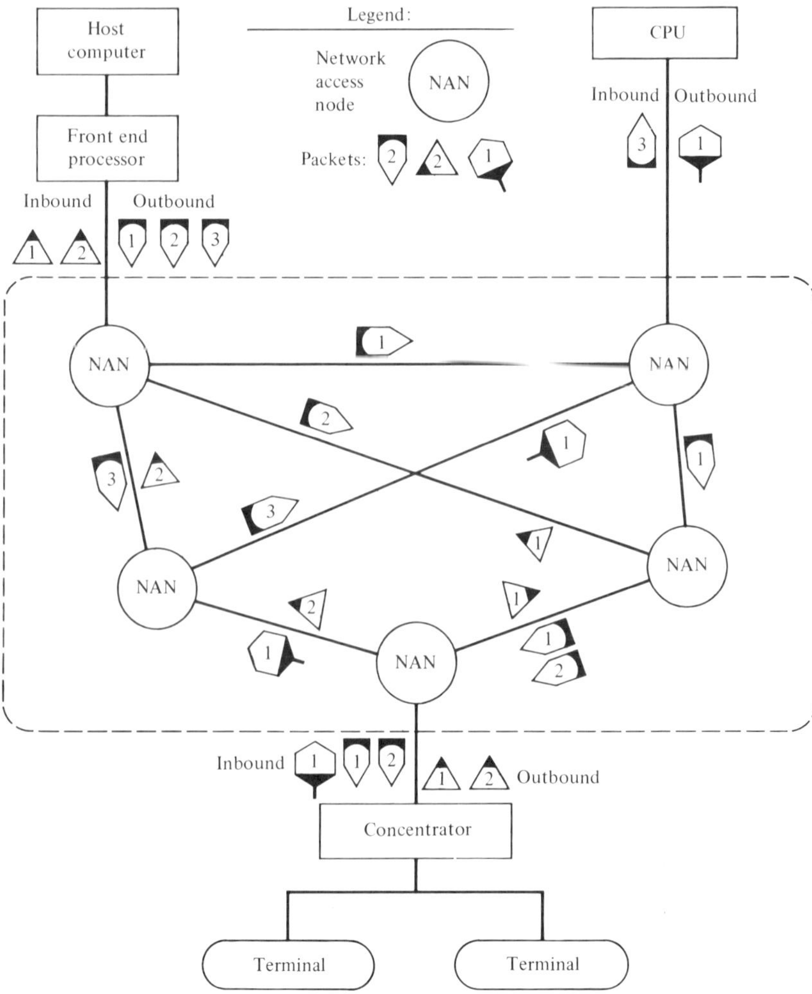

FIGURE 13-19. Most packet-switched networks do not choose a path between any two points in advance—the routes data travel are dynamic functions of the network conditions at any time. As a result, packet networking can overcome temporary localized congestion, failed links and faulty switches, and other transmission stumbling blocks.

tached to the access link. Consequently, the user's main concerns are the relatively simple procedures and structures for transmitting and receiving data to and from that "concentrator," and not its internal working.

13-5.3 Software Control Routines

Defined in X.25, these Level 2 and 3 procedures must be incorporated in the user's system as software routines. Typically, the routines reside in a programmable processor that serves as a front end to the host computer, as a terminal-cluster controller, or as a remote concentrator. New LSI chips for Level 2 of X.25 promise to off-load user software.

The virtual-circuit interface has been accepted by all packet-switching carriers throughout the world because it enables a teleprocessing computer to communicate with many devices simultaneously and efficiently. In packet-switched systems, in particular, the network is responsible for error control, flow control, path control, and internal management. What is more, CCITT X.25 is the first approved interface to describe explicitly three levels of data-communications architecture:

1. Level 1 specifies that X.21bis (equivalent to RS232C) applies for initial use, while the Level 1 characteristics of X.21 are intended as the standard.
2. Level 2 specifies bit-oriented link-access procedures (LAP or LAPB). LAPB provides proper alignment with the ISO HDLC balanced class of procedures (DIS 656). LAP and LAPB are point-to-point subsets of HDLC.
3. Level 3 defines the packet formats and call-control procedures between a DTE and the network for virtual-circuit packet-switched service.

A simpler interface—the Datagram service—allows users to send single-addressed packets through the network without prior-call setup or post-call clearing. Completely self-contained, the interface includes the network-control information and the intended receiver's complete address. (Datagram was accepted as an enhancement to X.25 at the CCITT meeting in April 1979.)

Each Datagram is handled independently by the network so that Datagrams are delivered with minimum delay. However, there is no guarantee against duplication, loss, or improper sequencing. As a result, the receiving DTE is responsible for such tasks as identifying each Datagram's source, detecting and recovering lost Datagrams, and reordering and processing undelivered Datagrams. Nevertheless, a Datagram interface is attractive in inquiry/response transactions such as point-of-sale, directory look-up, inventory control, and credit checking.

13-6 SATELLITE SERVICE TRENDS

The GTE Satellite Corp., the government of Australia, and the Arab Satellite Organization have become the latest parties to notify NASA of their intent to launch satellites from the Space Shuttle.

The GTE Satellite Corp. of Stamford, Connecticut, will use the Shuttle to launch its GSTAR-C communications satellite in May 1985. The satellite will be used to provide common carrier communication services to the continental United States, Alaska, and Hawaii. GSTAR-C will be built by RCA Astro Electronics Corp. and launched with the McDonnell Douglas Payload Assist Module. The PAM-D II to be used on that launch is an improved version of the original PAM.

Australia plans to use the Space Shuttle for the launching of two communication satellites in July and October 1985. The two satellites of the AUSSAT system will be used for direct television broadcast and voice, data, and telephone services to continental Australia and offshore territories. The satellites will be built by the Hughes Aircraft Co. AUSSAT PTY Ltd. is the Australian corporation responsible for development of the national satellite system

The Arab Satellite Organization, ARABSAT, signed a launch services agreement Nov. 23, 1982, for a satellite that will serve 22 member countries and organizations. The ARABSAT-A spacecraft, built by Aerospatiale and Ford Aerospace, is scheduled for launch from the Shuttle in October 1984. A McDonnell Douglas PAM-D will be used to boost the satellite into geosynchronous orbit. The Arab consortium will operate a public satellite system for telecommunications for its member entities.

PROBLEMS

13-1. A 1000-Hz 0.5-mW tone is present on a data communications line. Express this level in dBrn.

13-2. The interfering effect of noise is measured with a C-message filter. Express the signal level in Problem 13-1 in dBrnC.

13-3. The maximum data rate for a noiseless channel was shown to be given by

$$C = 2x \log_2 L \qquad \text{bits/s}$$

The channel capacity for a noisy channel is given by

$$C = x \log_2 (1 + S/N) \qquad \text{bits/s}$$

Assuming a 4-kHz line channel with four signal levels, compare these two expressions for a signal-to-noise ratio of 30 dB.

13-4. For the four-level coding scheme shown, what is the ratio between the baud rate and the bit rate?

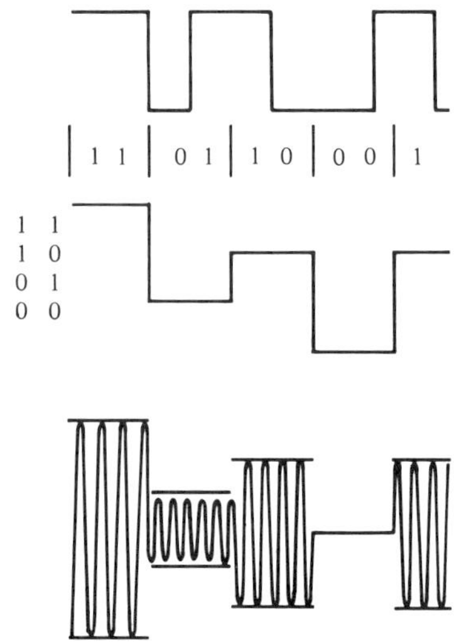

Four-level coding

PROBLEM 13-4

13-5. The block of data in the figure is transmitted using vertical and horizontal odd parity. Verify or correct the parity check bits.

Data flow ←

— Longitudinal check →

Vertical check ↓

1	0	1	0	0	1	1	0	0	
1	0	0	1	0	1	0	0	0	LRC
0	1	1	0	0	0	0	0	0	horizontal
0	0	0	1	1	1	0	1	1	parity
1	0	0	0	1	0	0	1	0	bits
0	0	0	1	1	0	1	0	0	(BCC)
1	1	1	0	0	1	1	0	0	
1	0	0	1	0	0	0	1	0	

Vertical parity bits

VRC/LRC checking (odd parity)

PROBLEM 13-5

13-6. A time-division-multiplex system uses 8-bit A/D converters and an 8-kbaud sampling rate for each of 10 PCM channels.

(a) Determine the equivalent baud rate per frame and the bandwidth required for transmission.

(b) The equivalent bandwidth that would be occupied by the 10 analog voice channels before PCM conversion.

13-7. An FSK data modem operates at 1200 baud with a 1200/2200-Hz frequency shift. Transmission is over a voice-grade telephone circuit.

(a) Determine the equivalent FM deviation of the FSK signal and the equivalent highest modulation frequency based on the baud rate.

(b) The equivalent FM modulation index.

13-8. Fill in the blanks for the circuit shown.

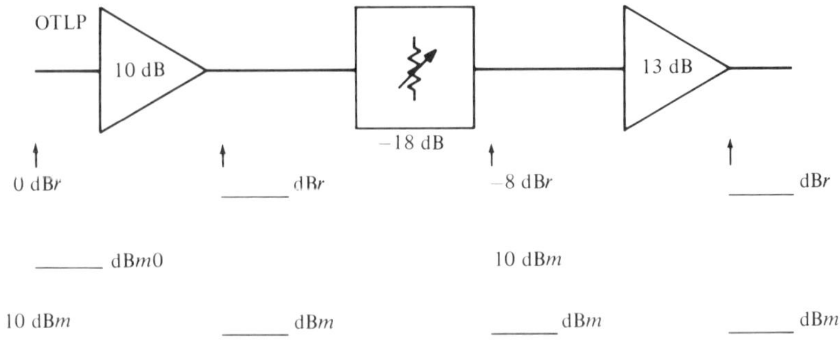

PROBLEM 13-8

13-9. What is the level in dBm of a 1000-Hz 0-dBrnC tone?

13-10. **(a)** Define the zero-level transmission point in a system.

(b) What is the definition of dBm0?

(c) What is the level in dBm0 of a +40-dBm tone at 20-dB relative transmission-level point?

13-11. If two noise sources produce 6 and 4 dBm of noise individually, what is the accumulative noise level in dBm?

13-12. Calculate the power and voltage gains of a +6-dB and a +8-dB amplifier. If the power level at the input of a 15-dB attenuator is 3 mW, what is the power level at the output?

13-13. Complete the following table:

$$\text{dBm} = 10 \log \frac{P(\text{watts})}{10^{-3}}$$

$$\text{dBrn} = 10 \log \frac{P(\text{watts})}{10^{-12}}$$

Power (W)	dBm	dBrn
10^1	40	130
10^0		
10^{-3}		
10^{-6}		
10^{-12}		

13-14. Describe simplex, half-duplex, and full-duplex transmission.

13-15. A telephone T1 carrier consists of 24 voice channels multiplex and PCM modulated. Each signal has a bandwidth of 300 Hz to 3.2 kHz. The encoder encodes to 8 bits, and the sampling rate meets the CCITT standard.

(a) Determine the bit rate of the T1 carrier.

(b) What baseband bandwidth is required to transmit this if the minimum bandwidth is half the channel capacity?

(c) Why does each voice channel require bandwidth limiting?

13-16. Why is companding used in voice transmission and not in data transmission?

13-17. What is the information capacity of a 3-kHz bandwidth channel that has a 30-dB SNR?

CHAPTER 14

Fiber Optics Communications

The fiber optics revolution is well under way. Optical fibers are tying together computers, instrumentation, telephones, and televisions all over the world. Tremendous advantages reside in the use of fibers. Fiber optics cables are smaller and lighter than copper conductor systems and offer total electrical isolation, widespread capability, and complete immunity to both noise and a broad spectra of interference. An additional attraction is the promise of low cost with volume. Best of all, fiber optics is unlocking the potential of integrated optics, which promises to process data in thousands of gigabits per second—literally, in a flash.

This chapter discusses the design aspects of using fibers. In fiber optics communication system design, several parameters enter into a designer's cost versus performance calculations. These may be grouped into three categories.

- The light source and associated drive circuitry.
- Optical-fiber and mechanical-cable construction.
- Photodetector and receiver circuitry characteristics.

A firm understanding of these parameters is mandatory for component selection. The topics in this chapter will be discussed in the following order:

- Optical-fiber transmission parameters.
- Detectors and sources.
- Systems considerations.
- Source drive and receive circuits.
- Data coding.

14-1 OPTICAL-FIBER TRANSMISSION PARAMETERS

The two parameters that introduce performance limitations of an optical-fiber system are (1) bandwidth (dispersion) and (2) optical power. These parameters define the systems repeater spacing. Because each repeater requires a back-to-back photodetector/transmitter arrangement, it is desirable to maximize repeater spacing or if possible eliminate it altogether. Dispersion and attenuation figures thus govern the bandwidth and power limitations of optical fibers. These factors may be mitigated somewhat by paying for higher-priced fibers. From an overall systems viewpoint, however, optical power and bandwidth limits also depend on the light source (modulation rates, radiance, optical coupling efficiency) and detector (responsivity, rise time, noise factors).

14-1.1 Dispersion

There are three basic types of fibers: (1) multimode step index, (2) single mode, and (3) graded index. A zero rise time in light intensity at the input of the fiber optic cable results in a nonzero rise time at the output. The mechanism of pulse dispersion is called *modal dispersion*. Dispersion traits vary with the fiber type. In considering the phenomenon of dispersion, the *index of refraction* (noted as n) is the best optical parameter to start with.

An optical fiber consists of a core surrounded by a cladding. A typical fiber is shown in Figure 14-1. Classically, for both glass and plastic, the index of refraction is defined as

$$n = \frac{c}{v} \tag{14-1}$$

where c is the speed of light in vacuum (3×10^8 m/s) and v is its velocity within the material.

Consider the behavior of light in the fiber from a light-ray approach. When a light ray inside the core meets the boundary between the core and the cladding

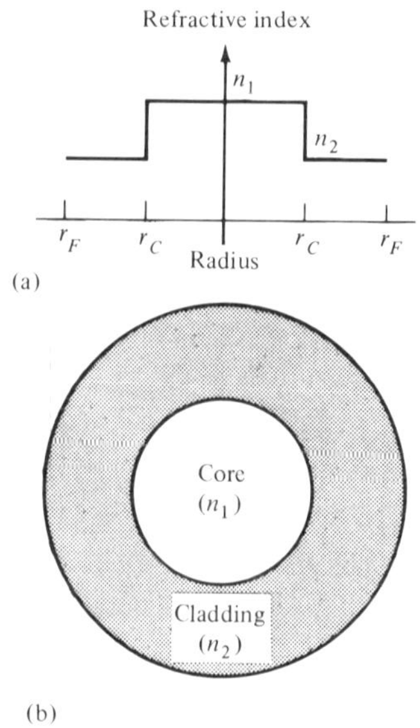

FIGURE 14-1. (a) The index profile takes a downward step at the core radius in a step-index optical fiber. (b) This step divides the fiber's cross-section into a central core with index n_1 and a surrounding cladding with index n_2.

at an angle (see Figure 14-2), total internal reflection occurs given that

$$\theta < \theta_c$$

where

$$\sin \theta_c = \frac{n_2}{n_1} \tag{14-2}$$

and $\theta = \theta_c$ is the critical angle. The critical angle is that angle of incidence beyond which the light will propagate into the cladding and be dissipated. Rays of light that enter the fiber at an angle less than θ_c will propagate down the fiber with little loss.

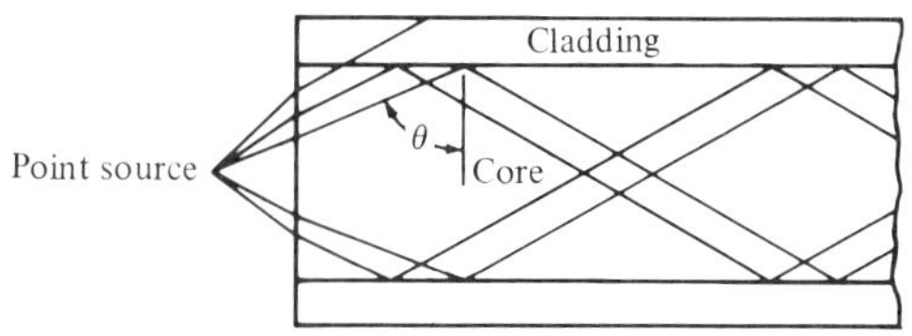

FIGURE 14-2. Light rays propagating down an optical fiber.

The *index profile* of a fiber refers to how the refractive index varies as a function of radial distance from the fiber's center. A fiber with a *step index* profile exhibits a refractive index that undergoes an abrupt change (step) in value at a radius r_f (Figure 14-1). In order that the optical energy propagate down the fiber, it is necessary that the core index of refraction, n_1, be greater than the cladding index, n_2. Thus air, with an index of refraction of 1.00, could serve as the cladding medium. Core support, however, is needed and this dictates the use of something other than air. The *cladding-to-core index ratio* is a prime factor in determining the dispersion of an optical pulse. Typical values for the indices of refraction are $n_1 = 1.48$ and $n_2 = 1.46$.

The geometrical propagation of light rays as indicated in Figure 14-2 is actually an approximate description of the method of optical energy propagation. Light is electromagnetic in nature and an exact analysis of the electromagnetic (EM) fields in the fiber involves a solution of Maxwell's equations in terms of the boundary conditions imposed by the fiber. This analysis shows that the optical energy is distributed among a discrete set of superimposed EM fields called *modes*. Differences in the propagation characteristics of these modes results in *modal dispersion*. As mentioned previously, this is one of the factors that limits the fiber's bandwidth.

EXAMPLE 14-1

An off-axis light ray enters a fiber at the minimum angle for total internal reflection (see the figure). Determine (a) the minimum (critical) angle, (b) the path length of the ray, and (c) the time delay of the off-axis ray compared to a ray that propagates straight down the fiber.

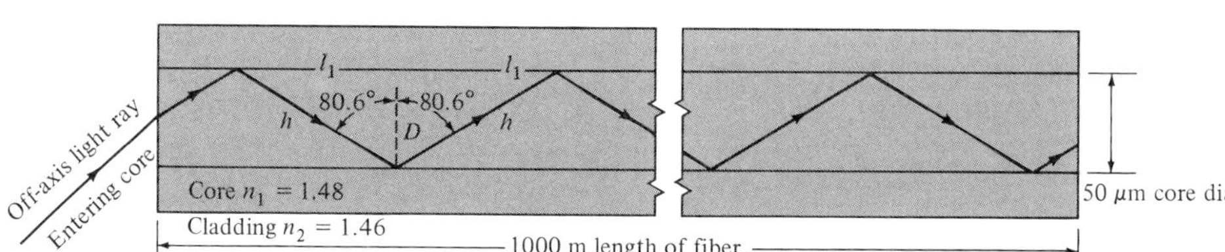

SOLUTION

(a) The minimum angle is obtained as follows:

$$\sin \theta_{\min} = \frac{n_2}{n_1} = \frac{1.48}{1.46}$$

$$\theta_{\min} = 80.6^\circ$$

(b) The total path length may be obtained by determining the distance $2h$ and then multiplying $2h$ by the number of distances $2l_1$ in the fiber.

$$\tan \theta_{min} = \frac{l_1}{D} \Rightarrow 2l_1 = 2D \tan \theta_{min}$$

$$2l_1 = (2)(50 \times 10^{-6}) \tan 80.6 = 6.0405 \times 10^{-4} \text{ m}$$

$$\frac{\text{no. } 2l_1}{1000 \text{ m}} = \frac{1000 \text{ m}}{6.0405 \times 10^{-4} \text{ m}}$$

$$= 1.655492 \times 10^6$$

$$\cos \theta = \frac{D}{h} \Rightarrow 2h = (2)\frac{50 \times 10^{-6}}{\cos 80.6°} = 612.27253 \times 10^{-6} \text{ m}$$

$$\text{path length} = \text{no.} \times 2h$$

$$= (1.655492 \times 10^{+6})(612.27253 \times 10^{-6})$$

$$= 1013.6123 \text{ m}$$

$$\approx 1014 \text{ m}$$

(c) The time delay is computed using equation (14-1):

$$n = \frac{c}{v} \Rightarrow v = \frac{c}{n_1} = \frac{3 \times 10^8}{1.48} \text{ m/s}$$

$$= 2.03 \times 10^8 \text{ m/s}$$

$$T_{\text{bouncing ray}} = \frac{1014 \text{ m}}{2.03 \times 10^8 \text{ m/s}} = 4995.0736 \text{ ns}$$

$$T_{\text{straight ray}} = \frac{1000}{2.03 \times 10^8} = 4926.1084 \text{ ns}$$

$$\Delta T = 69 \text{ ns}$$

Thus the bouncing ray reaches the end of the fiber 69 ns after the axial ray.

As indicated in Figure 14-2, there are many intermediate paths between the two ray paths considered in Example 14-1. The time delay (dispersion) in the arrival times of the rays produces bit smearing and consequently intersymbol interference. An order-of-magnitude estimate of the modal dispersion-limited bandwidth is often taken to be the reciprocal of the time delay. For the fiber in Example 14-1, this is

$$\text{BW} \approx \frac{1}{\Delta T} = \frac{1}{69 \times 10^{-9}} = 14.5 \text{ MHz}$$

The optical fiber of Figure 14-1 will support many modes. For this reason it is referred to as a multimode step-index type. From EM field theory an interesting parameter called the *normalized frequency*, V, is defined. Mathematically, it is expressed as

$$V = \frac{\pi d}{\lambda} \sqrt{n_1^2 - n_2^2} \tag{14-3}$$

where

d = fiber core diameter
λ = optical wavelength of the source

The number of propagating modes, N, in a step-index fiber may be estimated by the following relation:

$$N = \frac{V^2}{2} \tag{14-4}$$

EXAMPLE 14-2

Calculate the number of propagating modes of the fiber of Example 14-1. Assume red light, $\lambda = 6 \times 10^{-7}$ m.

SOLUTION

$$V = \frac{(\pi)(50 \times 10^{-6})}{6 \times 10^{-7}} \sqrt{1.48^2 - 1.46^2} = 63.45$$

$$N = \frac{V^2}{2} = \frac{(63.45)^2}{2} = 2{,}012.95 \text{ modes}$$

14-1.2 Single-Mode Fiber

From the preceding discussion we see that multiple modes results in dispersion with a consequent reduction in bandwidth. Obviously, a reduction in modes is desirable. Referring to equation (14-3), a reduction in mode volume can be accomplished in three ways: reduce the core diameter, increase the wavelength, or decrease the difference between the core and cladding indices.

Now, an increase in the wavelength results in greater absorption in the glass with attendant signal attenuation. For a single mode, $V = 1$, let us see what the core diameter would need to be for red light. Thus, from equation (14-3),

$$d = \frac{V\lambda}{\pi\sqrt{n_1^2 - n_2^2}} = \frac{(1)(6 \times 10^{-7})}{\pi\sqrt{1.48^2 - 1.46^2}} = 7.88\ \mu\text{m}$$

This fiber would offer no modal dispersion. Studies indicate that if $V < 2.405$, only a single mode (an axial ray) can propagate. This fiber, then, offers the ultimate in bandwidth. However, extreme difficulties are experienced in projecting light into such a small core. There is also a problem of alignment with connectors and fiber-to-fiber splices. Thus the final possibility is to make the ratio n_1/n_2 as small as practical, keeping in mind that $n_1/n_2 > 1$ must be maintained for total internal reflection.

14-1.3 Graded-Index Fiber

The third type of fiber mentioned was the graded index. This type provides greater bandwidths. In the graded index design (see Figure 14-3), the refractive index decreases continuously with radial distance from the center of the fiber. As demonstrated by Example 14-1, light propagation occurs through refraction. Thus, with a graded index of refraction there is a continuous bending of the ray toward the fiber's optical axis. The optimum graded-index profile for minimum modal dispersion is parabolic. The refractive index on the central axis, n_1, tapers off parabolically with radial distance to the value n_2 at the outer radius.

14-1.4 Fiber Optics Bandwidth Measurement

The response of a system to an impulse stimulus is a favorite technique for determining system bandwidth. The basis for this measurement technique lies in the fact that the frequency spectrum of an impulse is flat. The test setup for

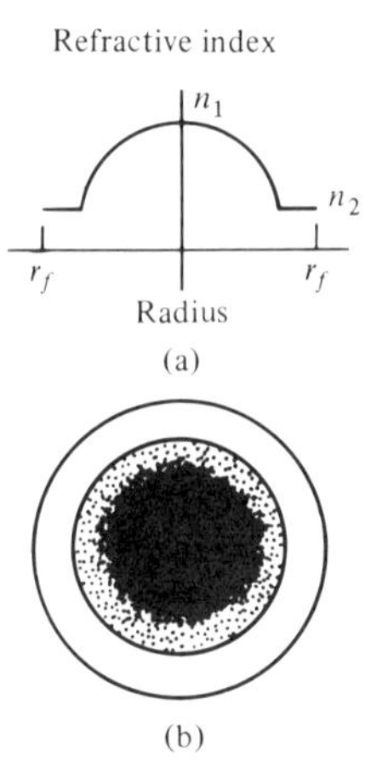

FIGURE 14-3. (a) The graded-index profile tapers off parabolically with radius from its central-axis value n_1 to a lower value n_2 at the fiber radius r_f. (b) The cross-section shows that light travels slower near the shaded center region than it does away from the center, resulting in more consistent arrival time and less dispersion.

this measurement is shown in Figure 14-4. The impulse response width is a function of the difference between the core and cladding indices. For a step-index fiber the modal impulse-response width varies linearly with the difference. For a graded-index fiber the width is proportional to the square of the difference. For example, for an index difference $n_1 - n_2 = 0.02$, the reduction is proportional to 0.0004 for a graded-index fiber. This results in a significant improvement. Typical values for 3-dB dispersion are:

- Step index, 15 ns/km
- Graded index, 2.5 ns/km

The specifications above imply that pulse broadening is proportional to fiber length. In actuality, it is proportional to the square root of the length for sufficiently long lengths. It is thought that coupling between the modes is responsible for this.

14-1.5 Material Dispersion

The refractive index is actually a function of the wavelength, λ. Thus in terms of the dispersive effect, this makes the spectrum of the light source assume immediate importance. Practical sources emit light over a band of frequencies. If the refractive index is not linear over this band of frequencies, material dispersion takes place.

Material dispersion's effect on the rms impulse response depends nonlinearly on the second derivative of the refractive index with respect to wavelength. The single-mode fiber is free from modal dispersion, but pulse broadening can still occur from material dispersion. Minimum material dispersion occurs in glass fibers at source wavelengths between 1.2 and 1.3 μm.

Figure 14-5 illustrates the bandwidth of two fibers 1 km in length. Note that these two fibers have bandwidths of 250 and 400 MHz, respectively. For a 5-km link, the 3-dB points would be 50 and 80 MHz, respectively.

14-1.6 System Losses

When designing an optical system it is necessary to compute a total loss budget for the system. The first order of business is to determine the optical power needed at the photodetector to ensure the desired signal-to-noise ratio (SNR) or bit error rate (BER). Various optical losses occur between the transmitter and the detector. We can divide these losses into the following:

- Input coupling loss.
- Connector/splice loss.
- Fiber attenuation.
- Output coupling loss.

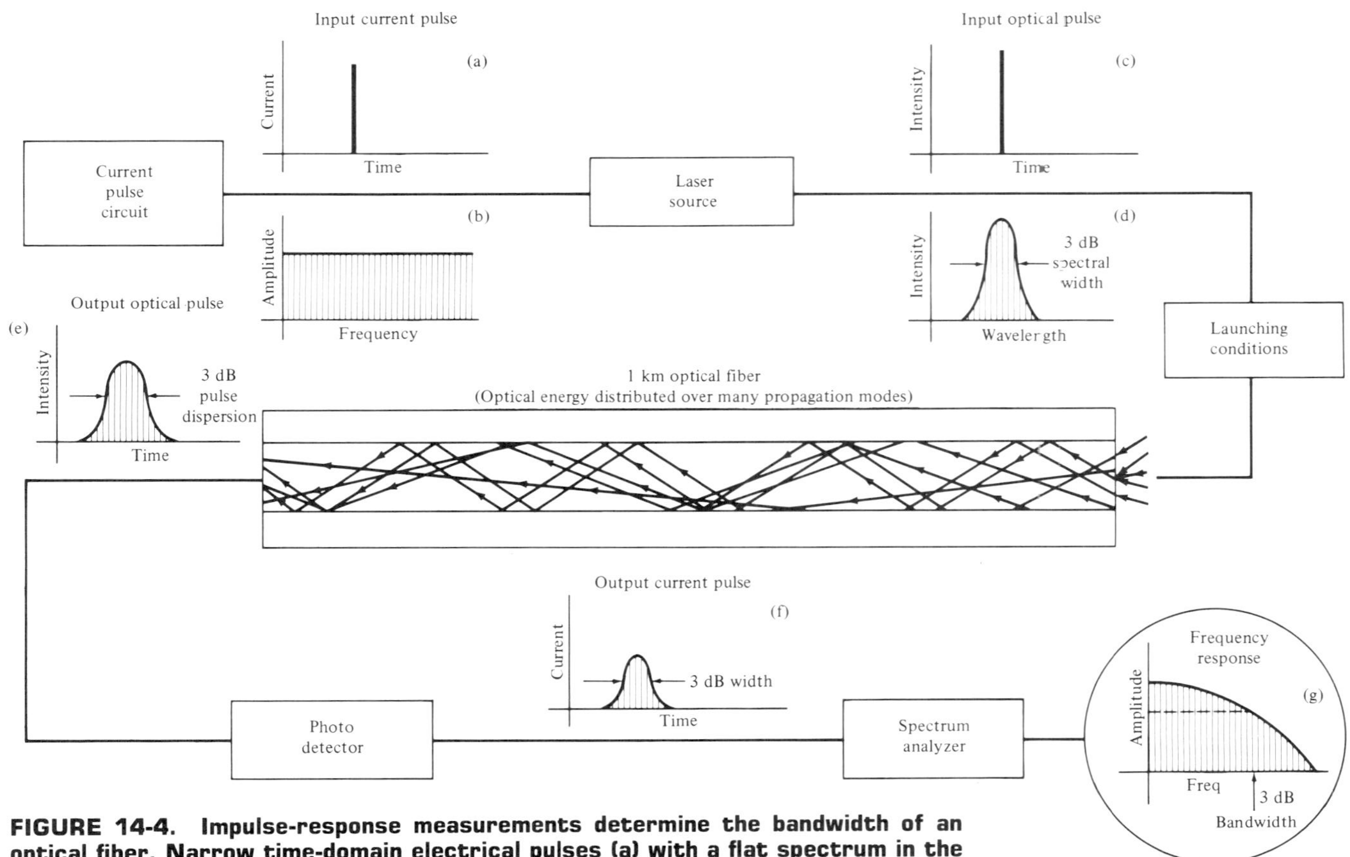

FIGURE 14-4. Impulse-response measurements determine the bandwidth of an optical fiber. Narrow time-domain electrical pulses (a) with a flat spectrum in the frequency domain (b) are converted to light pulses (c) with a wavelength distribution characteristic of the laser source (d). Depending on the input conditions, light energy propagates through many modes of the fiber under test. Because of modal and material dispersion, the output light pulse spreads with time (e) and is photodetected as a widened current pulse (f). A frequency-domain analysis then yields the fiber's 3-dB bandwidth (g).

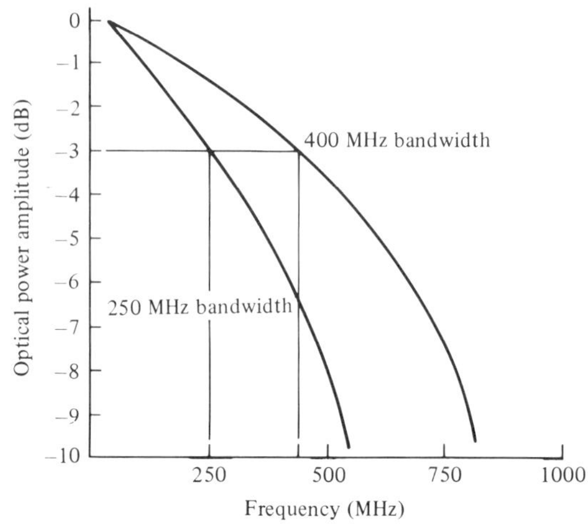

FIGURE 14-5. A fiber's 3dB frequency is identified by taking the Fourier transform of the output signal. Specs are usually quoted for 1-km fiber lengths.

Input coupling losses occur at the interface between the source and the fiber. A short length of fiber, called a "pigtail," is usually permanently attached to the source's emitting area in single-fiber communications. Loss at the interface can occur in two ways: (1) mismatch between the emitting area and the pigtail's core area results in input coupling loss, and (2) loss occurs if the core area is smaller than the source emitting area. An estimate of this loss is simply the ratio of the core area to the emitting area.

The light-gathering ability of the fiber also results in an input-coupling loss factor. The *numerical aperture* (NA) defines the *acceptance cone* half-angle, θ, of the fiber (see Figure 14-6). Light rays entering this cone will be propagated.

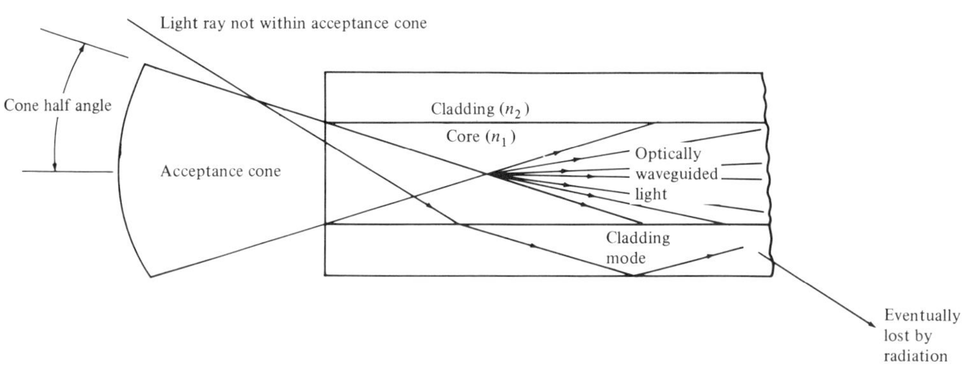

FIGURE 14-6. A fiber's acceptance cone half-angle derives from the fiber's numerical aperture. Light projected into the acceptance cone undergoes waveguiding in the cone, while rays outside the cone reflect into the cladding and are eventually lost.

Other rays, entering at a steeper angle, will be dissipated in the cladding. The numerical aperture is defined by

$$NA = \sin\theta \tag{14-5}$$

For step-index fibers, the numerical aperture is also given by

$$NA = \sqrt{n_1^2 - n_2^2} \tag{14-6}$$

Obviously, the source' emission profile must be known in order to evaluate the NA input-coupling loss

EXAMPLE 14-3

Determine the acceptance cone half-angle for the fiber in Example 14-1.

SOLUTION

$$NA = \sqrt{n_1^2 - n_2^2} = \sqrt{1.48^2 - 1.46^2} = 0.242$$

$$= \sin^{-1}(0.242) = 14.03°$$

An additional input-coupling loss occurs when a fiber bundle is used. That is, many fibers are grouped together and illuminated by one light source. This loss is termed the *packing loss* and is measured by the ratio of the collective core areas of the fibers to the bundle cross-sectional area. The packing loss in decibels is given by 10 times the log of this ratio.

An input-coupling loss also results from light reflection at the input end of the fiber. This loss is small and is usually on the order of 0.2 dB. Output-coupling losses occur at the receiving end of the link. These losses are not as severe as the input losses and are on the order of 1 dB.

Fiber attenuation losses occur between the system input and output together with connector losses and losses due to any splices made. The attenuation in the fiber is wavelength dependent and results from scattering and absorption. This loss may be minimized by selecting a source whose wavelength falls in an area of low fiber attenuation. An attenuation versus wavelength profile for two typical fibers is shown in Figure 14-7. At a typical source wavelength of 820 nm, these fibers exhibit a loss of 5 dB/km. At $\lambda = 1100$ nm, the loss is even less.

14-1.7 Single Fibers Versus Bundles

The developments in single-fiber optics has been so rapid that the trend is toward single-fiber optics for most applications. Even so, bundles are the most efficient

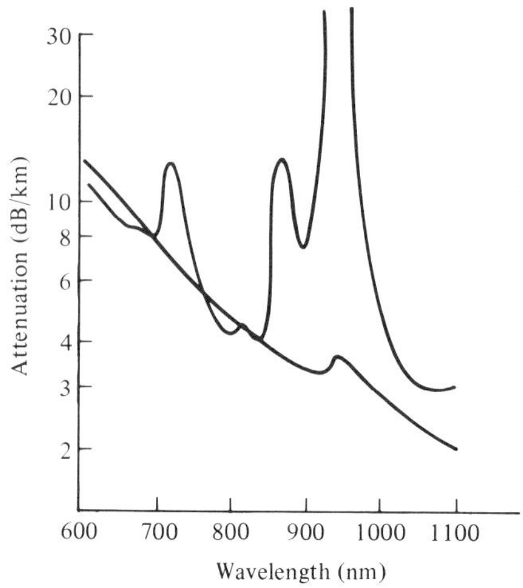

FIGURE 14-7. Fiber attenuation varies dramatically with wavelength, so good overall system performance requires a good match between source and fiber characteristics.

for either collecting or distributing light either at random or in a formatted mode. The bundles greater energy, then, is an advantage for applications such as scanners, card readers, and medical instruments. In the future, however, it appears that fiber optics bundles will become a custom design business.

14-2

DETECTORS AND SOURCES

In Section 14-1 we identified input-coupling losses. These losses may be categorized as follows:

- Unintercepted-illumination (UI) loss resulting from an area mismatch between the source's illumination spot (in the plane of the fiber end) and the fiber core area.
- Numerical-aperture (NA) loss resulting from light rays with angles of incidence not within the fiber's acceptance cone.
- Reflection (*R*) loss from the end of the fiber.

Thus we see that the magnitude of the input-coupling loss depends on the geometrical and optical characteristics of the source and the fiber. If the source's emitting area is larger than the core area, area mismatch occurs (see Figure 14-8). The solution to this problem is to choose a source whose emitter is smaller

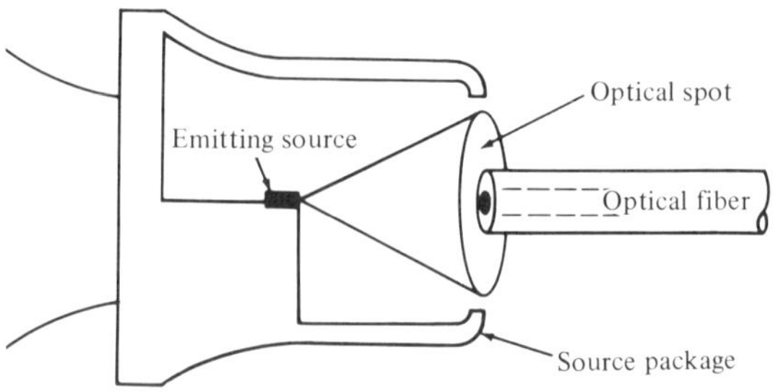

FIGURE 14-8. Unintercepted illumination loss.

than the core. There are available both LEDs and ILDs (injection laser diodes) that satisfy this requirement.

If the source is smaller than the core, problems with unintercepted illumination may still occur. Separation between the emitter and the core may allow light to miss the core. An approximation to this loss is

$$\mathrm{UI}_{\mathrm{loss}} = 10 \log \frac{A_c}{A_s} \tag{14-7}$$

where

A_c = fiber core area

A_s = area of the source's projected spot in the plane of the fiber's end

All small sources have rapidly divergent beams. To minimize these losses, the separation between the emitter surface and the fiber end must be no greater than two to four times the core diameter. For a core diameter of 50 μm, it is 100 to 200 μm. This small distance eliminates the use of glass-capped diodes. The simplest solution is to use a source with a pigtail already installed by the manufacturer.

Reduced performance occurs because of the numerical-aperture (NA) loss. The typical acceptance cone angle is on the order of 10 to 14° (Figure 14-9). Both LEDs and ILDs exhibit broad divergence. This loss can be estimated if the emission power profile of the light source is known.

14-2.1 Source-Beam Profile

Source-beam patterns are usually displayed on a polar diagram (Figure 14-10). These patterns are obtained by measuring the relative radiant intensity on a small detector that swings through a 180° arc. A source whose energy varies as the cosine of the angle, ϕ (Figure 14-10a) is referred to as *Lambertian*. That is,

$$P = P_0 \cos \phi \tag{14-8a}$$

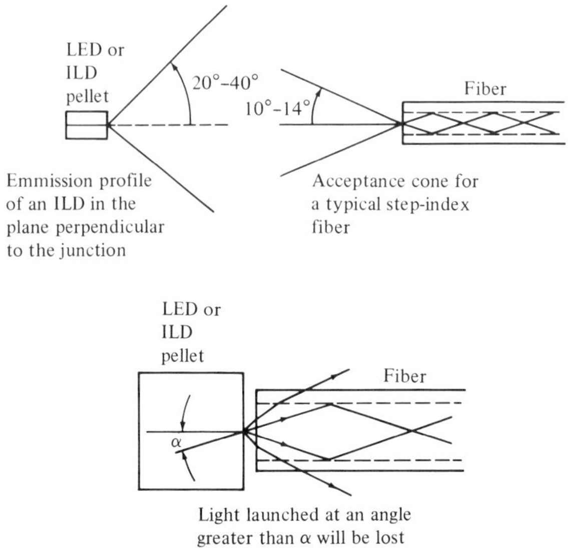

FIGURE 14-9. Lots of light is not necessarily a blessing at the source–fiber interface. Both LEDs and ILDs have broadly divergent emission beams, and all light radiated outside of the fiber's acceptance cone angle contributes to numerical-aperture loss.

where P_0 is the radiant intensity along the line $\phi = 0$. Some sources exhibit narrow beam patterns (Figure 14-10b). Mathematically, these patterns can be approximately described by

$$P = P_0(\cos \phi)^m \tag{14-8b}$$

The value of m can be obtained for a given source by comparing Figure 14-10b to the pattern on the manufacturer's data sheet. Some manufacturers specify the 50% intensity points. For example, if the 50% intensity point is specified at 25°, then from Figure 14-10b, m is approximately 7.

Now, the source power, P_c, coupled into the fiber is described by

$$P_c = P_T[1 - (\cos \theta)^{m+1}] \tag{14-9}$$

where θ is the fiber's acceptance cone half-angle and P_T is the total source power. The aperture loss, in decibels, can now be determined by evaluating equation (14-10).

$$\text{NA}_{\text{loss}} = 10 \log \frac{P_c}{P_T} \tag{14-10}$$

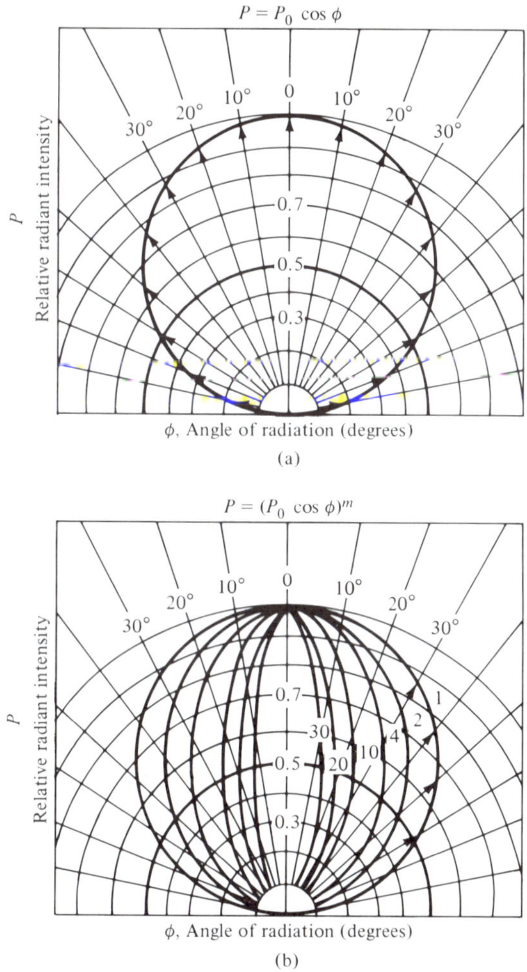

FIGURE 14-10. A wide variety of emission-beam patterns appear in sources available today. While uniform surface emitters typically have Lambertian type profiles (a), some sources exhibit much narrower beam profiles (b).

EXAMPLE 14-4

Calculate the aperture loss for the fiber in Example 14-1 if a Lambertian source is used.

SOLUTION

$$P_c = P_T[1 - \cos\theta)^{m+1}] = P_T(1 - \cos^2\theta)$$

$$= P_T \sin^2\theta = P_T(\text{NA})^2$$

From Example 14-2, NA = 0.242.

$$\frac{P_c}{P_T} = (\text{NA})^2 = 0.242^2$$

$$\text{NA}_{\text{loss}} = 10 \log \frac{P_c}{P_T} = 10 \log 0.242^2 = -13.15 \text{ dB}$$

14-2.2 Reflection Loss

Compared to the NA loss, the reflective (R) loss is almost negligible. However, this loss assumes a more important role in fiber splices. The light incident on the fiber core experiences some reflection because of a change in refractive index at the air/core interface. The reflection coefficient, ρ, which gives the fraction of reflected light, is approximately

$$\rho = \left(\frac{n_1 - 1}{n_1 + 1}\right)^2 \qquad (14\text{-}11)$$

This formula is based on the classical Fresnel formulas for normally incident light. The reflection loss in decibels is then

$$R = 10 \log_{10}(1 - \rho) \qquad (14\text{-}12)$$

For a core index of 1.5, the reflection loss is approximately 4% or −0.2 dB loss.

14-2.3 Source Requirements

System requirements can be very severe for sources used in long-haul fiber optics transmission systems. To minimize repeater spacing, the source should be as intense as the state of the art allows. As discussed previously, the source emitting area must be smaller than the fiber core area to launch optical power effectively into the fiber core. The fiber's acceptance cone dictates the emitter's beam pattern. The beam pattern must be very directional—almost collimated. In addition, it should be almost monochoromatic to avoid material dispersion. High-capacity digital systems require very fast (nanosecond range) rise and fall times, while optical systems require an optical output that is linearly related to the drive current or voltage over a large dynamic range.

Light-emitting diodes (LEDs) are particularly suited for analog applications. LEDs can be very nearly linearly modulated as a function of drive current. These devices have been in use for years as indicators and electro-optical isolators. Some refinements were necessary to make LEDs compatible with fiber optics. Primarily, this consisted of reducing the emitting area to more nearly match that of the fiber. One of the early designs (1971) for a high-radiance LED is shown in Figure 14-11. This configuration approximates a Lambertian source with consequent high NA loss. However, with an attached pigtail, the power out can approach 1 mW.

The power output for surface-emitting devices is usually specified in terms of *radiance,* the power per unit of solid angle (steradian) per unit area ($W/sr\text{-}cm^2$). An example will illustrate typical system calculations.

EXAMPLE 14-5

The radiance for an LED is specified to be 60 $W/sr\text{-}cm^2$. For Example 14-1, assuming a Lambertian source, calculate (a) total power emitted into a solid angle of 2π steradians, (b) P_c, and (c) NA loss.

SOLUTION

(a) $P_T = (60\ W/sr\text{-}cm^2)(2\pi sr)(\pi R^2)$

$$= (60)(2\pi)(\pi)(25 \times 10^{-4})^2 = 7.4\ mW$$

(b) $P_c = P_T[1 - (\cos\theta)^2]\ mW$

From Example 14-3, $\theta = 14.03°$.

$$P_c = 7.4[1 - (\cos 14.03)^2] = 0.43\ mW$$

(c) $10 \log \dfrac{P_c}{P_T} = 10 \log \dfrac{0.43}{7.4} = -12.3\ dB$

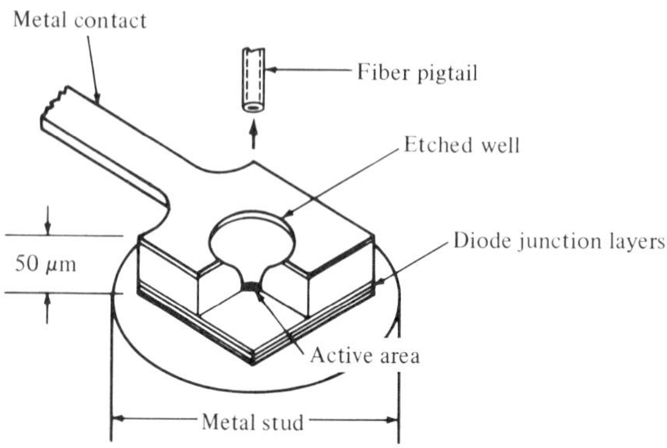

FIGURE 14-11. Fiber–source interfacing occurs in the etched well in this Burrus-configured surface emitter. The fiber pigtail, placed in the well in intimate contact with the emitting area, is epoxied into place.

Edge or side emitting diodes, in contrast to a surface design, are also available (Figure 14-12). In this configuration, guiding layers channel the light toward the fiber core. This produces a narrower beam source with reduced NA losses.

14-2.4 Source–Fiber Compatibility

The peak emission wavelength of the source should correspond to the fiber's minimum attenuation wavelength. Glass fibers exhibit two minimal areas—one around 800 to 850 nm and another around 1050 nm (see Figure 14-7). Most available LEDs have peak wavelengths of 800 to 850 nm with power levels on the order of 1 mW. Fewer sources are available at 1050-nm wavelengths but with reduced power: 0.1 to 0.2 mW.

Source purity must also be considered in matching the source and the fiber. We observed earlier that pulse spreading resulted from both modal and material dispersion. Material dispersion is wavelength dependent and thus depends on the spectral width of the source. The term *spectral width* refers to the band of wavelengths produced by the source.

Obviously, if material dispersion is to be contained, a reduction in spectral width of the source is necessary. Laser diodes provide just this feature (see Figure 14-13). The spectral width of laser diodes is so small that material dispersion is virtually eliminated. A spectral width of 4 nm is not uncommon

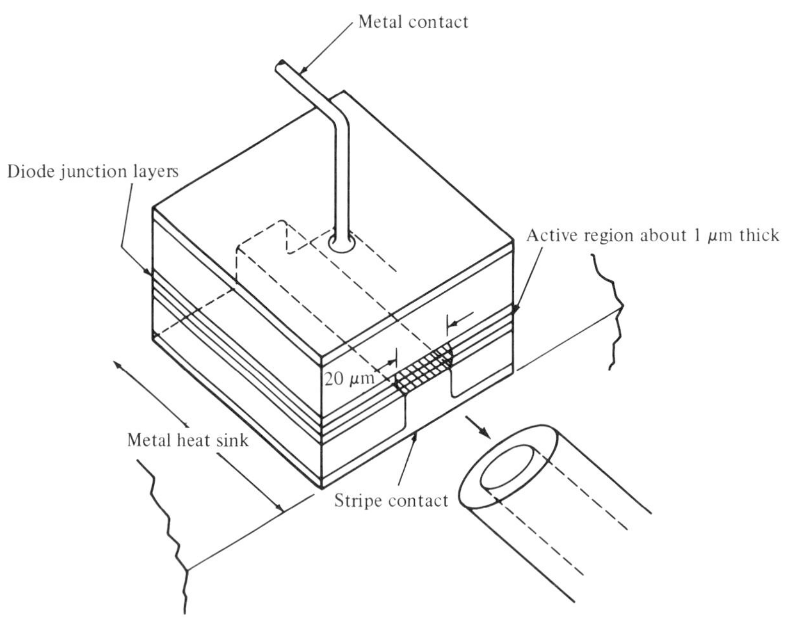

FIGURE 14-12. Easy fiber–source coupling occurs in stripe contact edge-emitting sources. However, some broad area (no stripe) edge emitters suffer from unintercepted-illumination loss.

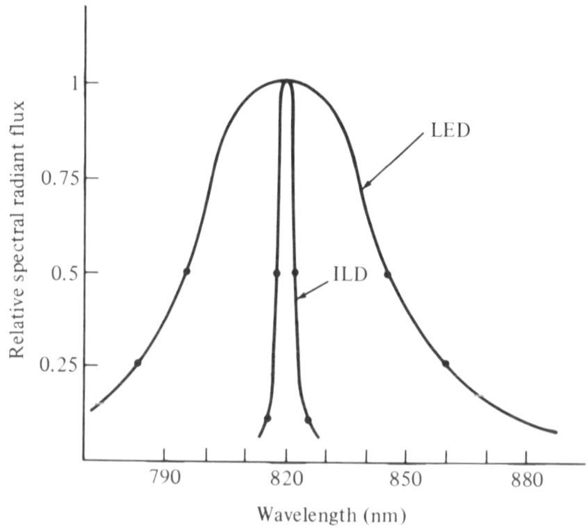

FIGURE 14-13. Fiber material dispersion is no problem with laser diodes because their spectral width is so narrow. Compared with LEDs, lasers can reduce time spreads by a factor of 40.

compared to a 50-nm spread for an LED. Material spread time for a typical LED source is on the order of 4 ns/km versus 0.1 ns/km for the laser diode.

Solid-state laser diodes are well suited for digital optical systems. The beam profile is illustrated in Figure 14-14. Emission through the narrow slit produces a rather broad beam in the plane perpendicular to the junction (20 to 40° half-

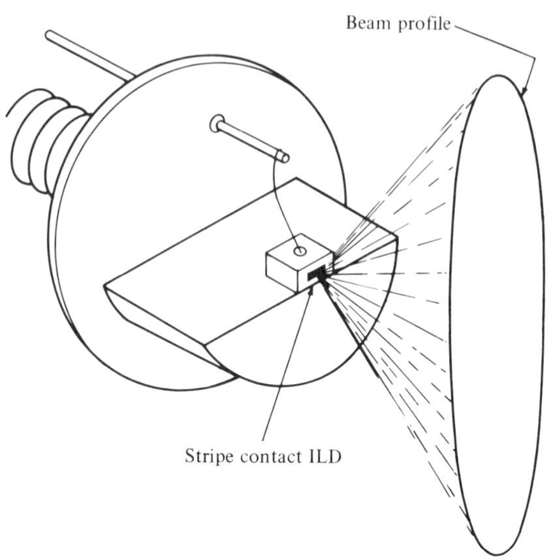

FIGURE 14-14. Significant NA loss, reduction results from the narrow beam profile (in the plane parallel to the junction) of stripe-contact-configured laser diodes.

angle). The beam in the plane parallel to the junction is much more narrowly confined. Current CW lasers have a power output in the range 5 to 10 mW. Rise times are less than 1 ns.

There are, however, several disadvantages associated with laser diodes. For example, laser diodes are more expensive than LEDs. Also, the characteristic curves are not uniform. This results in harmonic distortion when attempting to use analog modulation. To complicate the problem further, the power output is temperature dependent, thus making necessary some form of temperature control. Some designs employ temperature control of the heat sink.

14-2.5 Fiber Optic Receivers

The most commonly used receivers in fiber optics systems are photodiodes. These are usually either PIN or avalanche types, which convert incident light into electrical energy. Before considering receiver problems, however, it is of interest to examine the characteristics of diodes.

A PIN photodiode consists of a large intrinsic (very lightly doped) region sandwiched between *p* and *n* semiconducting regions. Photons of light energy absorbed in this region create electron–hole pairs which are then separated by an electric field. This generates an electric current in the load circuit (Figure 14-15). The efficiency of the optical photo/electron–hole conversion process is specified by the photodiode's quantum efficiency, η. This is a measurement of the average number of electrons released by each incident photon. A number near 1 is a highly efficient diode.

The response of the diode to a light simulus is a more meaningful design parameter. It is a ratio of the output current to the incident optical power and may be expressed as

$$r = \frac{\eta^2}{1.24}\ \text{A/W} \qquad (14\text{-}13)$$

The signal current generated when *P* watts of optical power are incident on the diode is

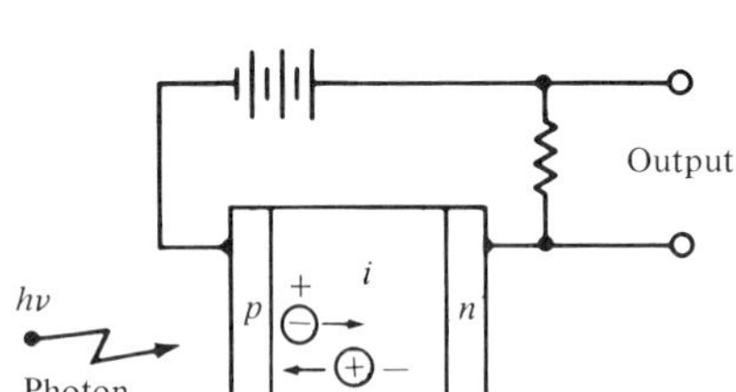

FIGURE 14-15. Electron–hole pairs, created by photons absorbed in the intrinsic region of a PIN diode, are separated by the diode's electric field to produce a current in the load circuit.

$$i_s = rP = \frac{\eta \lambda P}{1.24} \quad \text{amperes} \tag{14-14}$$

For a typical PIN diode, the peak wavelength response is less than 1 A/W. A typical response curve is shown in Figure 14-16.

For applications requiring greater sensitivity, an *avalanche photodiode* is preferred. As a result of external biasing, a strong electric field exists within the diode. This causes the APD to exhibit a gain mechanism. The primary electrons are accelerated by the field and undergo ionizing collisions with the surrounding atoms, resulting in an avalanche condition. As with PIN diodes, the APDs are reverse biased but with much higher voltages (up to 300 V). The higher gain results in a much higher responsivity with consequent greater signal current. The usable gain is limited by "excess noise." That is, the actual gain fluctuates about an average that creates the noise. Typical values of gain are on the order of 100 to 150. The gain is temperature dependent; thus some form of gain control is needed. In spite of these disadvantages, however, these diodes can yield up to 15 dB extra sensitivity, thus increasing the distance between repeaters.

In the receiver circuits, particularly the front end, the signal current must contend with noise currents. The primary noise component in PIN diodes is caused by *dark current,* the diode current that flows when no light is incident on the diode. Usually, the manufacturer specifies an average dc value that corresponds to a given temperature and bias voltage. As a general rule, dark current doubles for every 10°C increase in temperature.

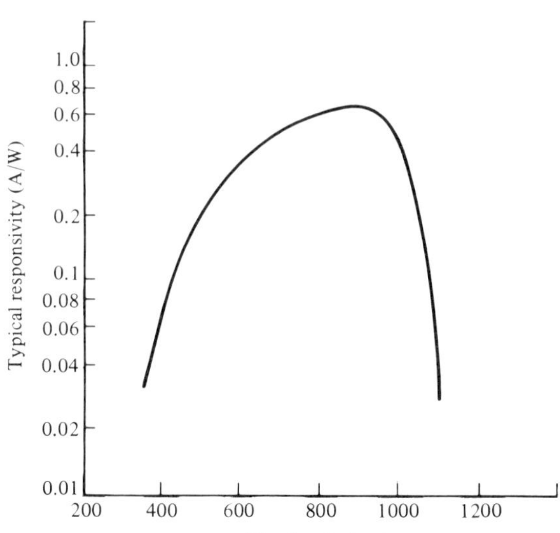

FIGURE 14-16. Typical photodiode response curve.

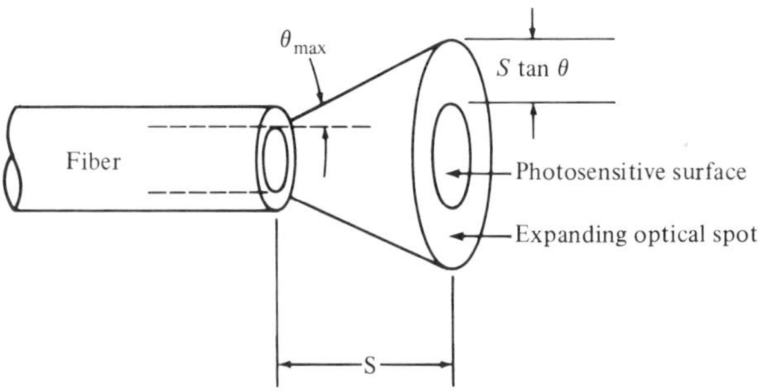

FIGURE 14-17. Unintercepted illumination is also a potential problem at the receiving end of a link. Using a photodiode with a large active area keeps this loss down; a better solution involves using an uncapped diode and moving the fiber as close as possible to it.

Another diode figure of merit is the *noise equivalent power* (NEP). This parameter is usually expressed in watts/$\sqrt{\text{Hz}}$. If we multiply NEP by the square root of the detector noise bandwidth, the result is a value for the absolute power and is referred to as the *minimum detectable signal* (MDS). The MDS defines the optical power incident on the diode that is required to generate a photocurrent equal to the total photodiode noise current. In other words, a zero signal-to-noise ratio at the output of the diode. Following the diode, a low-noise preamp is necessary to amplify the small-signal currents.

Data rates for a given diode are limited by the response time. The response time for a photodiode is limited by the *RC* time constant associated with the diode's load resistance and junction capacitance plus the load impedance into which the diode operates. Typical 10 to 90% rise times for photodiodes are on the order of a few nanoseconds. Response times for APDs are somewhat faster.

Light Reception. At the end of the fiber, the light must be coupled into the receiving diode. As the light exits from the end of the fiber, it will spread out with a divergence approximately equal to the acceptance angle (Figure 14-17). The photodiode is located a distance *s* behind a protective glass window. Any light that spreads beyond the active area of the photosensitive surface represents coupling loss.

EXAMPLE 14-6

A fiber has an exit angle $\theta = 14°$ and a core diameter of 0.05 mm. It is to be coupled to a photodiode whose circular sensitive area is 5 mm^2 and which is located 2.5 mm behind the glass window. Calculate the coupling loss, if any.

SOLUTION

The light area at the photodiodes surface is

$$A = \pi r'^2 = \pi\left(s \tan \theta + \frac{d}{2}\right)^2$$

$$= \pi\left(2.5 \tan 14° + \frac{0.05}{2}\right)^2 = 1.3 \text{ mm}^2$$

This figure is smaller than the 5-mm^2 detector area. Thus there is no coupling loss due to expansion of the light beam.

14-3
SYSTEM DESIGN PARAMETERS

At this point we have examined the characteristics of individual components that make up a fiber optic system. System parameters will now be examined. Generally speaking, system design involves trade-offs which are necessary to arrive at a final design. These include such factors as component selection, component compatibility, system configuration, multiplexing methods, type of modulation, and system performance itself. We see, then, that system design is the first step toward realizing a system.

14-3.1 Component Selection

The initial step in a system design involves evaluating the pros and cons of various components. For example, in an optical system, this requires an evaluation of the optical source. We first consider whether it is desirable to use an LED or an ILD. The ILD radiates more power and has a narrower beam profile. This allows a greater distance for repeater spacing. In addition, the ILD exhibits a narrower spectral width, which makes it possible to ignore fiber-material dispersion. ILDs have faster response times, which permit higher modulation rates than LEDs. As mentioned previously, there are some negative aspects. The ILDs exhibits output-power nonlinearities at and above the lasing current threshold. This is a definite drawback for linear analog-intensity modulation. The LED's linear power versus drive current characteristic is preferable.

The next item for consideration is selection of the fiber type. As indicated previously, the three available fiber types are: multimode step, graded index, and single mode. The single-mode fiber offers the ultimate in bandwidth capacity. Together with this, however, are several disadvantages. The extremely small core diameter makes input coupling and fiber splicing very difficult. In addition, it has poor numerical aperture. Step-index fiber has a higher NA then does graded-index fiber. Thus it provides better input-coupling efficiencies. In terms of bandwidth, however, graded index is superior.

For purposes of implementing the receiver, we can choose between a PIN photodiode and an avalanche photodiode (APD). The APD offers the advantage of greater bandwidth and lower power level detection. The disadvantages are temperature-dependent gain and large bias voltage. For moderately long links a PIN photodiode is preferable. When choosing components it is necessary to match the characteristics of the devices. For example, the source's peak emission wavelength should correspond to the lowest attenuation wavelength of the fiber (Figure 14-18). In addition, the photodetector must also have maximum response at this wavelength.

14-3.2 System Architecture

The simplest architecture for a fiber optic system is a point-to-point configuration. Bidirectional communications may require the use of multiplexing or possibly a two-fiber cable layout. A geographically distributed system may require a multiple-access system. This type of architecture requires the use of an additional component—a coupler. There are two coupler configurations: a T and a star. The *T coupler* is often referred to as an *in-line* data bus (Figure 14-19).

The *star configuration* offers an alternative coupling solution for multiterminal applications (Figure 14-20). The operation of the star coupler is as follows. A subscriber transmits a signal on a dedicated fiber toward the coupler. The light enters port 1, spreads out, and is reflected by the dielectric mirror into all other ports.

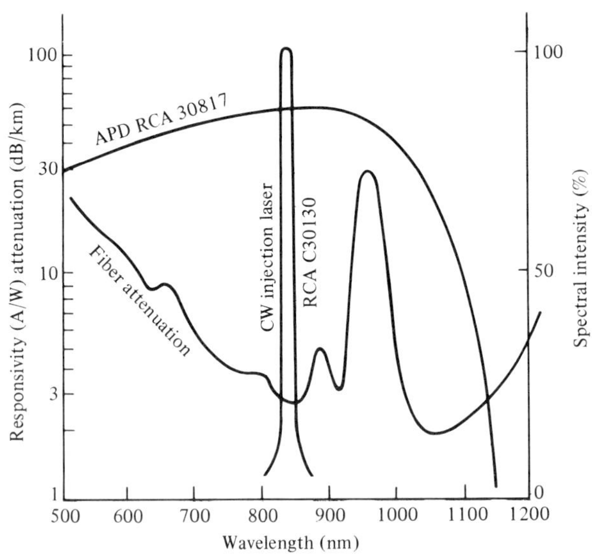

FIGURE 14-18. Wavelength compatibility is necessary for optimum system performance. Fortunately, today's fiber optic components provide this necessary compatibility.

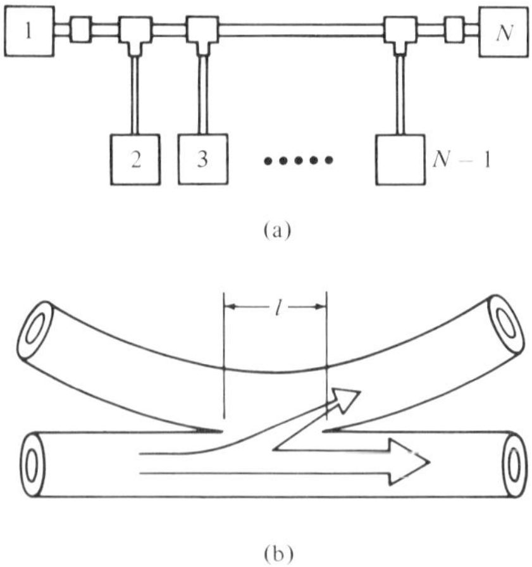

FIGURE 14-19. (a) Terminals tap off or inject optical energy onto the main trunk in the in-line bus configuration. (b) One approach to the T-coupler implementation of this configuration is to fuse the fibers together; the amount of light coupled varies with interaction length, 1, and core-to-core proximity.

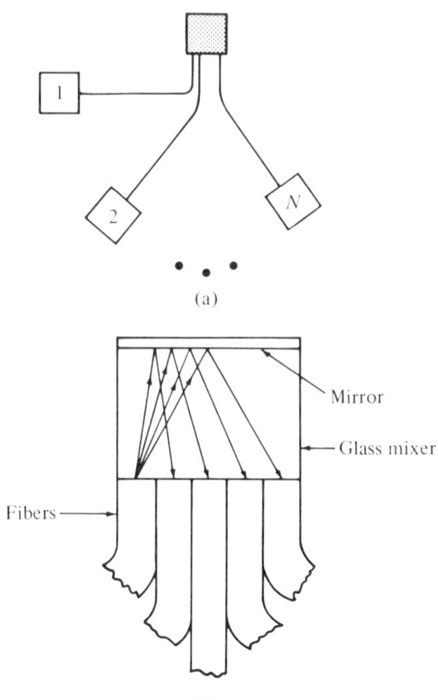

FIGURE 14-20. (a) One terminal can communicate with all others when you employ the star-coupler configuration (b) Light entering at one port spreads out in the glass mixer and reflects from the mirror into all other ports.

In choosing a T or star coupler, system losses may be the deciding factor. The optical signal suffers a loss at each T coupler on the bus. This loss increases linearly with the number of terminals. Comparatively, the star introduces only one coupler loss. Also, as the number of ports, *n*, increases, the star coupler's power splitting loss increases only as *n* log (*n*). Another disadvantage of T couplers is the dynamic range. In transmitting enough power to overcome losses at the most distant terminal, the nearest terminals may be overdriven. A comparison of the couplers is shown in Figure 14-21. This graph assumes a constant 10-dB tap loss, plus a 2-dB insertion loss for T couplers and a 7-dB insertion loss for the star designs. Losses of 1 and 3 dB were assumed for connector and I/O splitting losses.

14-3.3 Signal Multiplexing

Some form of multiplexing is required if several individual signals are to be transmitted. Frequency-division and time-division multiplexing (FDM and TDM) are familiar techniques that can readily be used. Space-division multiplexing (SDM) can be used if a separate fiber is available for each signal. A newer

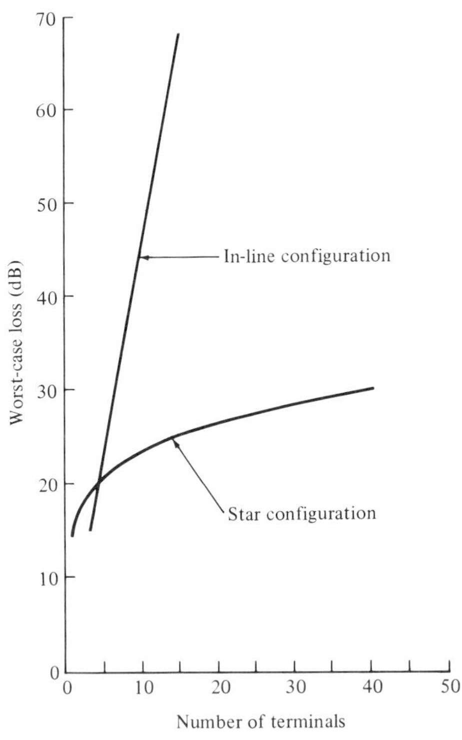

FIGURE 14-21. The star configuration proves superior as the number of terminals increases. Because there's only one coupler loss to contend with in this configuration, this result is not too surprising.

technique, *wavelength-division multiplexing* (WDM), uses sources of different wavelength. This technique requires filters at the photodetectors to select the appropriate signal. The advantage of this technique is that the bandwidth advantage of fibers is increased.

14-3.4 Modulation

Both LEDs and ILDs can be modulated simply by varying the current flowing through them. Consider first, methods for modulating with analog information. We may choose, for example, to use *baseband analog-intensity modulation*. The output optical power of the LED varies linearly with the input electrical signal. When using this method, the LED must be biased at a level such that overmodulation (clipping) does not occur. Signal detection occurs through a square law (optical power to electrical current) photodiode and a current-to-voltage preamplifier.

If a decision is made to frequency-division multiplex several signals, it is necessary to preprocess the baseband analog information before optical modulation. It is common practice with systems of the type to use *subcarrier analog-intensity modulation*. The baseband analog signal first modulates an RF subcarrier, which then intensity modulates the optical source (Figure 14-22). The modulation indices are selected to prevent overmodulation.

Pulse modulation is a feasible alternative with fiberoptic systems. Techniques such as pulse-position modulation (PPM) and pulse code modulation (PCM) were discussed in Chapters 11 and 12. The wide bandwidth of fiber optics communication systems make these attractive options, taking advantage of the wide bandwidth available. An improved SNR improvement can be achieved over baseband analog intensity modulation.

14-3.5 Signal-to-Noise Ratio

The usable SNR is determined by the photodiode and the front-end amplifier. This choice depends on the detector's bandwidth and the average optical power

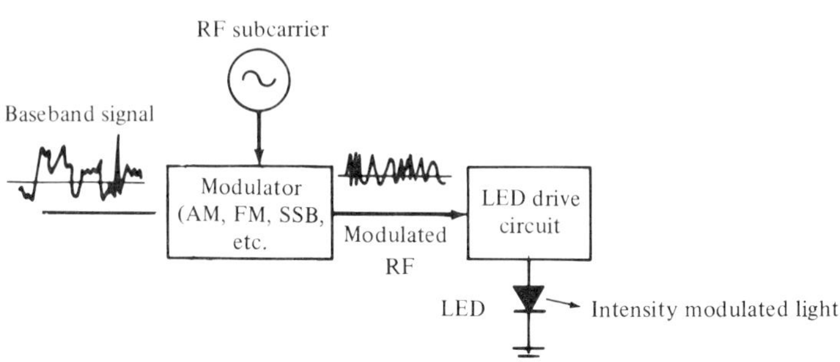

FIGURE 14-22. Using subcarrier analog modulation, you first modulate an RF carrier and then modulate the LED output.

that can be delivered. Figure 14-23 illustrates the average received optical power versus bandwidth performance curves for PIN and APD analog receivers.

We indicated previously that pulse modulation provided an improvement in SNR. For purposes of comparison with analog modulation, refer to Figure 14-24. For comparison, consider a PIN diode and fix the SNR at 60 dB. Note from the figure that PPM requires 30 dB less optical power than analog-intensity modulation to maintain an SNR of 60 dB. For a typical 20-dB/km fiber cable, this difference can increase the link length by as much as 1.5 km. PCM provides a gain margin of 18 dB. The penalty paid for this increase in margin is an increase in bandwidth. The ultimate limitation, then, for both techniques is the allowable system dispersion. The allowable dispersion is plotted in Figure 14-25.

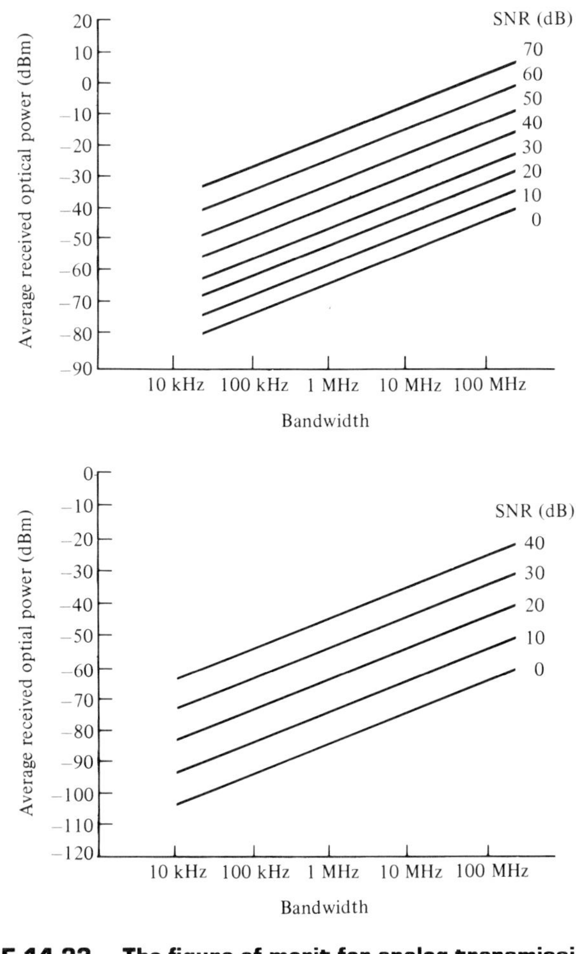

FIGURE 14-23. The figure of merit for analog transmission is signal-to-noise ratio (SNR). These plots of average received optical power vs bandwidth for several SNR values illustrate the capability of PIN (a) and APD (b) detectors.

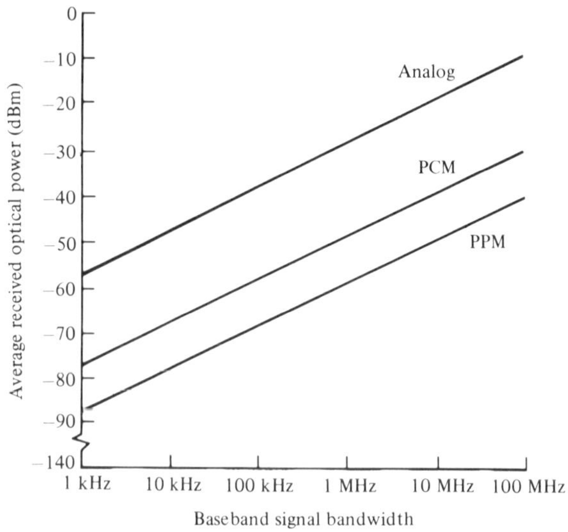

FIGURE 14-24. Pulse position modulation (PPM) requires less optical power to maintain a given SNR. This advantage can significantly increase link length or enhance signal quality.

14-3.6 Digital Modulation

The most straightforward scheme for transmitting binary digital data over a fiber optical link is to simply key the optical source "on" and "off." The BER replaces SNR as the measure of performance. The optical power required by state-of-the-art PIN and APD receivers is shown in Figure 14-26. At a data rate of 10 Mbits/s, a PIN diode receiver requires an average received optical power

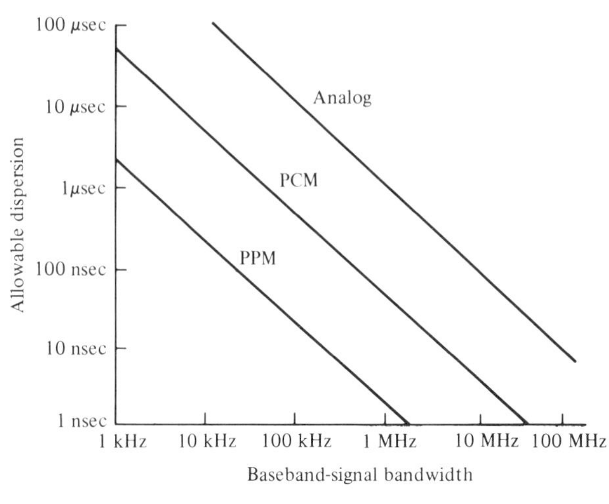

FIGURE 14-25. The shortcomings of PPM appear in these plots. Although PPM provides an extra power margin, it suffers with regard to allowable dispersion.

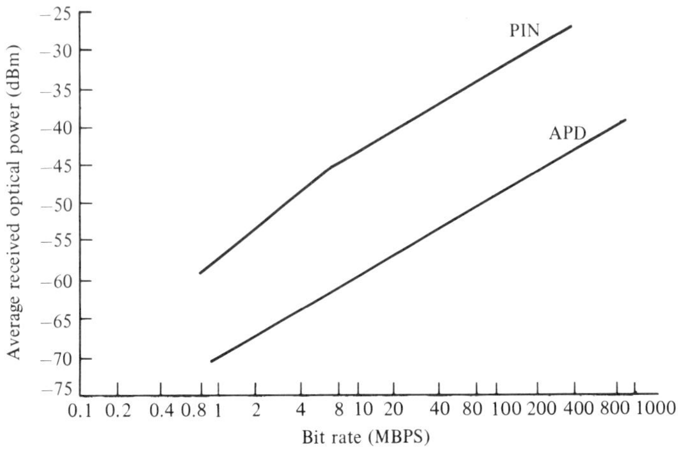

FIGURE 14-26. Enhanced sensitivity of the APD receiver appears when you consider the optical power required by state-of-the-art detectors.

of approximately −43 dBM. For comparison purposes, the APD requires −60 dBm, resulting in a margin of 17 dB. Another approach is first to modulate an RF carrier digitally and then intensity modulate the light source. Familiar techniques such as frequency, amplitude, or phase-shift keying may be used.

14-3.7 Loss Budgeting in System Design

To ensure system performance it is necessary to budget system losses. The curves in Figures 14-23 through 14-26 enable this to be accomplished. The optical power loss limit, L, for a particular SNR or BER is given by

$$L = P_s - P_r \qquad (14\text{-}15)$$

where

P_s = source power

P_r = received power

In a point-to-point system, the losses consists of (1) input coupling losses (ICL) between the source and fiber, (2) connector and/or splice losses (CSL), (3) fiber attenuation losses (AL), and (4) output coupling losses (OCL). In equation form, the link-loss budget is

$$L = \text{ICL} + \text{CSL} + \text{AL}(D) + \text{OCL} \quad \text{dB} \qquad (14\text{-}16)$$

where D is the "power-limited distance" of the link. Beyond this distance, D, the received power is not sufficient to maintain the required SNR or BER. Thus D dictates the repeater spacing. It is common practice to add two or three decibels to act as a safety margin for factors such as source-power temperature degradation.

EXAMPLE 14-7

It is desired to transmit a 100-MHz analog signal over a distance, D, of 2 km with a SNR spec of 40 dB. Use an analog intensity-modulated LED and calculate the loss budget using typical figures.

SOLUTION

From Figure 14-23, at 10 MHz, both the PIN and the APD need about the same amouont of received power at a 40 dB SNR. That is, $P_r = -35$ dBm. For cost considerations, choose the PIN diode and assume a Lambertian-emitter LED source with a total radiant power of 2 mW (+ 3 dBm). From equation (14-15), the loss limit, L, is

$$L = P_s - P_r = +3 - (-35) = 38 \text{ dB}$$

Next, consider the ICL loss. If the LED's emitting area is smaller than the core, the ICL is due primarily to the NA of the fiber plus 0.2 dB for reflection loss. Choose a fiber with an NA of 0.25. Then the ICL loss is

$$\begin{aligned} \text{ICL} &= 10 \log (0.25)^2 + 0.2 \\ &= 12.2 \text{ dB} \end{aligned}$$

Now, as an estimate, let the OCL loss be 1 dB and assume three connector losses (one at each end and one at the middle). Then CSL = 3 dB. From equation (14-16), the loss budget is

$$\begin{aligned} L &= \text{ICL} + \text{LSL} + \text{AL}(D) + \text{safety} \\ &= 12.2 + 3 + \text{AL}(D) + 1 + 3 \end{aligned}$$

Then

$$\text{AL}(D) = 18.8 \text{ dB}$$

The allowable cable loss for a 2-km link is then 18.8/2 = 9.4 dB/km. If a higher loss fiber is considered, it will be necessary to increase the source power.

14-3.8 Signal Degradation

Signal degradation is caused by noise and distortion. The receiver is the major source of noise in a fiber optic system. However, other parameters, such as nonlinearities in the power versus drive current characteristics of the light source can degrade system performance by introducing harmonic distortion. This is particularly true for an ILD. Material and modal dispersion cause distortion. This is evidenced as intersymbol interference for pulsed systems or bandlimited amplitude distortion for analog transmission. LED sources exhibit spectral widths between 30 and 50 nm. This results in material dispersion of about 4 ns/km. ILDs possess a spectral width of 2 to 4 nm. Thus there is negligible material dispersion.

Finally, system rise time can distort the shape of a signal. The system rise time is determined by the rise times of the source, detector, and the fiber. Rise times for LEDs range from 5 to 15 ns and from 0.1 to 2 ns for ILDs. At the other end of the fiber, PINs and APDs have rise times between 1 and 4 ns. A fiber's 10 to 90% rise time arising from modal dispersion measures about 70% of the 3-dB modal dispersion figure (in nanoseconds per kilometer). The total system rise time is given by

$$T_{\text{system}} = 1.1\sqrt{T^2_{\text{source}} + T^2_{\text{modal}} + T^2_{\text{material}} + T^2_{\text{detector}}} \qquad (14\text{-}17)$$

where the T's are 10 to 90% rise time. The system's 3-dB bandwidth may be expressed approximately as

$$\text{BW} = 0.35T_{\text{system}} \qquad (14\text{-}18)$$

Thus we see that the end-to-end data rate is determined by system rise time.

14-4 DRIVE AND RECEIVE CIRCUITS

The basic drive circuits for diode light sources are straightforward. Two such circuits are shown in Figure 14-27. In Figure 14-27a, the current through the diode is controlled by the bias on the transistor. Thus this circuit is suited to analog intensity modulation. Maximum forward currents for the LED or PPM are specified by the manufacturer. Typically, it is on the order of a few hundred milliamperes.

For pulsed-modulation techniques such as PPM or digital on/off keying, a FET can be used to switch an LED or ILD. If the ILD has a sharp "knee" at the lasing threshold current, a resistor R (Figure 14-27b) can be used to bias the ILD just below threshold. This action increases the switching speed since the switching action does not have to begin with the current at zero.

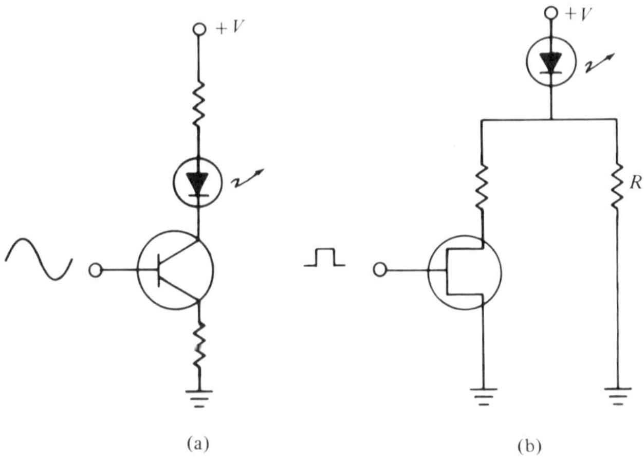

FIGURE 14-27. Straightforward basics suffice when designing source drive circuits. A simple transistor circuit (a) works well for analog-intensity modulation, and you can use an FET circuit (b) for pulsed-modulation schemes.

14-4.1 Photodiode Receiver Circuits

Two basic photodiode receiver circuits are shown in Figure 14-28. The transimpedance amplifier (Figure 14-28a) is designed for current sources like PIN or avalanche photodiodes. In the voltage-gain mode (Figure 14-28b), an op amp boasts the voltage developed across the load R_L.

In selecting the load resistance, it is necessary to make trade-offs in optical sensitivity versus response speed. For a given photodiode, large loads (R_L = 5 MΩ) reduce thermal noise currents from the resistor. This lowers the NEP

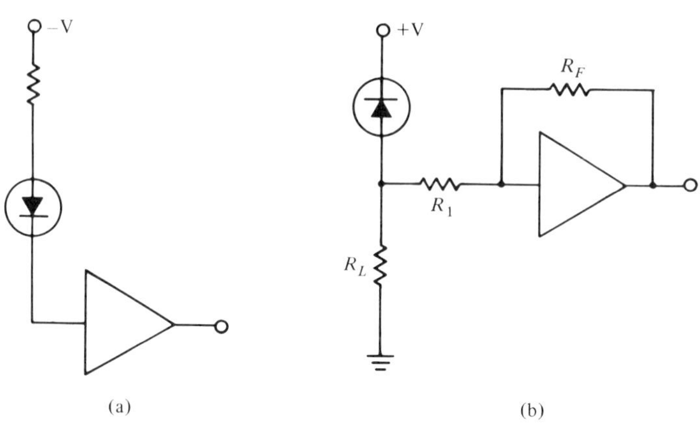

FIGURE 14-28. There is much flexibility when designing photodiode receiver circuits. The transimpedance amplifier (a) suits current sources like the PIN or avalanche photodiode. You can also operate the receiver in a voltage-gain mode (b) by using an op amp.

and allows detection at lower light levels. In this case the photodiode's dark-current shot noise limits signal detection. Conversely, small loads ($R_L = 50\ \Omega$) lower the *RC* time constant of the load and diode junction capacitance, leading to shorter rise times. Most photodiode manufacturers integrate a photodiode and preamp in one package. The responsivity figure of merit is specified in output volts per input optical watts. Temperature-compensated APD/preamp modules are available from RCA, Texas Instruments, and others.

14-5 PULSE-BIPOLAR DATA ENCODING

Coding must often permit a long string of all 1's or 0's to be transmitted without distortion. This can cause problems in high-speed fiber optic links, since all light pulses are unipolar (on or off) and the receiver must be ac coupled to eliminate dc-drift errors. Data transmission via fiber optics should be unipolar and contain a transition for each bit. Thus the conventional non-return to zero (NRZ) code may prove unsuitable, especially for high data rates.

Fiber optic receivers typically operate with detected signals in the range of a few millivolts, which must be compared with a reference level to decide whether the received signal represents a 1 or a 0. Because the signal level is low, a dc comparison can lead to errors, because of the difficulty of supplying a sufficiently stable dc reference voltage. In addition, the reference must normally be midway between the 1 and 0 states, but the peak signal will change as a function of many variables: fiber length, temperature and age of the LED, and so on. Thus the reference voltage cannot simply be stable, but must also be adjusted as a function of peak-to-peak signal amplitude.

If the signal is capacitively coupled to a comparator, the reference is set near ground potential so that the minima and maxima of the signal excursions go above and below the ground level symmetrically (provided that the duty cycle of the signal is 50%). Encoding schemes allow the use of ac coupling in the receiver by maintaining very close to 50% duty cycles in transmitted signals while allowing the receiver to decode the signals into a reproduction of the original data stream of arbitrary duty cycle. The use of ac rather than dc amplifiers in the receiver allows the designer to design circuits with greater sensitivity, less temperature dependence, and greater dynamic range.

14-5.1 Pulse-Bipolar and Manchester Encoding

Two encoding schemes are commonly used in industrial fiber optic systems: Manchester and pulse bipolar. Manchester encoding designates a 1 or a 0 by a transition from high to low or low to high, so that each bit cell has a 50% duty

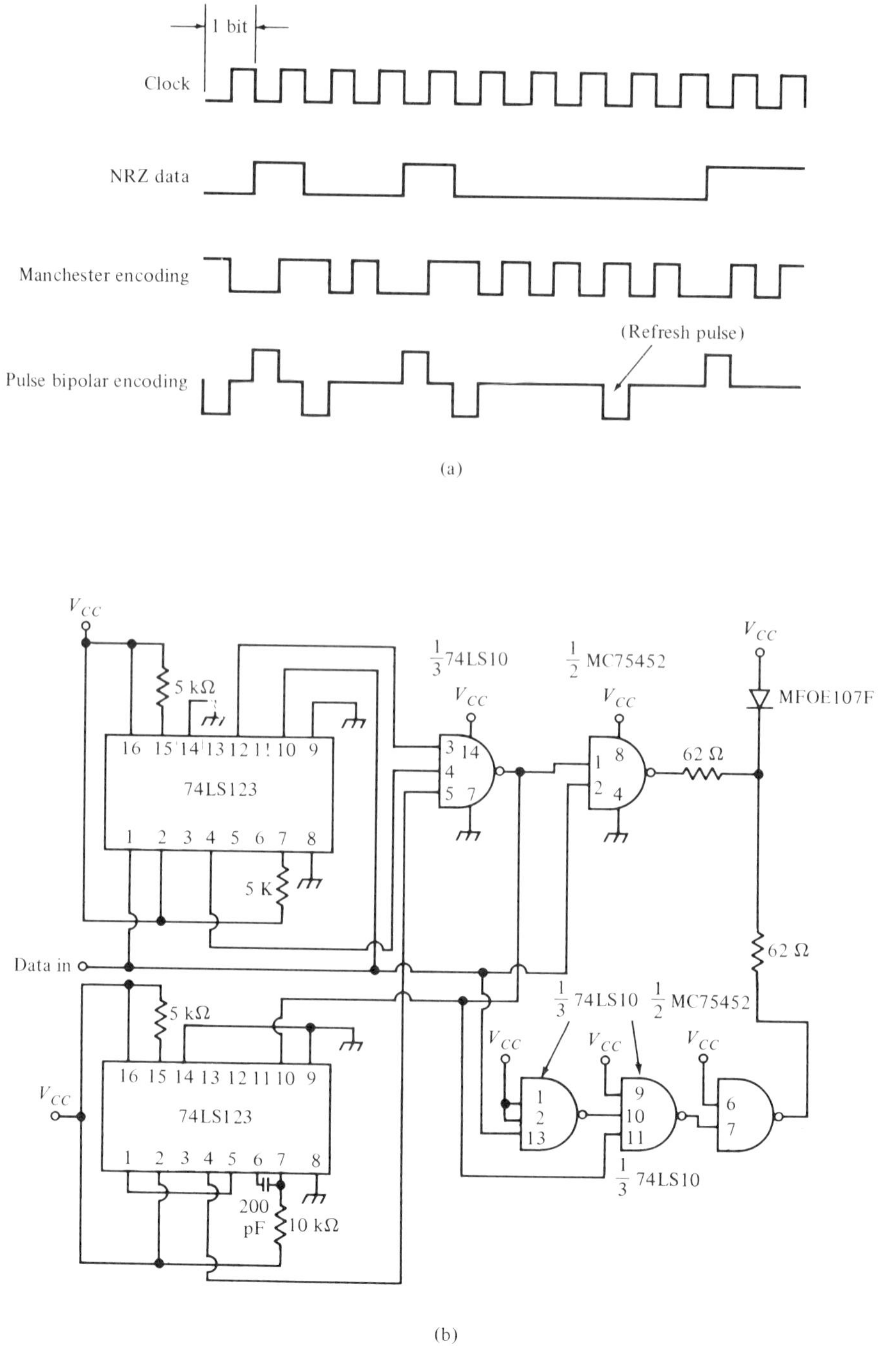

FIGURE 14-29. (a) Timing diagrams; (b) pulse–bipolar transmitter.

and the average transmitted duty cycle is 50%, an optimum for capacitively coupled systems. A disadvantage of Manchester is that it requires clock regeneration at the receiver; this is usually done with a phase-locked loop.

Pulse-bipolar encoding sends a constant average light signal through the fiber. This average signal is pulsed upward in intensity to indicate a transition to a 1 and pulsed downward in intensity to indicate a transition to a 0. This system also has a 50% average duty cycle. A refinement adds "refresh pulses" of the same polarity as the last transition so that automatic gain control can be maintained in the receiver. The refresh pulses must be of much lower repetition rate than the maximum data rate so that their impact on duty cycle is negligible.

Pulse-bipolar encoding has an advantage over Manchester encoding in that the receiving circuitry is very simple. Its disadvantage is that it does not facilitate clock regeneration, so its use is primarily for the transmission of asynchronous data. In many applications the simplicity of the circuitry and the lack of a requirement for a clock signal make pulse-bipolar encoding the more attractive mode of encoding a signal for transmission.

Figure 14-29 shows the timing diagram and transmitter for a pulse-bipolar-encoded asynchronous signal. The NRZ data are differentiated by the 74LS123 dual one-shot, yielding pulses that are steered to the appropriate output circuit by the 74LS10 triple three-input and gate. The output circuit is an MC75452 dual-LED driver. The remaining 74LS123 provides refresh pulses in the absence of any transition in the data.

A receiver is shown in Figure 14-30. It uses a special-purpose chip, the MFOC600, which includes the decoding circuitry to restore the pulse-bipolar received data stream back to NRZ.

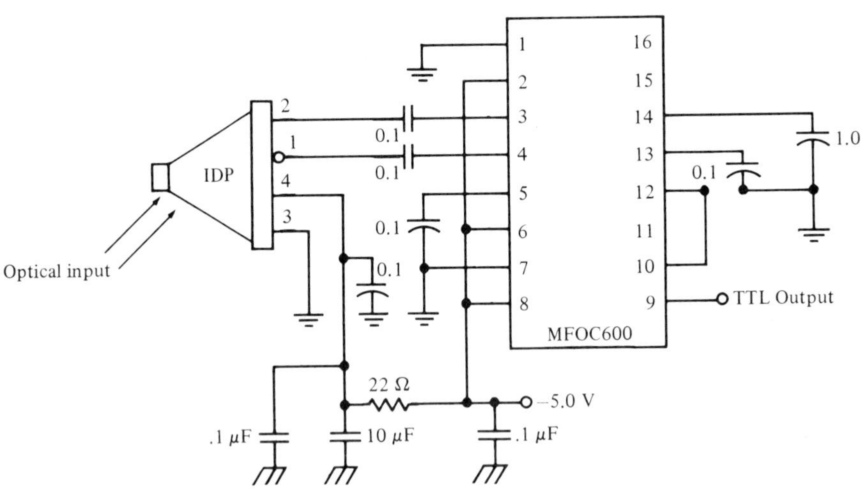

FIGURE 14-30. Pulse-bipolar decoding receiver.

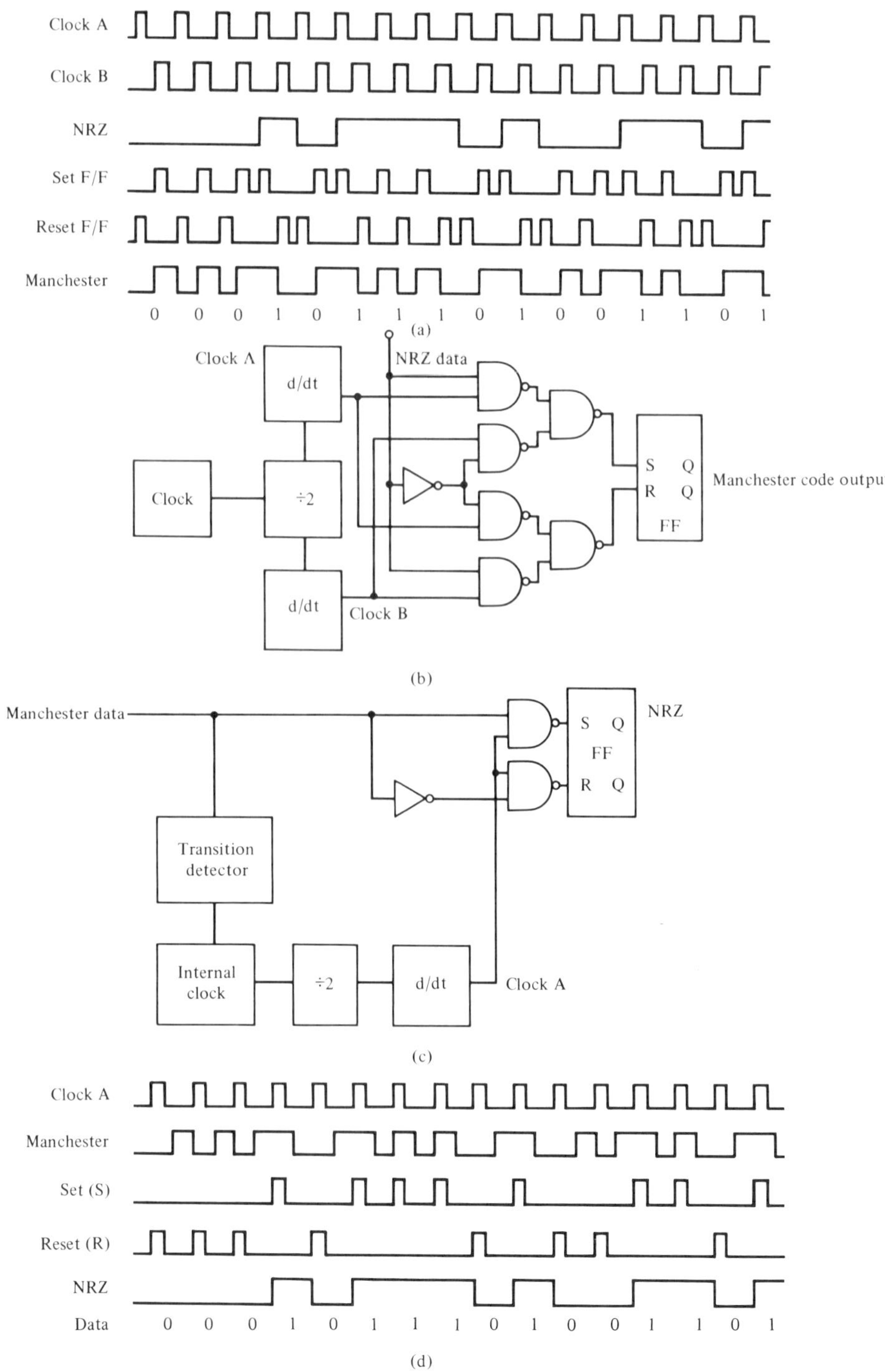

FIGURE 14-31. Manchester-coded data reverse polarity in each bit period independently of the bit value. A positive-to-negative reversal at the bit's center signifies ONE while negative-to-positive means ZERO (a). While you need a two-phase clock (b) to encode NRZ data into Manchester, a single-phase clock will do for decoding Manchester waveforms into NRZ (c). For synchronization, the decoder retimes its clock to received-signal transitions.

14-5.2 Manchester Encoding

In the Manchester code, polarity reverses in each bit period, regardless of the information. The transition occurs during the split-phase period at the bit's center. Positive-to-negative reversal signifies 1, negative-to-positive signifies 0. When all the bits in a string represent the same state, Manchester coding calls for additional transitions at the bit leading edges, which ensures proper polarity transitions at the bit centers.

To generate Manchester waveforms from NRZ input, two clock phases are needed, A and B (Figure 14-31a). Pulses in clock stream A coincide with the bit's leading edge, while those in B are in the bit's center. Output polarity always changes at clock time B: positive-to-negative for a 1 and negative-to-positive for a 0. Also, the output can change if the same-state bit occurs with clock A: negative to positive for a stream of 1's, positive to negative for a stream of 0's. Figure 14-31b shows the encoder logic based on an RS flip-flop whose

$$\text{SET} = A \cdot \text{data} + \text{B} \cdot \overline{\text{data}}$$
$$\text{RESET} = \text{A} \cdot \overline{\text{data}} + \text{B} \cdot \text{data}$$

Since, in NRZ code, the output changes only at the bit's leading edge, we can decode Manchester waveforms to NRZ using a single-phase clock with the circuit in Figure 14-31c. This decoder is based on an RS flip-flop whose

$$\text{SET} = \text{A} \cdot \text{data}$$

$$\text{RESET} = \text{A} \cdot \overline{\text{data}}$$

However, the decoder must be synchronized to the encoder, or the fiber optic link will not work properly. So the decoder circuit adjusts its clock frequency using received-signal transitions, then halves this frequency to regenerate clock phase A.

When synchronizing, use the pattern least likely to suffer distortion. With NRZ coding, that means avoiding an all-0's or an all-1's pattern, either of which contain no transitions, hence can lead to loss of decoder synchronization. Manchester code, with its guaranteed transition for each bit, is immune from this problem. Here alternating 1's and 0's will produce the lowest repetition rate and thus the surest synchronizing signal.

Data encoding is still an area of active research. Some of the more recently developed codes are simpler than Manchester and offer improved bandwidth. These new codes are based on digital frequency and phase modulation schemes—familiar subjects. The reader is referred to the literature for further reading.

PROBLEMS

14-1. Repeat Example 14-1 for a core diameter of 100 μm.

14-2. Calculate the modal dispersion-limited bandwidth for Problem 14-1.

14-3. Calculate the number of propagating modes for Problem 14-1.

14-4. A fiber has indices of refraction $n_2 = 1.49$ and $n_1 = 1.45$. Calculate the acceptance cone half-angle.

14-5. What is the value for m in equation (14-9) if the 50% intensity point is at 10°?

14-6. Calculate the aperture loss for the fiber in Problem 14-4. Use the source specified in Problem 14-5.

14-7. Repeat Example 14-4 for a fiber with a core diameter of 100 μm.

14-8. Calculate the signal current generated by a photodiode when 10^{-12} W of power is incident on the diode. Assume red light ($\lambda = 6 \times 10^{-7}$ m) and 100% efficiency.

14-9. Repeat Example 14-7 for an analog signal with frequency content: (a) 1 MHz and (b) 100 MHz.

14-10. Sketch the NRZ and Manchester codes for the following binary signal: 1101001101.

14-11. Light is launched into an optical fiber having refractive indices for the core and cladding of 1.5 and 1.49, respectively.
(a) What is the acceptance cone half-angle?
(b) What is the numerical aperture of the fiber?

14-12. If the index of refraction of glass is 1.5, what is the critical angle, θ_c, for an electromagnetic wave in glass?

14-13. Calculate the numerical aperature of a fiber optics cable if the following indices of refraction apply: core, 1.51; cladding, 1.30.

14-14. A fiber optic system requires 4 μw of received power. The cable is 3 km long and exhibits 3.0 dB/km loss. There is a 1.5 dB loss at both the source and the receiver. Calculate the required level of optical power at the optical source.

CHAPTER 15

Free-Space Optical and Radio-Frequency Communications

In Chapter 14 we considered in detail the design of fiber optics communication systems. The radiation channel was the optical fiber. We now consider optical communication systems where free space constitutes the radiation channel. The elements of a system such as this was introduced in Chapter 1 and a block diagram is shown in Figure 1-3. Since we are still dealing with the transmittance and reception of light, the general problem of defining the required transmitter power to ensure a given received signal-to-noise ratio remains. That is, we must determine the link-loss budget.

Consider the communication system in Figure 15-1. The transmitter power, P_T, is applied to the transmitting antenna, where it is directed to a receiving antenna located at distance, L, away. Now, as indicated in Chapter 1, light is simply electromagnetic radiation (EMR). We know from the study of electromagnetics that the intensity of radiation (both electric and magnetic fields) decreases by a factor of $1/R$, where R is the distance from the source. Thus a *free-space loss* exists between the transmitting and receiving antennas. To find the received power, P_r, it is necessary, then, to compute the path loss and also take into account any gain factors associated with the antennas.

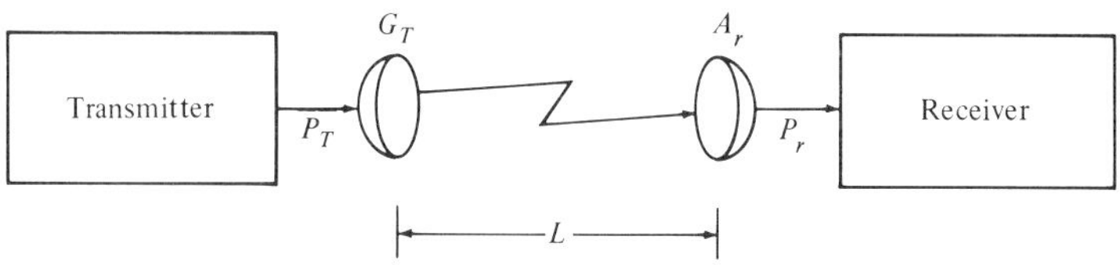

FIGURE 15-1. Typical free-space communication link.

15-1 ANTENNA GAIN

We defined directive antenna gain [equation (1-3)] as

$$G_d = \frac{\text{maximum power density of the radiation pattern}}{\text{power density of an isotropic antenna}}$$

where an isotropic antenna is one that radiates power in all directions. This definition is reflected by equation (15-1).

$$G_d = \frac{P_m}{P_T/4\pi r^2} = 4\pi r^2 \frac{P_m}{P_T} \qquad (15\text{-}1)$$

where P_m is the maximum power density of the radiation pattern, P_T is the total radiated power, and r is the distance from an isotropic antenna radiating total power P_T. Expressed in decibels, the directive gain, G_d, is

$$G_d(\text{dB}) = 10 \log_{10} G_d \qquad (15\text{-}2)$$

Now note that the directive gain, G_d, is based on the power transmitted. It does not take into account losses associated with the antenna. The antenna power gain is given by

$$G_a = \frac{P_m}{P_d/4\pi r^2} = 4\pi r^2 \frac{P_m}{P_d} \qquad (15\text{-}3)$$

where P_d is the power delivered to the antenna. If we divide equation (15-3) by (15-2), we see that the resulting ratio is a measure of the efficiency of the antenna:

$$\frac{G_d}{G_a} = \frac{P_t}{P_d} \qquad (15\text{-}4)$$

15-2

EFFECTIVE AREA OF AN ANTENNA

The *effective area* of an antenna is defined as the ratio of power available at the antenna output to the power density of the incident wave, that is,

$$A = \frac{P_r}{P} \tag{15-5}$$

where

P_r = received power
A = effective area of the antenna
P = power density of the incident wave

From equation (15-5) we may state that the received power, P_r, is equal to the product of the effective area or aperture times the power density of the incident wave. It can be shown that the effective area, A, is given by

$$A = \frac{\lambda^2}{4\pi} G_{dr} \tag{15-6}$$

where G_{dr} is the receiving antenna directive gain. This assumes that there are no losses in the antenna and that all available power is delivered to the load.

15-3

FREE-SPACE PATH LOSS

Now, referring back to Figure 15-1, we can compute the received power, P_r, for the free-space communication link. From equation (15-1), the power density, P_m, at a distance L from the transmitter is given by

$$P_m = \frac{G_d P_t}{4\pi L^2} \qquad \text{W/m}^2 \tag{15-7}$$

The receiving antenna, has an effective area, A_r, and intercepts a total power of

$$P_r = A_r P_m \tag{15-8}$$

Using equations (15-6) and (15-7), equation (15-8) becomes

$$P_r = \frac{\lambda^2}{4\pi} G_{dr} \frac{G_{dt}P_T}{4\pi L^2} = \frac{\lambda^2 P_T G_{dr} G_{dt}}{(4\pi L)^2} \tag{15-9}$$

The *free-space loss* is defined as the ratio of transmitted power to received power. From equation (15-9) this is

$$\frac{P_T}{P_r} = \frac{(4\pi L/\lambda)^2}{G_{dr}G_{dt}} \tag{15-10}$$

In decibels the free-space loss is

$$\text{free-space loss} = 10 \log \frac{(4\pi L/\lambda)^2}{G_{dr}G_{dt}} \tag{15-11}$$

For the frequency in megahertz and L in miles, equation (15-11) becomes

$$\text{FSL} = 36.6 + 20 \log L + 20 \log f - G_{dt} - G_{dr} \tag{15-12a}$$

If isotropic antennas are used ($G_{dt} = G_{dr} = 0$ dB), equation (15-12a) reduces to

$$\text{FSL} = 36.6 + 20 \log L + 20 \log f \tag{15-12b}$$

Equation (15-12b) is the free-space path loss that occurs over the path length, L. Thus

$$L_s = 36.6 + 20 \log L + 20 \log f \tag{15-13}$$

and the received power may be computed as

$$P_r = P_T + G_{dr} + G_{dt} - L_s(\text{dB}) \tag{15-14}$$

Using equation (15-14), for a required received power, P_r, we may readily determine the transmitter power.

EXAMPLE 15-1

A PCM signal with a bandwidth of 1 MHz is to be transmitted over an optical length of 10 miles. Coherent red light ($\lambda = 6 \times 10^{-7}$ m) is to be used and a required received SNR = 60 dB must be maintained. The transmitting and receiving antennas both have 100-dB gain. Determine (a) the required transmit power, P_T, and (b) the diameter of the receiving antenna lens.

SOLUTION

(a) From Figure 14-24, for a 60-dB SNR at a PIN diode input:

$$P_R = -50 \text{ dBm}$$

$$P_T = P_r - G_{dr} - G_{dt} + L_s$$

$$= -50 - 100 - 100 + 36.6 + 20 \log 10 + 20 \log \frac{300}{6 \times 10^{-7}}$$

$$= -19.42 \text{ dBm} \Rightarrow 0.0114 \text{ mW}$$

$$\text{(b)} \quad A = \frac{\lambda^2}{4\pi} G_{dr} = \frac{(6 \times 10^{-7})^2}{4\pi} (10^{10}) = 2.866 \times 10^{-4} \text{ m}^2$$

$$\pi r^2 = 2.866 \times 10^{-4} \text{ m}^2$$

$$r = 0.39 \text{ in.} \Rightarrow \text{diameter} = 0.78 \text{ in.}$$

The size of the receiving lens illustrates the pointing problems inherent in optical communications. A direct detection receiver is shown in Figure 15-2. In designing receivers such as this, we must make allowances for sources of radiation that are not present in receivers attached to a fiber optic cable. Specifically, this is "background" radiation which contributes to noise in the receiver.

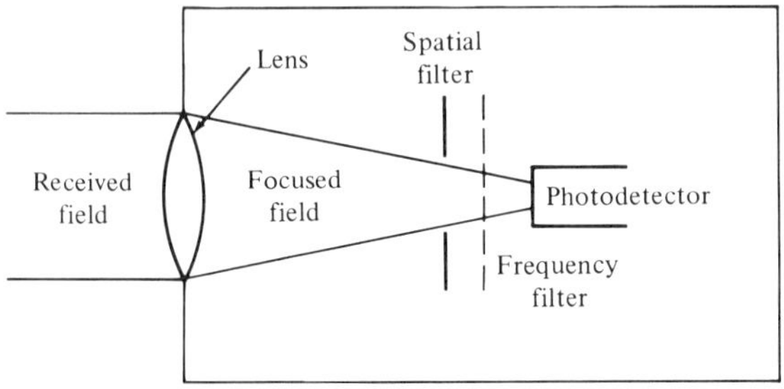

FIGURE 15-2. Direct detection receiver.

NOISE CONSIDERATIONS

In addition to receiver noise, a receiving system that views an atmosphere's background also collects undesirable background radiation that falls within the spatial and frequency ranges of the detector. This radiation is processed along with the signal and represents an overall degradation in system performance. The direct detection receiver with noise sources can be modeled as shown in Figure 15-3. The incident field results in a current through the *p-n* junction of the photodiode. The detected field produces a shot noise process at the photodetector output. This shot noise process is governed by the received intensity arising from both the signal and the background radiation collected in the optical front end. In addition to shot noise, circuit (thermal) noise is added to the detector output prior to receiver postdetection processing.

Consider the receiver in Figure 15-4. The incident light results in current, I_p, through the diode, with this current representing the detected signal. The signal voltage developed across R_L is amplified by the preamp with gain A. Now, for wide-band amplifiers, thermal noise and shot noise are the only noise components that must be taken into consideration. As indicated previously, the photodiode produces shot noise and we know that R_L will produce thermal noise. Consequently, we can model the preamplifier input as shown in Figure 15-5.

From Chapter 5 we know that thermal noise is caused by the random thermal motion of charge carriers. Shot noise is caused by the random generation of

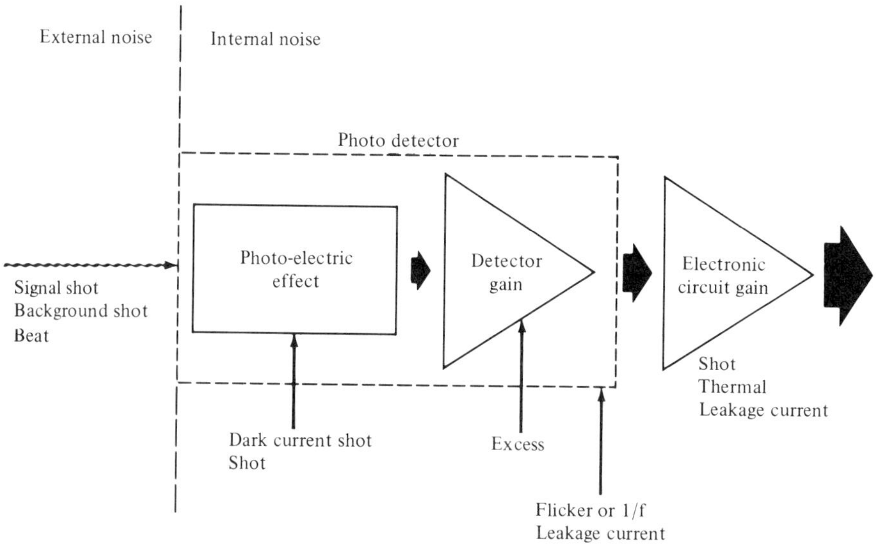

FIGURE 15-3. Noise sources (direct detection optical communication receiver).

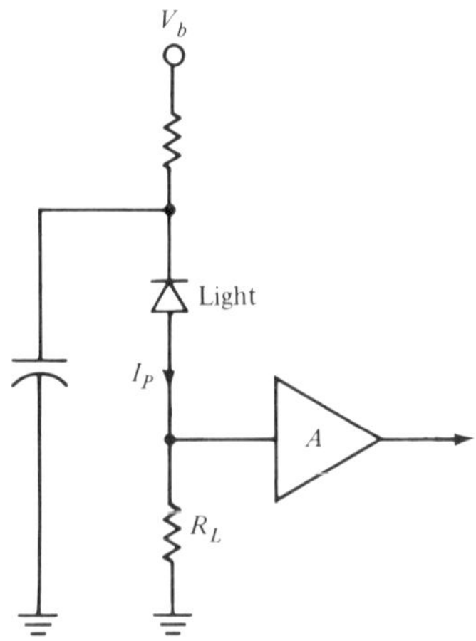

FIGURE 15-4. Preamplifier input.

carriers. The mean-square value of noise current for a resistor of R ohms is [see equation (5-12a)].

$$\bar{I}_{tn}^2 = \frac{4kT}{R} = \frac{4eV_T}{R} \qquad \text{A/Hz} \qquad (15\text{-}15)$$

where $V_T = kT/e$ and q is the charge of an electron, 1.6×10^{-19} A-s. If the noise equivalent bandwidth of the amplifier is B, the mean-square value of thermal noise current is

$$\bar{I}_{tn}^2 = \frac{4qV_TB}{R} \qquad \text{amperes} \qquad (15\text{-}16)$$

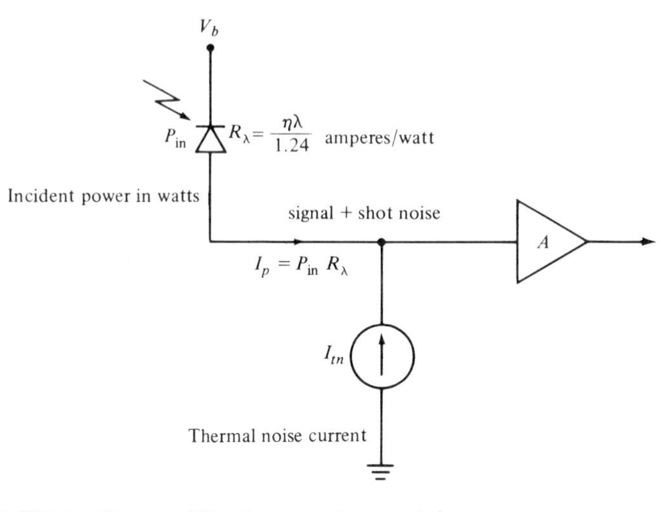

FIGURE 15-5. Preamplifier input noise model.

Note that the thermal noise current increases with the bandwidth and decreases with the resistance, R.

Now, from equation (5-2), the mean-square shot noise current is given by

$$\bar{I}_{sh}^2 = 2qI_{dc}B \tag{15-17}$$

The mean-square value of the total noise current is the sum of the mean squares of the individual noise currents. That is,

$$\begin{aligned} \bar{I}_n^2 = \bar{I}_{tn}^2 + \bar{I}_{sh}^2 &= \frac{4qV_TB}{R} + 2qI_{dc}B \\ &= 2q\left(I_{dc} + \frac{2V_T}{R}\right)B \end{aligned} \tag{15-18}$$

Now, at room temperature, $T = 300°K$, $V_T = 0.025$ V. Thus, if $2V_T/R = 50\ \text{mV}/R >> I_{dc}$, the shot noise contribution in equation (15-18) can be ignored. To get an idea of the magnitudes involved, for $B = 10$ MHz and $I_{dc} = 10$ nA, the shot noise current, I_{sh}, from equation (15-17) is

$$\begin{aligned} \bar{I}_{sh} &= \sqrt{2qI_{dc}B} = \sqrt{2(1.6 \times 10^{-19})(10 \times 10^{-9})(10 \times 10^6)} \\ &= 1.8 \times 10^{-9}\ \text{A} = 1.8\ \text{nA} \end{aligned}$$

Referring to Figure 15-5, at the input to the preamp, the signal-to-noise ratio is given by

$$\text{SNR} = 20 \log \frac{I_p}{I_n} \quad \text{dB} \tag{15-19}$$

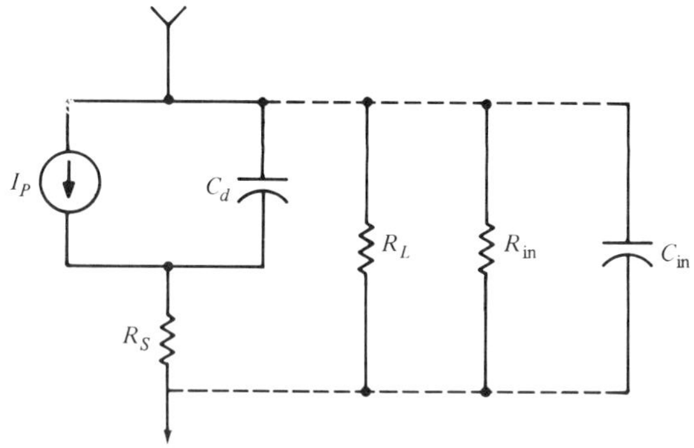

FIGURE 15-6. Input preamplifier equivalent circuit.

The minimum detectable signal will depend on the noise level and the shot noise depends on the signal level. Now, the thermal noise is independent of the signal and is dependent on the temperature. Thus, to make the noise current as low as possible, we must make the input resistance of the amplifier as large as possible. However, as we observed in Chapter 14, a high input resistance decreases the system data rate. High data rates require a wide-bandwidth amplifier with fast response. The time constant, $\tau = RC$, of the amplifier then places an upper limit on R. The equivalent circuit for Figure 15-4 is shown in Figure 15-6, and some typical circuits for the receiver input are shown in Figure 15-7.

15-5 BASEBAND MODULATION

Receiver performance in analog systems is normally specified as a signal-to-noise ratio or a carrier-to-noise ratio. For baseband intensity modulation, the transmitted optical signal is given by

$$P(t) = P_0[1 + mf(t)] \tag{15-20}$$

where

P_0 = average optical power
$m \leq 1$ = modulation index
$f(t)$ = normalized signal with $|f(t)|_{max} = 1$, $\overline{f}(t) = 0$, and $\overline{f^2}(t) \approx \frac{1}{2}$.

The instantaneous photocurrent, I_p, is proportional to the incident optical signal. That is,

$$I_p = I_0[1 + mf(t)] \tag{15-21}$$

where $I_0 = \bar{I}_p$, the average photocurrent. The mean-square value of photocurrent associated with the signal is

$$\begin{aligned} \bar{I}_s^2 &= m^2 I_0^2 \overline{f^2(t)} \\ &\approx \frac{1}{2} m^2 I_0^2 \end{aligned} \tag{15-22}$$

We can now directly define the SNR. Specifically, the SNR is the ratio of signal current to noise current.

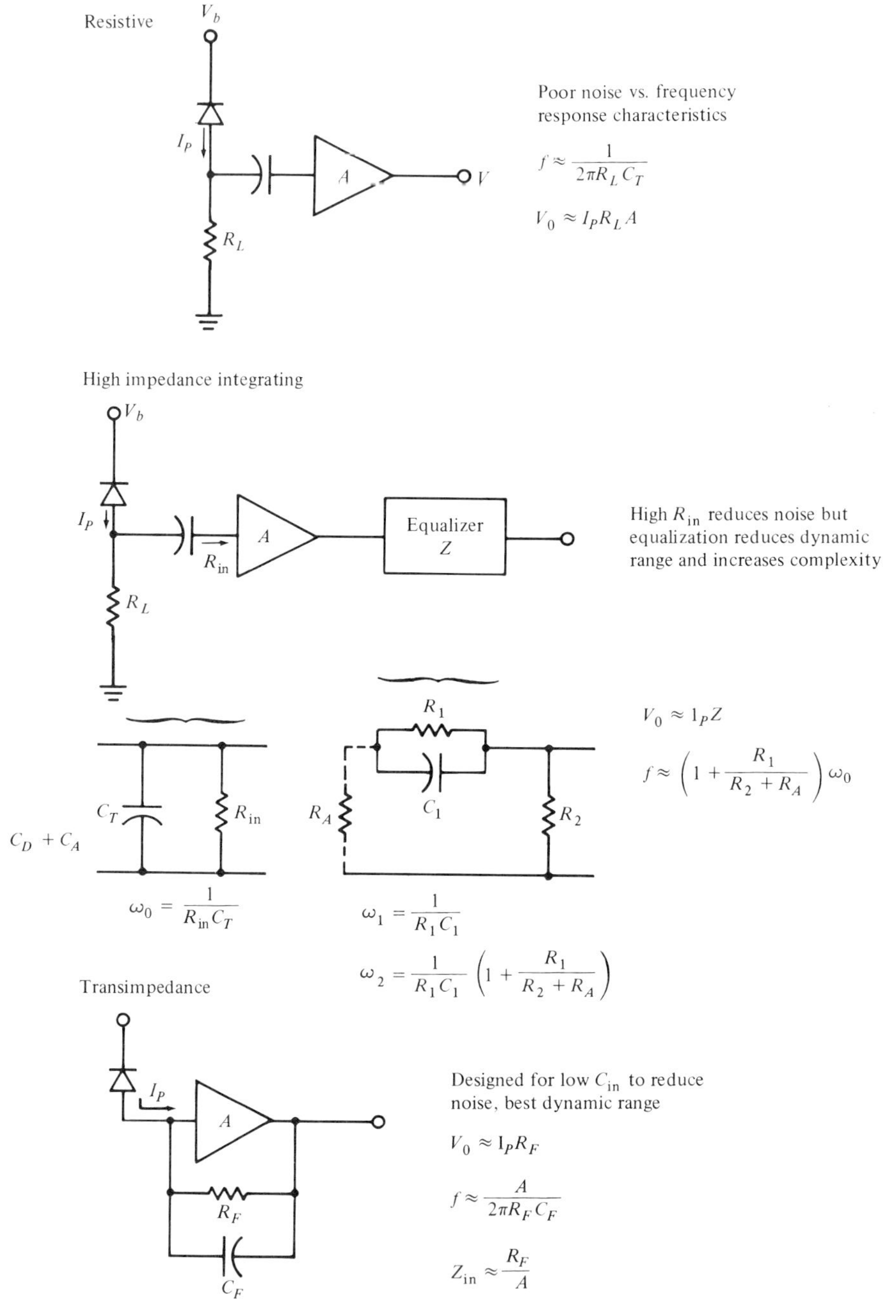

FIGURE 15-7. Receiver electronics: resistive, high-impedance integrating, and transimpedance.

$$\text{SNR} = \frac{\overline{I}_s^2}{\overline{I}_n^2} = \frac{\frac{1}{2}m^2I_0^2}{2qI_0B + (4qV_T/R)B}$$
$$= \frac{1}{2}\frac{m^2I_0^2}{2qB(I_0 + 2V_T/R)} \tag{15-23}$$

From equation (15-23) it is obvious that the maximum SNR occurs when the modulation index, $m = 1$. For a preamplifier with -6 dB/octave roll-off, the noise bandwidth, B, is, from equation (5-8),

$$B = \frac{1}{4RC} \tag{15-24}$$

Now if the basebandwidth of the signal, $f(t)$, is equal to the -3-dB bandwidth of the amplifier, equation (15-24) may be written as

$$B = \frac{\omega_1}{4} = \frac{\pi}{2}f_1 \tag{15-25}$$

and the maximum SNR can be expressed as

$$\text{SNR}_{\text{max}} = \frac{I_0^2}{2\pi f_1 q(I_0 + 2V_T/R)} \tag{15-26}$$

To maximize the expression in (15-26) we need to make R as large as possible. However, the required -3-dB bandwidth of the amplifier sets an upper limit on R. Specifically, $R = 1/2\pi f_1 C$. Thus we may write equation (15-26) as

$$\text{SNR}_{\text{max}} = \frac{I_0^2}{2\pi f_1 q(I_0 + 4\pi V_T f_1 C)} \tag{15-27}$$

From this equation, for any desired SNR, we can determine the minimum detectable signal current, I_0, and the power level, P_0, at the receiving end. Specifically,

$$I_0 = \pi f_1 q \,\text{SNR}\left(1 + \sqrt{1 + \frac{8V_T C}{q\,\text{SNR}}}\right) \tag{15-28}$$

The optical power is related to the photocurrent by

$$\eta P_0 = \frac{1.24}{\lambda/\mu\text{m}} I_0 \quad \text{W/A}$$
$$= \frac{1.24\ \mu\text{m}}{\lambda} I_0 \quad \text{W/A} \qquad (15\text{-}29)$$

where η is the quantum efficiency of the diode.

EXAMPLE 15-2

It is desired to transmit black-and-white television (bandwidth = 4.5 mHz) using red light ($\lambda = 6 \times 10^{-7}$ m). Assuming a photodiode quantum efficiency of 100%, calculate the minimum detectable signal power. Assume an effective feedback capacitance across R of 0.5 pF and a required SNR of 47 dB (5×10^4).

SOLUTION

$$I_0 = (1.601 \times 10^{-19})(3.14)(4.5 \times 10^6)(5 \times 10^4) \times \left[1 + \sqrt{1 + \frac{(8)(0.025)(0.5 \times 10^{-12})}{(1.601 \times 10^{-19})(5 \times 10^4)}}\right]$$
$$= 528.83 \text{ nA}$$
$$P_0 = \frac{1.24 \times 10^{-6}}{6 \times 10^{-7}}(528.83 \times 10^{-9})$$
$$= 1.09\ \mu\text{W}$$

This is a relatively large signal. Consequently, it may be desirable for a given design to investigate the use of an avalanche photodiode. For low signal levels, the shot noise can be ignored. The presence of thermal noise is the significant parameter. For the design in Example 15-2 the link margin requirements may be determined as in Example 15-1 to determine transmitter power for a given communication path.

15-6 FREE-SPACE RF COMMUNICATIONS

One of the more interesting characteristics of communications theory is the variety of design approaches allowable when considering an initial design. This variety precludes a complete discussion of the techniques available. However, it is informative at this point to consider an actual communications system design example. To accomplish this it is necessary to choose a particular technique to serve as a design vehicle. A telemetry system utilizing FM/PM modulation is a widespread technique in use. Prior to considering a design example, we will first develop some useful system design curves.

15-6.1 FM/PM Telemetry System

The block diagram for an FM/PM telemetry system is shown in Figure 15-8. The VCOs constitute the subcarriers with the RF carrier provided by the PM

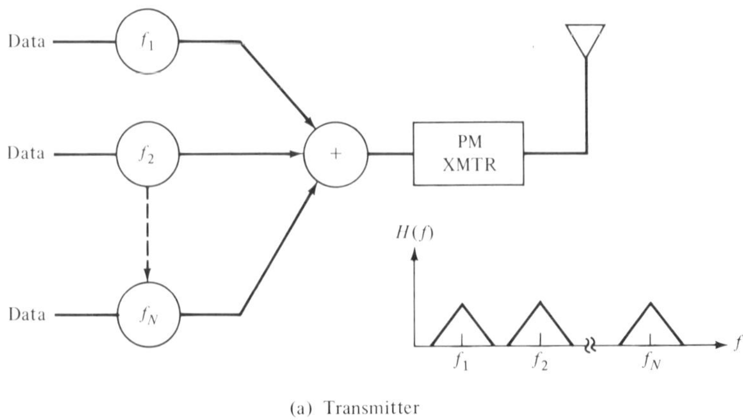

FIGURE 15-8a. Functional block diagram of an FM-PM system: transmitter.

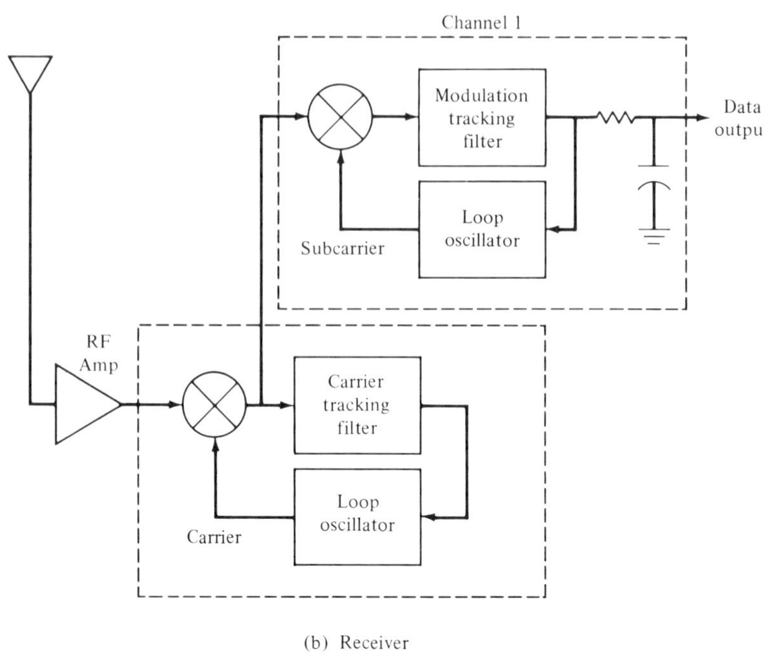

FIGURE 15-8b. Functional block diagram of an FM-PM system: receiver.

transmitter. The receiver (Figure 15-8b) consists of a narrowband carrier tracking loop with additional subcarrier tracking loops. For simplicity, only one subcarrier tracking loop is shown. Referring to equation (9-24), the output of the multiplier in the carrier tracking loop contains the phase modulating signal $g(t)$, which in this case is the sum of the FM subcarriers. The subcarrier spectrum is indicated in Figure 15-8a. The carrier tracking filter is made narrow, on the order of 20 Hz, so that the subcarrier spectrum is rejected and only the carrier is tracked. With each subcarrier loop locked on its respective VCO frequency, the output of the multiplier in the subcarrier loop contains the data. The loop filter is designed with sufficient bandwidth to pass these data. Consequently, we refer to the PLL discriminator as a modulation tracking loop. The system design problem is to allocate the available RF power between the carrier and the sideband frequencies corresponding to each subcarrier (VCO). Typically, we allocate this power so that the carrier tracking loop will not lose lock before the subcarrier tracking loops. The design of an FM/PM link then begins with a specification as to the fraction of the total RF power that must remain in the carrier channel to ensure proper carrier threshold performance. Prior to considering the power budget, let us first consider the carrier tracking loop as a phase demodulator.

15-6.2 Carrier Tracking Loop as a Phase Demodulator

The receiver in Figure 15-8 may be reduced in its simplest form to a carrier tracking-phase coherent loop, as shown in Figure 15-9. Now, if the subcarrier spectrum does not extend into the tracking range of the loop, the receiver can be further simplified to the equivalent operation of multiplication (Figure 15-10). In this figure $a(t)$ is the total received PM spectrum and $b(t)$ is the recovered subcarrier spectrum, and ω_c is the carrier frequency. The general expression for the case of k subcarriers phase modulating a carrier signal is similar to equation (9-20) for multitone FM modulation. The exact expression is

$$a(t) = \sum_{n_1=-\infty}^{\infty} \sum_{n_2=-\infty}^{\infty} \cdots \sum_{n_k=-\infty}^{\infty} J_{n1}(m_{p1}) J_{n2}(m_{p2}) \cdots J_{nk}(m_{pk}) \times \sin\left(\omega_c + \sum_{i=1}^{k} n_i \omega_i\right)t \quad (15\text{-}30)$$

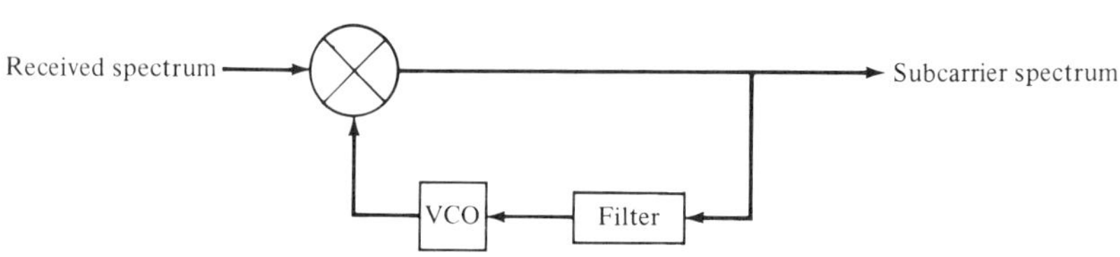

FIGURE 15-9. Carrier tracking-phase coherent loop receiver.

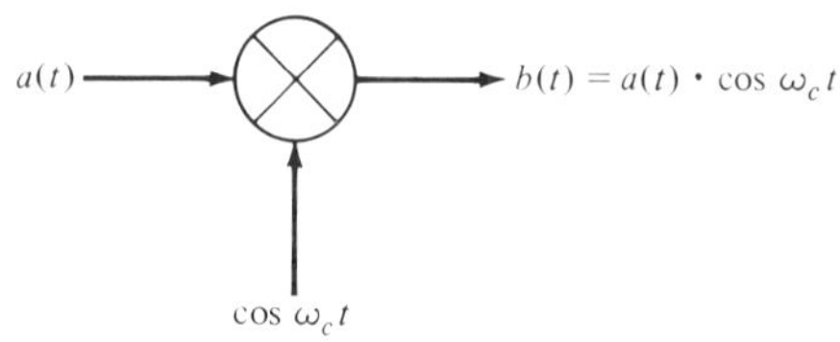

FIGURE 15-10. Basic receiver operation.

where the $J_{nk}(m_{pk})$ are the nth-order Bessel coefficients of the kth subcarrier.

Performing the indicated multiplication in Figure 15-10 and substituting the trigonometric expression

$$\sin x \sin y = \tfrac{1}{2} < \sin (x + y) + \sin (x - y)$$

yields the following expression for the received subcarrier spectrum:

$$b(t) = \frac{1}{2}\left[\sum_{n_1=-\infty}^{\infty} \sum_{n_2=-\infty}^{\infty} \cdots \sum_{n_k=-\infty}^{\infty} J_{n1}(m_{p1}) J_{n2}(m_{p2}) \cdots J_{nk}(m_{pk}) \times \sin \left(\sum_{i=1}^{k} n_i \omega_i \right) t \right] \quad (15\text{-}31)$$

Note that terms in $2\omega_c$ are omitted since they are many times higher than the highest subcarrier.

By expanding equation (15-31) into positive and negative n_i and applying the identities

$$J_{-n}(m_p) = (-1)^n J_n(m_p) \quad (15\text{-}32)$$

$$\sin -\theta = -\sin \theta \quad (15\text{-}33)$$

it can be shown that the recovered spectrum will contain only modulation products which satisfy the following conditions:

$$\sum_{i=1}^{k} n_i \text{ is odd}$$

As a simple example, consider the case of a single subcarrier. The incoming RF wave, $a(t)$, becomes

$$a(t) = \sum_{n=-\infty}^{\infty} J_n(m_p) \sin(\omega_c + n\omega_{sc})t \quad (15\text{-}34)$$

The recovered spectrum, $b(t)$, is then

$$b(t) = J_1(m_p) \sin \omega_{sc}t + J_3(m_p) \sin 3\omega_{sc}t + J_5(m_p) \cdots \qquad (15\text{-}35)$$

As a second example, consider the case of two subcarriers, with center frequencies ω_1 and ω_2 and modulation indicies m_{p1} and m_{p2}. The recovered subcarrier spectrum, $b(t)$, becomes

$$\begin{aligned} b(t) = {} & J_1(m_{p1})J_0(m_{p2}) \sin \omega_1 t + J_0(m_{p1})J_1(m_{p2}) \sin \omega_2 t \\ & + J_3(m_{p1})J_0(m_{p2}) \sin 3\omega_1 t + J_0(m_{p1})J_3(m_{p2}) \sin 3\omega_2 t \\ & + J_1(m_{p1})J_2(m_{p2}) \sin (\omega_1 \pm 2\omega_2)t \\ & + J_2(m_{p1})J_1(m_{p2}) \sin (\omega_2 \pm 2\omega_1)t \\ & + J_2(m_{p1})J_3(m_{p2}) \sin (2\omega_1 \pm 3\omega_2)t \\ & + J_3(m_{p1})J_3(m_{p2}) \sin (3\omega_1 \pm 2\omega_2)t + \cdots \end{aligned} \qquad (15\text{-}36)$$

If the multiplier is perfect, the following modulation products are theoretically absent:

$\omega_1 \pm \omega_2$	$2\omega_1$
$2(\omega_1 \pm \omega_2)$	$2\omega_2$
$3(\omega_1 \pm \omega_2)$	$\omega_1 \pm 3\omega_2$
$4(\omega_1 \pm \omega_2)$	$3(\omega_1 \pm \omega_2)$

Referring to equation (15-36), the useful subcarrier sidebands are given by

$$J_1(m_{p1})J_0(m_{p2}) \sin \omega_1 t \qquad (15\text{-}37)$$

and

$$J_0(m_{p1})J_1(m_{p2}) \sin \omega_2 t \qquad (15\text{-}38)$$

The remaining terms are cross-modulation products and, as such, are wasted power.

For the case of a single subcarrier, then, the recoverable sideband power resides in the first-order sidebands $J_1(m_p)$, and the remaining components appear as distortion products and, therefore, lost power. The case of two or more sidebands presents an additional problem to the system designer in that the odd-order sum and difference products can fall in or near the desired channel pass-bands and possibly create an interference problem. Fortunately, these products are often so small as to be negligible, except in the case of a large number of subcarriers.

15-6.3 Carrier Channel Power

Refer to equation (9-21). The power in the carrier is given as the product of the zero-order Bessel coefficents. That is,

$$\frac{P_c}{P_{tot}} = \prod_{n=1}^{N} [J_0(m_{pn})]^2 \tag{15-39}$$

where m_{pn} are the modulation indices of the N subcarriers on the carrier. If the simplifying assumption is made that all subcarriers have equal modulation indices, it is possible to plot a family of curves as shown in Figure 15-11. Experience with FM/PM system designs has shown these curves to be valid even when the modulation indices are not equal, but fall within a range of 4 or 5 to 1, if three or more subcarriers are used.

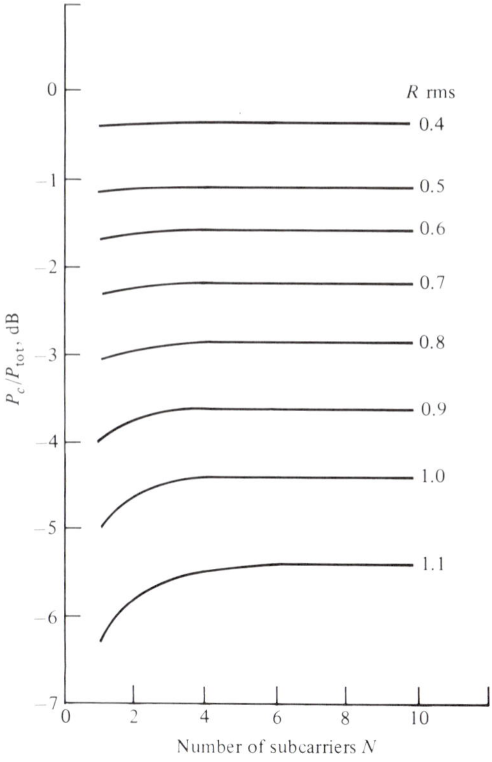

FIGURE 15-11. Carrier power versus *N* for various modulation levels.

If, for example, a five-channel system were needed and the P_c/P_{tot} ratio had to be kept at -3 dB, the total allowable rms modulation of the carrier by the subcarriers would be 0.82 rad rms. We see that for a given total modulation index, the carrier power as a fraction of the total power can readily be determined. It should be mentioned that in a properly designed system, this rms modulation on the carrier will not appear as a modulation error in the carrier loop, nor contribute to the carrier threshold. Only when the subcarriers fall within the tracking bandwidth of the loop will the carrier threshold be altered. Under this condition there be an effective loss of subcarrier power, and the entire system performance will be degraded.

The remaining power must be distributed among the subcarriers. This process is examined in the next section.

15-6.4 Subcarrier Channel Power

We saw [equations (15-37) and (15-38)] that for the case of two subcarriers the usable sideband power is given by the $J_1(m_{p1})J_0(m_{p2})$ product for the first subcarrier and by the $J_0(m_{p1})J_1(m_{p2})$ product for the second subcarrier. If we extend this to the case of N subcarriers, we have the following expression for the nth subcarrier:

$$\frac{P_{sc}^{(N)}}{P_{tot}} = \left[2 \prod_{n=1}^{N} J_0(m_{pn}) J_1(m_{pn}) \right]^2 \tag{15-40}$$

This expression is plotted in Figure 15-12.

When using the design curves in Figure 15-12, note that the curves correspond to the power in each of the N subcarriers and not to the total subcarrier sideband power. Examining the curves, we see that for a given rms modulation of the carrier, the total usable subcarrier power decreases as the number of subcarriers increases. As the number of subcarriers, N, increases, more of the total received power appears in the unusable odd-order cross-modulation products. This is shown in Figure 15-13, where the usable percent of the total system power is plotted as a function of the rms modulation level and the number of subcarriers.

A common criticism of the phase coherent FM/PM system is that only a relatively small percentage of the total power can be placed in the usable sidebands, which results in a rather inefficient communication link. To establish an upper bound on the amount of usable sideband power, let us consider the most efficient case: a single subcarrier modulating the carrier. A study of the Bessel coefficients quickly reveals that the $J_1(m_p)$ term will be a maximum for a modulation index of 1.84 rad peak. Use of this modulation index would place 68% of the total power in the usable sidebands and 10% in the carrier. The remaining 22% would be wasted in the higher-order modulation products. At this modulation level, however, a small change in modulation index results in a relatively

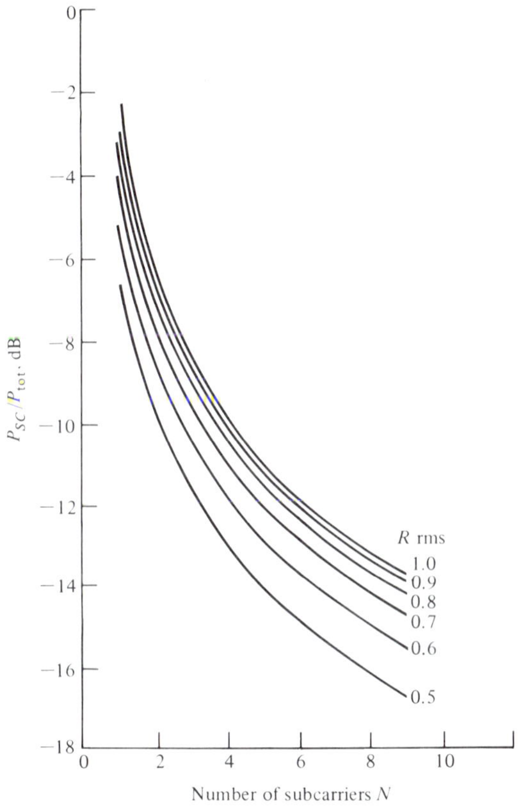

FIGURE 15-12. Individual subcarrier power versus *N* for various modulation levels.

large change in the carrier power and therefore a rather unstable design point. If the peak modulation index is chosen to be $\sqrt{2}$, or 1.0 rad rms, 59% of the power will appear in the subcarrier and 31% in the carrier. Only 10% of the power will be wasted. Reducing the modulation index from 1.84 to 1.41 rad peak therefore increases the carrier power by 4.0 dB while decreasing the subcarrier power by only 0.6 dB. It is therefore suggested that 1.0 rad rms is the highest practical modulation level in an FM/PM system, and in this case the ratio of subcarrier to carrier power is nearly 2/1.

Experience has shown that when 5 to 10 subcarriers are used in a system with information rates considerably less than the maximum defined by the IRIG standards, a total modulation level of 0.9 rad rms is about the largest practical. If a full 18 IRIG channels are required, the modulation level must be reduced to 0.5 rad rms or less to avoid excessive odd-order modulation products in the spectrum. Clearly, the coherent FM/PM link is basically a narrowband technique.

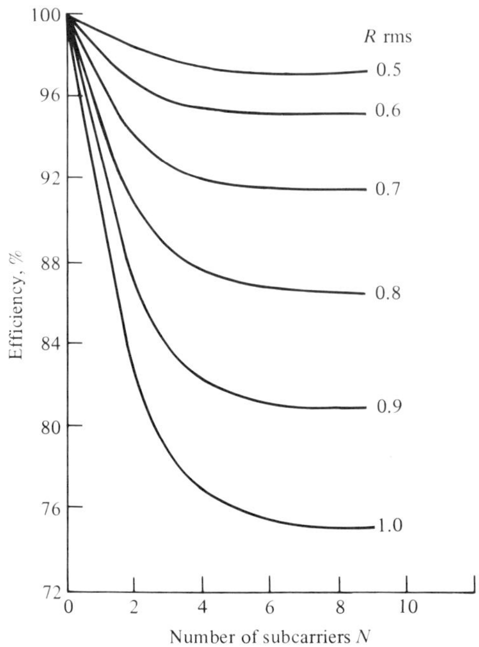

FIGURE 15-13. System power efficiency versus *N* for various modulation levels.

15-6.5 Choice of Modulation Indices

Since the modulation indices determine the sideband power, it is of interest at this point to consider a procedure for choosing the modulation indices. The power in the significant sidebands of the Nth subcarrier is given by an expansion of equation (15-40). That is,

$$P_{sc} = P_{tot}2[J_0(m_{p1})J_0(m_{p2}), \ldots, J_0(m_{pN-1})J_1(m_{pN})]^2 \qquad (15\text{-}41)$$

Now equation (15-41) may be rewritten as

$$P_{sc} = P_{tot}2\left[J_0(m_{p1})J_0(m_{p2}), \ldots, J_0(m_{pN-1})J_0(m_{pN})\,\frac{J_1(m_{pN})}{J_0(m_{pN})}\right]^2$$
$$\frac{P_{sc}}{P_{tot}} = 2\,\frac{P_c}{P_{tot}}\left[\frac{J_1(m_{pN})}{J_0(m_{pN})}\right]^2 \qquad (15\text{-}42)$$

Equation (15-42) is a very useful design equation. The bracketed ratio is a function of the modulation index, m_p, only, and the ratio P_c/P_{tot} is usually specified as a system design constraint. This equation is plotted in Figure 15-14 as P_{sc}/P_{tot} versus the modulation index. The process of using these curves

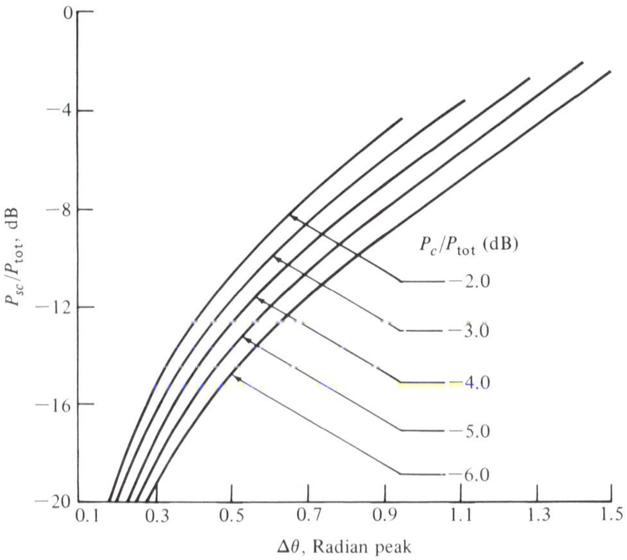

FIGURE 15-14. Individual subcarrier power versus modulation index.

reduces first to calculating the required sideband power and then referring to the proper curve in Figure 15-14 for P_c/P_{tot} and then reading the exact modulation index.

Now, once the ratio P_c/P_{tot} is chosen and the number of subcarriers determined, the total allowable rms modulation of the carrier may be read directly from the curves in Figure 15-11. We next determine all N modulation indices from Figure 15-14 and compute the rms value. This number is compared with the allowable value. Obviously, for the design to be realizable the computed rms value must be equal to or less than the allowable. If this is not true, the required sideband power is not available and the communication system is not capable of transmitting the information the required threshold distance.

The preceding design description assumed that the system was to transmit a given amount of data to a required threshold distance. Many times the design approach is considerably simpler in that either the available communication distance is the parameter to be computed, or the number of subcarrier channels and threshold distance are fixed parameters and the characteristics of the subcarrier channels are to be calculated. In the latter case the data would be conditioned so as to keep the loop modulation error within allowable bounds.

15-6.6 Channel Design Equation

Before considering a design example, it is useful to introduce a version of the free-space transmission equation developed earlier [equation (15-9)]. The received signal power is given by

$$P_r = \frac{P_t G_t G_r}{L_s L_m L_p} \tag{15-43}$$

In this equation L_p is a design margin, usually 6 dB, and L_m is losses associated with coaxial cables, connectors, and so on. This is generally on the order of 1.5 to 2 dB. The quantity, L_s, is free-space loss, previously discussed.

The range equation permits calculating the necessary power for a specified signal-to-noise ratio at the receiver. For our particular receiver we need to design for a particular signal-to-noise ratio in the phase-locked loop. The concept of an equivalent noise bandwidth (ENBW) was introduced in Chapter 5 in terms of a low-pass filter. A PLL is itself a filter and the ENBW is given by

$$\text{ENBW} = 2\beta_{\text{Lo}} = \frac{1}{2\pi j} \int_{-j\infty}^{j\infty} \left| \frac{\phi_o(s)}{\phi_i(s)} \right|^2 ds \tag{15-44}$$

where $\phi_o(s)/\phi_i(s)$ is the loop transfer function [equation (7-33)]. For a second-order loop, $\beta_{\text{LO}} = 3.33\omega_n$, where ω_n is the loop natural frequency [see equation (7-37)].

Now the received noise spectral density is given by

$$\phi_{kT} = kT \qquad \text{W/Hz} \tag{15-45}$$

Thus the total noise power, N, is

$$N = 2\beta_L \phi_{kT} \tag{15-46}$$

where the factor of 2 arises from considering a two-sided noise spectrum. Using equations (15-43) and (15-46), the resulting SNR

$$\frac{P_r}{N} = \text{SNR} = \frac{P_t G_t G_r}{L_s L_m L_p 2\beta_L \phi_{kT}} \tag{15-47}$$

We can express this equation logarithmically as

$$P_t + G_t + G_r = 2\beta_L + (S/N)_{2\beta L} + \phi_{kT} + L_s + L_m + L_p \tag{15-48}$$

Note that since bandwidths are also expressed logarithmically, a slightly different concept in usage arises. For example, 2 Hz becomes 3 dB · Hz, 5 Hz becomes 7 dB · Hz, and so on. Equation (15-48) lends itself readily to the design of FM/PM systems since the carrier and each subcarrier channel can be considered separately as having a given fraction of the total transmitted power.

EXAMPLE 15-3

A rocket containing a low power beacon is to be launched, with its destination the moon. It is desired to track the beacon to determine a trajectory and at the same time telemeter a half-dozen relatively static vehicle functions. The required communication distance is one lunar distance (250,000 miles) and the transmitting frequency is chosen to be 960 MHz. Assume the following system boundary conditions:

$$P_{\text{tot}} = 50 \text{ mW, } 17.0 \text{ dBm}$$

$$G_t = 0.0 \text{ dB (omnidirectional)}$$

$$G_r = 45.0 \text{ dB (85-ft-diameter dish)}$$

$$\Sigma L_m = 2.0 \text{ dB}$$

$$\phi_{kT} = -174.0 \text{ dBm/Hz (effective system temperature of 290°K)}$$

$$L_s = 204.2 \text{ dB (250,000 miles)}$$

$$L_p = 6.0 \text{ dB}$$

$$2\beta_{\text{LO}} = 20 \text{ Hz}$$

$$\text{threshold (SNR)}_{2\beta_{\text{LO}}} = 3.0 \text{ dB}$$

SOLUTION

We first solve for the required carrier power using equation (15-48):

$$\begin{aligned} P_c &= 2\beta_L + (S/N)_{2\beta_L} + \phi_{kT} + L_s + L_m + L_p - G_t - G_r \\ &= 13 + 3 - 174 + 204.3 + 2 + 6 - 0.0 - 45 \\ &= +9.2 \text{ dBm} \end{aligned}$$

Normalizing the carrier power by the total power yields

$$\frac{P_c}{P_{\text{tot}}} = 9.2 - 17 = -7.8 \text{ dB}$$

Referring to Figure 15-11, we see that this is more carrier suppression than is practical. From this same figure it is evident that if a maximum of 1 rad rms modulation of the carrier is permitted the carrier channel will be overdesigned by approximately 3 dB.

A variety of choices are available for design at the subcarrier level. We could time multiplex the information onto a single VCO or we could use a VCO for each channel of information. Assume first a PAM multiplexed signal modulating a single VCO. From Figure 15-12 the normalized subcarrier power for a single channel and a 1.0-rad rms modulation level is

$$\frac{P_{sc}}{P_{tot}} = -2.3 \text{ dB}$$

or

$$P_{sc} = 14.7 \text{ dBm}$$

Now, if the modulation error in the ground discriminator is constrained to be within 30° peak, a satisfactory design threshold SNR in the subcarrier $2\beta_{LO}$ is on the order of 5.6 dB. Using this information, the channel design equation can be used to find the available subcarrier, $2\beta_{LO}$. Thus

$$\begin{aligned} 2\beta_{LO} &= P_t + G_t + G_r - (S/N)_{2\beta_L} - \phi_{kT} - L_s - L_m - L_p \\ &= 14.7 + 0.0 + 45 - 5.6 - (-174) - 204.2 - 2 - 6 \\ &= 15.9 \text{ db} \cdot \text{Hz} \end{aligned}$$

or

$$2\beta_{LO} = 38.6 \text{ Hz}$$

The corresponding loop natural frequency (loop information bandwidth) is

$$f_n = \frac{\beta_{LO}}{3.33} = \frac{19.3}{3.33} = 5.8 \text{ Hz}$$

A little thought will convince us that if we are to limit the phase error in the subcarrier discriminator to 30° peak, we must limit the rate at which the data deviate the VCO frequency. Specifically, if the subcarrier PLL is to track the incoming signal with less than 30° phase error, we must limit the rise time of the data pulses before it is applied to the VCO. A scheme for accomplishing this is shown in the figure.

Applying a ramp input to the VCO results in a frequency ramp applied to the subcarrier PLL. We wish the PLL to track the frequency ramp with a phase error (lag) not to exceed 30° peak. Referring to equation (7-32), we see that if $\theta_i(s)$ is the incoming phase associated with a frequency ramp, an application of

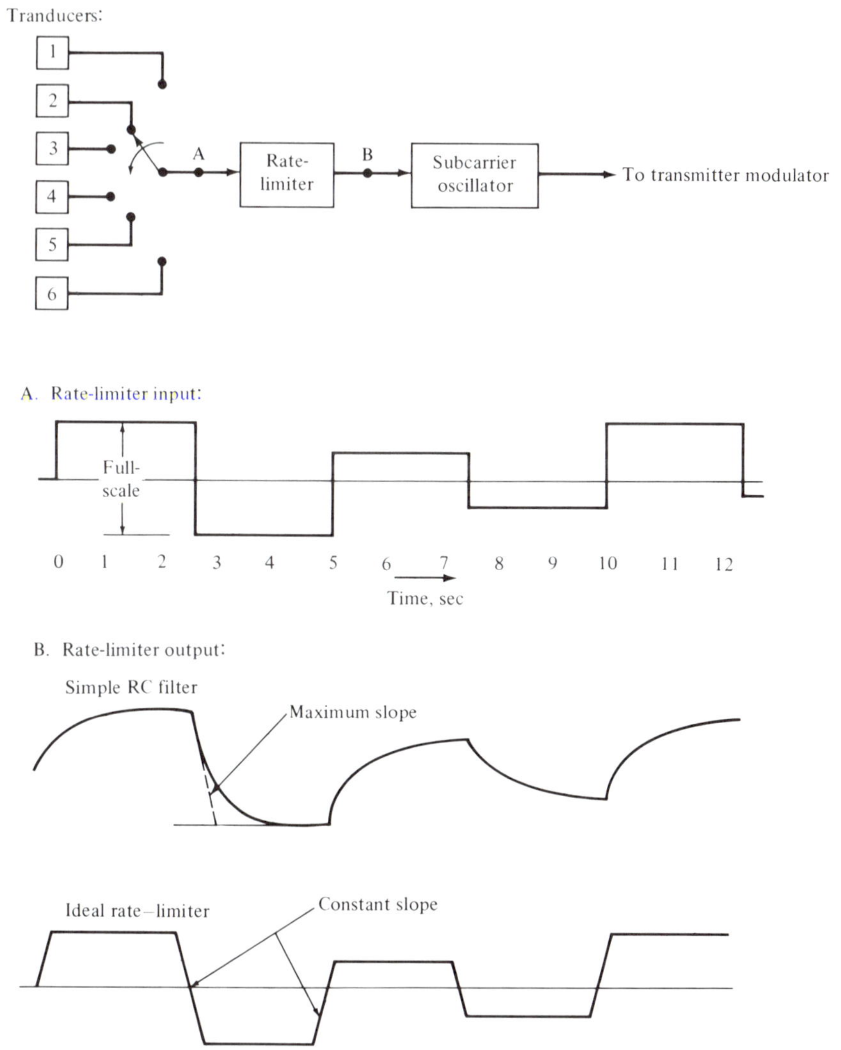

the final-value theorem to this equation will show that the steady-state output error for a second-order PLL is given by

$$\Delta\theta_{ss} = \frac{\dot{\omega}}{\omega_n^2}$$

or

$$\dot{f} = \Delta\theta_{ss}\ \omega_n^2/2\pi$$
$$= \Delta\theta_{ss}\ 2\pi f_n^2$$

For $\Delta\theta_{ss} = 30°$ this becomes

$$\dot{f} = 3.28 f_n^2$$

$$= (3.28)(5.8)^2 = 110 \text{ Hz/s}$$

That is, the allowable rate of change of VCO frequency is 110 Hz/s. Now for IRIG channels 1 or 2, a full-scale channel deviation on the order of 55 Hz is reasonable. Thus the rate-limiting time constant must be 0.5 s. If we allow five time constants per sample, the allowable sampling rate becomes 24 samples per minute. Thus we would sample each channel every 15 s. If an ideal rate limiter were used instead of a simple *RC* filter, the sampling rate could be increased by a factor of 5.

As an alternative approach, assume that two, or possibly three, subcarriers are desired instead of a single time-multiplexed channel. In this case, Figure 15-12 shows that two channels would each have a normalized power ratio P_{sc}/P_{tot} of -6.2 dB if the modulation level is held to 1 rad rms, and as in the preceding example, the information bandwidths may be calculated to be 2.4 Hz each. Each subcarrier would have a modulation index of 1 rad peak in this case.

If three subcarriers were used, each would have an information bandwidth of 1.4 Hz and a modulation index of 0.82 rad peak. Note from Figure 15-11 that if the total modulation level is held constant, the carrier power increases slightly as the number of subcarriers is increased, up to about four channels.

Many other variations occur in FM/PM designs. The design curves and equations are applicable to most situations that will be encountered.

EXAMPLE 15-4

Four channels of data, each with well-defined bandwidths, are to be transmitted. Two of the subcarriers are required to transmit essentially sinusoidal information with a maximum frequency, $f_m = 5$ Hz, and a deviation $\Delta f = 30$ Hz. The remaining two channels are also identical and require only a 1.0-Hz loop information bandwidth. This would be typical for quasi-static temperature measurements. The system parameters are

$$P_{tot} = 250 \text{ mW, or } +24 \text{ dBm}$$

$$G_t = 0.0 \text{ dB}$$

$$G_r = 45.0 \text{ dB}$$

$$L_m = 1.5 \text{ dB}$$

$$\phi_{kT} = -175.6 \text{ dBm/Hz } (200^\circ)$$

$$L_s = 204.2 \text{ dB } (250{,}000 \text{ miles})$$

$$L_p = 6.0 \text{ dB}$$

SOLUTION

The approach taken is first to calculate the required loop information bandwidths, then the required channel powers, and finally arrive at a choice of modulation indices for each of the channels. The carrier channel power is

$$
\begin{aligned}
P_c &= 2\beta_L + (S/N)_{2\beta_L} + \phi_{kT} + L_s + L_m + L_p - G_t - G_r \\
&= 13 + 3 - 175.6 + 204.2 + 1.5 + 6 - 0.0 - 45 \\
&= 7.1 \text{ dBm}
\end{aligned}
$$

$$
\frac{P_c}{P_{\text{tot}}} = 7.1 - 24 = -16.9 \text{ dB}
$$

As in Example 15-3, this is more carrier suppression than is practical. Since there is more than enough power in the carrier, let us hold $P_c/P_{\text{tot}} = -4$ dB. From Figure 15-11, for four subcarriers, the allowable rms modulation level is approximately 0.95 rad rms.

Two of the subcarriers transmit essentially sinusoidal information. As in Example 15-3, let us constrain the modulation error to 30° maximum. It may be shown as discussed in Example 15-3 that for sinusoidal modulation and f_m/f_n small, the maximum phase error is given

$$
\Delta\theta_{\max} \approx \frac{\Delta f \, f_m}{f_n^2}
$$

$$
f_n = \sqrt{\frac{(30)(5)(57.3)}{30}} = 16.92 \text{ Hz}
$$

The subcarrier loop noise bandwidth is then

$$
\begin{aligned}
2\beta_{\text{LO}} &= 2(3.33\, f_n) \\
&= 2(3.33)(16.92) = 113 \text{ Hz}
\end{aligned}
$$

The required subcarrier power

$$
\begin{aligned}
P_{\text{sc}} &= 2\beta_{\text{L}} + (S/N)_{2\beta_L} + \phi_{kT} + L_s + L_p - G_t - G_r \\
&= 20.5 + 5.6 - 175.6 + 204.2 + 1.5 + 6 - 0.0 - 45 \\
&= 17.2 \text{ dBm}
\end{aligned}
$$

or

$$\frac{P_{sc}}{P_{tot}} = -6.8 \text{ dB}$$

From Figure 15-14 we see that the modulation index for each of these channels is 0.91 rad.

The remaining two channels are also identical and require only a 1.0-Hz loop information bandwidth. In this case, performing the calculations as above, we find the following:

$$2\beta_{LO} = 6.66 \text{ Hz or } 8.2 \text{ dB} \cdot \text{Hz}$$

$$P_{sc} = 4.9 \text{ dBm/channel}$$

$$\frac{P_{sc}}{P_{tot}} = -19.1 \text{ dB}$$

From Figure 15-14 the required modulation indices are 0.26 rad for each channel. We determined earlier that the total allowable rms modulation level would be set at 0.95 rad. The rms sum of the individual modulation indices are

$$\text{rms} = \frac{1}{\sqrt{2}} \sqrt{(0.91)^2 + (0.91)^2 + (0.26)^2 + (0.26)^2} = 0.946 \text{ rad}$$

Thus the proposed design is realizable.

15-7 SATELLITE COMMUNICATIONS

Satellite communications was invented in the 1940s by the noted British scientist and science fiction writer, Arthur C. Clarke. It remained for the advent of the U.S. space program to implement the concept. Prior to discussing specifics relating to electronic communications via satellites, let us consider some mechanics of the system.

A communications satellite system consists of a satellite, a ground-based station to control its operation, and a network of earth stations providing transmission (uplink) and reception (downlink) facilities for communications traffic relayed through the satellite. The satellite itself can be considered as being composed of two major parts: a payload and a bus. The payload consists of the antennas, receivers, and transmitters that perform the communications signal transponding function. The bus includes attitude control, thermal control, and command and telemetry systems. Essentially, then, the bus provides the support (housekeeping) functions needed for payload operation.

After launch, the satellite is assigned an orbital position, or slot, above the equator at a geostationary altitude. These slots are boxes with 0.1 to 0.2° sides at 3 or 4° spacings. Due to forces that act on the satellite (solar and lunar gravity, etc.), the satellite in geostationary orbit wanders. The ground station orders stationkeeping maneuvers to keep the satellite in its assigned slot. On-board thrusters using hydrazine (N_2H_4) as the propellent are fired to accomplish this. Typically, 20 to 30 lb of propellant is used each year.

The satellite is normally powered by solar cells. However, it will be eclipsed by the earth once each day around the time of the vernal and autumnal equinoxes. The eclipses can last as long as 70 minutes and battery power is used during this time.

15-7.1 Satellite Classes

Two main classes of satellites are used (Figure 15-15). These are spinners and three-axis stabilized satellites. A spinner uses the angular momentum of its spinning body to provide attitude stabilization in yaw and pitch. The body of a three-axis stabilized satellite remains fixed relative to the earth's surface. An internal momentum wheel provides roll/yaw stabilization.

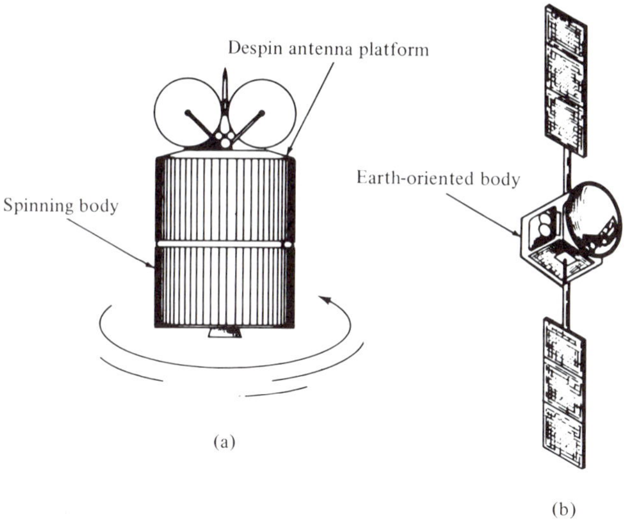

FIGURE 15-15. Principal classes of communications satellite: (a) spinner; (b) three-axis stabilized. (© 1983 IEEE)

15-7.2 Stationkeeping

The communications functions of a geostationary communications satellite demands accurate control of the orbit and the attitude. Attitude control is needed to point the antennas.

Each satellite is assigned a longitude in the geostationary arc located 22,300 miles above the equator (Figure 15-16). The common carriers exercise tight control over the satellite's orbit. This is necessary because many of the ground-based antennas are of the nontracking type. Significant satellite motion would move the satellite's beam.

The size of the acceptable tolerance on movement depends on the communications frequency band. Typically $\pm 0.1°$ is allowed in both longitude and latitude for the popular C-band systems (6-GHz uplink and 4-GHz downlink). For the increasingly used Ku-band systems (14-GHz uplink, 12-GHz downlink), the tolerance bands are $\pm 0.05°$ in longitude and latitude. Ku-band antennas exhibit a much smaller beamwidth than C-band antennas.

15-7.3 The Intelsat System

The International Telecommunications Satellite system provides worldwide, international telephone, television, telegraph, and data transmission service, as well as being available for domestic use. As of mid-1982, there were 424 earth station antennas with over 55,000 combined full-time channels using Intelsat. Needless to say, a system of this magnitude requires considerable software to

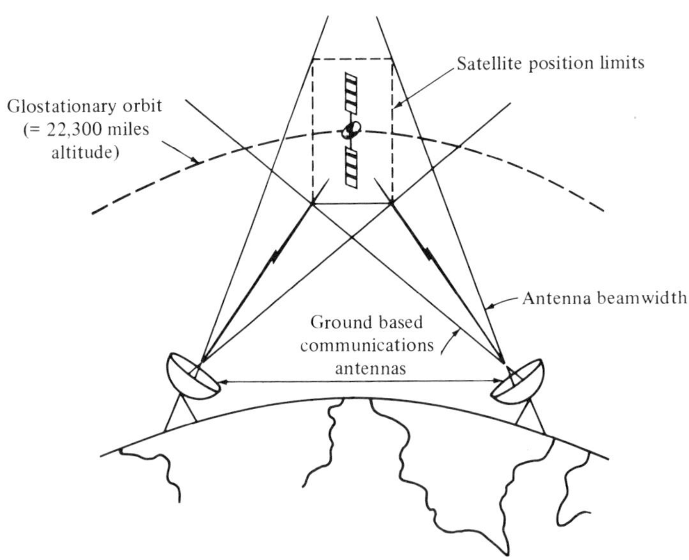

FIGURE 15-16. Satellite latitude and longitude must be tightly controlled to meet communications system requirements. (© 1983 IEEE)

control channel routing, monitor performance, and to optimize transmission planning.

Transmission planning begins with the annual global traffic meeting at which representatives from each participating country establishes the number, connectivity, and types of circuits that will exist between the countries in subsequent years. Using the results of this meeting, a traffic data base is derived to allocate the channel requirements between earth stations to designated satellites. The actual frequency plan for each transponder of a given satellite is then developed with the aid of computer programs that indicate how channels could be allocated to satellite beams. Other software programs are then used to evaluate and optimize the initial frequency plans which must meet Intelsat's transmission-quality standards. Once a frequency plan has been judged acceptable for implementation, it is presented to the participating countries, allowing sufficient time to acquire suitable earth-station equipment. After a plan is implemented, its performance is monitored and these data are compared with system model predictions. Results from these programs are used to detect discrepancies in operating conditions and to establish performance trends.

Intelsat V Spacecraft Antenna Coverage. The Intelsat V spacecraft has four types of antenna coverage. A global beam provides coverage over the entire earth's disk from a geosynchronous orbit in the 6/4-GHz band. Four times frequency reuse is obtained in this band, with the use of both east and west hemisphere beams in one polarization sense, and zone beams in the opposite sense. Two times frequency reuse in the 14/11-GHz band is accomplished by means of an east and west steerable spot beam. A typical hemispheric and spot coverage on the earth's disk as seen by a satellite in the Atlantic region is shown in Figure 15-17. The contour lines represent the amount of satellite power which is available to earth stations located in those regions.

Frequency Plans. Intelsat uses the frequency-division multiple-access (FDMA) method for most services. In an FDMA system, each frequency carrier occupies a discrete frequency allocation and is assigned a specific location within a multicarrier transponder. Two FDMA techniques are employed: (1) FDM/FM/FDMA, where the transmitting earth station frequency-division multiplexes (FDM) several single-sideband carrier telephone channels into one carrier baseband assembly, which frequency modulates (FM) a carrier which is then applied to a FDMA satellite network, and (2) single channels per carrier, where each telephone channel modulates a separate radio-frequency carrier. A typical frequency plan for an Intelsat V transponder is shown in Figure 15-18. The actual signal routing through the satellite can be switched by ground command. This capability provides a larger number of combinations of paths, which provides various receive and transmit antennas connectivity with many of the transponders. The transponders are numbered 1 through 12.

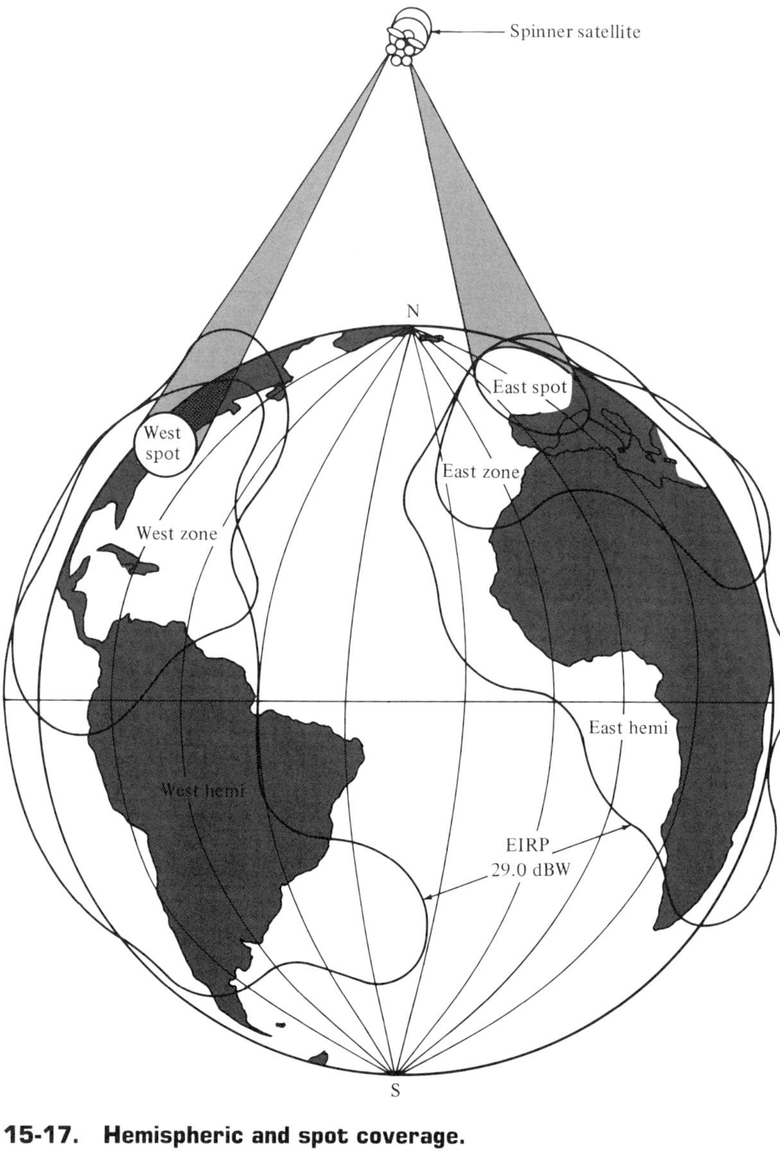

FIGURE 15-17. Hemispheric and spot coverage.

Time-Division Multiple Access (TDMA). Efforts are under way to convert major portions of Intelsat's traffic to TDMA, beginning in 1984. In the TDMA mode, each user accesses the satellite transponder using the same carrier frequency during specified time slots relative to a reference time. A time interval called a TDMA frame is established during which all users within the network transmit bursts of information during their designated time slots. This time frame is 2 ms or 120,832 symbols long. Each network user is assigned burst time slots

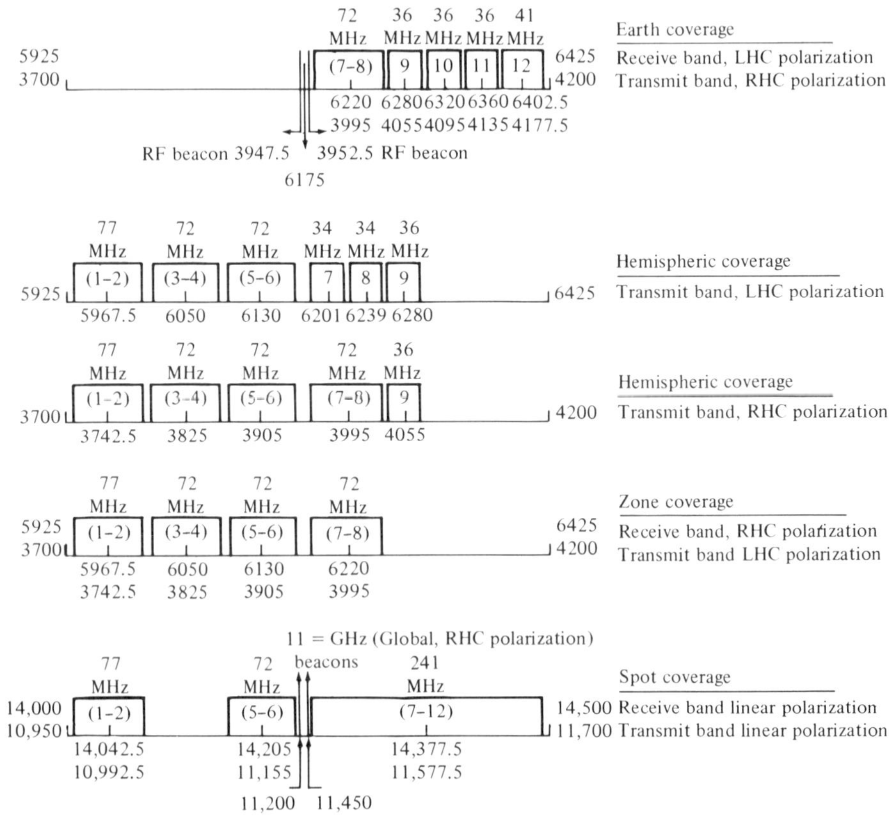

FIGURE 15-18. Intelsat V transmitter frequency plan.

on the basis of how much traffic is to be transmitted and received during the frame. A typical TDMA frame is shown in Figure 15-19.

Crosstalk is generated in the baseband of an FDM/FM carrier by other carriers when two or more FDM/FM signals are simultaneously amplified by a nonlinear filter. When this occurs, amplitude modulation/phase modulation characteristics

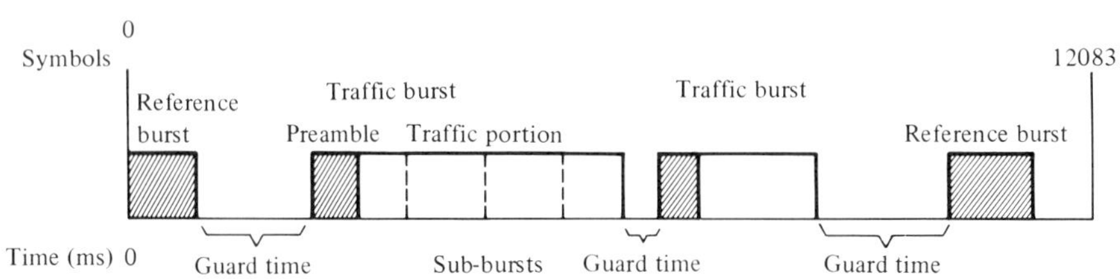

FIGURE 15-19. Frame format for Intelsat's time-division multiple access. (© 1983 IEEE)

of the nonlinear amplifier produce an amplitude modulation in a carrier, which in turn produces a phase modulation in other carriers. The TDMA scheme eliminates this problem. The satellite power amplifier can be operated in the saturated (maximum power) mode.

15-7.4 Satellite Link Budgets

Uplink power budgets are not a critical factor for satellite communications since powerful earth-based stations can supply all the power that is needed. Downlink considerations are another story. Power is limited and also the real possibility exists that there will be interference in ground communication systems by signal radiation from spaceborne communication satellites. Guidelines have been established by the CCIR (International Consultative Committee) of the ITU (International Telecommunications Union) on the maximum flux density produced at the earth's surface from space communication systems.

At the satellite, the signal is received by the antenna, filtered, and then amplified by a low-noise amplifier. A frequency translater then translates the frequency before retransmission. This prevents inband interference between the high power output of the satellite and its input. Without this frequency translation an exceptionally high isolation would be required (100 to 150 dB).

Link Equations. The familiar range equation is used to compute the uplink and downlink power requirements. For digital transmission we are generally interested in the ratio of carrier power to noise power per hertz (noise density). This, in turn, can be related to the bit energy-to-noise density ratio required for the entire system. The basic uplink evaluation is given by

$$(C_u/N_{ou})_{\mathrm{dB}} = \underbrace{10 \log P_T G_T}_{\substack{\text{earth-station} \\ \text{EIRP}}} - \underbrace{20 \log \frac{4\pi R_u}{\lambda_u}}_{\substack{\text{free-space} \\ \text{uplink loss}}} + \underbrace{10 \log \frac{G_{su}}{T_s}}_{\substack{\text{satellite} \\ \text{G/T}}} - \underbrace{10 \log L_u - 10 \log k}_{\substack{\text{additional} \\ \text{uplink losses}}} \tag{15-49}$$

Two terms in this form of the range equation require an explanation. The efficiency of a satellite or a ground station is frequently specified in terms of the gain to equivalent noise temperature, G/T_e:

$$T_e = N/kW' \tag{15-50}$$

where

N = total noise power
k = Boltzmann's constant
W' = double-sided bandwidth of the system being evaluated

From Chapter 5 we remember that the equivalent noise temperature, T_e, is also related to the noise figure F by

$$T_e = (F - 1)T_0$$

The effective isotropic radiated power (EIRP) is the product of the transmitted power and the antenna gain. That is,

$$\text{EIRP} = PG$$

or

$$\text{EIRP(dB)} = P(\text{dBW}) + G(\text{dB}) \tag{15-51}$$

where

P = transmitted power
G = antenna gain

The ratio C/N_0 is the average wide-band carrier power-to-noise density ratio, where the noise is measured in a 1-Hz bandwidth. From equation (15-50) the noise density, N_0, is given by

$$N_0 = \frac{N}{W'} = kT_e \tag{15-52}$$

For the downlink, the basic range equation is

$$(C_d/N_{od_{\text{dB}}} = \underbrace{10 \log (P_s G_{sd})}_{\substack{\text{satellite}\\ \text{EIRP}}} - \underbrace{20 \log \frac{4\pi R_d}{\lambda_d}}_{\substack{\text{free-space}\\ \text{downlink}\\ \text{loss}}} + \underbrace{10 \log G_d/T_d}_{\substack{\text{earth-}\\ \text{station}\\ G/T}}$$

$$+ \underbrace{10 \log L_d - 10 \log k}_{\substack{\text{additional}\\ \text{downlink}\\ \text{losses}}} \tag{15-53}$$

We note that the form is the same as the uplink. Now, obviously, the C/N_o of the uplink will carry through and contribute to determining the C/N_o of the received downlink. In a frequency-translating satellite, the overall end-to-end carrier-to-noise density ratio, $(C/N_o)_T$, at the receiver earth station is given by

$$(C/N_o)_T = \frac{1}{N_{ou}/C_u + N_{od}/C_d} \tag{15-54}$$

The bit energy, E_b, is obtained by multiplying the carrier power, C, by the bit duration, T_b. That is,

$$E_b = CT_b \tag{15-55}$$

Thus equation (15-54) may be written as

$$(E_b/N_o)T = \frac{1}{(N_{ou}/E_{bu}) + (N_{od}/E_{bd})} \tag{15-56}$$

Equation (15-56) is very useful in system calculations for digital communication systems since the probability of error is a function of the total energy per bit-to-noise density ratio $(E_b/N_o)T$. A noise link budget calculation for a frequency-translating satellite transponder is shown in Table 15-1.

EXAMPLE 15-5

The output power of an earth-station transmitter is 2.2 kW and the diameter of its antenna is 15 m. If the transmitting frequency is 14 GHz, calculate the EIRP of the station. Assume losses between the transmitter and antenna of 7 dB.

SOLUTION

From equation (1-7), the antenna gain is

$$G = K\left(\frac{4D}{\lambda}\right)^2 = 0.55\left[\frac{(4)(15)}{(300/14{,}000)}\right]^2$$

$$= 431{,}200 = 66.3 \text{ dB}$$

In this expression $K = 0.55$ is the antenna efficiency. Using equation (15-51), we have

$$\text{EIRP} = P(\text{dBW}) + G(\text{dB})$$

$$= (33.42 - 7) + 66.3 = 92.72 \text{ dBW}$$

TABLE 15-1 Link Budget of the 14/11-GHz European Communications Satellite (ECS) System[a]

	Uplink		
1.	Transmitter output power at saturation, 2 kW	33	dBW
2.	Backoff and combining losses	7	dB
3.	Transmitting antenna gain (15 m, 14 GHz)	64	dB
4.	EIRP	90	dBW
5.	Free-space loss (14 GHz)	207.5	dB
6.	Atmospheric loss (14 GHz, clear weather)	0.6	dB
7.	Satellite G/T	−5.3	dB/°K
8.	C/T at repeater input	−123.4	dBW/°K
9.	Boltzmann's constant	−228.6	dBW/°K/Hz
10.	Bit rate, 120 Mbits/s	80.8	dBHz
11.	E_b/N_o at repeater input	24.4	dB
	Downlink		
1.	EIRP at beam edge (unmodulated carrier, saturation)	40.8	dBW
2.	Modulation backoff and bandlimiting losses	0.6	dB
3.	Free space loss (11.7 GHz)	205.6	dB
4.	Atmospheric loss (11.7 GHz, clear weather)	0.4	dB
5.	Power at receive antenna	−165.8	dBW
6.	Receiving antenna gain (15 m, 11.7 GHz)	62	dB
7.	Receiving system noise temperature (clear weather), 270°K	24.3	dBK
8.	Earth-station G/T	37.7	dB/°K
9.	C/T at receiver input	−128.1	dBW/°K
10.	Boltzmann's constant	−228.6	dBW/°K/Hz
11.	Bit rate, 120 Mbits/s	80.8	dBHz
12.	E_b/N_o at receiver input	19.7	dB

[a]This system will provide 17,000 two-way digital telephone circuits by 1990. Individual satellite transponders will carry 120 Mbits/s of traffic.

EXAMPLE 15-6

Using the results from Example 15-5 and Table 15-1, find E_b/N_o at the satellite's repeater input if the bit rate is 100 Mbits/s (70 dBHz).

SOLUTION

Using equation (15-49), we have

$$\begin{aligned}(C_u/N_{ou})_{\text{dB}} &= 92.72 - 207.5 + (-5.3) + (-0.6) - (-228.6)\\ &= 107.92 \text{ dBW/Hz}\\ &= 137.92 \text{ dBm/Hz}\end{aligned}$$

The uplink bit energy is given by equation (15-55)

$$E_{bu} = CT_b$$

or

$$(E_{bu})_{\mathrm{dBm}} = C_u(\mathrm{dBW}) + 10 \log T_b(\mathrm{seconds})$$

$$= 107.92 + 10 \log (1/100 \times 10^6)$$

$$= 107.92 - 70 = 37.92 \text{ dB}$$

Then

$$\frac{E_b}{N_0} = \frac{C_u T_b}{N_0}$$

$$= 107.92 \text{ dBW/Hz} + 10 \log (1/100 \times 10^6)$$

$$= 107.92 - 70 = 37.92 \text{ dB}$$

15-7.5 Satellite Television

The previous discussion regarding satellite communications is oriented toward data transmission. Satellites are also used as links in a direct broadcast system (DBS) for television. The satellite television system in use by the United States operates at C-band frequencies (4 to 6 GHz). This system was set up for cable TV (CATV) systems. The installation of home TV reception was not envisioned at the time. It was thought that the large size of the receiving antenna would preclude this. At present, though, there are thousands of ''pirate'' installations in operation. There are proposals for extensive use of Ku-band (12 to 14 GHz) satellites as transponders for television. This would be favorable to the consumer in that the antenna size would be reduced from 6 to 10 ft to an order of 2 or 3 ft. Two Ku-band satellites were launched for Canada in November 1982 on board the NASA Space Shuttle *Columbia*.

Satellite Television Broadcast Stations. A block diagram of a typical TV broadcast station is shown in Figure 15-20. Referring to the figure, we see that the baseband video is applied to a baseband processor with preemphasis and energy dispersal features. The need for preemphasis has already been discussed in regard to FM modulation. In addition to preemphasis, an ''energy'' dispersal waveform is added to the baseband video signal. The dispersal waveform is a triangular waveform with frequency equal to the video frame rate, 30 Hz. This signal is synchronized to the video frame rate and is used to broaden the spectrum of the frequency-modulated RF signal. Without this feature, problems of interference can arise with respect to terrestial microwave links. For example, without energy dispersal, with low-average picture levels, the primary component of the transmitted Fm signal would be the horizontal sync and 3.58-MHz color subcarrier, both of which are present in the composite video

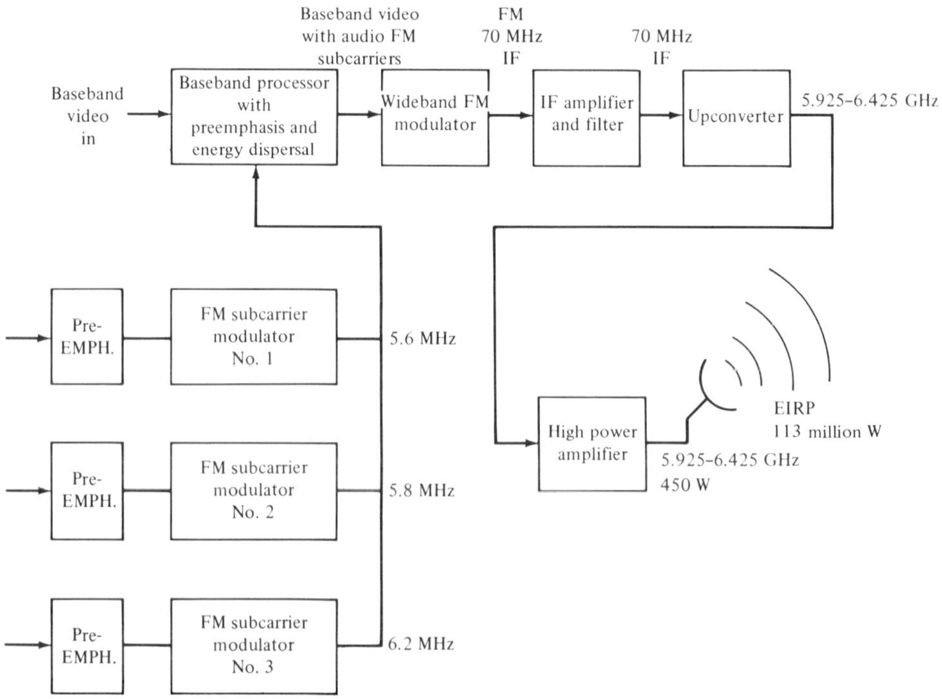

FIGURE 15-20. Block diagram of a satellite TV broadcast transmitter.

signal. This concentration of energy can result in unwanted interference with other systems. The spectrum of the transmitted signal should be at least 2 MHz.

Note also that a number of subcarriers are available in the transmitter. These are used to transmit the audio. Since the subcarriers are FM, preemphasis networks are necessary. With the same picture we have the option of transmitting English on one channel and perhaps Spanish on another. Of course, the television receiver must be tunable to the preferred audio channel.

The audio subcarriers are summed with the processed video signal and applied to a wide-band FM modulator. The output is an FM signal centered at 70 MHz with a bandwidth on the order of 36 MHz. Finally, this signal is applied to an upconverter which produces the C-band signal. The EIRP of the transmit antenna is approximately 113 million watts. Uplink and downlink frequency plans are illustrated in Figure 15-21.

Satellite Transponder. A block diagram of the satellite transponder is shown in Figure 15-22. Note that each transponder is allowed a bandwidth of 40 MHz with 4 MHz serving as a guard band to separate transponders. A study of the frequency plan reveals that overlap exists in frequency assignments for transponders. As discussed earlier, frequency reuse is allowed by making use of orthogonal polarizations for adjacent channel transponders.

TABLE 15-2 North American Domestic Satellites

Name	Operator	Launch Date(s)	Number of Transponders per Satellite	Bandwidth (MHz)	Uplink (GHz)	G/T_0 (dBi/K) at edge	Downlink (GHz)	EIRP (dBW) at edge	Lifetime (years)	Orbit Location (longitude)
Alascom 1 (Satcom V)	Alascom, Inc.	82	24	36	5.925–6.425	−6	3.7–4.2	34 SSPA	10	143W
American Satellite Co. 1 to 3	American Satellite Co.	84+	12 and 6 6	36 and 72 72	5.925–6.425 14–14.5	−4 −2	3.7–4.2 11.7–12.2	34 and 36 42	7.5	
Anik–A1 to A3	Telesat Canada	72,73,75	12	36	5.925–6.425	−6	3.7–4.2	33 TWTA	7	104, 114, and 114W
Anik–B1	Telesat Canada	78	12 6	36 72	5.925–6.425 14.0–14.5	−6 +0	3.7–4.2 11.7–12.2	36 47 TWTA	7	109W
Anik–C1 to C3	Telesat Canada	82+	16	54	14.0–14.5	+3	11.7–12.2	48 TWTA	8	112.5, 116, and 109W
Anik–D1 to D2	Telesat Canada	83+	24	36	5.925–6.425	−6	3.7–4.2	36	8	104, 109 and 114W
Comstar D1 to D4	Comsat General for AT&T/GTE	76, 76, 78, 81	24	34	5.925–6.425	−4.5	3.7–4.2	33 TWTA	7	95, 95, 87, 127.25W
Galaxy I to III	Hughes	82+	24	36	5.925–6.425	−7	3.7–4.2	34.5 TWTA	9	135, 74W, + ?
GSTAR–1 and 2	GTE Satellite	84+	16	54	14–14.5	+1.6	11.7–12.2	43.1 TWTA	10	103 and 106W

Satcom I and II	RCA Americom	75, 76	24	34	5.925–	−6	3.7–4.2	32 TWTA	8	135 and 119W
Satcom IIIR and IV	RCA Americom	81, 82	24	34	6.425	−6	3.7–4.2	32 and 34 TWTA	8	131 and 83W
Satcom IIR, IIR and "Sixth"	RCA Americom	82–85	24	36	5.925–6.425	−6	3.7–4.2	34 SSPA	10	139, 66W +
Satmex	Mexico	83	24	36	5.925–6.425	−9	3.7–4.2	approx. 35	10	85?
SBS–1 to 4	Satellite Business Systems	80, 81 +	10	43	14–14.5	+1.8	11.7–12.2	37 to 43.8 TWTA	7	100, 97, and 94W
Spacenet 1 to 3	SP Communications Co.	83 +	12 and 6	36 and 72	5.925–	−5	3.7–4.2	34 and 36 SS and TWTA	7.5	119, 70W
			6	72	6.426					
					14–14.5	−5	11.7–12.2	39 TWTA		
Telstar 3–A to 3D	AT&T	83 +	24	34	5.925–6.425	−5	3.7–4.2	32 to 34 TWTA	10	87, 95, + one more
Westar I to III	Western Union Telegraph	74, 75, 79	12	36	5.925–6.425	−7.4	3.7–4.2	33 TWTA	7	99, 123.5, and 91W
Westar IV to VI	Western Union Telegraph	82 +	24	36	5.925–6.425	−6	3.7–4.2	34	10	99, 123W
Advanced Westar	Space Comm. Co.	83 +	12	36	5.925–6.426	−7.4	3.7–4.2	33 TWTA	10	91W, spare at 79W
			6	225	14–14.5	−5 to +4.4	11.7–12.2	43 to 53	10	
USAT–1 and 2	U.S. Satellite Systems, Inc.	85 +	10	43	14–14.5		11.7–12.2			

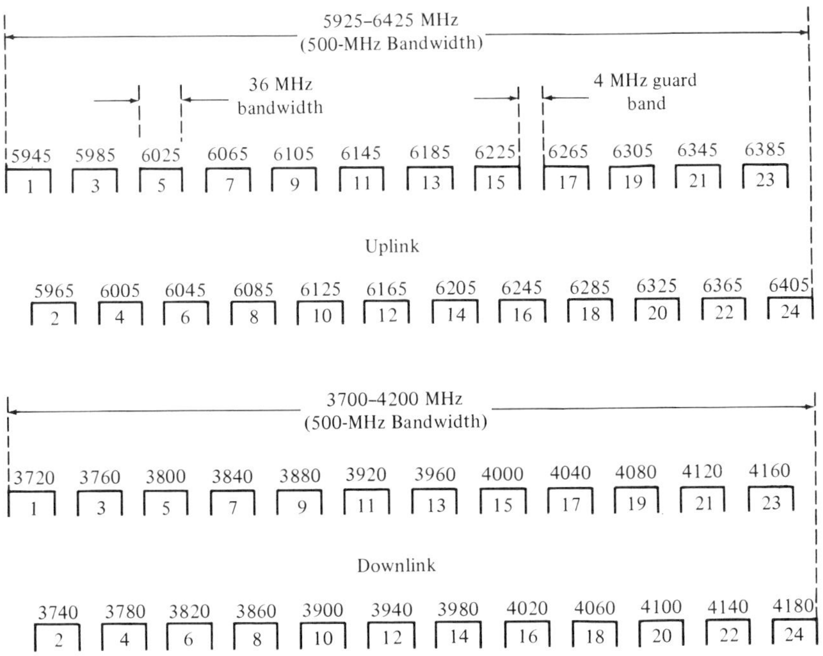

FIGURE 15-21. Uplink and downlink frequency plans.

We can, at this point, relate the satellite directly to the operation of our cable-connected television set. As far as the satellite is concerned, HBO, Showtime, Cinemax, the Super Station, and so on, are all just a signal that passes through one of the transponders. The owners of the television service lease the transponder from the owner of the satellite. Table 15-2 illustrates a number of satellites currently in use which carry television.

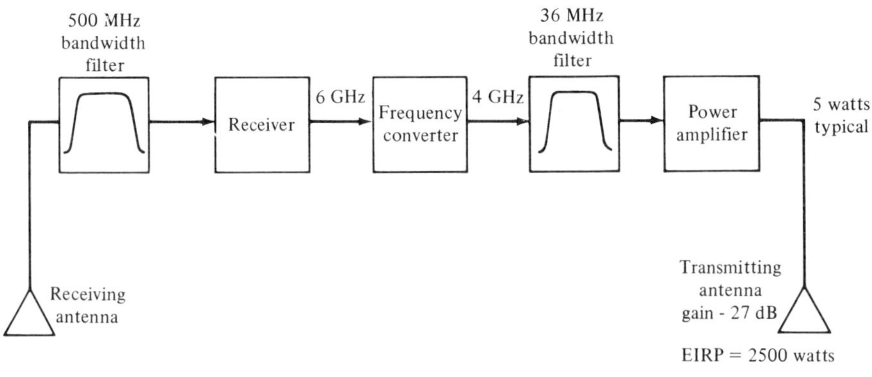

FIGURE 15-22. Satellite transponder.

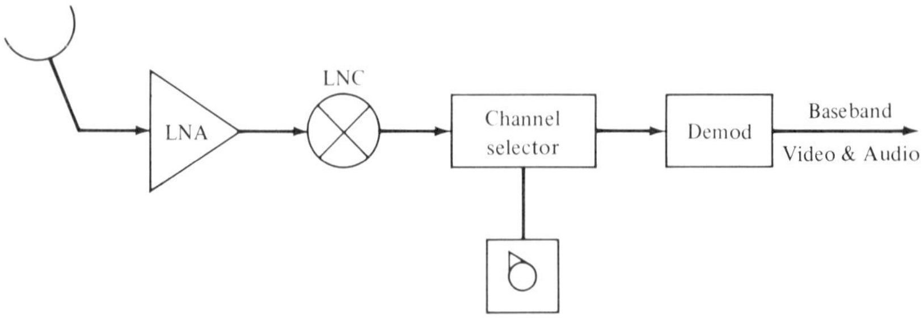

FIGURE 15-23. Satellite TV receiver.

DBS Receiver. The block diagram for a satellite TV receiver is shown in Figure 15-23. The received 4-GHz signal is on the order of −135 dBm. Thus at the input a low-noise amplifier (LNA) is required. Generally speaking, a GaAs FET transistor is used. Following the LNA is a low-noise frequency converter (LNC). In many cases both the LNA and LNC is located near the antenna. The channel selector allows one transponder at a time to be selected. The channel selector is essentially a tunable converter that converts the selected channel to 70 MHz.

A more detailed block diagram of a TV receiving system is shown in Figure 15-24. The EIRP from the satellite's antenna is +65 dBm. After experiencing a 200-dB path loss, the received signal at the antenna is approximately −135 dBm. If the antenna gain is 100 dB, the signal input to the LNA is −35 dBM. Losses in the LNC can be on the order of 10 to 30 dB with an additional 30-dB loss due to filtering. Typically, the input to the demodulator must be in the vicinity of 0 dBm; thus the required gain of the IF is 55 to 100 dB.

Note that the output of the video is clamped to remove the energy dispersal waveform (Figure 15-25). Finally, the clamped video and demodulated audio is

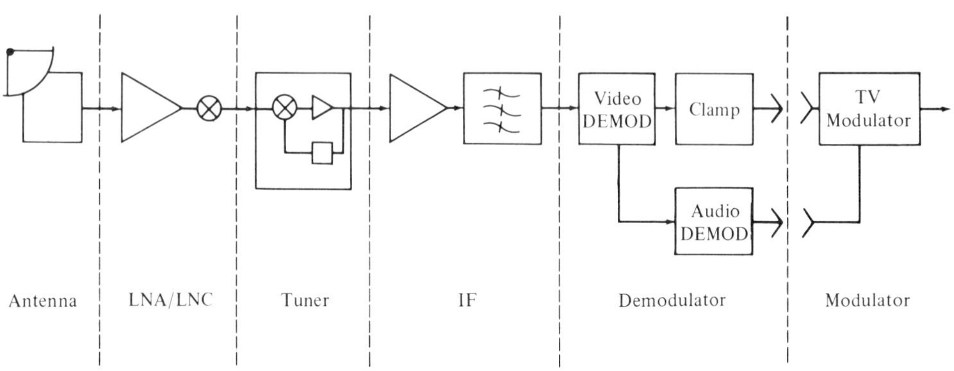

FIGURE 15-24. Detailed block diagram of a TV receiving system.

applied to a modulator so that the baseband video and audio is converted to a standard TV signal. Typically, this is either channel 3 or 4.

Pay TV Descramblers. The growth of cable TV has been extremely rapid in the last decade. The legalities concerned in having private citizens own their own cable TV equipment appears to be somewhat vague. Attitudes toward this can vary from city to city and the FCC has been reluctant to rule on the matter. The fact is, however, that it would probably be difficult to justify requiring telephone companies to allow consumers to use their own equipment and at the same time exempt cable companies.

The same comments could be made regarding descrambling devices. It appears that as long as you pay for the service, the customer should be allowed to use his or her own descrambling device. Most cable TV companies use a scrambling system known as the sync suppression method. This system consists of two types of sync suppression: gated sync suppression and sine-wave sync suppression. An interference signal method is also used.

Gated sync suppression uses a pulse train to suppress the horizontal and vertical blanking and sync pulses by a factor of 6 dB (Figure 15-26). Normally, the TV receiver's timing circuits respond to the sync pulses because they are the highest level in the video signal. If these pulses are suppressed, the receiving circuits attempt to lock on random video information and a scrambled picture results.

For purposes of descrambling, the scrambling pulses are inverted and transmitted on a separate pilot frequency. These pulses are detected by the descrambler and used either to attenuate the video 6 dB or to amplify the sync and blanking pulses 6 dB. The most commonly used pilot frequencies are 50.5 and 114 MHz. Other frequencies that may be used are 108, 110.5, 93, and 99 MHz.

The sine-wave system is similar to the gated sync suppression method (Figure

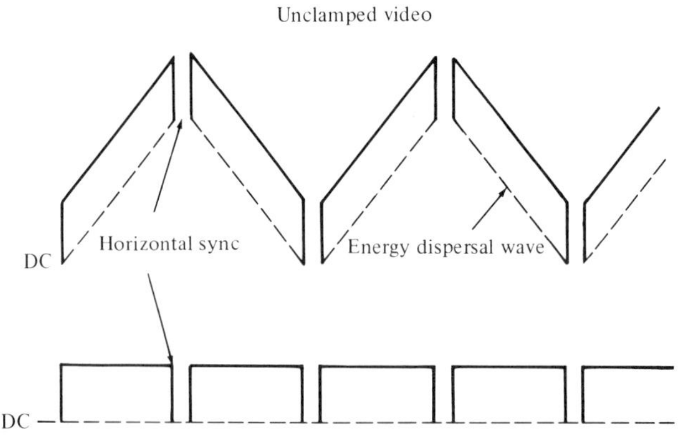

FIGURE 15-25. Clamped and unclamped video.

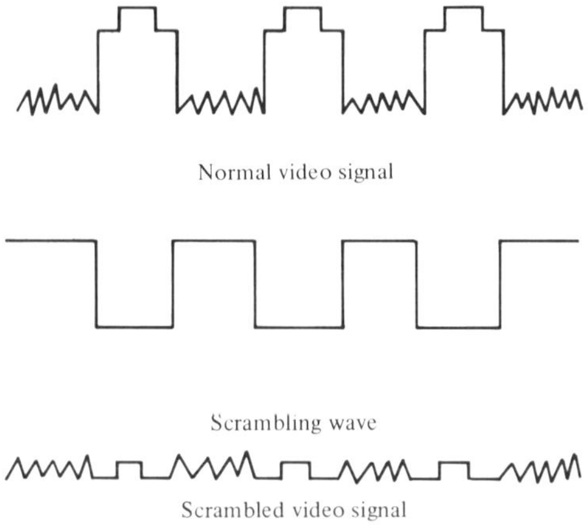

FIGURE 15-26. Video signal scrambling sequence.

15-27). The sine wave is also transmitted on the audio channel. Unscrambling consists of extracting the sine wave from the audio carrier, inverting it, and applying it as AGC to the video signal path.

The interference method is exactly as it sounds. A frequency-modulated signal is transmitted with the video. The descrambler is a notch filter tuned to eliminate the interfering signal. A simple parallel tuned circuit may also be used as a wave trap.

Cable TV channels: The following channels are allocated by the FCC for cable TV used:

- *Channels 2 through 13.*
- *Channels A through I:* These are the midband channels and range in frequency from 120 to 174 MHz with the standard 6-MHz channel bandwidth.

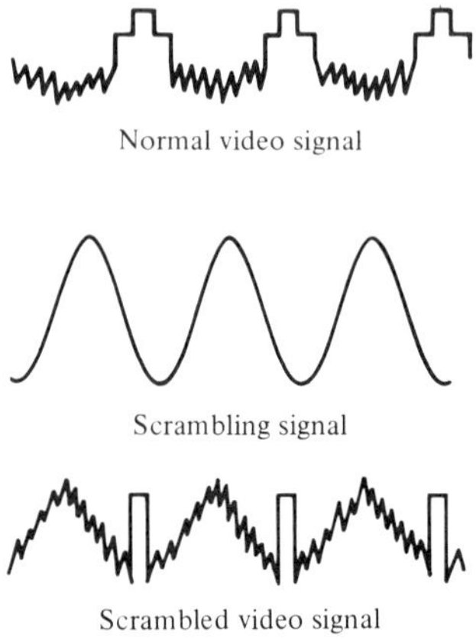

FIGURE 15-27. Sine-wave sync suppression.

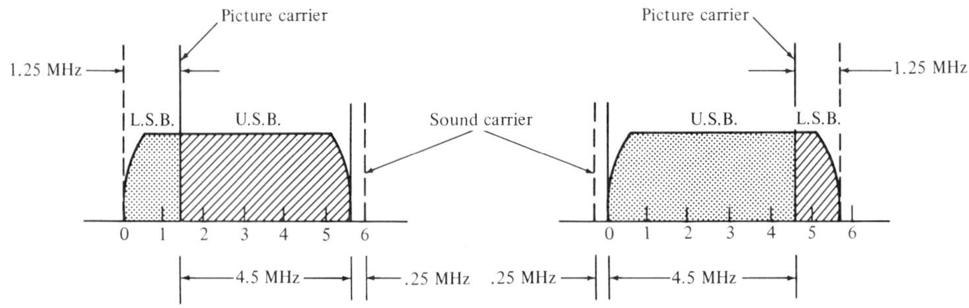

FIGURE 15-28. Midband and superband spectrums.

- *J through W:* These are the superband channels and range from 216 to 300 MHz with the standard 6-MHz channel bandwidth.

For the midband and superband channels, the cable companies can choose to transmit either normal or inverted spectra (Figure 15-28). Referring to the figure, we see that the inverted signal is a mirror image of the normal video spectrum. If the inverted signal is converted to a lower frequency (channel 3 or 4) by a simple one-stage converter, a normal signal is produced that can be received by any television set.

Television Receiver Antennas. The most popular antenna is the dish antenna because of its high gain (Figure 15-29). For example, if the antenna

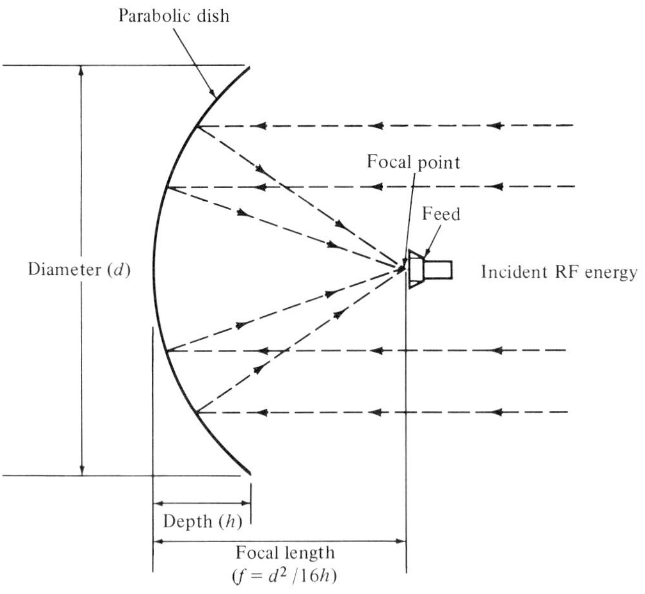

FIGURE 15-29. Parabolic dish antenna.

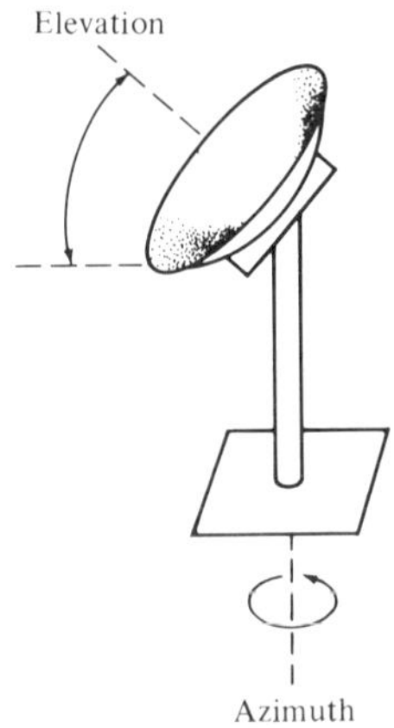

FIGURE 15-30. AZ-EL antenna mount.

diameter ranges from 3 to 11 m, the corresponding gain is 39.4 to 54 dB. Prior to aiming the antenna at a particular satellite, some advanced work is necessary. Referring to Table 15-2, the orbital location is given for a number of satellites. For example, Satcom IV is at 83° west longitude.

As discussed previously, the satellites are located along the equatorial belt at an altitude of 22,300 miles. In order to locate a satellite, it is necessary to first locate due south. This can be done by locating magnetic north with a compass and then correcting for magnetic deviation in the immediate area. Once true north is located, a 180° reversal yields true south. In the northern hemisphere this is the direction that the antenna must point. For a polar-mounted antenna (Figure 15-30), the angle of elevation must be determined. This can be accomplished using the following equation:

$$E_L = 90° - \tan^{-1} \frac{3974L}{22{,}300 + 3974(1 - \cos L)} \tag{15-57}$$

where L is the latitudinal position in degrees of the antenna and 3974 is the equatorial radius of the earth in miles. Once the antenna has been positioned correctly, it will easily sweep the arc of satellites over the equator.

PROBLEMS

15-1. Using results from Example 15-1, determine the transmitted power required if the transmitter is located 10 miles away. Assume antenna gains of 120 dB. What is the required diameter for the receiving antenna?

15-2. Repeat Problem 15-1 using an APD diode with a 15-dB improvement factor in sensitivity.

15-3. For the transimpedance amplifier in Figure 15-7, show that the following is true: (a) $V_o \approx I_p R_F$, (b) $f_{-3\text{dB}} \approx A/2\pi R_F C_F$, and (c) $Z_{\text{in}} \approx R_i/A$

15-4. For the equalizer network in Figure 15-7, verify the expressions for the break-point frequencies, ω_1 and ω_2. Sketch the voltage transfer function and phase response versus frequency for the network.

15-5. Design a three-channel FM/PM system. The modulation is sinusoidal, with one channel having a deviation of 30 Hz and a message frequency of 5 Hz. The remaining two channels have a deviation of 20 Hz and f_m of 2 Hz. Use the system parameters of Example 15-4.

15-6. What is the efficiency of the communication system of Example 15-4?

15-7. Using results from Example 15-4, calculate the range at which the carrier loop begins to lose lock. Does this occur before or after the subcarrier loops begin to lose lock? Do all of the subcarrier loops begin to lose lock at the same range? Numerical results are required.

15-8. In Example 15-3, which is the more efficient communication system: single or multiple subcarriers? Which system will transmit the most data?

15-9. The noise figure of an earth station is 2 dB. What is the equivalent noise temperature for the earth terminal?

15-10. For Table 15-1, calculate $(C_u/N_{ou})_{\text{dB}}$ at the satellite transponder input.

15-11. What is the E_b/N_o ratio at the input to a receiving earth station if the transmission rate is 100 Mbits/s and C/T at the receiver terminals is -130.2 dBW/°K?

15-12. Use equations (15-49), (15-52), and (15-53) to verify the end results of Table 15-1.

15-13. A communications satellite located 22,300 miles above the earth is to transmit a red light beam to cover a 1-mile radial distance on the earth. What is the approximate transmitting lens size needed on the satellite? What is its gain?

PART V

RADIO TRANSMISSION

CHAPTER 16

Transmission Lines

In basic circuit theory we neglect the finite amount of time it takes for a circuit to respond to an input stimulus. Specifically, we assume that an output voltage or current appears at about the time the input is applied. There are situations, however, where the travel time of an electrical wave through a network must be considered. In these situations, analysis demands that traveling-wave theory be employed. In general, if the physical size of a network (structure) is $\frac{1}{10}\lambda$, traveling-wave theory is applicable. We generally characterize structures of this type as *distributed-parameter networks* versus *lumped-constant networks,* where propagation time can be ignored. A simple example of a distributed-parameter network with which all of us are familiar is the television lead-in (transmission line) from the antenna to the television receiver.

TRANSIENTS ON A LOSSLESS TRANSMISSION LINE

In common with other networks, an understanding of transmission lines dictates that we understand its response to both transient and steady-state signals (sine waves). Unlike lumped-constant networks, transient solutions for transmission lines are easiest to visualize and are considered first. The circuit conventions used are indicated in Figure 16-1. In the figure the subscripts S and R refer to the sending end and the receiving end, respectively.

We shall see that when a signal is applied to a line such as this, two distinct waves may be present. These are called the *incident wave* and the *reflected wave*. These waves are indicated in Figure 16-2. The incident wave travels from the sending end to the receiving end and the reflected wave from the receiving end to the sending end. By convention the current wave traveling toward the generator end is taken to be negative.

Now consider an infinitely long transmission line with incident voltage and current pulses, e^+ and i^+ traveling down the line. These traveling waves "see" a constant impedance. By Ohm's law this impedance is described by

$$Z_0 = \frac{e^+}{i^+} \tag{16-1}$$

where Z_0 is characterized as the "surge" or "characteristic" impedance. Since a transmission line is bilinear, any reflected wave must see the same impedance. That is,

$$Z_0 = -\frac{e^-}{i^-} \tag{16-2}$$

where the negative sign results from the negative current pulse.

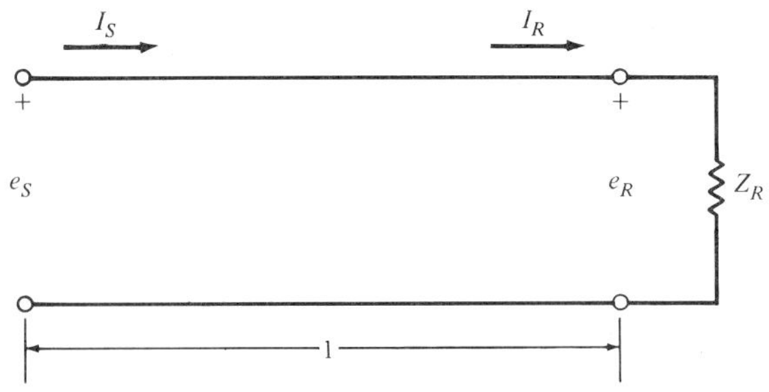

FIGURE 16-1. Conventions used in transmission lines.

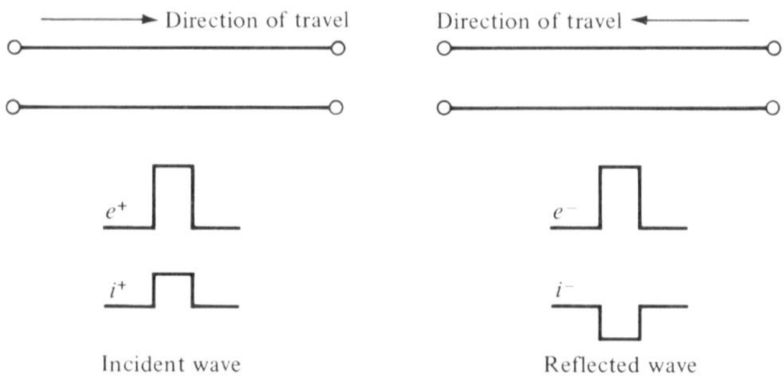

FIGURE 16-2. Traveling voltage-current waves.

Now an obvious question is: What causes a reflected wave to exist? Consider a transmission line with an incident traveling wave. Ohm's law must be satisfied at all points on the line. By equation (16-1) the wave traveling down the line sees impedance Z_0. If the line is terminated with impedance $Z_R = Z_0$, this relation holds true at all points on the line. There is no need for a reflected wave. At the receiving end of the line we see that

$$\frac{e_R^+}{i_R^+} = Z_r = Z_0$$

Now terminate the line in an impedance other than Z_0. Thus

$$\frac{e_R^+}{i_R^+} = Z_0 \neq Z_R$$

and a reflection must occur if Ohm's law is to be satisfied. At the end of the line, by Ohm's law,

$$\frac{\text{total } v}{\text{total } i} = Z_R \tag{16-3}$$

If $Z_R \neq Z_0$, the total voltage and current must include a reflected wave. Thus at the load end of the line,

$$\text{total } v = e_R^+ + e_R^- \tag{16-4}$$

$$\text{total } i = i_R^+ + i_R^- = \frac{e_R^+}{Z_0} - \frac{e_R^-}{Z_0} \tag{16-5}$$

Using these relations, equation (16-3) may be expressed as

$$\frac{\text{total } v}{\text{total } i} = Z_R = Z_0 \frac{e_R^+ + e_R^-}{e_R^+ - e_R^-} \tag{16-6}$$

Solving for the ratio of reflected to incident voltage, we have

$$\frac{e_R^-}{e_R^+} = \frac{Z_R - Z_0}{Z_R + Z_0} = \Gamma_R \tag{16-7}$$

The ratio Γ_R is termed the *voltage reflection coefficient*. Thus we see that the incident and reflected voltages are related by

$$e_R^- = \Gamma_R e_R^+ \tag{16-8}$$

It readily follows that the *current reflection coefficient* is the negative of the voltage reflection coefficient. If $Z_R = Z_0$, the reflection coefficient is

$$\Gamma_R = \frac{Z_R - Z_0}{Z_R + Z_0} = \frac{Z_0 - Z_0}{Z_0 + Z_0} = 0$$

and no reflected voltage exists.

EXAMPLE 16-1

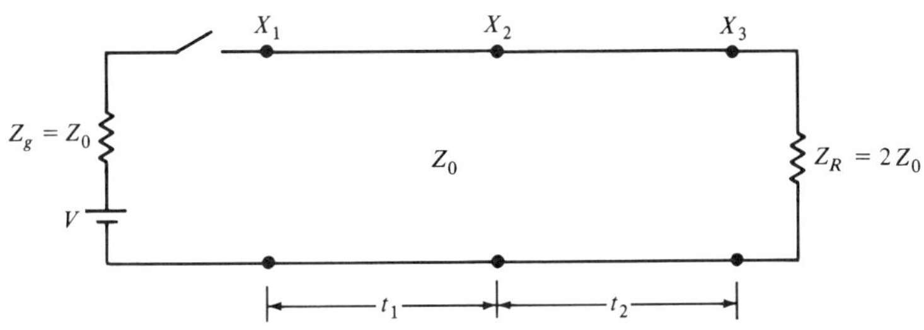

A battery, having a voltage V and an internal impedance of Z_0, is applied to a transmission line at time $t = 0$. Determine the waveforms that will be present at locations X_1, X_2, and X_3. Assume t_1 to be the time taken for the wave to arrive at point X_2 and t_2 to be the time taken to travel the distance from X_2 to X_3.

SOLUTION

When the switch closes, a voltage wave moves down the transmission line, seeing an impedance Z_0. Since there will be a voltage drop across the generator impedance, the magnitude of the voltage moving down the line is not equal to the battery voltage. The equivalent circuit for the wave as it just enters the line consists of a generator with internal impedance Z_0 and characteristic line impedance Z_0:

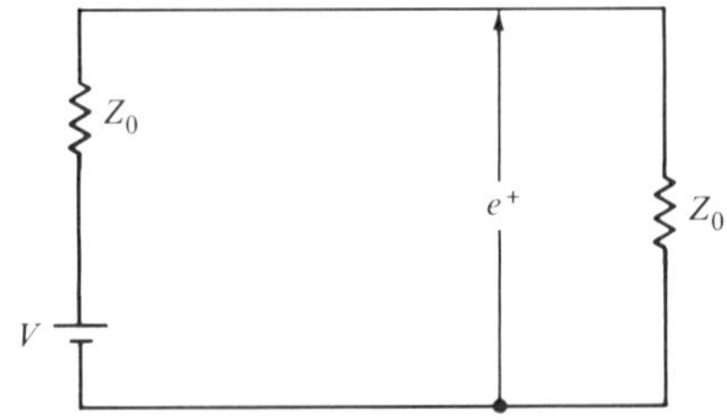

Step wave applied to a mismatched transmission line

The incident voltage has a magnitude of

$$e^+ = V \frac{Z_0}{Z_0 + Z_0} = \frac{V}{2}$$

The reflection coefficient at the receiving end has a value of

$$\Gamma_R = \frac{Z_R - Z_0}{Z_R + Z_0} = \frac{2Z_0 - Z_0}{2Z_0 + Z_0} = \frac{1}{3}$$

The reflected voltage has a magnitude of

$$e^- = \Gamma_R e^+ = \left(\frac{1}{3}\right)\frac{V}{2} = \frac{V}{6}$$

The voltage waveforms are shown below.

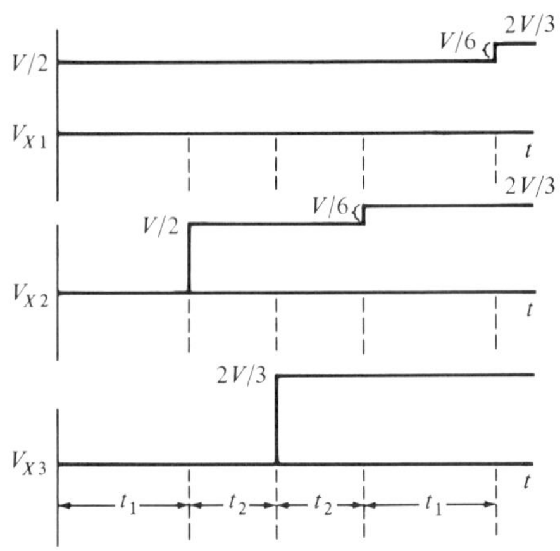

DIFFERENTIAL EQUATIONS FOR VOLTAGE AND CURRENT ON A TRANSMISSION LINE

A differential length of transmission line is shown in Figure 16-3. Note that a lumped-constant model is employed. The concepts of circuit analysis will be employed through the introduction of the principles of distributed parameter networks. The differential length of transmission line is characterized by four parameters, assumed to be constant. These are: R = series resistance, L = series inductance, G = shunt conductance, and C = shunt capacitance, all per unit length of line.

Kirchhoff's voltage law applied to the outside loop results in

$$\begin{aligned} v(x, t) = {} & \frac{R\,\Delta x}{2} i(x, t) + \frac{L\,\Delta x}{2} \frac{d}{dt} i(x, t) + v(x + \Delta x, t) \\ & + \frac{R\,\Delta x}{2} i(x + \Delta x, t) + \frac{L\,\Delta x}{2} \frac{d}{dt} i(x + \Delta x, t) \end{aligned} \tag{16-9}$$

Transposing $v(x + \Delta x, t)$ and collecting terms yields

$$\begin{aligned} -[v(x + \Delta x, t) - v(x, t)] = {} & \frac{R\,\Delta x}{2} [i(x, t) + i(x + \Delta x, t)] \\ & + \frac{L\,\Delta x}{2} \frac{d}{dt} [i(x, t) + i(x + \Delta x, t)] \end{aligned} \tag{16-10}$$

Now a Taylor series expansion may be written for a term of the form $i(x + \Delta x, t)$. The power series expansion allows equation (16-10) to be expressed in terms of only one current, $i(x, t)$:

$$i(x + \Delta x, t) = i(x, t) + \frac{di(x, t)}{dt} \Delta x + \frac{1}{2!} \frac{d^2 i(x, t)}{dt^2} \Delta x^2 + \cdots \tag{16-11}$$

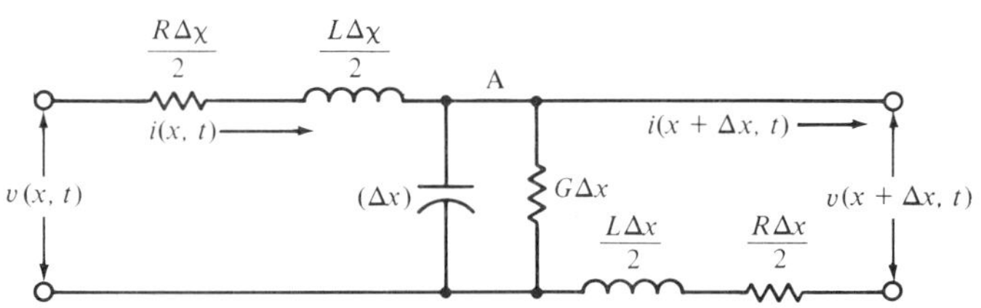

FIGURE 16-3. Circuit representation of a Δx length of transmission line.

Substituting this expression into equation (16-10) and taking the limit as $\Delta x \to 0$ results in the final equation:

$$-\lim_{\Delta x \to 0} \frac{[v(x + \Delta x, t) - v(x, t)]}{\Delta x} = -\frac{\partial v(x, t)}{\partial x}$$
$$= Ri(x, t) + L\frac{\partial i(x, t)}{\partial t} \qquad (16\text{-}12)$$

The dual of this equation is a relation between the distance derivative of the current and the time derivative of the voltage. It is obtained in an analogous way by writing the nodal equation at point A in Figure 16-3. The result is

$$-\frac{\partial i(x, t)}{\partial x} = Gv(x, t) + C\frac{\partial v(x, t)}{\partial t} \qquad (16\text{-}13)$$

The physical interpretation of equations (16-12) and (16-13) is that the voltage decreases as the distance increases because of the voltage drop across the series resistance and inductance. The shunting effect of the conductance and the capacitance reduces the current with increasing distance.

16-2.1 Differential Steady-State Equations for a Transmission Line

Let the driving source for a transmission line be a sinusoid. That is,

$$f(t) = A\cos\omega t = Re(Ae^{j\omega t}) \qquad (16\text{-}14)$$

where A is the peak phasor of $f(t)$. From Chapter 4 we know that differentiation with respect to time becomes one of multiplication by $j\omega$, and integration is equivalent to division by $j\omega$. Substituting equation (16-14) into equations (16-12) and (16-13) then reduces these equations to ordinary differential equations in space. Thus

$$\frac{dE}{dx} = -(R + j\omega L)I \qquad (16\text{-}15)$$

$$\frac{dI}{dx} = -(G + j\omega C)E \qquad (16\text{-}16)$$

where E and I are the peak phasor values of the voltage and current, respectively.

It is usual practice to denote the ac series impedance per unit length $(R + j\omega L)$ by the symbol Z and the shunt admittance per unit length by Y. That is,

$$Z = R + j\omega L \qquad \text{ohms/unit length}$$

$$Y = G + j\omega C \qquad \text{siemens/unit length}$$

The steady-state differential equations may be written as

$$\frac{dE}{dx} = -ZI \tag{16-17}$$

$$\frac{dI}{dx} = -YE \tag{16-18}$$

16-2.2 Steady-State Solutions for a Transmission Line

In arriving at steady-state solutions for voltage and current on a transmission line, it is helpful to have an expression relating the distance derivative of voltage or current to the time derivative of voltage or current. This relation can be easily obtained from equations (16-17) and (16-18). Differentiating equation (16-17) with respect to x results in the following:

$$\frac{d^2E}{dx^2} = -Z\frac{dI}{dx} \tag{16-19}$$

Substituting this expression into equation (16-18) for dI/dx yields

$$\frac{d^2E}{dx^2} = (YZ)E \tag{16-20}$$

Equation (16-20) is known as the one-dimensional *wave equation*.

The general solution to this equation is

$$E(x) = A_1e^{-\sqrt{ZY}x} + A_2e^{\sqrt{ZY}x} \tag{16-21}$$

where the A's are complex constants of integration which must be determined from the boundary conditions associated with a particular problem. The corresponding expression for the current is found by substituting this expression into equation (16-17). Thus

$$\begin{aligned} I(x) &= -\frac{1}{Z}\frac{dE(x)}{dx} \\ &= \frac{1}{\sqrt{Z/Y}}(A_1e^{-\sqrt{ZY}x} - A_2e^{\sqrt{ZY}x}) \end{aligned} \tag{16-22}$$

The quantity $\sqrt{Z/Y}$ has units of ohms and represents a formal way of calculating the characteristic impedance, Z_0, of the line. From equations (16-21) and (16-22) we

$$Z_0 = \frac{E(x)}{I(x)} = \sqrt{\frac{Z}{Y}} = \sqrt{\frac{R + j\omega L}{G + j\omega C}} \tag{16-23}$$

From these same equations, we see that the quantity $\sqrt{ZY}$ governs the way the amplitude and phase of the voltage and current varies along the line. This term is, then, aptly called the *propagation constant*, γ.

$$\gamma = \sqrt{ZY} = \sqrt{(R + j\omega L)(G + j\omega C)} \tag{16-24}$$

The propagation constant is generally a complex number. The real part, α, is called the *attenuation constant* and the imaginary part, β, is the *phase constant*.

$$\gamma = \alpha + j\beta \tag{16-25}$$

where

α = attenuation constant (nepers/unit length)
β = phase constant (radians/unit length)

EXAMPLE 16-2

An open-wire transmission line has constants of

$$R = 13\ \Omega/\text{mi}$$

$$L = 4 \cdot 2\ \text{mH/mi}$$

$$C = 0.015\ \mu\text{F/mi}$$

$$G = 0.25 \times 10^{-6}\ \text{S/mi}$$

and operates at a frequency of 10,000 Hz. Calculate (a) Z_0, (b) α, and (c) β.

SOLUTION

(a) $$Z_0 = \sqrt{\frac{R + j\omega L}{G + j\omega C}} = \sqrt{\frac{13 + j263.7}{0.25 \times 10^{-6} + j942 \times 10^{-6}}}$$

$$= \sqrt{\frac{264.02\angle 87.1^\circ}{942 \times 10^{-6}\angle 90^\circ}} = 0.53\angle -1.45^\circ\ \Omega$$

$$\gamma = \sqrt{ZY} = \sqrt{(264.02\angle 87.1^\circ)(942 \times 10^{-6}\angle 90^\circ)}$$

$$= 0.498\angle 88.55^\circ = 0.0123 + j0.498$$

Then

(a) $\alpha = 0.0123$ Np/mi

(c) $\beta = 0.498$ rad/mi

16-2.3 The Neper and the Decibel

The attenuation coefficient α was introduced as having units of nepers per mile. Often it is desirable to be able to relate this quantity to the more familiar unit, decibels. Consider the section of line, Δx, in Figure 16-4. Consider that both the input and output of the line are terminated in resistive impedances of Z_0. It is possible, then, to describe the gain characteristics of this line by the decibel. If, in addition, the line is distortionless, it has an exponentially decaying response. Thus

$$|E_1| = |E_s|e^{-\alpha x_1} \tag{16-26}$$

$$|E_2| = |E_s|e^{-\alpha x_2} \tag{16-27}$$

The decibel gain for the length of line Δx is then

$$\begin{aligned} \text{dB} &= 20 \log_{10} \frac{|E_2|}{|E_1|} = 20 \log_{10} \frac{|E_s|e^{-\alpha x_2}}{|E_s|e^{-\alpha x_1}} \\ &= 20 \log_{10} e^{-\alpha x_2} - 20 \log_{10} e^{-\alpha x_1} \\ &= 20(-\alpha x_2 \log_{10} e + \alpha x_1 \log_{10} e) \\ &= 8.686[-\alpha(x_2 - x_1)] \end{aligned}$$

where $-\alpha(x_2 = x_1) = -\alpha\,\Delta x$ represents the total loss in nepers. Furthermore, then, 1 Np represents 8.686 dB.

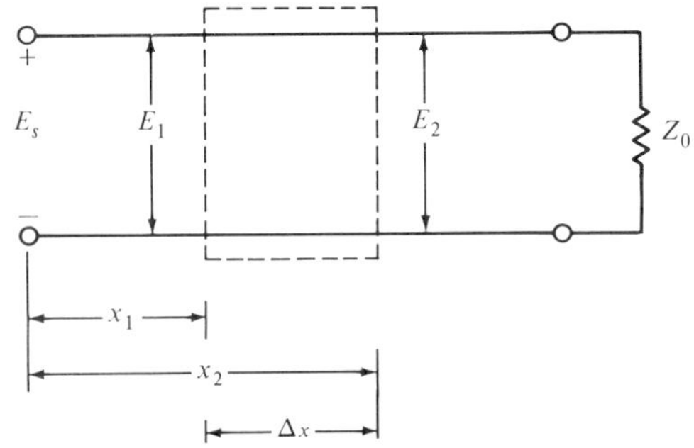

FIGURE 16-4. Matched transmission line.

EXAMPLE 16-3

It is desired to construct a long-distance transmission system. A low-loss, high-power transmission cable is selected which has an attenuation of 1 dB/100 ft. A 1000-V source is connected to the sending end and the sensitivity of the detector is 1 mV. Determine the longest distance between the transmitting and receiving station.

SOLUTION

The ratio between sending and receiving stations is

$$\frac{V_S}{V_R} = \frac{10^3}{10^{-3}} = 10^6$$

Then the maximum attenuation is

$$\text{maximum attenuation} = 20 \log 10^6 = 120 \text{ dB}$$

$$\text{maximum cable length} = 120 \times 100 \approx 2\tfrac{1}{4} \text{ mi}$$

This indicates that even in low-loss systems it is necessary to use repeaters at regular intervals to compensate for line loss.

16-2.4 Matched Transmission Line

The general solution for voltage along a transmission line is expressed by equation (16-21); that is,

$$E(x) = A_1 e^{-(\alpha + j\beta)x} + A_2 e^{(\alpha + j\beta)x} \tag{16-28}$$

Now, consider again the case of an infinitely long transmission line. Since the exponent for the second term is positive, this term would tend to infinity. This is a physical impossibility, thus, a "boundary" condition for the line implies that $A_2 = 0$. Then

$$E(x) = A_1 e^{-(\alpha + j\beta)x} \tag{16-29}$$

To obtain a value of A_1, we use the boundary condition at the source. At the source ($x = 0$), the applied voltage is E_s:

$$E(0) = E_s = A_1 e^{-0} = A_1$$

Then equation (16-29) becomes

$$E(x) = E_s e^{-(\alpha + j\beta)x} \tag{16-30}$$

and the corresponding expression for the current is, from equation (16-22),

$$I(x) = \frac{E_s}{Z_0} e^{-(\alpha + j\beta)x} \tag{16-31}$$

Next, suppose that we cut the line and terminate it in its characteristic impedance, $Z_R = Z_0$. From equations (16-7) and (16-8) there is no reflected wave and the solutions for voltage and current on the line indicated by equations (16-30) and (16-31) still hold. A transmission line such as this is termed a *matched line*, a *flat line*, or a *properly terminated line*.

The input voltage, E_s, to the line must still be solved for. Refer to Figure 16-5. From the figure we see that E_s is given by

$$E_s = \frac{Z_0}{Z_0 + Z_g} E_g \tag{16-32}$$

16-2.5 Incident Traveling Wave

At this point it is instructive to verify that the voltage waveform described by equation (16-30) is indeed a traveling wave. We may rewrite equation (16-30) as

$$E(x) = E_s e^{-\alpha x} e^{-j\beta x} \tag{16-33}$$

The factor $e^{-\alpha x}$ simply indicates that the amplitude of the wave decreases with increasing x. If we set $\alpha = 0$, the line is lossless and equation (16-33) reduces to

$$E(x) = E_s e^{-j\beta x} \tag{16-34}$$

If we now reinsert the $j\omega t$ factor and take the real part, equation (16-34) becomes

$$E(x, t) = E_s \cos(\omega t - \beta x) \tag{16-35}$$

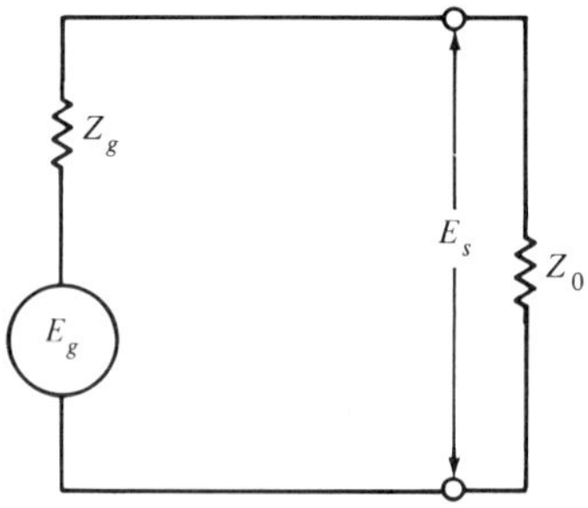

FIGURE 16-5. Equivalent circuit of sending end.

That this wave represents a traveling wave can be verified by plotting this equation for increasing values of ωt (Figure 16-6). To determine the velocity of wave travel (phase velocity), we can imagine an observer riding the crest of the wave. Thus the phase $\omega t - \beta x$ is a constant. Taking the derivative of x with respect to time, we find the phase velocity to be $v_p = dx/dt = \omega/\beta$.

16-2.6 Terminated Transmission Line

A wave launched on a terminated line propagates down the line as if the line were of infinite extent, until it reaches the termination. At that point, if conditions are not correct, a reflection occurs. The total voltage at any point x along the line is given by

$$E(x) = E^+(x) + E^-(x) \tag{16-36}$$

where $E^+(x)$ is the incident voltage at point x and $E^-(x)$ is the reflected voltage at point x.

The incident voltage $E^+(x)$ may be written as

$$E^+(x) = E^+(0)e^{-\gamma x} \tag{16-37}$$

where $E^+(0)$ is the incident sending-end voltage. Now the reflected voltage can be obtained by following the incident wave to the end of the line, allowing a reflection, and then traveling back up the line to point x (Figure 16-7). Thus

$$E^-(d) = E^+(0)e^{-\gamma l}\Gamma_R e^{-\gamma d} \tag{16-38}$$

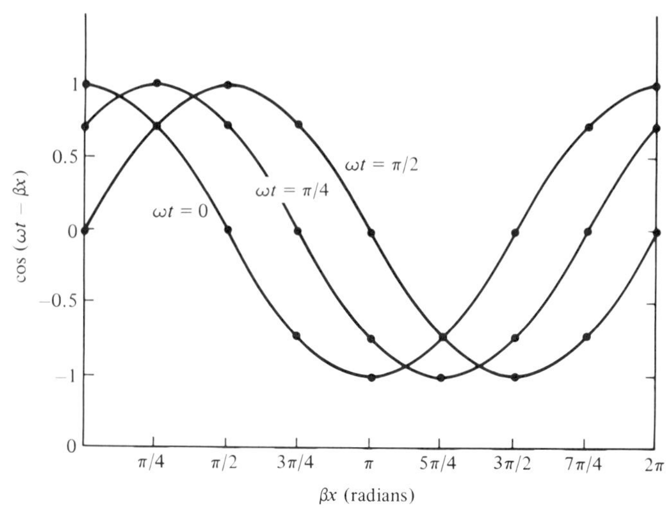

FIGURE 16-6. Traveling waves on a lossless line.

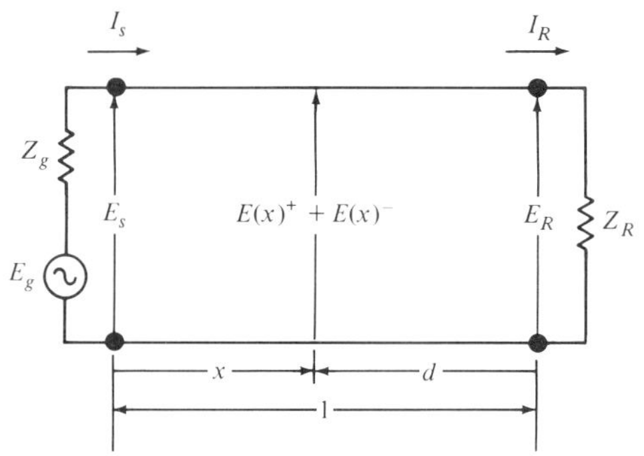

FIGURE 16-7. Relationship between the distance variables for a terminated transmission line.

The points $E^-(x)$ and $E^-(d)$ are physically the same, with the relationship $d = 1 - x$ prevailing. Thus equation (16-38) may be written as

$$E^-(x) = \Gamma_R E^+(0) e^{-\gamma(2l-x)} \tag{16-39}$$

Using equations (16-36), (16-38), and (16-39), the total voltage at any point x is then

$$E(x) = E^+(0)[e^{-\gamma x} + \Gamma_R e^{-\gamma(2l-x)}] \tag{16-40}$$

It is convenient to express the incident sending-end voltage $E^+(0)$ in terms of the sending-end voltage E_s since it cannot be easily measured. Thus from equation (16-40) with $x = 0$,

$$E(0) = E_s = E^+(0)[1 + \Gamma_R e^{-2\gamma l}] \tag{16-41}$$

or

$$E^+(0) = \frac{E_s}{1 + \Gamma_R e^{-2\gamma l}} \tag{16-42}$$

Using this result, equation (16-40) may be written as

$$E(x) = E_s \frac{e^{-\gamma x} + \Gamma_R e^{-\gamma(2l-x)}}{1 + \Gamma_R e^{-2\gamma l}} \tag{16-43}$$

A similar expression may be found for the current at point x by noting that the expressions for incident and reflected current are

$$I^+(x) = \frac{E^+(x)}{Z_0} \tag{16-44}$$

and

$$I^-(x) = \frac{E^-(x)}{Z_0} \tag{16-45}$$

where the negative sign for the reflected current was introduced earlier. The total current at point x is then

$$\begin{aligned} I(x) &= I^+(x) + I^-(x) \\ &= \frac{E^+(0)e^{-\gamma x}}{Z_0} - \frac{E^+(0)\Gamma_R e^{-\gamma(2l-x)}}{Z_0} \\ &= \frac{E^+(0)}{Z_0}[e^{-\gamma x} - \Gamma_R e^{-\gamma(2l-x)}] \end{aligned} \tag{16-46}$$

If we now use equation (16-42), the expression for current becomes

$$I(x) = \frac{E_s}{Z_0}\left[\frac{e^{-\gamma x} - \Gamma_R e^{-\gamma(2l-x)}}{1 + \Gamma_R e^{-2\gamma l}}\right] \tag{16-47}$$

16-2.7 Driving-Point Impedance of a Terminated Line

The driving source impedance is the ratio of the voltage to the current at the driving terminals. At any point on the line the ratio of voltage to current at that point is the driving-point impedance. We may obtain the driving-point impedance at any point x from equations (16-43) and (16-47). Thus

$$Z(x) = \frac{E(x)}{I(x)} = Z_0 \frac{1 + \Gamma_R e^{-2\gamma(l-x)}}{1 - \Gamma_R e^{-2\gamma(l-x)}} \tag{16-48}$$

This is a very interesting and useful result. It states that the impedance at any point on the line depends on the characteristic of the line, the terminating impedance, and the distance from the termination.

EXAMPLE 16-4

Find the input impedance of a transmission line 37.5 m long with a 50-Ω characteristic impedance and terminated in a short circuit. Assume a lossless line. The operating frequency is 3 MHz.

SOLUTION

At the termination, the reflection factor is

$$\Gamma_R = \frac{Z_R - Z_0}{Z_R + Z_0} = -1$$

At the input ($x = 0$), the reflection factor is

$$\Gamma(0) = \Gamma_R e^{-j2\beta l} \qquad l = 37.5 \text{ m}$$
$$= -e^{-j75\beta} = -e^{-j75(2\pi/\lambda)}$$
$$= -e^{-j(150\pi/100)} = e^{-j(\pi/2)}$$

where use has been made of the relations $\beta = 2\pi/\lambda$ and $\lambda = 300/f_{\text{MHz}}$ meters. The sending-end impedance from equation (16-48) is

$$Z(0) = 50 \frac{1 + e^{j(\pi/2)}}{1 - e^{-j(\pi/2)}} = 50 \frac{1 - j}{1 + j} = -j50 \ \Omega$$

16-3 VOLTAGE STANDING-WAVE RATIO

The voltage along the line varies as a function of position unless the line is properly terminated. From equation (16-40) the normalized voltage at any point on the line may be written as

$$\overline{E}(x) = \frac{E(x)}{E^+(0)e^{-\gamma x}} = 1 + \Gamma_R e^{-2\gamma(l-x)} \tag{16-49}$$

The voltage has a maximum value

$$E_{\max} = 1 + |\Gamma_R| \tag{16-50}$$

and a minimum value

$$E_{\min} = 1 - |\Gamma_R| \tag{16-51}$$

The ratio of the maximum to the minimum is known as the *voltage standing-wave ratio* (VSWR):

$$\text{VSWR} = \rho = \frac{1 + |\Gamma_R|}{1 - |\Gamma_R|} \tag{16-52}$$

The reflection factor can be expressed in terms of the VSWR:

$$|\Gamma_R| = \frac{1 - \rho}{1 + \rho} \tag{16-53}$$

Note that the VSWR is a measure of the degree of mismatch. The VSWR for a properly terminated line is unity.

A disadvantage of a high VSWR is the fact that it limits the power that can be transmitted. This is because all lines have at least two effective conductors separated by a dielectric. The maximum electric field strength must not be exceeded or breakdown occurs.

EXAMPLE 16-5

Consider a transmission line with a maximum permissible voltage of 5000 V, terminated in a 200-Ω resistance. The characteristic impedance is 50 Ω. Determine the maximum voltage that can be applied to the line.

SOLUTION

$$|\Gamma_R| = \frac{Z_R - Z_0}{Z_R + Z_0} = \frac{200 - 50}{200 + 50} = 0.6$$

The voltage VSWR is

$$\rho = \frac{1 + |\Gamma_R|}{1 - |\Gamma_R|} = \frac{1.6}{0.4} = 4$$

The maximum applied voltage is

$$V_{\max} = \frac{5000}{4} = 1250 \text{ V}$$

16-4

QUARTER-WAVELENGTH MATCHING SECTION

A quarter-wavelength of line may be used to match two impedances. From equation (16-48) we see that the maximum and minimum values of impedance occur along the line when $e^{-j2\beta d} = -1$, and $e^{-j2\beta d} = +1$, respectively, where

$d = l - x$. The maximum impedance occurs when

$$e^{-j2\beta d} = +1 \Rightarrow 2\beta d = 2k\pi$$

or

$$d = \frac{2k\pi}{2\beta}$$

The minimum impedance occurs when

$$e^{-j2\beta d} = -1 \Rightarrow 2\beta d = (2k + 1)\pi$$

or

$$d = \frac{(2k + 1)\pi}{2\beta}$$

Thus the positions of maximum and minimum impedance are

$$d_{max} - d_{min} = \frac{2k\pi - (2k + 1)\pi}{2\beta} = \frac{\pi}{2\beta} = \frac{\lambda}{4} \tag{16-54}$$

The positions of maximum and minimum impedance along the line are separated by a quarter-wavelength. From equation (16-48), the maximum and minimum impedances are then

$$Z_{max} = Z_0 \frac{1 + |\Gamma_R|}{1 - |\Gamma_R|} \tag{16-55}$$

and

$$Z_{min} = Z_0 \frac{1 - |\Gamma_R|}{1 + |\Gamma_R|} \tag{16-56}$$

Using equation (16-52), these two equations may be expressed in terms of the VSWR:

$$Z_{max} = \rho Z_0 \tag{16-57}$$

and

$$Z_{min} = \frac{1}{\rho} Z_0 \tag{16-58}$$

It is of interest to formulate the product of the maximum and minimum impedances. Thus

$$Z_{max} Z_{min} = Z_0^2$$

and

$$Z_0 = \sqrt{Z_{max} Z_{min}} \tag{16-59}$$

This relation indicates a method of matching two impedances. Essentially, the impedance transforming properties of the line can be used to match impedances.

EXAMPLE 16-6

A 100-Ω load is to be matched to a 50-Ω line. Determine the characteristic impedance of a quarter-wavelength matching section.

SOLUTION

$$Z_0 = \sqrt{Z_{max} Z_{min}} = \sqrt{(100)(50)}$$
$$= 70.71\ \Omega$$

16-5 THE SMITH CHART

We have seen that the circuit quantities voltage, current, impedance, and standing-wave ratio all depend on the reflection coefficient, Γ. With the exception of the VSWR, all these quantities vary with distance. This occurs because the reflection coefficient is a function of distance. If distances are specified, we can compute any of the foregoing quantities by using the proper equations. However, if the distances are not specified, the resulting equations are transcendental and difficult to solve analytically. Graphical procedures are often most convenient for solving transcendental functions. The Smith chart was developed to provide convenient solutions for transmission-line problems.

The expression for impedance on a lossless line normalized with respect to the characteristic impedance is

$$\frac{Z}{Z_0} = \frac{1 + \Gamma_R e^{-j2\beta d}}{1 - \Gamma_R e^{-j2\beta d}} \tag{16-60}$$

This is a complex quantity and can be expressed as a real and an imaginary part:

$$\frac{Z}{Z_0} = r + jx \tag{16-61}$$

Now the complex term in equation (16-60) may be written as

$$\begin{aligned} \Gamma_R e^{-j2\beta d} &= |\Gamma_R| e^{j\phi} e^{-j2\beta d} \\ &= |\Gamma_R| e^{j(\phi - 2\beta d)} = u + jv \end{aligned} \tag{16-62}$$

where ϕ is the angle of the reflection coefficient. Using equations (16-61) and (16-62), equation (16-60) may be expressed as

$$\begin{aligned} \frac{Z}{Z_0} = r + jx &= \frac{1 + u + jv}{1 - u - jv} \\ &= \frac{1 - u^2 - v^2 - 2jv}{(1 - u)^2 + v^2} \end{aligned} \tag{16-63}$$

From equation (16-63) we may identify the real and imaginary parts as

$$r(1 - 2u + u^2 - v^2) = 1 - u^2 - v^2 \tag{16-64}$$

$$x(1 - 2u + u^2 - v^2) = 2v \tag{16-65}$$

Equation (16-64) involves r only; thus we can obtain the loci of constant r in the Γ plane. We can similarly obtain the loci of constant reactance from equation (16-65). These two equations may be rewritten as

$$u^2 - \frac{2r}{r+1} u + v^2 = \frac{1 - r}{1 + r} \tag{16-66}$$

$$u^2 - 2u + v^2 - \frac{2v}{x} = -1 \tag{16-67}$$

and are readily recognized as representing circles. To determine the centers and radii, it is necessary to complete the square. After performing this operation, the expressions are

$$\left(u - \frac{r}{r + 1}\right)^2 + v^2 = \frac{1}{(r + 1)^2} \tag{16-68}$$

$$(u - 1)^2 + \left(v - \frac{1}{x}\right)^2 = \left(\frac{1}{x}\right)^2 \tag{16-69}$$

The first equation represents a circle with radius $1/(r + 1)$ and centered at $u = r/(r + 1)$ and $v = 0$. The second equation represents a circle with radius $1/x$ and centered at $u = 1$ and $v = 1/x$. The equations are plotted in Figure 16-8. A chart of this type was first described by P. J. Smith and is known as a *Smith chart*. The reflection coefficient [equation (16-62)] is also plotted on the chart. As can be seen from the chart, as we move along a transmission line (increase d), a clockwise rotation is effected on the Smith chart. On the chart this corresponds to an angular distance of $2\beta d$ radians.

16-5.1 Use of the Smith Chart

Referring to Figure 16-8, the loci of constant Γ are concentric circles centered at the origin of the $u-v$ plane. For a lossless line the reflection factor is a complex quantity of constant magnitude whose phase varies linearly with distance. Located around the rim of the chart are three angle scales: one is calibrated in degrees and the other two in wavelengths toward the generator and toward the load. This scale arises because the argument of the reflection factor is

$$\text{Arg } \Gamma = 2Bd = \frac{4\pi}{\lambda} d$$

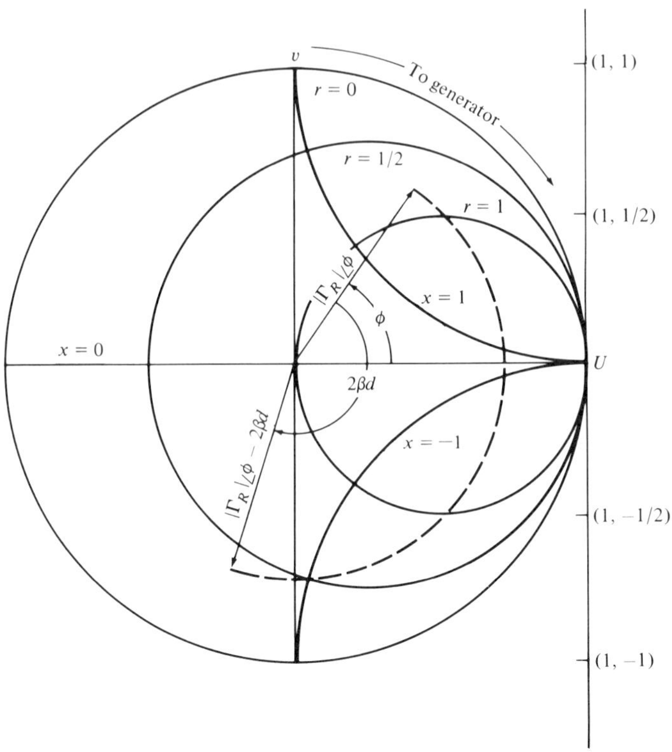

FIGURE 16-8. Smith chart.

Note that when d changes by $\lambda/2$, the angle of the reflection coefficient changes by 2π radians. Thus moving along the line $\lambda/2$ results in a complete rotation about the chart. This means that all functions of the line are periodic with period $\lambda/2$.

We observed earlier that the distance between the maximum and minimum impedance along the line was $\lambda/4$. This corresponds to a rotation of π radians on the Smith chart. From equations (16-57) and (16-58), this transforms a normalized impedance of ρ ohms to $1/\rho$ ohms. This, then, offers the possibility that the Smith chart can be used on an admittance basis as well as for impedance. Some examples will illustrate the use of the Smith chart.

EXAMPLE 16-7

A transmission line with characteristic impedance 50 Ω is terminated by a complex impedance of 50 − j50 Ω. Find the impedance at a point 0.45λ from the load.

SOLUTION

Step 1: The reflection factor is

$$\Gamma_R = \frac{Z_R - Z_0}{Z_R + Z_0} = \frac{(50 - j50) - 50}{50 - j50 + 50} = 0.446\underline{/-63.5^\circ}$$

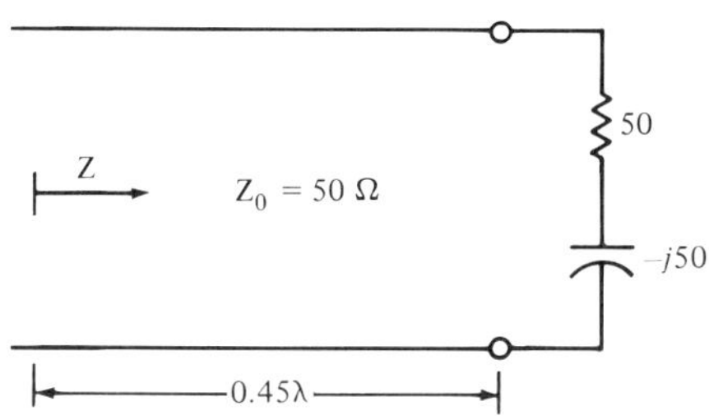

Step 2: From the Smith chart (Figure 16-9) at the bottom of the chart, determine—using the voltage-reflection scale—the radius that corresponds to a magnitude of 0.446. Use the "angle of reflection coefficient" scale on the rim of the chart to locate the desired point.

Step 3: Locate a point 0.45λ from the load by proceeding along the circle of constant Γ in the "wavelengths toward generator" direction. This point is located on the "wavelengths toward generator" scale at

$$[(0.342 + 0.45) - 0.5]\lambda = 0.292\lambda$$

Step 4: Read the normalized impedance from the curves of constant r and constant x.

$$Z_n = 2.0 - j1.0\ \Omega$$

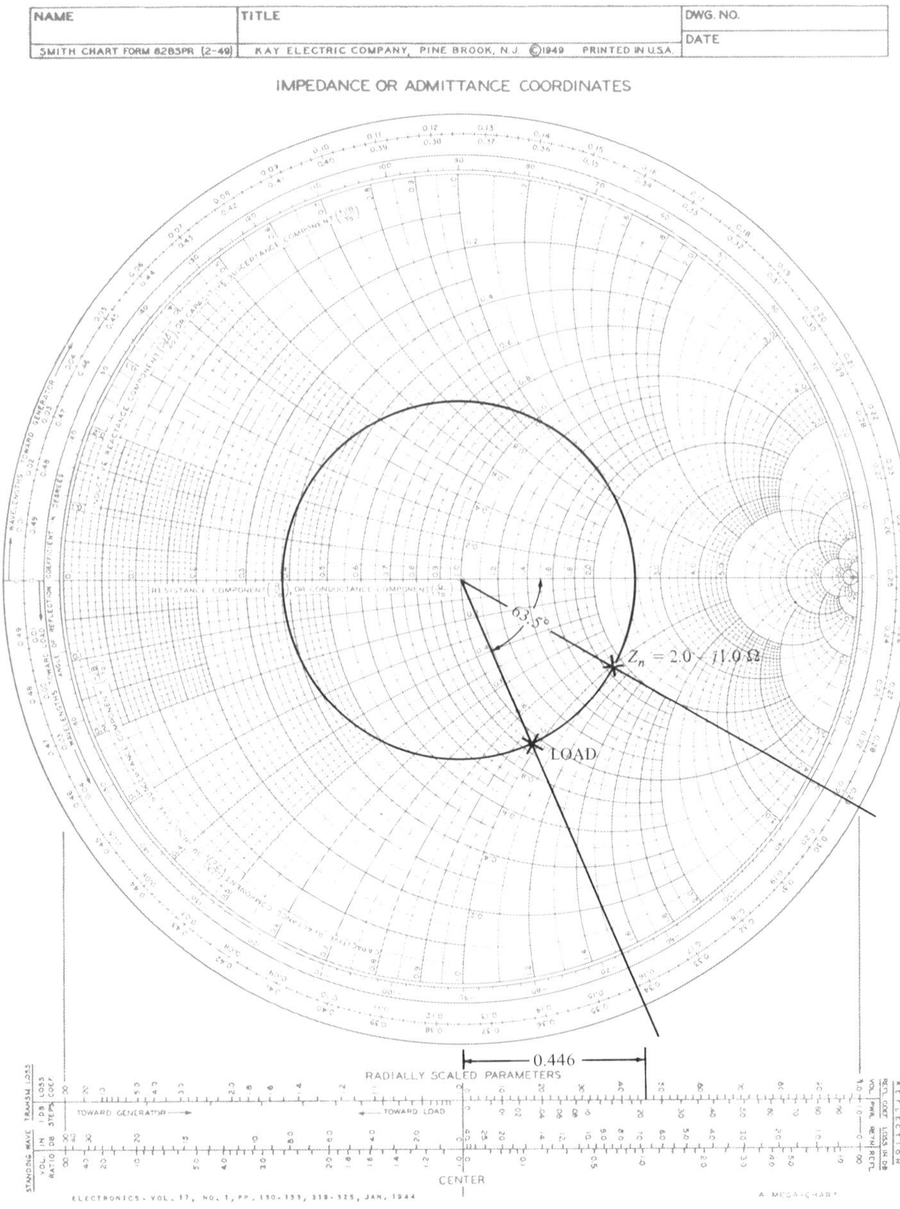

FIGURE 16-9. Smith chart solution for Example 16-7.

Scale the impedance by multiplying by Z_0.

$$Z = Z_0 Z_n$$

$$= 50(2 - j1.0) = 100 - j50\ \Omega$$

EXAMPLE 16-8

The VSWR on a transmission line is 4. At a point d_1 along the line, Z_n is $0.3 + j0.42\ \Omega$. Determine the location of the nearest point, d_2, along the line where the impedance is purely real and is a minimum.

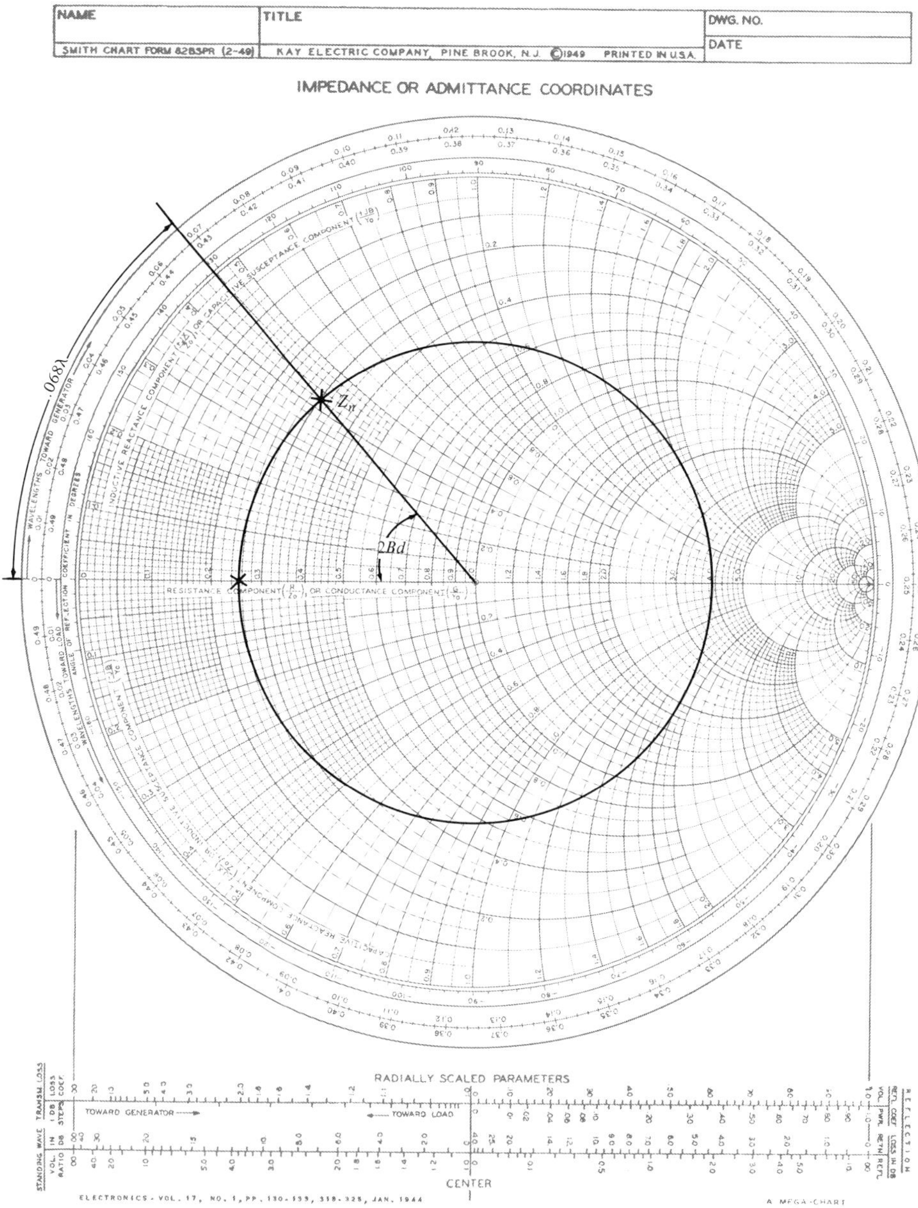

FIGURE 16-10. Smith chart solution for Example 16-8.

SOLUTION

Step 1: Locate the point $Z_n = 0.3 + j0.42$ on the Smith chart (Figure 16-10) and draw the circle of constant standing-wave ratio (constant Γ). Note the point on the "wavelengths toward generator" scale.

Step 2: The impedance is purely real when $x = 0$. This occurs two places on the ρ = constant circle: $r = 4$, $r = 0.25$. The point of minimum real impedance is at $r = 0.25$. We see that this point is located 0.068λ toward the load.

16-5.2 Impedance Matching and Stubbing Using the Smith Chart

The maximum utilization of a transmission line occurs when the standing-wave ratio is unity. From preceding discussions this implies a matched termination which may not always be practical. There are several procedures available for making a voltage appear to the line as a matched termination. One technique using a quarter-wavelength matching section has already been discussed (see Example 16-6). This technique is of limited practical importance because coaxial lines do not come in many different characteristic impedances. Generally, these lines are available with characteristic impedances between 50 and 100 Ω. This technique is more practical with "strip-line" techniques. An impedance-matching technique that removes the need for a section of line with particular characteristic impedance is *stubbing*.

The *single-stub method* is often used. It offers the advantage that the complete transmission line has the same characteristic impedance. Using this procedure, an auxiliary reactive element is added to the line so as to cancel the imaginary component of the input admittance. This auxiliary line is placed at a point where the real part of the input admittance is equal to the characteristic admittance. Thus, canceling the reactive part of the input admittance achieves a matched or "flat" line. In this technique, the unknown quantity is distance. Thus the Smith chart is very useful. An example will illustate the technique.

EXAMPLE 16-9

A 50-Ω line is terminated in an impedance of $7.5 + j60$ Ω. Determine the characteristics and position of a stub so that the line is matched.

SOLUTION

The reactive element is usually placed in parallel with the line. Thus it is easiest to solve the problem using admittances.

Step 1: Find the normalized load admittance.

$$Z_n = \frac{7.5 + j60}{50}$$

$$Y_n = \frac{1}{Z_n} = \frac{50}{7.5 + j60} = 0.1 - j0.83 \text{ S}$$

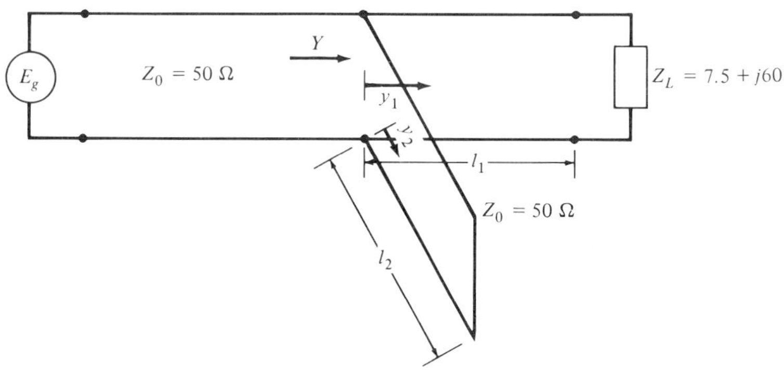

Solution of the problem dictates that we find distances l_1 and l_2.

Step 2: Mark the point $Y_L = 0.1 - j0.83$ S on the Smith chart (Figure 16-11). Through this point draw a circle of constant-voltage standing-wave ratio. Note the position of Y_L on the distance scales at the rim of the chart.

Step 3: Proceed along the constant VSWR circle toward the generator until the real part is 1.0 S. The distance form the load to this point is l_1.

$$l_1 = (0.500\lambda - 0.394\lambda) + 0.209\lambda = 0.315\lambda$$

Step 4: The stub must have a susceptance equal to $-j3.8$ S. A stub terminated in a short circuit is generally used to produce the required admittance. The VSWR of a shorted stub is infinite; thus the circle of constant VSWR for this segment of line is the rim of the chart. From the chart $l_2 = 0.07\lambda$. At the stubbing point

$$Y = Y_1 + Y_2$$

$$= 1.0 + j3.8 - j3.8 = 1.0 \text{ S}$$

and

$$Z = Z_0 Z_n = Z_0 \frac{1}{Y} = Z_0$$

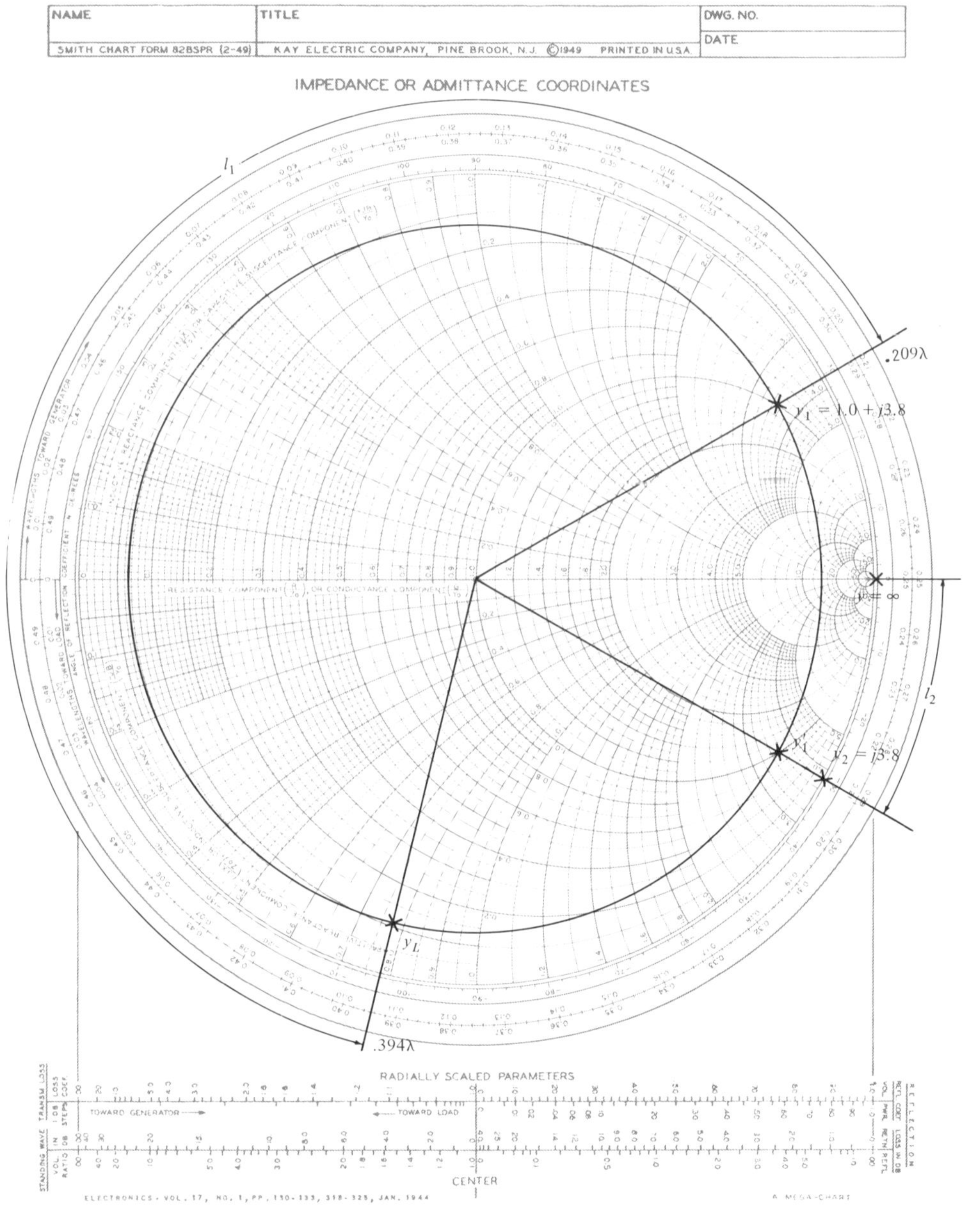

FIGURE 16-11. Smith chart solution for Example 16-9.

Note that there are two points at which an impedance match can be effected. If we continue on up the line to $Y_1' = 1.0 - j3.8$, an impedance match can be effected at this new point on the line if a stub with admittance $+j3.8$ S is parallel with the line at this stubbing point.

Double-Stub Matching. *Double-stub tuning* is a matching technique which consists of two adjustable stubs fixed in position on the transmission line. The spacing between the stubs is not critical. A common practice is to use an odd number of eighth-wavelengths between the stubs, since this will permit a match to be achieved for a wide range of frequencies. An example will illustrate the procedure.

EXAMPLE 16-10

A 50-Ω line feeds an inductive load, $Z = 35 + j35\ \Omega$. Design a double-stub tuner to match this load to the line.

SOLUTION

Step 1: Normalize the impedance and find its reciprocal.

$$Z_n = \frac{Z}{50} = \frac{35 + j35}{50} = 0.7 + j0.7\ \Omega$$

$$Y_n = \frac{1}{Z_n} = \frac{1}{0.7 + j0.7} = 0.71 - j0.71\ \text{S}$$

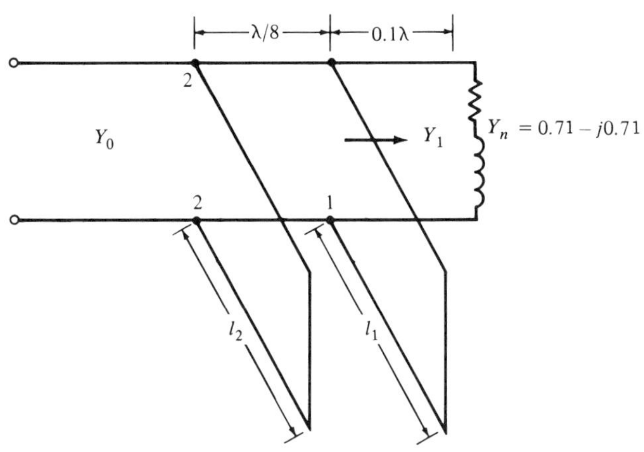

The configuration is as shown.

Step 2: Construct two circles, *A* and *B*, on the Smith chart (Figure 16-12). The transformer formed by the eighth-wavelengths of the line between 2-2 and 1-1 transforms all admittances that lie on circle *A* to circle *B*. Now, stub 1 must transform the input admittance of the line and load to the right of junction 1-1 into an admittance that will plot on the circle *B* locus. The λ/8 section will transform this impedance into a value just to the right of 2-2 which will lie on the locus of circle *A*. This admittance is $1 + jb$ and stub 2 must cancel the susceptance component $+ jb$.

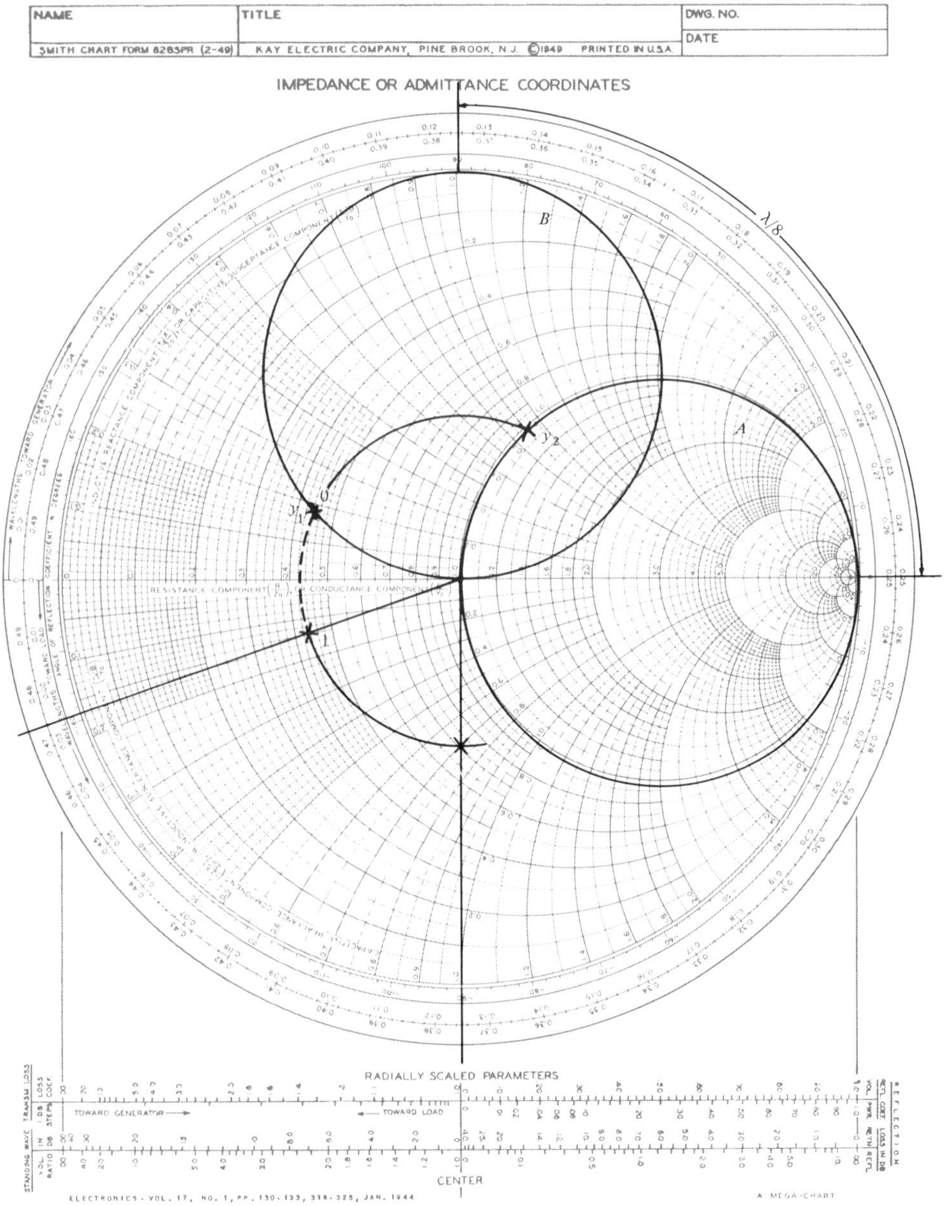

FIGURE 16-12. Smith chart solution for Example 16-10.

Step 3: Enter the Smith chart at $0.71 - j0.71$ and rotate 0.1λ toward the generator on a constant VSWR circle. The normalized admittance to the right of 1-1 is $0.43 - j0.11$ S (point 1 on the chart). Stub 1 adds a susceptance to the line, which must change the admittance to a point on circle *B*. Since stub 1 does not change the conductance, a constant conductance circle is followed from point 1 to point 0 on circle B, where

$$Y_1 = 0.43 + j1.9 \text{ S}$$

Stub 1 must add a normalized susceptance of

$$(0.43 + j1.9) - (0.43 - j0.11) = j2.01$$

From the Smith chart,

$$l_1 = 0.43\lambda$$

Step 4: Starting at point 0, follow a circle of constant VSWR an eighth-wavelength, to circle B. The admittance of the line at 2-2 without stub 2 connected is

$$Y_2 = 1 + j0.9$$

Stub 2 must then have a susceptance of $-j0.9$ S. From the Smith chart,

$$l_2 = 0.126\lambda$$

16-6

LOSSY-LINE SOLUTIONS USING THE SMITH CHART

The previous solutions for lossless lines used constant VSWRs and reflection coefficients. For a lossy line, the propagation constant was noted to be

$$\gamma = \alpha + j\beta$$

As a consequence of being nonzero, the amplitude of an applied waveform decreases as a function of distance. When a reflection occurs, the reflected wave also decreases in amplitude as it progresses backward along the line. Consequently, the VSWR decreases as a function of distance from the load. On the Smith chart, the circles of constant Γ becomes spirals, but the basic principles of the solution are unchanged.

PROBLEMS

16-1. Find the voltage and current reflection coefficients at the load.

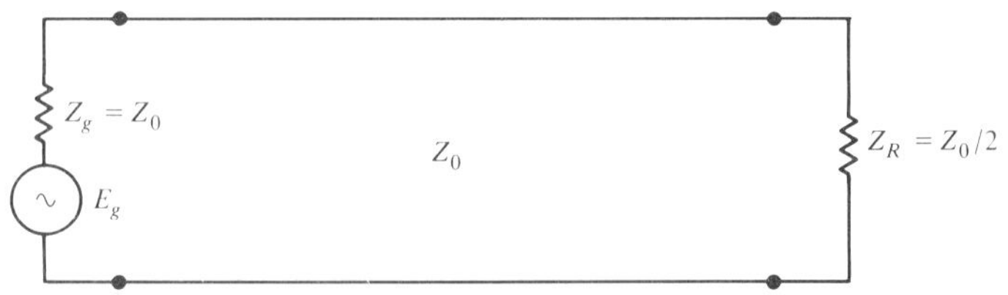

PROBLEM 16-1

16-2.

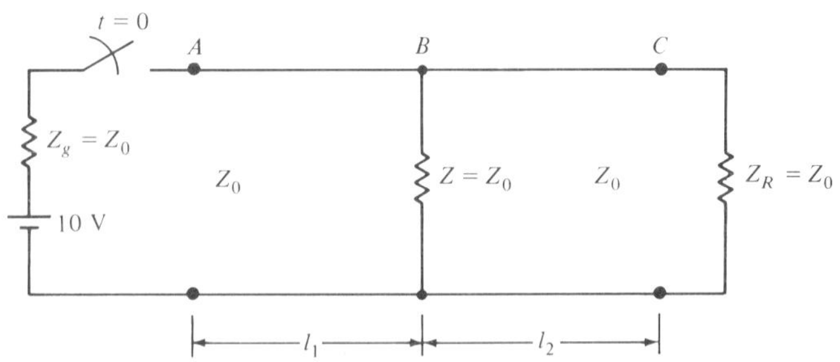

PROBLEM 16-2

Sketch the voltages at points A, B, and C as a function of time.

16-3. An open-wire line has the following constants:

$$R = 10\ \Omega/\text{mi} \qquad G = 0$$
$$L = 10\ \text{mH/mi} \qquad f = 2\ \text{kHz}$$
$$C = 0.01\ \mu\text{F/mi}$$

Find Z_0, γ, α, and β.

16-4. A lossless open-circuited line has an output voltage E_R of 10 V. If β = 1.256 rad/mi, and the line length is 50 mi, what is the voltage amplitude at a distance of 1 mi from the load end? At 2.5 mi?

16-5. For the circuits shown, find Z_a and Z_b.

16-6. Determine the location of the nearest voltage maximum and minimum from the load. What is the VSWR? λ = 5 cm and the normalized load is $1 + j1\ \Omega$.

16-7. The terminating impedance of a transmission line is $Z_R = 100 + j100\ \Omega$ and $Z_0 = 50\ \Omega$. Match the line using double-stub matching. The first stub is to be 0.4λ from the load and the spacing between stubs is to be $\frac{3}{8}\lambda$. Use short-circuited stubs.

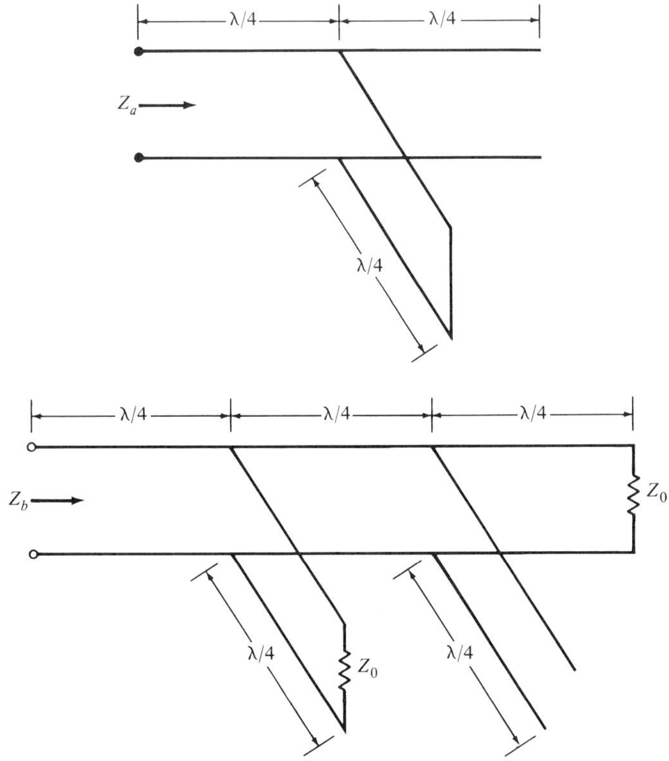

PROBLEM 16-5

16-8. Match the line in Problem 16-7 using a single stub.

16-9. Determine the length of a 75-Ω short-circuited line to produce a reactance of $j125\ \Omega$.

16-10. Using the Smith chart, find the load impedance at the end of a λ/8 line if the sending end impedance is $50 + j20\ \Omega$. Z_0 is 50 Ω.

16-11. For Problem 16-10, what are the reflection coefficient and the VSWR?

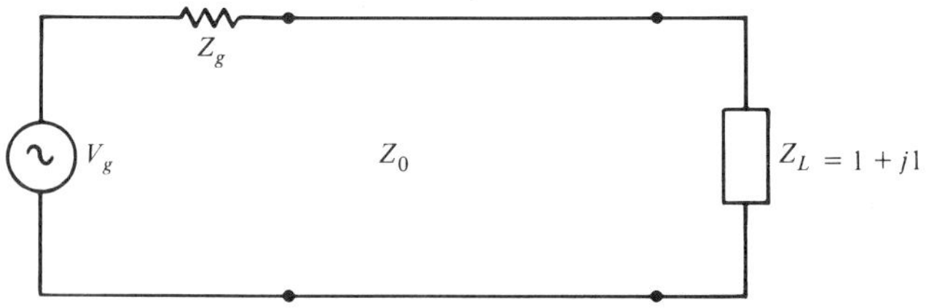

PROBLEM 16-6

16-12. A directional coupler can be used to determine incident and reflected power. If the incident power is measured to be 10 W and the VSWR on the main line is 2.5, what power is being absorbed by the load?

16-13. A load is fed by a properly terminated line. The sending end current is 5 A. If the characteristic impedance of the line is 500 ohm, how much power is being delivered to the line?

16-14. What should the characteristic impedance of a 1/4λ matching section of line be if it is to be used to match a 500 ohm feeder to a 72 ohm load?

16-15. Use the Smith chart to calculate the input impedance of a line 600 electrical degrees long if $Z_0 = 75$ ohm and $Z_L = 50 + j75$ ohms.

16-16. Match a load of $25 + j35$ ohms to a 50 ohm line using: (a) single stub matching, and (b) double stub matching.

16-17. What is the length of a short circuited stub necessary to provide an inductance of 0.5 nh at 260 MHz? Assume a 50 ohm line.

16-18. What is the length of a short circuited line to provide a capacitance of 25 pf at 600 MHz? Assume a 50 ohm line.

CHAPTER 17

Antennas and Radio-Wave Propagation

The design of radio systems requires an estimate of the power loss between the transmitter and the receiver. In Chapter 16 we studied transmission lines used for guided-wave propagation of RF energy. Between free-space propagation and guided-wave propagation, an interface is needed. An antenna is a structure that transforms the energy contained in a guided wave to that of free space or, on the receiving end, free space to guided wave. The parameters that describe the properties of an antenna in accomplishing this transformation are the input impedance, directivity, radiation efficiency, radiation pattern, and polarization.

Once the wave is launched, many factors affect propagation of the signal. Some of these are reflections, fading, refraction in the atmosphere, and diffraction over the earth's surface. These factors are frequency dependent. For completeness we will discuss ground-wave and sky-wave propagation and also both line-of-sight and beyond-horizon transmission. In the following we consider first the basic relation between a radio wave and an antenna.

RADIO-WAVE GENERATION

At this point it is instructive to introduce antennas as a generator of radio waves. The basic antenna or aerial is simply a wire suspended above ground with insulators at both ends (Figure 17-1).

In Chapter 16 we observed that stubs (short-circuited or open-circuited) could be used as a reactance (inductive or capacitive) to match the line and load on a transmission line. In fact, a quarter-wavelength long short-circuited line appears as an open circuit and can be considered parallel resonant. At a length of one half-wavelength, the line appears to be series resonant. Consequently, if the line is cut to, for example, a half-wavelength, it will act as an oscillating *LC* circuit. The impedance characteristic is illustrated in Figure 17-2. At $2f_0$ the line acts like a series resonant circuit. An open-circuited line will act like a series resonant circuit at f_0.

Based on the discussion above, if we cut the open wire line of Figure 17-1 at a half wavelength, it will act as an *LC* oscillating circuit. If the line is excited by an RF source, the free electrons in the wire will oscillate back and fourth from one end to the other. For a high Q, this oscillation will continue for some time. There are, however, two sources of loss. The first is I^2R loss due to resistance in the wire. The second and larger loss is due to energy radiated into space. This energy is manifested as a *radio wave*.

From basic electromagnetic field theory we recognize that current flow in a wire produces time-varying electric and magnetic fields. In fact, as the current along the wire builds to a maximum, decreases to zero, and then reverses direction, this produces corresponding expanding and contracting electric and magnetic fields. A portion of these fields radiate far enough into space that they

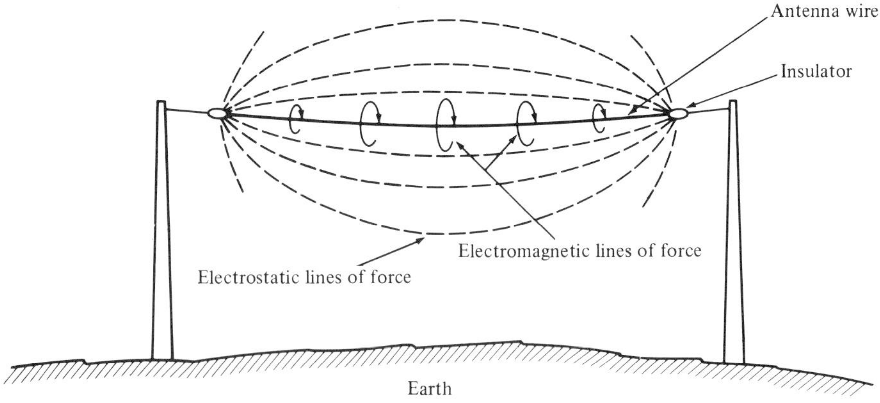

FIGURE 17-1. Basic antenna.

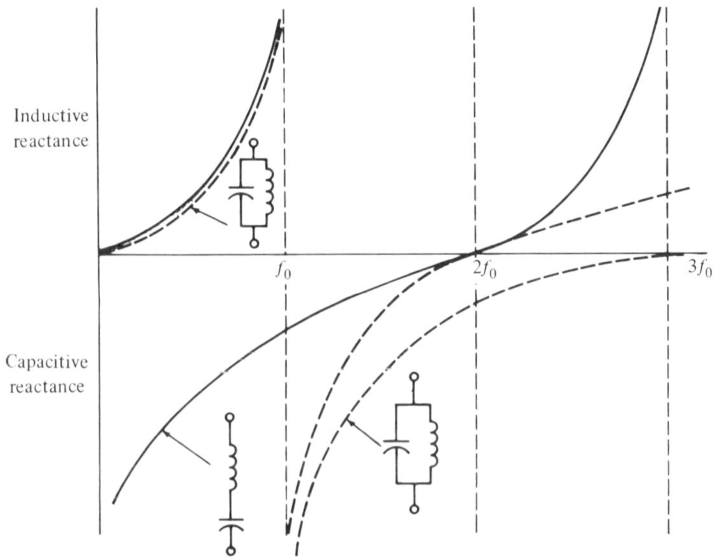

FIGURE 17-2. Short-circuited line as a resonant circuit.

cannot return to the antenna when the current alternates direction. They radiate into space and are lost to the antenna.

For resonance to occur, it is necessary that the antenna be long enough so that the oscillating current does not reach the end and return to the driving point before the potential of the driving source reverses. At low frequencies, this requires a rather long antenna. For example, at 1000 kHz an antenna length of 468 ft is required. A technique known as *loading* is used to reduce this length. This consists of placing an inductor in series with the line. This has the effect of increasing its electrical length.

Once launched, the radio wave must propagate to its destination, where a receiving antenna must intercept a portion of the radiated energy. The paths that the radiation may take are illustrated in Figure 17-3. Lower-frequency radio waves tend to travel along the surface of the earth without attenuation. The ''ground wave'' travels just above the surface of the ground and is usable for hundreds to thousands of miles, day or night. At higher frequencies (500 to 1500 kHz), the earth absorbs energy from the ground wave. Consequently, for these frequencies usable ranges may vary from 40 to 600 miles. Above 5 to 10 MHz the ground wave exists for only a few miles. At these frequencies it is possible to transmit further using the ''direct'' wave. In order to use this wave, the transmitting and receiving antennas must be within *line of sight* of each other. At high frequencies, signals are routinely beamed into space to satellites and deep-space probes.

Near the earth the air is quite dense. Sixty to 600 miles above the earth, the

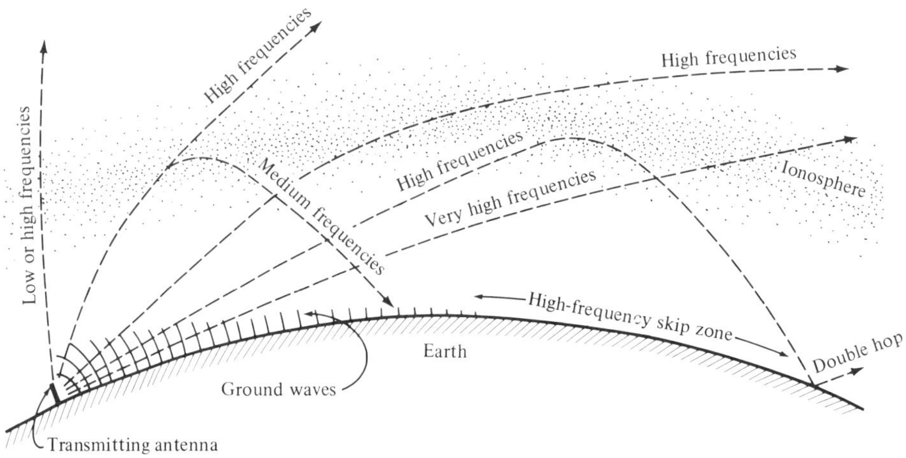

FIGURE 17-3. RADIO-WAVE PATHS.

atmosphere becomes quite thin. The widely spaced air molecules are susceptible to radiation from the sun which ionizes the air molecules. In fact, several different layers of ionization are recognizable at different heights above the earth's surface. The ionized atmosphere allows the radiated wave to travel through it faster than the more dense lower atmosphere. Thus the top part of the wavefront travels faster than the lower part. The wavefront bends and may be "refracted" downward. At the earth's surface this results in the phenomenon of skip distances, which may be used for long-distance communication at high frequencies. The topic of radio-wave propagation is discussed in detail in Section 17-12.

17-2 SHORT DIPOLE ANTENNA

Insight into antenna operation may be gained by considering a short dipole. We may define a short dipole antenna as consisting of a short linear conductor. The importance of this antenna lies in the fact that any linear antenna may be considered to be composed of a large number of short dipoles connected in series. Thus a knowledge of the properties of short dipoles leads to an understanding of antennas of more complex shape.

The relation of a short dipole to a coordinate system is shown in Figure 17-4. The dipole may be energized by a balanced transmission line connected to equal halves of the antenna at its center. Solutions for the fields everywhere around the dipole are given in numerous books on antennas. These solutions

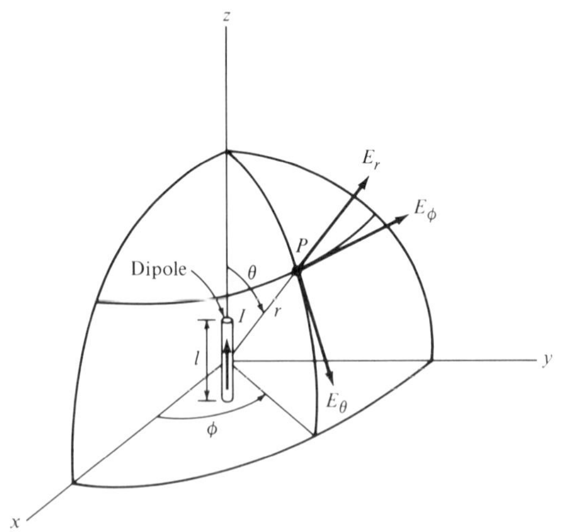

FIGURE 17-4. Relation of dipole to coordinates.

are tabulated in Table 17-1. Note that the electric and magnetic fields from the dipole have three components: E_r, E_θ, and H_ϕ. For r very large the terms involving $1/r^2$ and $1/r^3$ can be neglected relative to terms involving $1/r$. We see that in the *far field*, E_r is negligible and only two field components, E_θ and H_ϕ, remain. Note that these terms are at right angles and in the far field propagate as a plane wave (Figure 17-5). Taking the ratio of E_θ to H_ϕ, we obtain

$$\frac{E_\theta}{H_\phi} = \eta = 120\pi = 376.7\ \Omega$$

This is the intrinsic or characteristic wave impedance of free space.

TABLE 17-1 Fields of a Short Dipole

Component	General expression	Far field	Quasi-stationary
E_τ	$\dfrac{I_0 l e^{j(\omega t - \beta r)} \cos\theta}{2\pi\epsilon_0}\left[\dfrac{1}{cr^2} + \dfrac{1}{j\omega r^3}\right]$	0	$\dfrac{q_0 l \cos\theta}{2\pi\epsilon r^3}$
E_θ	$\dfrac{I_0 l e^{j(\omega t - \beta r)} \sin\theta}{4\pi\epsilon_0}\left[\dfrac{j\omega}{c^2 r} + \dfrac{1}{cr^2} + \dfrac{1}{j\omega r^3}\right]$	$\dfrac{j60\pi I_0 e^{j(\omega t - \beta r)} \sin\theta}{r}\dfrac{l}{\lambda}$	$\dfrac{q_0 l \sin\theta}{4\pi\epsilon r^3}$
H_ϕ	$\dfrac{I_0 l e^{j(\omega t - \beta r)} \sin\theta}{4\pi}\left[\dfrac{j\omega}{cr} + \dfrac{1}{r^2}\right]$	$\dfrac{jI_0 e^{j(\omega t - \beta r)} \sin\theta}{2r}\dfrac{l}{\lambda}$	$\dfrac{I_0 l \sin\theta}{4\pi r^2}$

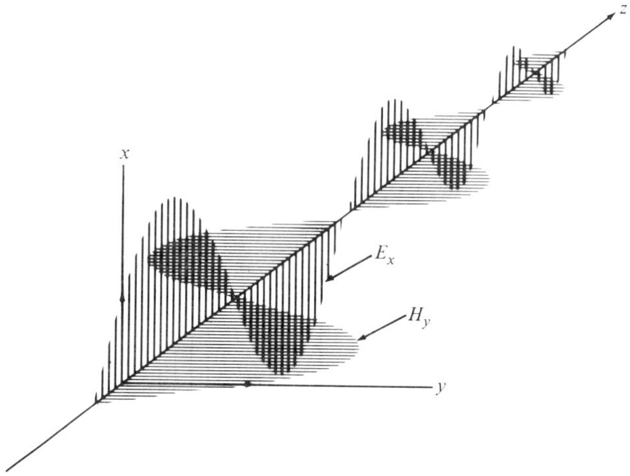

FIGURE 17-5. Plane electromagnetic wave.

17-2.1 Radiation Resistance of a Short Dipole

When an antenna is driven by a source (caused to oscillate), power is radiated. As far as the space surrounding the antenna is concerned, the antenna is a source of power and any source must have an internal impedance. We know from Ohm's law that power can be computed as $P = I^2R_r$. Thus we can compute the resistance of $R_r = P/I^2$. This value of R_r is the *radiation resistance*. It is the ratio of the radiated power, P, to the antenna terminal current, I. The radiation resistance for an antenna can be formally expressed as

$$R_r = \frac{120\pi}{I^2} \int_S |H|^2 \, ds \qquad \text{ohms} \tag{17-1}$$

where

$|H|$ = amplitude of the far field (amperes/meter)
I = amplitude of terminal current

The surface integral is performed over a surface surrounding the antenna. For a short dipole, the radiation resistance is

$$R = 80\pi^2\left(\frac{l}{\lambda}\right)^2 \qquad \text{ohms} \tag{17-2}$$

EXAMPLE 17-1

Find the radiation resistance of a dipole $\frac{1}{10}$-wavelength long.

SOLUTION

$$R = 80\pi^2(\tfrac{1}{10})^2 = 7.9\ \Omega$$

Radiation resistance is also the ratio of the voltage at any point on an antenna to the current flowing at that point ($R = E/I$). Since the current decreases toward the end of the dipole and the voltage increases toward the end of the dipole, the minimum resistance point will be at the center of the antenna.

17-3 HALF-WAVELENGTH DIPOLE

We can deduce from the small value of radiation resistance of Example 17-1 that the short dipole is not an efficient radiator. To improve the radiation efficiency, a longer dipole is used. Refer to Figure 17-6. The half-wave dipole can be thought of as a section of open-circuit transmission line with the wires bent to 90°. Note that the antenna is fed at the center and the current distribution is essentially sinusoidal.

The far fields of a half-wavelength dipole at a distance r from the antenna are

$$E_\theta = \frac{jI\eta}{2\pi r}\,\frac{\cos[(\pi/2)\cos\theta]}{\sin\theta}\,e^{-j\beta r} \tag{17-3}$$

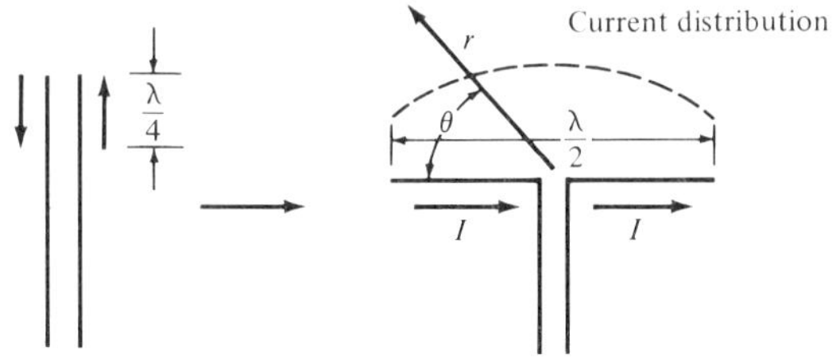

FIGURE 17-6. Half-wavelength dipole.

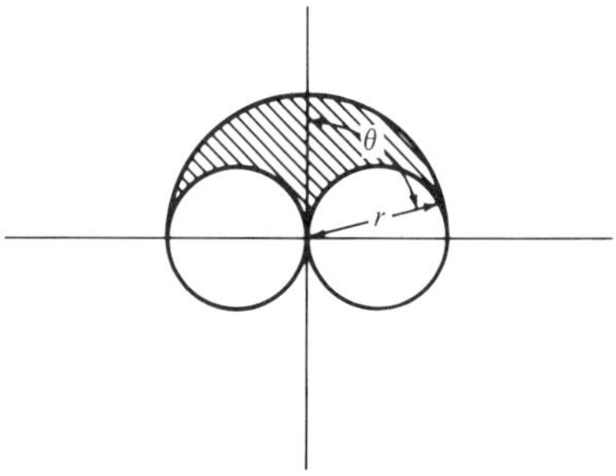

FIGURE 17-7. Three-dimensional view of the far-field pattern (E_θ or H_θ) from a dipole.

and

$$H_\phi = \frac{E_\theta}{\eta} \tag{17-4}$$

The far-field pattern is shown in Figure 17-7.

EXAMPLE 17-2

A half-wave dipole radiating in free space is driven with a current of $\frac{1}{2}$ A at the terminals. Calculate the E and H field strengths 1000 m from the antenna at angles of (a) $\theta = 45°$ and (b) $\theta = 90°$.

SOLUTION

(a) At $\theta = 45°$:

$$|E_\theta| = \frac{I\eta \cos[(\pi/2) \cos \theta]}{2\pi r \sin \theta}$$

$$= \frac{(0.5)(376.7) \cos [(\pi/2) \cos 45°]}{(2\pi)(1000) \sin 45°}$$

$$= 0.0188 \text{ V/m}$$

$$|H_\phi| = \frac{|E_\theta|}{\eta} = \frac{0.0188}{376.7} = 0.00005 \text{ A/m}$$

(b) At $\theta = 90°$:

$$|E_\theta| = \frac{(0.5)(376.7) \cos 0°}{(2\pi)(1000) \sin 90°} = 0.02999 \text{ V/m}$$

$$|H_\phi| = \frac{|E_\theta|}{\eta} = \frac{0.02999}{376.7} = 0.0000796 \text{ A/m}$$

It is shown in basic texts on field theory that the power density in the electromagnetic field is given by a cross-product of the electric and magnetic fields. Specifically,

$$p = \mathbf{E} \times \mathbf{H} = \frac{E_\theta^2}{\eta} \tag{17-5}$$

For example 17-2 the power density at 1000 m and $\theta = 90°$ is

$$p = \frac{E_\theta^2}{\eta} = \frac{(0.02999)^2}{376.7}$$
$$= 2.4\ \mu\text{W/m}^2$$

Theoretically, the radiation resistance of a half-wave dipole is 73 Ω. This is true, however, only if it is constructed of infinitely thin wire and is located away from any reflecting surfaces. When the dipole is placed a quarter-wavelength above a ground plane it is also 73 Ω. As it is lowered toward the ground plane the radiation resistance decreases almost linearly to zero. At one-third of a wavelength above ground the resistance is about 95 Ω. It returns to 73 Ω at one-half wavelength. At two-thirds wavelength it drops to 58 Ω and returns to 73 Ω at three quarters wavelength. As the antenna is raised higher above ground, the deviation about 73 Ω becomes less and less. This variation is plotted in Figure 17-8. Physically, the variation in impedance is due to a reflected wave from the ground which induces currents in the antenna. The phase of these currents relative to the antenna currents produced by the excitation source varies depending on the height above ground.

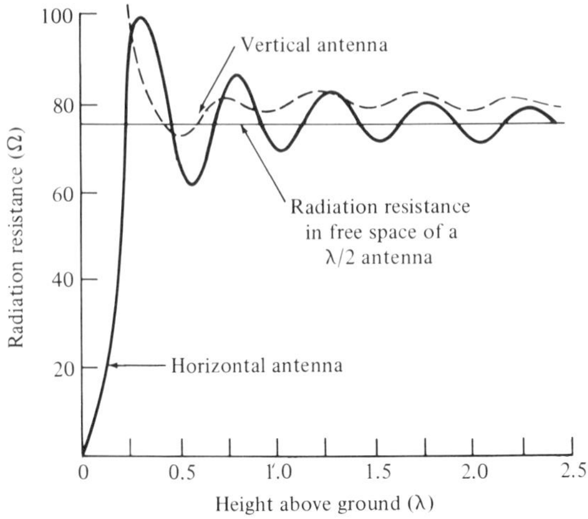

FIGURE 17-8. Radiation resistance of half-wavelength antennas at various heights.

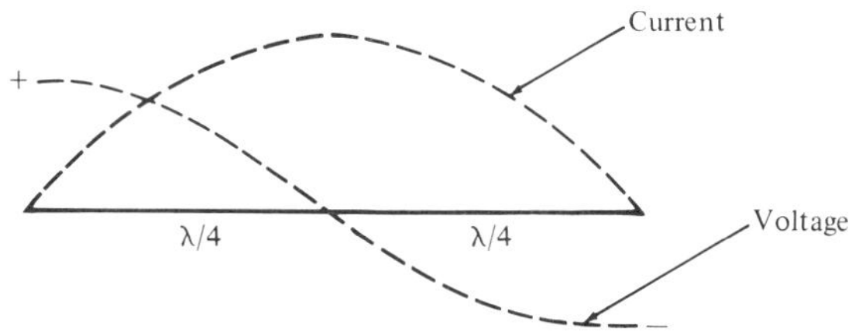

FIGURE 17-9. Voltage and current distribution on half-wavelength dipole.

The radiation efficiency depends not only on the radiation resistance but also on the ohmic resistance. Specifically, we can define the radiation efficiency as

$$\% \text{ efficiency} = \frac{R_r}{R_r + R_a} \times 100 \tag{17-6}$$

when R_a is the antenna ohmic resistance. We can think of the antenna in terms of an equivalent circuit where the radiation resistance and the ohmic resistance are in series. Thus at the terminals

$$Z_{\text{in}} = R_r + R_a \tag{17-7}$$

Generally speaking, the ohmic resistance for a dipole is usually no greater than 1 Ω. Besides the resistive component indicated by equation (17-7), an antenna may be reactive. It is capacitive if it is less than a half-wavelength and inductive if more than a half-wavelength or multiple of a half-wavelength.

We have discussed the half-wavelength antenna to this point without considering certain practical effects associated with the length. A phenomenon referred to as *end effect* must be considered with any single half-wavelength antenna. A dielectric effect exists at the end of the antenna which effectively lengthens it. This has the effect of making the antenna appear about 5% longer than it is. Thus, to make the antenna resonate properly, it is necessary to make the physical length approximately 95% of the electrical length.

The voltage and current distribution on a half-wavelength dipole are shown in Figure 17-9. The impedance varies along the antenna depending on the voltage–current distribution. The farther from the center, the higher the impedance, with the highest impedance being at the ends.

17-4 POLARIZATION

The polarization of the radiated electromagnetic wave is considered to be the direction of the electric field. Depending on the antenna type, the polarization may be linear, circular, or elliptical (Figure 17-10). The waveform in Figure 17-10 traveling in the positive z direction (out of the page) may be described by

$$E_y = E_2 \sin(\omega t - \beta z) \tag{17-8}$$

This expression describes the electric field produced by a linear antenna such as a half-wavelength dipole. The electric field is collinear with the antenna. More complex antenna structures may produce fields with both x and y components (Figure 17-10b). At a fixed value of z, the field vector E rotates as a function of time and the tip of the vector (resultant of x and y components) traces out an ellipse called the *polarization ellipse*.

The instantaneous total vector field **E** resulting from the x and y components may be expressed as

$$\begin{aligned} \mathbf{E} &= \mathbf{i}E_x + \mathbf{j}E_y \\ &= \mathbf{i}E_1 \sin(\omega t - \beta z) + \mathbf{j}E_2 \sin(\omega t - \beta z + \phi) \end{aligned} \tag{17-9}$$

where **i** and **j** are unit vectors in the direction of the x and y axes, respectively, and ϕ is the time phase angle by which E_y leads E_x. Here the x component is taken as the reference for the time phase angle. That the field vector **E** traces out an ellipse is easily demonstrated by fixing the value z. Take $z = 0$. The x and y components are then

$$E_x = E_1 \sin \omega t \tag{17-10}$$

$$E_y = E_2 \sin(\omega t + \phi) \tag{17-11}$$

First we expand equation (17-11) and then use (17-10) to eliminate ωt, the independent variable. That is,

$$E_y = E_2(\sin \omega t \cos \phi + \cos \omega t \sin \phi) \tag{17-12}$$

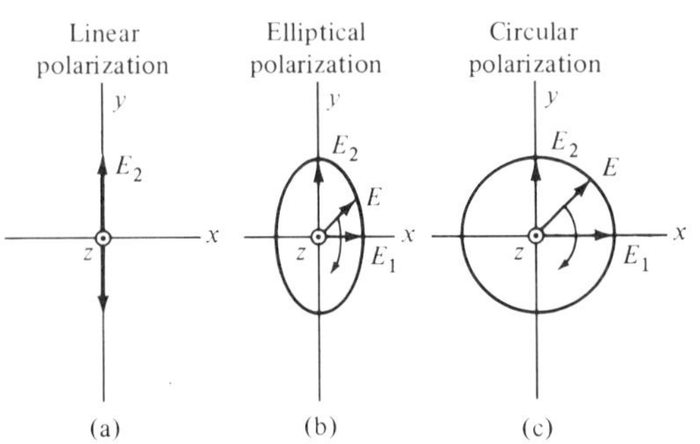

FIGURE 17-10. Linear, elliptical, and circular polarization.

and

$$\sin \omega t = \frac{E_x}{E_1}$$

We may also write

$$\cos \omega t = \sqrt{1 - \sin^2 \omega t} = \sqrt{1 - \left(\frac{E_x}{E_1}\right)^2}$$

Substituting these relations into equation (17-12) and rearranging results in

$$\frac{E_x^2}{E_1^2} - \frac{2E_xE_y \cos \phi}{E_1E_2} + \frac{E_y^2}{E_2^2} = \sin^2 \phi \qquad (17\text{-}13)$$

Finally, dividing by $\sin^2 \phi$, we have

$$aE_x^2 - bE_xE_y + cE_y^2 = 1 \qquad (17\text{-}14)$$

where

$a = 1/(E_1^2 \sin^2 \phi)$
$b = (2 \cos \phi)/(E_1E_2 \sin^2 \phi)$
$c = 1/(E_2^2 \sin^2 \phi)$

Equation (17-14) has the general form of an ellipse. Furthermore, if $\phi = 90°$ and $E_1 = E_2$, it is easily shown that this expression reduces to an equation describing a circle (Figure 17-10c).

As a final comment, the polarization may be right circular or left circular, depending on which field component, x or y, leads the other in phase. If an observer stands "behind" the radiation, clockwise rotation of the **E** field vector is termed *right circular* and counterclockwise is *left circular*. On some antenna types, the "feed" structure determines left or right polarization, and for other antenna types it is determined solely by the antenna structure.

Television and FM broadcasts are made with horizontally polarized antennas. Thus the receiving antenna must be mounted horizontally. Most transmissions below 2 MHz are vertically polarized. AM broadcast transmitters (550 to 1650 kHz) have vertically polarized antennas. Most transmitters below 500 kHz have horizontal wire antennas. However, at these frequencies, there may be a significant vertical component. Above 30 MHz, most communications, with the exception of TV and FM, are vertically polarized. Due to reflections from the ionosphere or from buildings, the polarization of the received wave may be

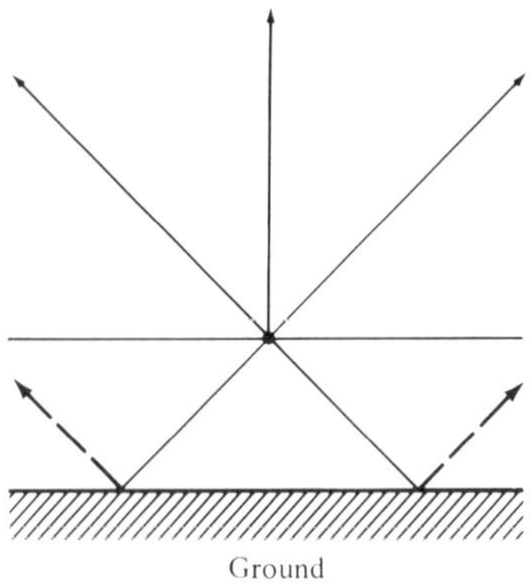

FIGURE 17-11. Reflected energy near ground.

altered. For maximum signal reception, stations located some distance away must have an antenna oriented to this new polarization.

17-5

ANTENNA DIRECTIVITY

We saw (Figure 17-7) that the half-wavelength antenna radiates energy at right angles to the wire. Thus, if the antenna is oriented in a north-south direction, it radiates in an east-west direction. As the radiation pattern for an antenna that is receiving energy is the same as the pattern when the antenna is transmitting energy, to receive from an east-west direction, the antenna must be aligned in a north-south direction.

The antenna pattern illustrated in Figure 17-7 is accurate only if the antenna is in free space. When located near the ground, reflections will alter the antenna pattern (Figure 17-11). Vertical radiation patterns for a horizontal antenna at various heights above ground are shown in Figure 17-12. Note that additions and subtractions of the radiated and reflected fields result in a distinct lobing structure.

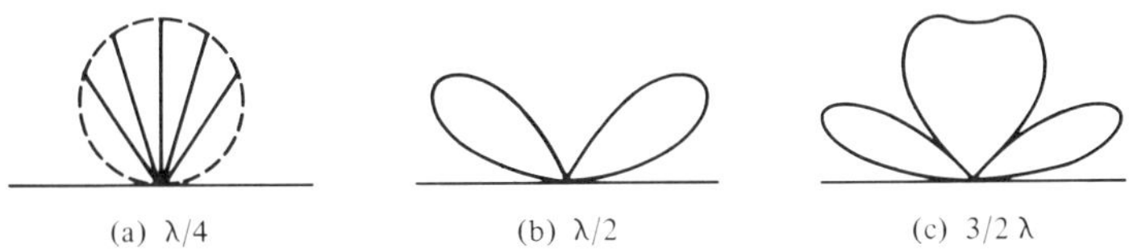

FIGURE 17-12. Vertical radiation patterns of a horizontal antenna located at different distances above ground.

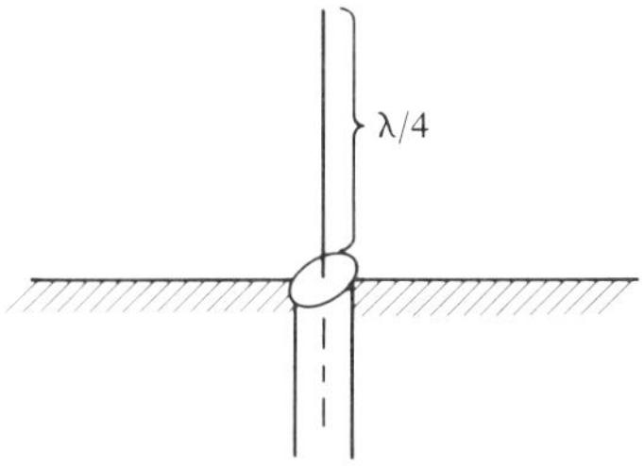

FIGURE 17-13. Quarter-wavelength antenna.

17-6

QUARTER-WAVELENGTH ANTENNA

For medium-frequency and low-frequency transmissions, the antenna most commonly used is the quarter-wavelength vertical antenna (Figure 17-13). This antenna consists of a quarter-wavelength wire or mast extending above a ground plane. In operation, we can visualize a quarter-wavelength "image" antenna below the ground plane. Thus the quarter-wavelength antenna will resonate as a half-wavelength antenna. The radiation patterns for a vertical quarter-wavelength antenna and a vertical half-wavelength antenna are shown in Figure 17-14.

The radiation resistance of a quarter-wavelength antenna between ground and the lower end of the wire (Figure 17-13) is, theoretically, 36.5 Ω. This is exactly half that of the dipole. This antenna is often fed by a 50-Ω coaxial cable.

EXAMPLE 17-3

Calculate (a) the VSWR on a 50-Ω line feeding a quarter-wavelength vertical antenna and (b) the impedance of a quarter-wavelength matching section.

SOLUTION

(a) Using equation (16-52), we have

$$\text{VSWR} = \rho = \frac{1 + |\Gamma_R|}{1 - |\Gamma_R|}$$

and

$$\Gamma = \frac{Z_R - Z_0}{Z_R + Z_0} = \frac{36.5 - 50}{36.5 + 50} = -0.16$$

$$\rho = \frac{1 + 0.16}{1 - 0.16} = 1.38$$

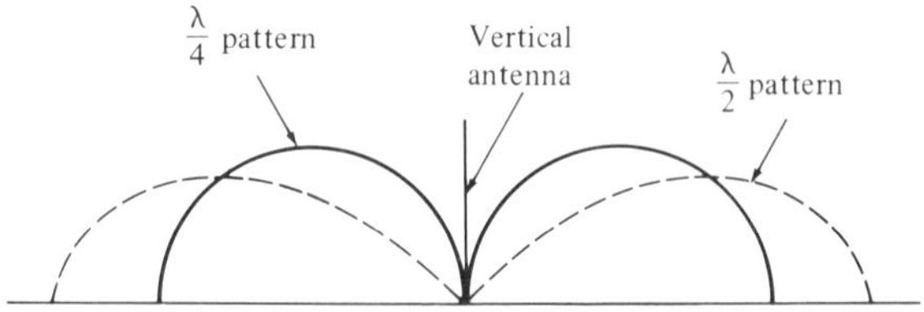

FIGURE 17-14. Vertical radiation patterns for a half- and quarter-wavelength vertical antenna.

$$\text{(b) } Z_0 = \sqrt{Z_{max} Z_{min}} \text{ [equation (16-59)}$$
$$= \sqrt{(50)(36.5)} = 42.72\ \Omega$$

In this example a voltage standing-wave ratio of 1.38 exists on the line. A quarter-wavelength section with characteristic impedance $Z_0 = 42.72\ \Omega$ would achieve an impedance match and permit maximum energy transfer between the line and the antenna. As an alternative, the single-stub or double-stub technique discussed in Chapter 16 could be used.

The quarter-wavelength antenna is often referred to as a *whip antenna*. When mounted on top of a car, the roof of the car serves as the ground plane. If it is mounted near the edge of the car, the ground and roadway in conjunction with the car make up the ground plane. Quarter-wavelength radials can also be used to make up the ground plane (Figure 17-15). This technique makes the ground plane portable with the antenna.

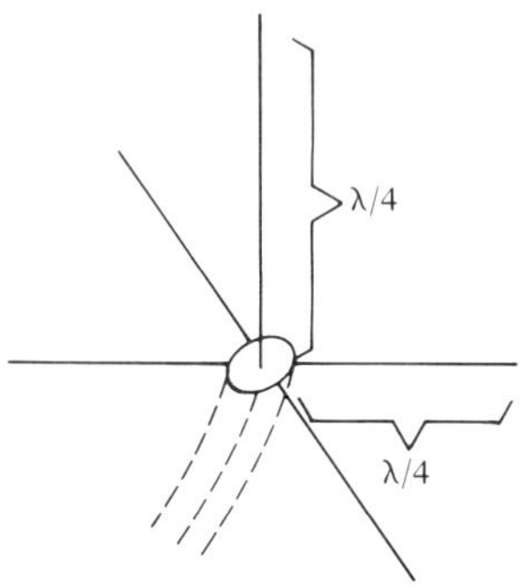

FIGURE 17-15. Quarter-wavelength antenna with radials forming the ground plane.

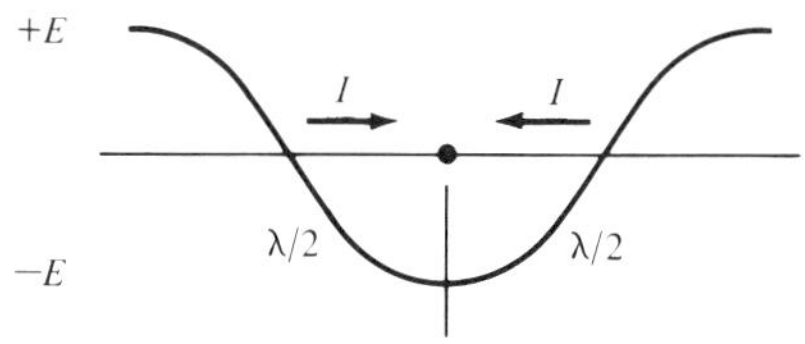

FIGURE 17-16. Full-wavelength antenna.

17-7

FULL-WAVELENGTH ANTENNA

A full-wavelength antenna is shown in Figure 17-16. At resonance the current flow is from the ends to the center during one-half of the excitation cycle and away from the center during the other half-cycle. A voltage standing wave is set up such that the maximum occurs at the center of the antenna. Since the polarity of the current in each half is 180° out of phase, the radiated field is zero at right angles to the wire at the center. The radiation pattern is shown in Figure 17-17. Note that the maximum radiation is at a 45° angle to the wire. This occurs because radiation from mirror-image points such as A, A' arrive at point P' in phase.

A $1\frac{1}{2}$-wavelength antenna has radiation cancellation at right angles to the center of the wire for two of the half-wavelengths. The remaining half-wavelength is free to radiate at right angles. The overall lobing structure becomes more complex. The radiation pattern for an antenna that is eight half-wavelengths long is shown in Figure 17-18. Note that a lobe exists for each half-wavelength.

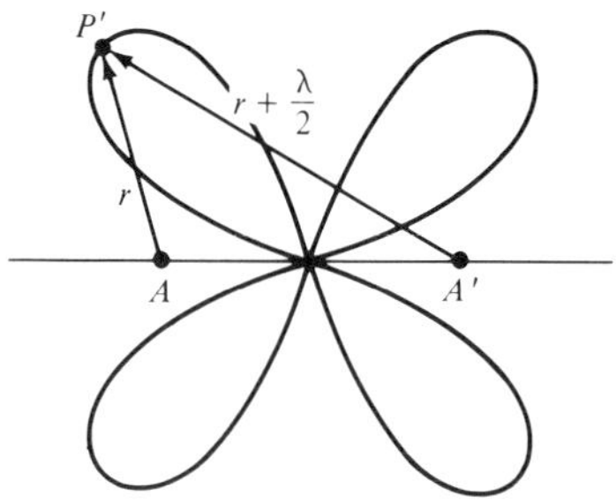

FIGURE 17-17. Radiation pattern of a full-wavelength antenna located in free space.

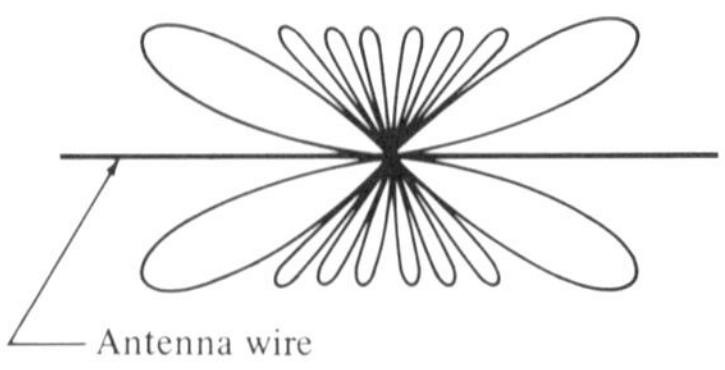

FIGURE 17-18. Radiation pattern of an antenna four wavelengths long.

17-8

ANTENNA FEED STRUCTURE

Methods of feeding energy to the antenna vary. We previously discussed using a 50-Ω coax to feed the whip antenna. The dipole can be cut at the center and fed by a 50- to 73-Ω line. This method is known as *center feeding* or *current feeding*. This name derives from the fact that current is maximum at the center of a dipole. A full-wavelength antenna cut at the center has an impedance on the order of 2000 to 3000 Ω. The impedance of a two-wire line may be calculated by

$$Z = 276 \log \frac{d}{r} \tag{17-15}$$

where

Z = impedance (ohms)
d = separation distance, center to center of the conductors
r = radius of conductors

Now, a two-wire line with an impedance greater than 700 or 800 Ω is not practical. Matching can be achieved, however, by deliberately producing standing waves on a line with resultant high- and low-impedance points (Figure 17-19). The high impedance of the antenna is repeated a half-wavelength down the line. Impedance matching is achieved using the parallel tuned circuit.

EXAMPLE 17-4

Calculate the characteristic impedance of a two-wire line consisting of two No. 14 wires (diameter 0.064 in.) held 1 inch apart center to center.

SOLUTION

$$Z = 276 \log \frac{1}{0.032} = 412.57\ \Omega$$

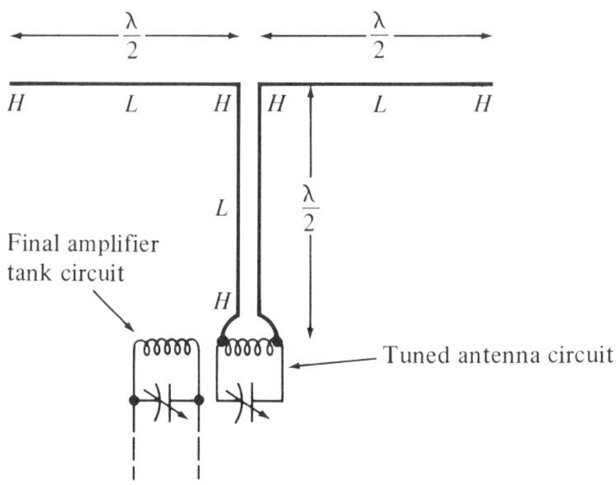

FIGURE 17-19. Full-wavelength antenna fed with a tuned circuit. H and L indicate high- and low-impedance points.

Many additional schemes are available for antenna feed structures. The interested reader is referred to the *Radio Engineers Handbook* for additional information. We see that the essential problem is impedance matching to achieve maximum transfer of energy between the line and the antenna. Also, in common with other reactive circuits, the impedance of the antenna's terminal impedance will vary with frequency. Consequently, the impedance-matching network together with the antenna must have a bandwidth wide enough for the signal frequencies of interest. Broadband antennas have received considerable emphasis in development.

17-9 ARRAYS

If two half-wavelength elements are spaced one half-wavelength apart and one element is excited, the second element has a voltage induced in it due to the radiated field from the first element. This element is said to be parasitically excited. By Lenz's law, the current in the second element will be 180° out of phase with the current in the excited element (Figure 17-20). At the parasitic

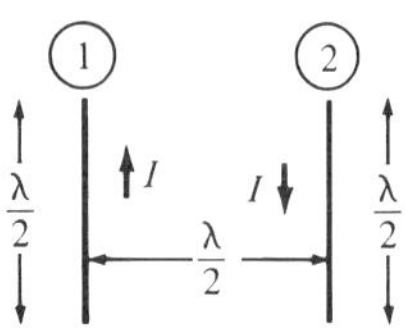

FIGURE 17-20. Two-element parasitic array.

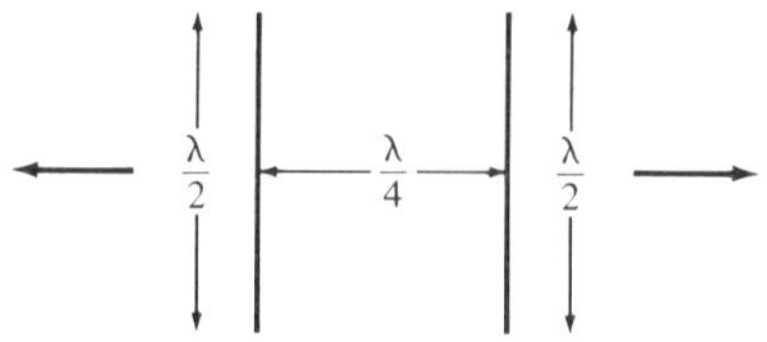

FIGURE 17-21. Array with quarter-wavelength spacing.

element the current produces a radiated wave that tends to cancel the wave from the driven element. This reduces the radiation in the plane of the page. At right angles to the plane of the page, the radiation from the elements tends to add, so that the radiation is in this direction. Since the radiated wave cannot be as strong as the original radiated wave, there will not be a complete null in any direction. If the spacing between the elements is reduced to a quarter-wavelength, the radiated wave travels 90° to the parasitic element, reverses phase 180°, and returns 90° to the excited element. Thus the returning wave is in phase with the original radiated wave and the maximum radiation is to the left and right in the plane of the page (Figure 17-21). This antenna is said to "end fire."

In Figure 17-21, the parasitic element acts as both a reflector and a director. The two-element array exhibits about 5 dB gain in both the forward and backward directions. The antenna in Figure 17-21 has the greatest forward gain when the parasitic element is approximately 0.2λ behind the driven element. With this spacing, the center of the driven element has an impedance of about 40 Ω. As more elements are added, the impedance decreases.

17-9.1 Yagi Antenna

A three-element yagi antenna is shown in Figure 17-22. It consists of a driven element, a reflector, and a director. If a parasitic element one half-wavelength

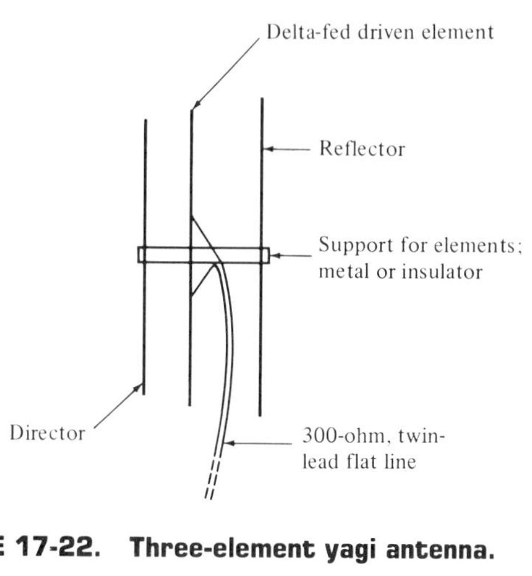

FIGURE 17-22. Three-element yagi antenna.

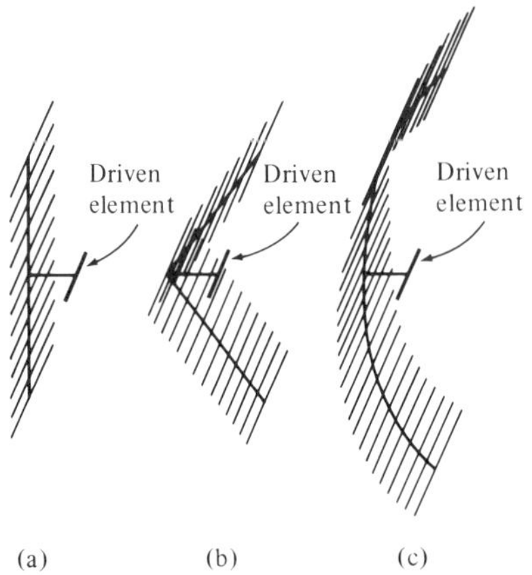

FIGURE 17-23. Beam antennas: (a) plane reflector; (b) 90° corner reflector; (c) parabolic reflector.

long is placed within 0.1λ of the driven element, the induced voltage and current cause it to act more as a director than as a reflector. This produces approximately 6 dB of gain in the forward direction. For the yagi shown in Figure 17-20, if the reflector is 1.05 (λ/2) and the director is 0.95 (λ/2), the gain is approximately 8 dB. The radiation resistance is on the order of 10 Ω.

Variations of parasitic-type antennas are shown in Figure 17-23. The plane reflector antenna uses a plane surface about one wavelength behind the driven element. The gain is approximately 7 dB. If the plane is bent into a 90° corner with each side approximately two wavelengths long, the antenna becomes move directive and has a gain of about 12 dB. If the corner reflector is now made to take on a parabolic shape, a beamwidth of 2 or 3° is attainable with gains in excess of 30 dB.

Antennas of this type are used in radars. The parabolic reflector can be extended to form the familiar parabolic "dish" antennas used with deep-space probes and satellite communications. These antennas feature high gains with narrow beamwidths.

17-10

LOOP ANTENNA

A loop antenna is often used at low frequencies. The reason for this is that its smaller size at low frequencies may make it the only feasible solution. The loop itself may be round, square, and so on. Essentially, the loop forms the inductance, L, of a resonant LC circuit (Figure 17-24).

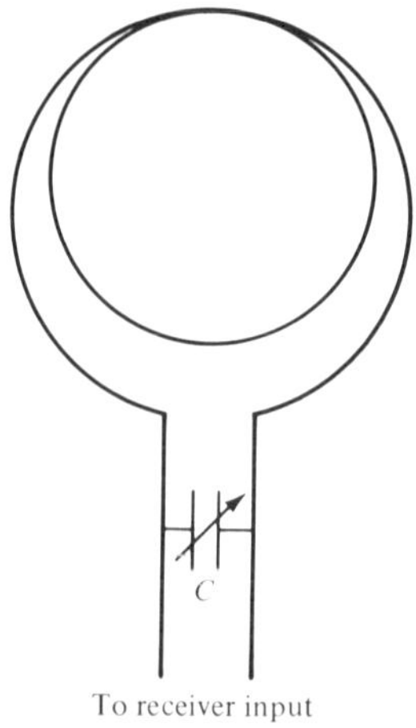

FIGURE 17-24. Loop antenna.

The radiation efficiency of a loop antenna is very low. Consequently, it is rarely used as a transmitting antenna. The loop antenna often finds use as a direction finder. As the loop is rotated 360°, there are two points where little or no voltage is induced and intermediate points in between where maximum voltage is induced.

The efficiency of the antenna can be raised by constructing the antenna around a ferrite rod loop. Essentially, this reduces the reluctance of the magnetic circuit. A typical ferrite loop antenna is shown in Figure 17-25. The maximum voltage

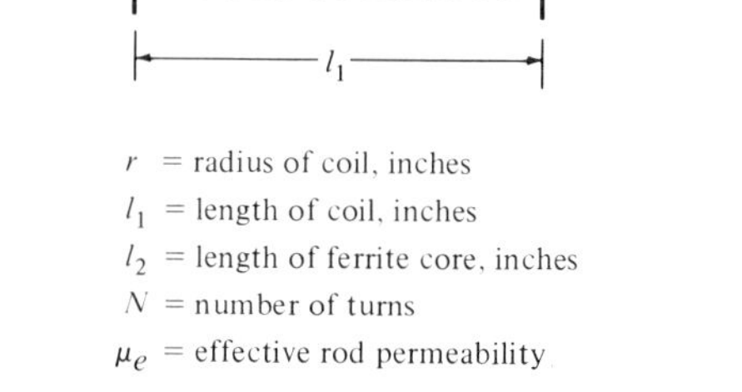

r = radius of coil, inches
l_1 = length of coil, inches
l_2 = length of ferrite core, inches
N = number of turns
μ_e = effective rod permeability

FIGURE 17-25. Ferrite-rod loop antenna.

induced in the loop, E, is

$$E = 13.55 \times 10^{-12} \mu_e e N A f F_A \qquad \text{volts} \tag{17-16}$$

where

e = field strength (volts/meter)
μ_e = effective rod permeability
N = number of turns
A = area of loop, πr^2 (in.^{2})
f = frequency (hertz)
F_A = averaging factor

The averaging factor, F_A, is a function of loop configuration. It may be approximated by

$$F_A = 1 - 0.17\left(\frac{l_1}{l_2}\right) \tag{17-17}$$

Since it is necessary to tune the loop to maximize the voltage induced in it, two possibilities exist: series or parallel tuning (Figure 17-26).

17-10.1 Parallel Tuning

The equivalent circuit for parallel tuning is shown in Figure 17-26. From the figure the circuit transfer function is

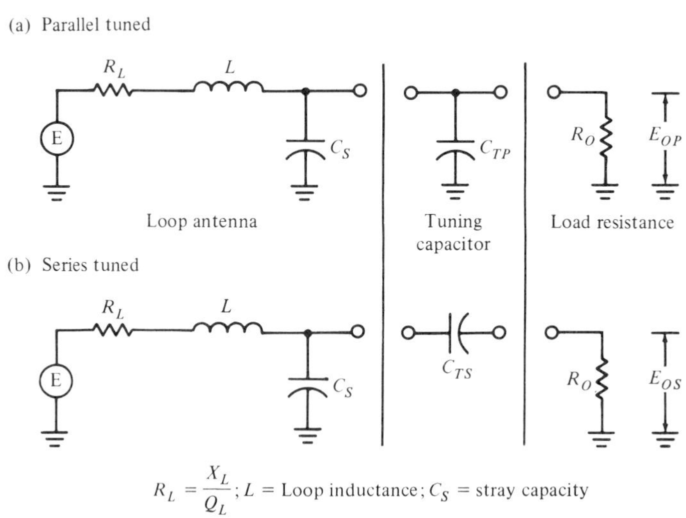

FIGURE 17-26. Loop-antenna equivalent circuit.

$$H(j\omega) = \frac{E_{op}}{E} = \frac{I(R_1 + jx_1)}{E} = \frac{R_1 + jx_1}{R_2 + jx_2} \tag{17-18}$$

where the terms are defined on the figure.

EXAMPLE 17-5

For the following design, calculate (a) the induced loop voltage and (b) the circuit gain.

$R_0 = 10{,}000\ \Omega$	$l_1 = 7.5$ in.
$f = 20{,}000$ Hz	$l_2 = 8.0$ in.
$L = 0.037$ H	$r = 0.625$ in.
$Q_L = 50$	$N = 640$ turns
$C_s = 100 \times 10^{-12}$ F	$\mu_e = 17$
$e = 10 \times 10^{-6}$ V/m	$F_A = 0.85$ [calculated using equation (17-17)]

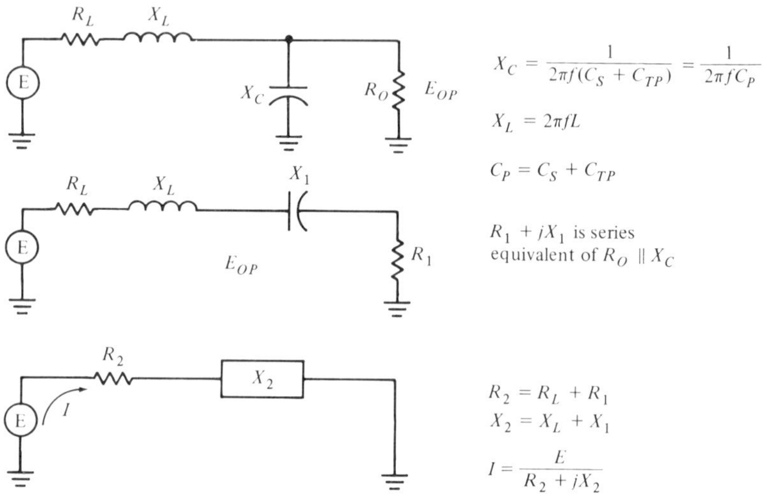

SOLUTION

(a) $E = 13.55 \times 10^{-12} \mu_e e N A f F_A$

$= (13.55 \times 10^{-12})(17)(10 \times 10^{-6})(640)(\pi)(0.625)^2(2 \times 10^4)(0.85)$

$= 0.0308 \times 10^{-6}$ V

(b) At resonance $X_L = X_C$ or $\frac{1}{\omega C_P} = X_L$:

$$C_P = \frac{1}{\omega X_L} = \frac{1}{(2\pi)(2 \times 10^4)(2\pi)(2 \times 10^4)(0.037)}$$

$$= 1710 \times 10^{-12}\text{ F}$$

$$C_p = C_{tp} + C_s \Rightarrow C_{tp} = C_p - C_s$$

$$= 1710 - 100 = 1610\text{ pF}$$

$$R_1 + jx_1 = R_0 \| X_C \Rightarrow R_1 = \frac{X_c^2 R_0}{R_0^2 + X_C^2} \qquad X_1 = -\frac{X_c R_0^2}{R_0^2 + X_C^2}$$

$$R_1 = \frac{(4580)^2(10^4)}{(10^4)^2 + (4580)^2} = 1733$$

$$X_2 = -\frac{(4580)(10^4)^2}{(10^4)^2 + (4580)^2} = -3786\ \Omega$$

$$R_2 = R_L + R_1 = \frac{\omega L}{Q} + R_1$$

$$= \frac{4580}{50} + 1733 = 1824\ \Omega$$

$$X_2 = X_L + X_1 = 4580 - 3786 = 795\ \Omega$$

The gain is computed using equation (17-18):

$$\frac{E_{\text{op}}}{E} = \left|\frac{R_1 + jx_1}{R_2 + jx_2}\right| = \frac{|1733 - j3786|}{|1824 + j795|} = 2.091$$

17-10.2 Series Tuning

The antenna may be series tuned (Figure 17-27). For this circuit the gain is given by

$$\frac{E_{os}}{E} = R_0 \frac{R_2 + jx_3}{(R_3 + jx_4)(R_0 + jx_{TS})} \tag{17-19}$$

where the parameters are defined in Figure 17-27. Circuit gain is maximized by proper choice of the tuning capacitor, C_{TS}.

We note from equation (17-16) that the induced voltage in the loop is a function of frequency. This is exhibited as a high-pass effect. From the tuned equivalent circuits we see that a low-pass effect is also exhibited. The combined

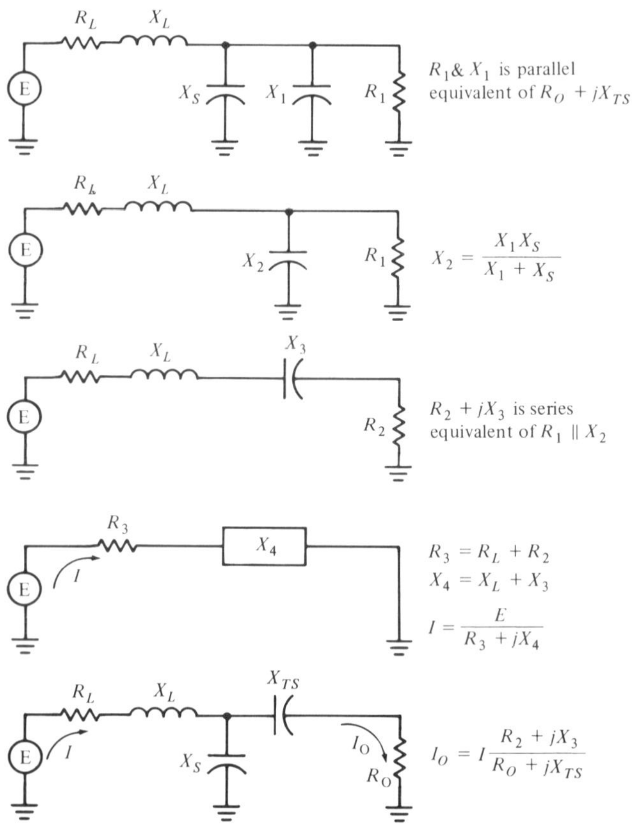

FIGURE 17-27. Series tuned equivalent circuit.

effect is a bandpass characteristic. Thus the circuit Q can be calculated by equation (2-17):

$$Q = \frac{f_0}{f_2 - f_1} \tag{17-20}$$

For maximum sensitivity, the load, R_0, should determine the bandwidth, Q, of the circuit. A transformer can be used to raise or lower the load impedance as needed.

17-11

ANTENNA CHART FOR SYSTEM DESIGNERS

Actual antenna design is generally left to specialists. Specialized equipment is needed and many hours of tedious work may be involved with the simplest of structures. However, antenna parameters may be tabulated so that the systems designer can intelligently select an antenna for a particular design. Antennas may be classified as linear radiators, apertures, arrays, and traveling-wave types. The chart at the front of the book for a few well-known antennas represents the foregoing types. For each antenna type the chart gives the antenna name, its physical size in wavelengths, a line drawing superimposed on the coordinate axes, and the impedance R in ohms. At the resonant frequency, f_r, the half-power bandwidth (3 dB) is given in percent. The gain in decibels above an isotropic radiator is given and also relative to a half-wavelength dipole. An isotropic radiator is, by definition, an ideal antenna that radiates equally well in all directions with unity gain. Also given is the polarization and a set of far-field patterns for each of the three orthogonal planes, of the X-Y-Z axis system.

17-12 RADIO-WAVE PROPAGATION

We indicated the nature of radio-wave propagation in Figure 17-3. The power radiated from an antenna is ordinarily spread over a large area. Consequently, the power at the receiver may be only a small percentage of the radiated power. The ratio of radiated power to received power is called the *transmission loss* and may be large as 10^{15} to 10^{20} (150 to 200 dB).

The transmission loss determines whether or not the received signal is useful. Each communication system is designed for maximum transmission loss. From equation (15-10), the free-space transmission ratio at a distance, d, is given by

$$\frac{P_r}{P_t} = \left(\frac{\lambda}{4\pi d}\right)^2 G_t G_r \tag{17-21}$$

where

P_r = received power
P_t = transmitted power
λ = wavelength
G_t (or G_r) = power gain of transmitting or receiving antenna

A nomogram for free-space loss between isotropic antennas is given in Figure 17-28. Using results from equation (15-9), equation (17-21) may be expressed as

$$\frac{P_r}{P_t} = \frac{A_t A_r}{(\lambda d)^2} \tag{17-22}$$

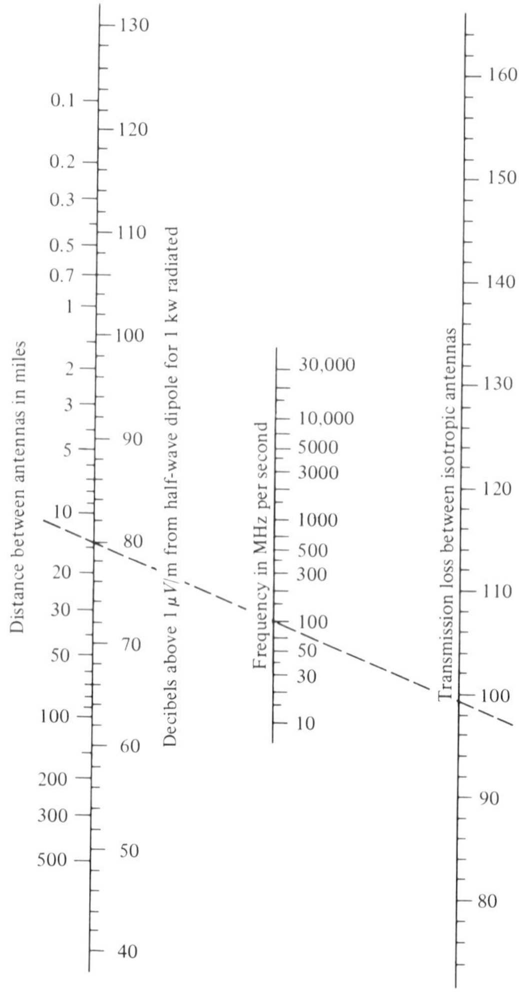

FIGURE 17-28. Free-space transmission.

where A_t and A_r represent the effective area of the transmitting and receiving antennas, respectively.

The concept of free-space transmission assumes that the atmosphere is perfectly uniform and nonabsorbing and that there are no reflections from any object. In practice it is necessary to consider the effect of the atmosphere, the ionosphere, and the earth. Free-space field intensity, E_0, is given by

$$E_0 = \frac{\sqrt{30 P_t G_t}}{d} \quad \text{volts/meter} \qquad (17\text{-}23)$$

where d is in meters and P_t is in watts.

17-12.1 Line-of-Sight Transmission

The generation and propagation of radio waves is modified by the ground. The practical effect of this is that the received power is usually less than would be received in free space. The effect of a plane earth on the propagation of radio wave is given by

$$\frac{E}{E_0} = \underbrace{1}_{\text{direct wave}} + \underbrace{Re^{j\Delta}}_{\text{reflected wave}} + \underbrace{(1 - R)Ae^{j\Delta}}_{\text{``surface wave''}} + \underbrace{\cdots}_{\text{induction field and secondary effects of the ground}} \tag{17-24}$$

where

R = reflection coefficient of the ground
A = surface-wave attenuation factor
Δ = $4\pi h_1 h_2/\lambda d$
$h_{1,2}$ = antenna heights measured in same units as the wavelengths and distance

Both the reflection coefficient, R, and the attenuation constant, A, vary with the electrical constants of the ground. In most practical situations, however, considerable simplification of equation (17-24) is possible. For near grazing paths, R is approximately -1. The factor A can be neglected as long as both antennas are elevated more than a wavelength above ground (5 to 10 wavelengths above seawater). With these conditions we can neglect the effect of polarization and ground constants and equation (17-24) reduces to

$$\left|\frac{E}{E_0}\right| = \sqrt{\frac{P_r}{P_0}} = 2 \sin \frac{\Delta}{2} = 2 \sin \frac{2\pi h_1 h_2}{\lambda d} \tag{17-25}$$

where P_0 is the received power expected in free space.

Equation (17-25) is the sum of the direct and reflected waves, and shows the lobe structure as the signal oscillates around the free-space value. In many radio applications (air to ground excepted) the principal interest is in the lower part of the first lobe. This occurs where $\Delta/2 < \pi/4$. For this situation $\sin(\Delta/2) \approx \Delta/2$ and the transmission loss over plane earth is given by

$$\frac{P_r}{P_t} = \left(\frac{\lambda}{4\pi d}\right)^2 \left(\frac{4\pi h_1 h_2}{\lambda d}\right)^2 G_t G_r = \left(\frac{h_1 h_2}{d^2}\right) G_t G_r \tag{17-26}$$

This expression is independent of frequency and is plotted in Figure 17-29. The results obtained from Figure 17-28 for transmission loss are not valid if they

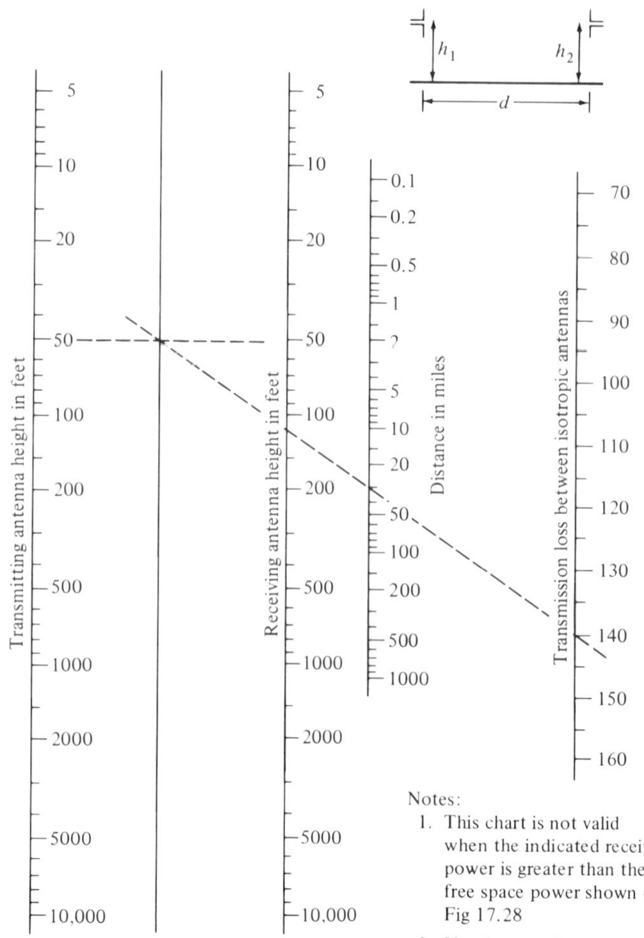

Notes:

1. This chart is not valid when the indicated received power is greater than the free space power shown on Fig 17.28
2. Use the actual antenna height or the minimum effective height shown on Fig. 17.30, whichever is the larger

FIGURE 17-29. Transmission loss over plane earth.

are greater than the free-space value obtained from Figure 17-29. This means that the approximation for Δ leading to equation (17-26) is not valid. Because of soil conditions the effective height of the antenna is not necessarily equal to the physical height. The preceding results for transmission loss were derived based on optical concepts. For antenna heights of less than a few wavelengths, approximate results can be obtained by using h_1 (or h_2) as the larger of either the actual antenna height or the minimum effective antenna height (Figure 17-30).

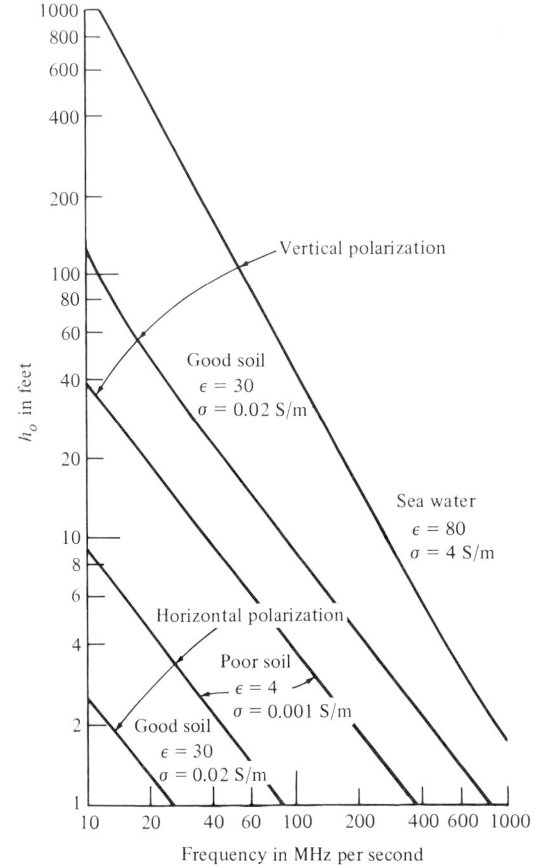

FIGURE 17-30. Minimum effective antenna height.

17-12.2 Fresnel Zones

We noted previously from equation (17-25) that the received field intensity oscillates around the free-space value as the antenna heights are increased. A maximum will occur when the difference path length between the direct and ground reflected wave is one half-wavelength. The signal maxima have a magnitude of $1 + |R|$ and signal minima of $1 - |R|$. Thus the received signal can be twice as large (6 dB) as the free-space value.

Often, the amount of clearance is described in terms of Fresnel zones. All points for which a wave could be reflected with a path difference of one half-wavelength form the boundary of the first Fresnel zone. The boundary of the nth Fresnel zone consists of all points for which the path difference is $n/2$ wavelengths. The nth Fresnel zone clearance H_n at any distance d_1 is given by

$$H_n = \sqrt{n\lambda \frac{d_1 d_2}{d_1 + d_2}} \qquad (17\text{-}27)$$

The first Fresnel zone, H_1, is indicated in Figure 17-31.

17-12.3 Diffraction over a Smooth Spherical Earth

In the troposphere region, the temperature, pressure, and water vapor cause variations in the dielectric constant. This changes the index of refraction, which causes a downward bending of the wave in the vertical plane. The effect of diffraction around the earth's curvature makes possible transmission beyond the line of sight. The magnitude of the loss caused by obstructions (earth) increases as the distance or wavelength is increased. Two techniques are often employed when plotting profiles of the microwave beam and the earth when investigating microwave paths. In one method, the microwave beam is considered to be straight and the curvature of the earth is changed to compensate for beam bending. The change in the earth's radius is denoted by, K, where

$$K = \frac{\text{effective earth radius}}{\text{actual earth radius}} \qquad (17\text{-}28)$$

This method is illustrated in Figure 17-32. In the second method, a flat earth is used with the microwave beam having a curvature of $K \times$ the actual earth radius (Figure 17-33). The most commonly used value for K is $\frac{4}{3}$. This value of K is based on a "standard atmosphere," that is, a stable atmosphere during daylight.

The "earth bulge" indicated in Figure 17-32 is given by

$$h = \frac{d_1 d_2}{1.5K} \quad \text{feet} \qquad (17\text{-}29)$$

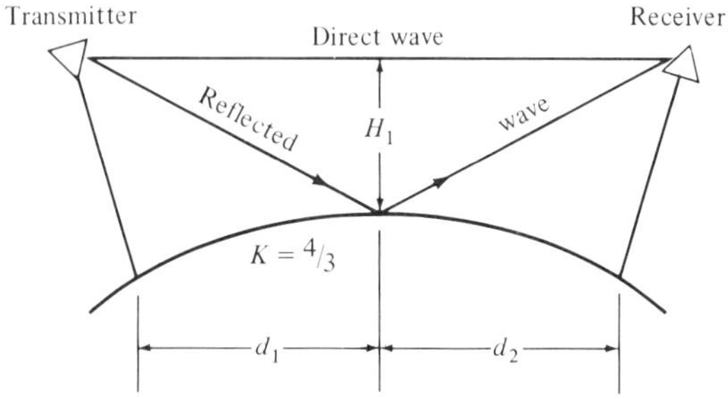

FIGURE 17-31. First Fresnel zone around direct wave.

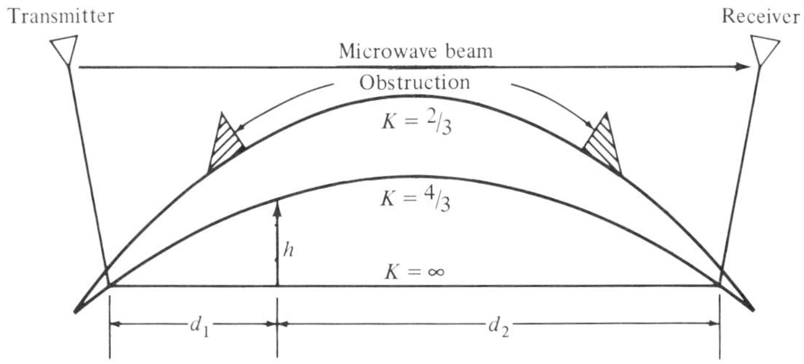

FIGURE 17-32. Straight-line microwave beam.

where d_1 and d_2 are in miles. The criterion used to determine whether the earth is sufficiently removed from the radio line-of-sight ray to allow mean free-space propagation conditions to apply is to have the first Fresnel zone clear all obstacles in the path of the ray. Figure 17-34 shows the effect of path clearance on radio transmission.

17-13

MICROWAVE SYSTEM DESIGN

In this section we discuss techniques for microwave path design. Basically, the path is initially designed assuming free-space propagation and then the effects of terrain on the path are determined by plotting path profiles.

The goal of a microwave path designer is to provide the most reliable path possible considering the constraints on the design. For the designer, this reliability or system quality is usually determined by the number of times the microwave signal "fades" to the point of being useless and the total time the system is useless during any given interval. To provide the best performance, the system designer normally provides as much *fade margin* as possible. Fade margin is the amount of attenuation that the path can tolerate and still provide

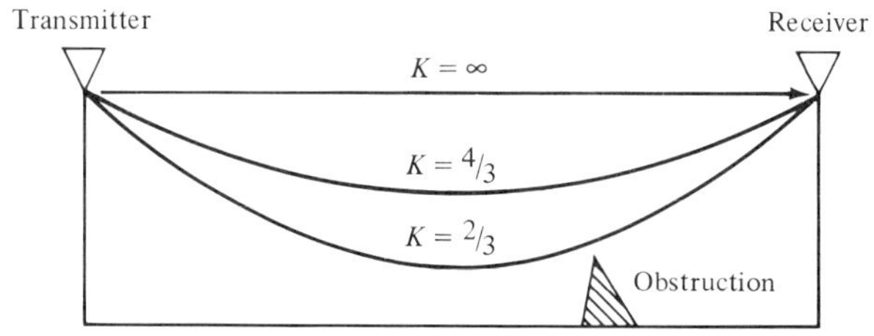

FIGURE 17-33. Microwave beams compensated for a flat earth.

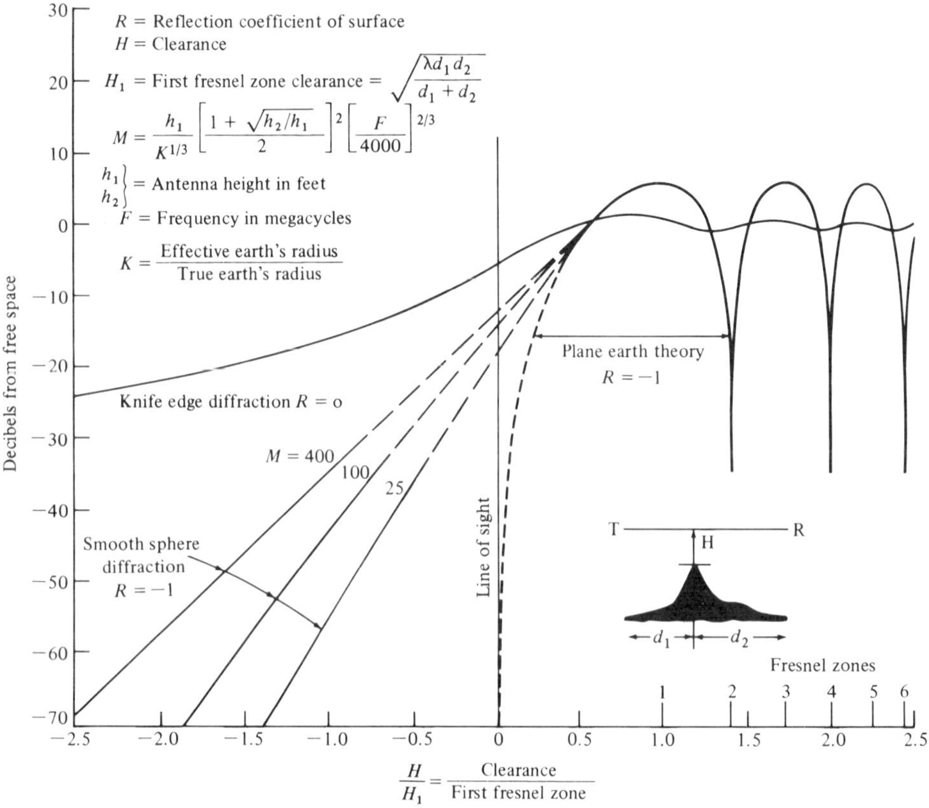

FIGURE 17-34. Transmission loss versus clearance.

an adequate signal to the receiver. More precisely, fade margin is defined as the difference between the nominal receiver level and the receiver threshold.

The fade margin of a line-of-sight path is determined as shown in Figure 17-35. The receiver threshold is a characteristic of the receiver and is fixed. The nominal receiver level is derived by adding system gains (antenna gains) and then subtracting system losses (coaxial or waveguide losses, connector losses, and free-space loss) from the transmitter output power. All system gains and losses are constant as long as the path is in a free-space propagation mode (i.e., it is not affected by blockage or reflections off adjacent bodies). When non-free-space propagation is involved, we must determine the effects on system calculations.

The most obvious change to path losses is caused by obstructions. These obstructions can take the form of hills, trees, buildings, or other structures. Also, as mentioned previously, they can result from the curvature of the earth's surface (Figure 17-36). The first practical step, then, is to raise the antennas high enough to clear all obstructions, including earth curvature.

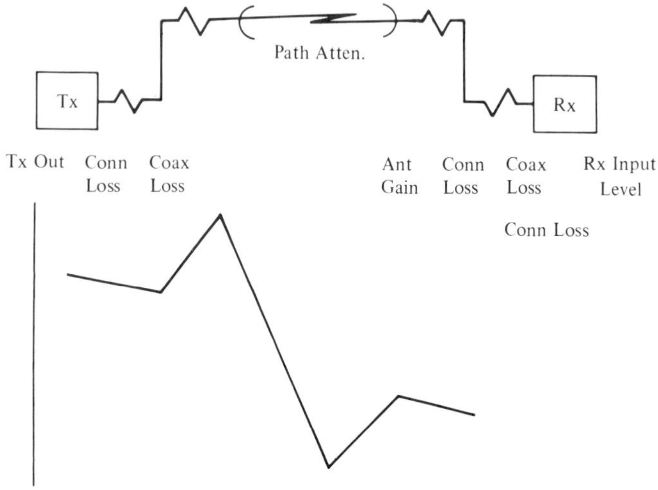

Tx Out − Coax Loss + Ant Gain − Path Loss + Ant. Gain − Coax Loss = Rx Input Level

Rx Input Level − Rx Threshold = Fade Margin

Free Space Path Loss = 96.6 + 20 Log_{10} FMHz + 20 Log Distance Miles

Tx = transmitter
Rx = receiver

FIGURE 17-35. Path gains/losses, with free-space path loss.

Figure 17-37 shows a nonobstructed path over smooth earth with the earth's surface drawn to scale. The horizontal axis represents the distance between the towers in miles, while the scale of the vertical axis is in feet to give an exaggerated representation of the heights. The RF path is drawn as a straight line, indicating that radio waves travel in a straight line. However, instead of a straight line, radio waves tend to "bend" over the horizon normally as shown by the RF path $K = \frac{4}{3}$ in Figure 17-38. The bending is caused by the refractivity of the atmosphere, which in turn is related to such atmospheric conditions as temperature, pressure, and water vapor content.

The "K" in $K = 1$ and $K = \frac{4}{3}$ is defined as *effective earth radius*. As used by radio engineers, this indicates that if we assume the radio wave travels in a straight line, we must effectively "change" the earth's curvature to observe the known ability of the waves to be received over the horizon. For years, $K = \frac{4}{3}$ was used in the design of most radio paths, and for mobile radio system design it is still used because it is an "average" value. However, on a given path

FIGURE 17-36. Obstruction losses, with earth's surface drawn to scale.

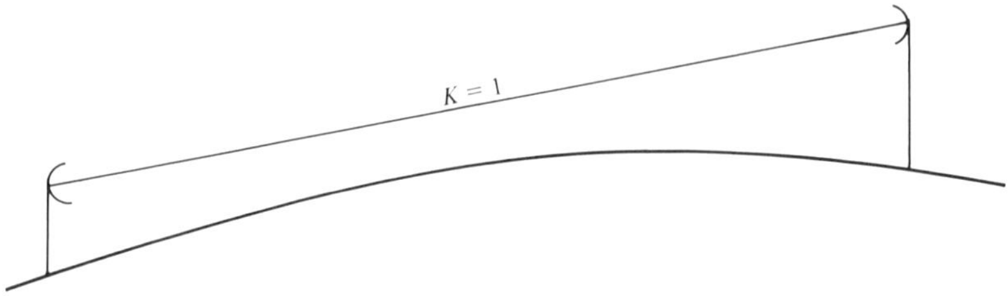

FIGURE 17-37. Nonobstructed path, with earth's surface drawn to scale.

between two fixed points, we must look at any changes that can take place in the path. Figure 17-39 illustrates some representative K factors that can occur in a given path over a period of time. Note that our nonobstructed path at $K = 1$ becomes an obstructed path if $K = \frac{5}{12}$ or less. We must therefore locate the antennas high enough on the towers to ensure that the path will not be blocked at the lowest K factor that we expect to see, or we must expect system outages for those times that the K factor exeeds our design.

Since it is inconvenient to draw both the earth's surface and the RF path as curved lines, radio engineers years ago decided to draw the RF path as a straight line and effectively "change" the earth curvature. Figure 17-40 shows the data in the previous figure redrawn for every K factor desired. This drawing can be used to demonstrate three terms commonly used by microwave path designers:

1. The term *effective earth radius* becomes obvious.
2. *Earth bulge* is used to define propagation where the K factor decreases, causing the earth's surface to bulge up into the RF path.
3. *Earth flattening* is used when the K factor increases, causing the earth to appear flatter. Note that the K factor can exceed infinity, at which time the curvature becomes negative.

Graph paper is available drawn to scale for $K = \frac{4}{3}$, which allows the mobile radio designer to check specific locations in a mobile radio coverage area. However, for microwave designers to design a fixed path considering all the K factors

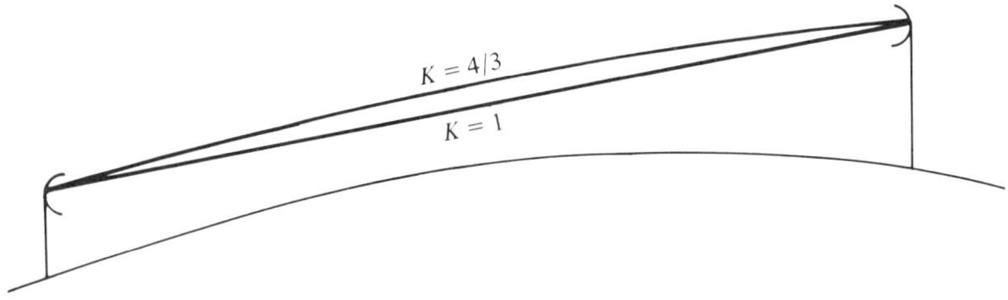

FIGURE 17-38. $K = 4/3$ versus $K = 1$, with earth's surface drawn to scale.

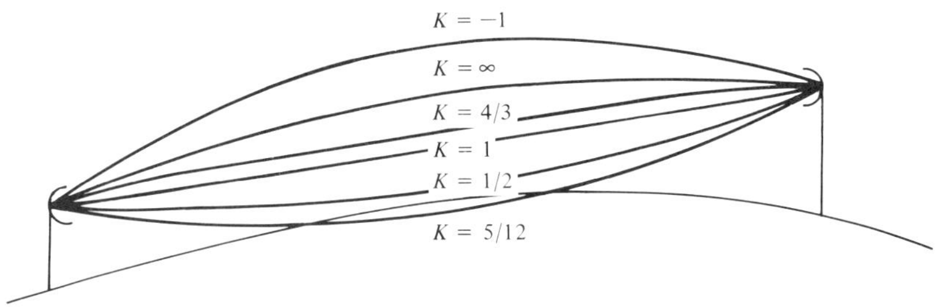

FIGURE 17-39. Selective values of *K*.

necessary, it may be necessary to redraw the path profile five or six times. With the smooth earth used in the preceding illustrations, that would not be a difficult task, but for onshore paths which include human-made obstructions, trees, and terrain irregularities, the effort becomes a bit more tedious. Microwave designers therefore use the second technique, making the earth's surface flat with various *K*-factor changes incorporated into the RF path (Figure 17-41). This method allows the designer to draw the irregular terrain and other physical features along the path only once. The curves for the various *K* factors can then consist of templates drawn to scale, which can be superimposed on the drawing. If the scale of all paths is the same, the same *K*-factor template can be used for all without redrawing.

Thus far we have discussed problems encountered when the RF path is too low. Is there any concern that the path may be too high above ground? Do we really care if the *K* factor takes us into the earth-flattening mode? The answer to both questions is "yes."

To see how a path may be too high, we reintroduce Fresnel zones (Figure 17-42). The first Fresnel zone is the boundary of all reflection points that cause a reflected wave traveling from one antenna to another to exceed the length of

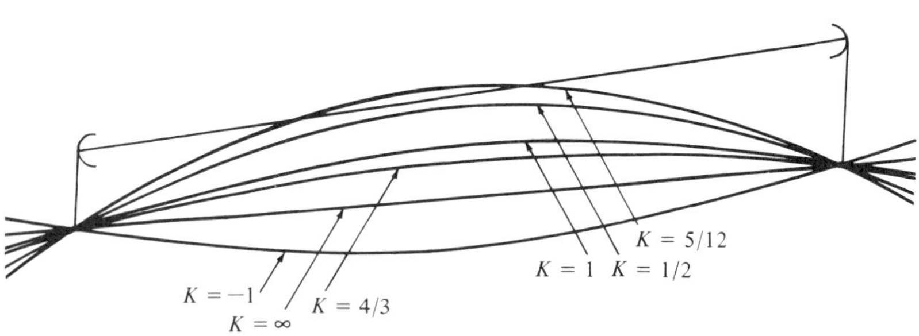

FIGURE 17-40. Selective values of *K* with RF path drawn as a straight line and the earth's surface modified.

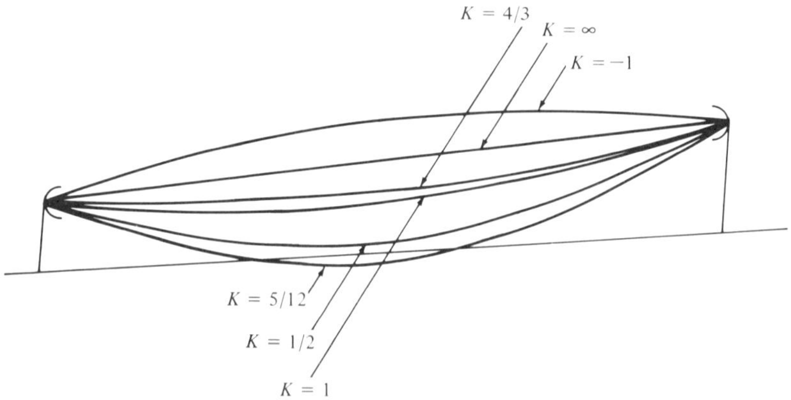

FIGURE 17-41. Selective values of *K* with earth drawn flat and RF path modified.

a direct wave between the two antennas by

$$\frac{n = 1}{2} \times 360° = 180°$$

or one half-wavelength. The second Fresnel zone would be 360° or one wavelength longer than the direct wave. Figure 17-43 illustrates a typical representation of the first four Fresnel zones. From the cross-sectional view at the bottom, we can see that the Fresnel zones are three-dimensional, indicating that we must be concerned about reflections above and to the sides of our microwave path in addition to reflections off objects below the path. In practice, we are interested primarily in reflections off highly reflective surfaces below the path, such as large bodies of calm water and smooth ground.

When a microwave signal is reflected from a smooth surface, it experiences a 180° phase shift. When this is added to the 180° delay in the first Fresnel zone reflected wave, the reflected wave and the direct wave arrive 360° apart, or in

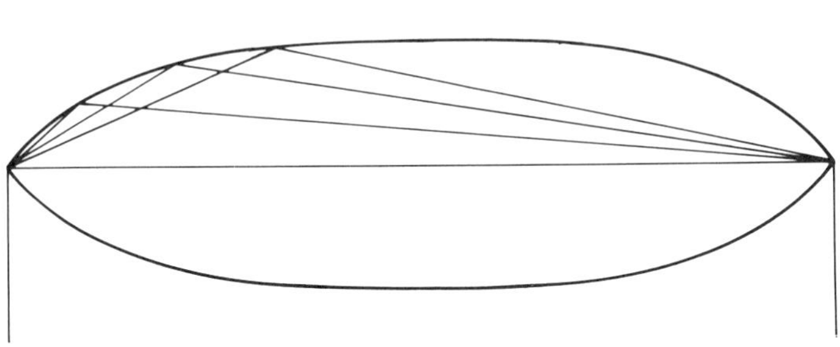

FIGURE 17-42. Definition of Fresnel zones.

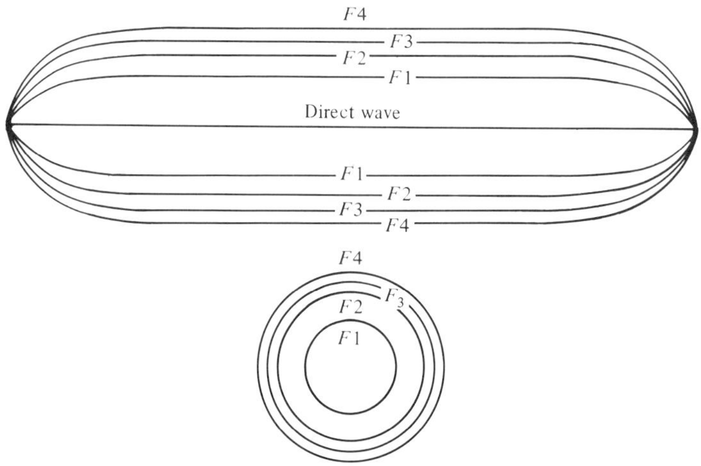

FIGURE 17-43. Fresnel zones 1 through 4.

phase. This results in an increase of up to 6 dB in signal strength, depending on the apparent smoothness of the surface that is reflecting the RF. For the second Fresnel zone, the 180° phase shift on reflection added to the 360° delay of the reflected wave results in a 540° difference, or 180° out of phase. Again depending on the amount of signal reflected back toward the antenna, the effect could vary from almost none to complete cancellation. Extending this analysis to higher-order Fresnel zones, it can be shown that the odd-numbered Fresnel zones are additive and the even-numbered ones cancel (Figure 17-34). As would be expected on paths that are too high, the variations in signals for nonreflective paths ($R = 0$) is negligible, while on highly reflective paths ($R = 1$), wide variations occur. In addition, paths with clearance less than 0.6 times the first Fresnel zone drop in signal strength much more over highly reflective surfaces than over nonreflective surfaces.

It is necessary to consider all the undesirable Fresnel zone values (all even-numbered zones and any clearance less than zero) for all values of K that we expect to encounter. The interaction between Fresnel zones and the K factor will be different for each path that is analyzed. Figure 17-44 illustrates this relationship for the path previously discussed. It is derived by combining Figure 17-39 with Figure 17-43. As K varies from $\frac{5}{12}$ to -1, the path can change from a totally obstructed path all the way through the fourth Fresnel zone.

Specific charts are available commercially which provide a cumulative distribution of K factors for specific geological areas. The charts provide information concerning temperature extremes, precipitation, and an actual description of the terrain. The charts also provide propagation data useful in determining the value of K factors that must be considered within a given location. These particular charts show the percentage of time that a given K factor is exceeded

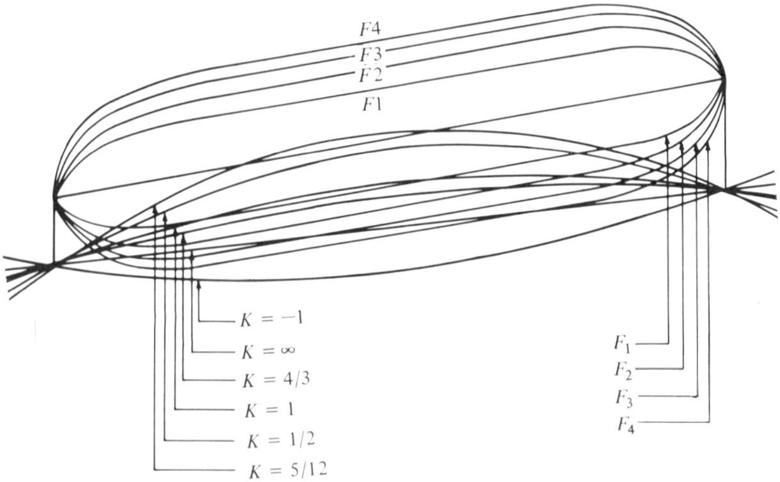

FIGURE 17-44. ***K* versus *F*.**

or not exceeded. As with any statistical data such as this, care must be taken in their use. The most serious problem with data taken from the charts is that they represent data taken only two times each day, and these times may not coincide with the times when most serious problems occur on microwave paths. The other notable problem is that averaging readings taken 12 hours apart may give a result which does not closely represent either of the actual readings. However, the radiosonde graphs are the best published data currently available.

The first step in plotting a path profile is to perform a map survey using the topographic map. From this we get elevations of the terrain along the path and any major obstacles that may constitute path obstacles. After the map survey has been completed, a field survey is necessary to locate new structures and tree heights. Using these results, we can plot a profile of the terrain along the path. A field survey is essential because geological maps available may be years old and new obstructions such as buildings, lakes, and other human-made structures may not be on the original maps. At the most obvious controlling point or points (those that extend upward closest to the path) we calculate or determine from charts or tables the different Fresnel zone values at the frequency of interest. Next we note that it is necessary to protect paths in this area against $K = \frac{1}{2}$ (see Figure 17-41).

Using the tower heights selected, we next determine path performance at other K factors. Path performance can be determined mathematically and computer programs are available for both computers and programmable calculators.

There are several methods available for profiling a path using curved earth paper, curved beam paper, or flat paper. This is a personal preference on the designer's part, but all will result in the same answer. If 10-by-10-in. graph paper is used for the profile, the scale can be: vertical scale, 1 in. = 400 ft and horizontal scale, 1 in. = 4 mi [see equation (17-29)]. Plotting the parameters on paper proceeds as follows. Always place the low antenna on a major division of the graph paper because we calculate the reflection points from the low antenna and this makes measuring easier. We calculate points along the path that will influence system performance. The reflection points extend from immediately below one antenna to below the next antenna. The point of interest to us is that point which will reflect power into the associated antenna. For any specified K factor there is only one point along the path. Other reflections will not be received by the receiving antenna because the energy either goes over or under it. These reflections add vectorially to the direct wave and if in phase will provide a 6-dB increase (double the signal) or completely cancel the signal received if it arrives out of phase. This, of course, assumes a perfectly reflecting surface that does not absorb any of the signal. This is never true; even seawater absorbs some signal, so that the signal being reflected is always weaker after reflection. However, it still may be strong enough to reduce the signal received to the point of receiver squelch, with resultant system fade. The system designer can influence the location of critical reflection points on a path by selecting different antenna heights. When one antenna is raised, the reflection points for all values of K will travel toward the low antenna. It should be noted that the critical reflection point for all values of K will be the exact center of the path when the antennas are at the same height. In this special case the reflection point calculation has to be done only once. If the antennas are at different heights, the points must be calculated for all values of K that are of interest. As mentioned previously, the K factor of $\frac{4}{3}$ is called *radio earth* and it is this effective earth curvature that the system sees a great majority of the time. There are, however, an infinite number of K factors that influence a path. By convention, only particular ones are normally investigated. Some examples will illustrate the principles involved in path design for a microwave system:

EXAMPLE 17-6

(a) A transmitting antenna is 1000 ft in height. Determine the maximum line-of-sight distance. (b) A receiving antenna is used which is 100 ft in height. What is the total line of sight distance? Use $K = \frac{4}{3}$ for both (a) and (b).

SOLUTION

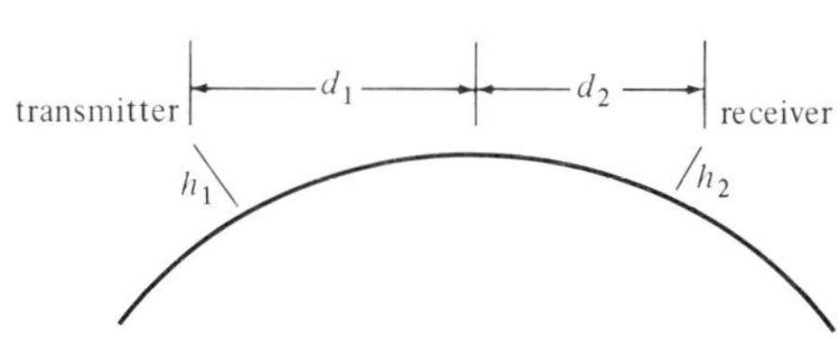

(a) $d_1 = d_2 = d$. Then, from equation (17-29),

$$h_1 = \frac{d^2}{(1.5)(\frac{4}{3})} \Rightarrow d_1 = \sqrt{2h_1} = \sqrt{(2)(1000)} = 44.72 \text{ mi}$$

(b) $d_2 = d_1 = d$, or

$$d_2 = \sqrt{2h_2} = \sqrt{(2)(100)} = 14.14 \text{ mi}$$

$$\text{total path} = d_1 + d_2 = 44.72 + 14.14 = 58.86 \text{ mi}$$

EXAMPLE 17-7

(a) Two points are located 30 mi apart on a smooth earth ($K = \frac{4}{3}$). If the frequency of transmission is 6 GHz, obtain the radius of the first Fresnel zone at a point midway between the antennas.
(b) Draw a topographic profile of the $K = \frac{4}{3}$ earth over the 50-mi path.
(c) If a clearance of $\frac{1}{4}$ Fresnel zone is desired, what antenna heights are required? Assume equal antenna heights and superimpose the microwave beam.
(d) For smooth-sphere diffraction, what is the loss in decibels from the free-space conditions?

SOLUTION

(a) Using equation (17-27), $n = 1$:

$$H_1 = \sqrt{\lambda \frac{d_1 d_2}{d_1 + d_2}}$$

$$\lambda = \frac{300}{f_{\text{MHz}}} = \frac{300}{6000} = 0.05 \text{ m} = 31.1 \times 10^{-6} \text{ mi}$$

$$H_1 = \sqrt{\frac{(31.1 \times 10^{-6})(25)^2}{30}} = 25.45 \times 10^{-3} \text{ mi}$$

$$= 134.39 \text{ ft}$$

(b) Using equation (17-29), we have

$$h = \frac{d_1 d_2}{2}$$

d_1	d_2	h
5	25	62.5
10	20	100
15	15	112.5

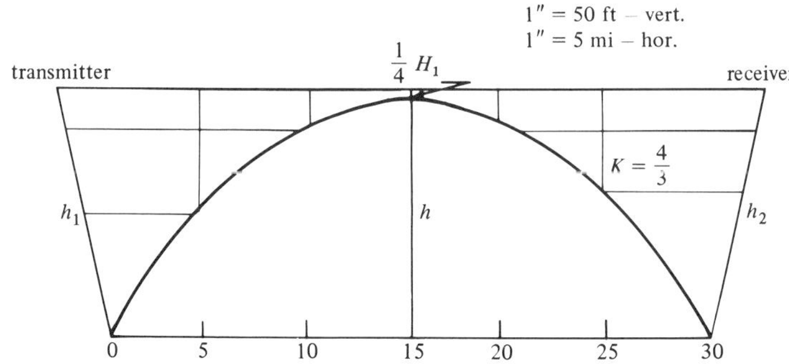

(c) $h_1 = h_2 = h + \frac{1}{4}H_1 = \dfrac{d^2}{2} + \frac{1}{4}H_1$

$$= \frac{15^2}{2} + \frac{134.39}{4} = 146.05 \text{ ft}$$

(d) From Figure 17-34, $H/H_1 = 0.25 \Rightarrow -6$ dB loss relative to the free-space value.

EXAMPLE 17-8

The figure shown can be used to convert from midpoint Fresnel zone clearance for other than midpoint obstacle points. For Example 17-7, if the obstacle is located 5 mi from the transmitting antenna, what must the clearance be?

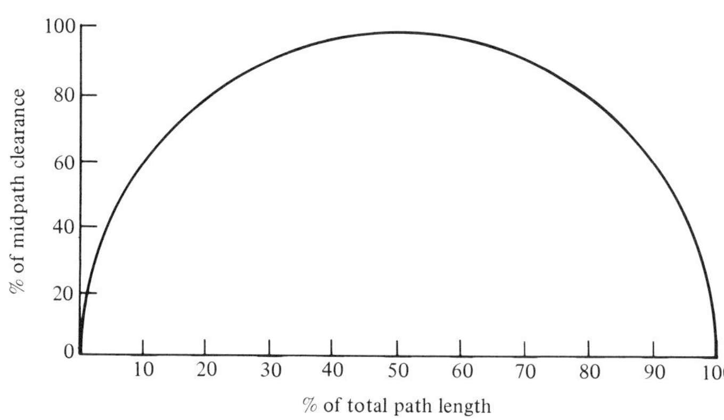

SOLUTION

In this particular case, $5/30 = 16.6\%$ and $25/30 = 83\%$. From the chart,

$$\text{required clearance} = (0.73)(\tfrac{1}{4}H_1) = 0.73 \times \frac{134.39}{4} = 24.52 \text{ ft}$$

PROBLEMS

17-1. Explain the difference between gain and directivity.

17-2. Find the aperture area of a 10-cm-wavelength antenna having a directivity of 1000.

17-3. The input to a 30-dB gain transmitting antenna is 1000 W. A receiver 100 km distant has a 20-dB antenna. The frequency is 300 MHz. Assuming free-space transmission, find the maximum receiver power.

17-4. The input power to a 95% efficient antenna is 100 W. The measured radiation intensity has a maximum value of 500 W/sr. What are the directivity and the gain of the antenna?

17-5. The electric field intensity of a wave in free space is 2.7×10^{-3} V/m. (a) What is the power density in volts per meter? (b) What is the magnetic field intensity in ampere-turns/meter?

17-6. A transmitting antenna is located on top of a mountain overlooking the sea. A ship is steaming directly away from the antenna. (a) Neglecting the height of the ship's antenna, what is the maximum distance that the ship will be able to receive line-of-sight propagation? (b) If the ship's antenna is 30 ft above sea level, what is the maximum distance the ship will be able to receive signals?

17-7. A transmitter radiates 1000 W of power and has an antenna gain of 10 dB. If free-space conditions exist, what is the electric field intensity in volts per meter at a distance of 1000 km?

17-8. An airplane is 100 mi away from a transmitting antenna whose frequency is 6 GHz. How high must the airplane be to guarantee a clearance of $0.6H_1$ at the midpoint?

17-9. An antenna whose input impedance is 80 Ω is fed from a transmitter by a line whose characteristic impedance is 400 Ω. Calculate the impedance of a quarter-wavelength matching section of line to ensure that the VSWR on the line is unity.

17-10. A center-fed dipole has an input impedance of 75 Ω resistive. The ohmic resistance of the conductor is 4 Ω. (a) What is the radiation resistance if the efficiency of the dipole is 90%? (b) If the directivity of the dipole is 1.64, what is the power gain?

17-11. Repeat Example 17-7 for a separation distance of 60 mi and a midpoint clearance of $0.6H_1$.

17-12. Repeat Example 17-7 for $K = 1$ and for $K = \frac{1}{2}$.

17-13. Repeat Example 17-6 for $K = 1$ and for $K = \frac{1}{2}$.

17-14. Using the parameters for the loop antenna of Example 17-5, calculate the induced loop voltage and circuit gain at frequencies of 10,000 Hz and 30,000 Hz. Assume that the circuit Q remains constant.

17-15. Calculate the directional gain of a parabolic antenna if radiated power of 100 mW can produce a signal level equal to that of an 800-W transmitter with an isotropic radiator. Express the gain in decibels.

17-16. A dipole antenna has a resistance of 73 Ω, of which 4 Ω is due to losses. When parasitic elements are added, the input resistance drops to 30 Ω, and 6 Ω of this is due to losses. Calculate the radiation efficiency.

17-17. What is the maximum communication distance if the transmitting and receiving antennas are at heights of 100 and 50 ft, respectively.

17-18. For line-of-sight transmission, what transmitting antenna height will be needed for reliable communication at a location 100 mi distant and a receiving antenna height of 200 ft?

17-19. The antenna current for a center-fed dipole is 7.5 A. Neglecting losses, how much power does this antenna radiate?

17-20. Calculate the length of a half-wave diole to receive a 250 MHz broadcast.

17-21. If the power of a 1000 KHz transmitter is changed from 100 to 500 watts, what is the percent change in electric field intensity at a given distance from the transmitter?

17-22. A half-wave dipole is used as the transmitting antenna. Can a loop antenna be used as the receiving antenna? Explain.

Answers to Selected Problems

Chapter 2

2-1. 795.775 kHz

2-5. **(a)** 499.806 kHz
(b) 40.02
(c) 6.49 $\angle 39^\circ$ Ω

2-7. **(a)** 20 Ω
(c) 0.5 $\angle 0^\circ$ amps
(d) 499.5 volts

2-9. **(a)** 0.5
(c) 2.0

2-11. 0.05 amps

2-13. **(a)** 367.6 Ω
377.5 + j1250 Ω

(b) $2.82 + j31.07\ \Omega$
$12.82 + j1241\ \Omega$
(c) $0.34 - j13.37\ \Omega$
$10.34 + j1368.88\ \Omega$

2-15. Overcoupled

2-17. **(a)** 51.16 kΩ
(b) 3.1
(c) 205.572 kHz
67.489 kHz
(d) 511.6
(e) 2.56 volts

2-19. **(a)** $R\ \Omega$
(b) $R\ \Omega$
(c) $L/RC\ \Omega$

2-21. 25.60 Hz

2-23. **(a)** 6.58 *nh*
(b) 0.19 F

2-25. **(a)** 3.8 amps
(b) 152.4 watts/sec

Chapter 3

3-1. Choose $R = 1$ kΩ
$c = 0.5$ μF
$\omega_c = 3003$ rad/s

3-3. 37.5 MHz
0.637 μH
63.6 pF

3-5. Choose R = 1 kΩ
$C_1 = 0.035$ μF
$C_2 = 0.0139$ μF
$C_3 = 0.002$ μF

Chapter 4

4-1. 0.418

4-3. $a_0 = 7.5$

$$a_1 = -\frac{10}{\sqrt{2}\pi}$$

$$a_2 = -\frac{5}{\pi}$$

$$a_3 = -\frac{30}{\sqrt{2}\pi}$$

$$a_4 = \frac{2.5}{\pi}$$

$$a_5 = \frac{2}{\sqrt{2}\pi}$$

4-5. $f(t) = 60.9 \sin \omega t + 58.5 \sin 2\omega t + 137.9 \sin 3\omega t - 36.8 \sin 4\omega t + 2.76 \sin 5\omega t + \ldots$

4-7. $v_R(t) = 0.45 \cos (\omega t + 38.9°) - 0.19 \cos (3\omega t + 18.7°) + 0.125 \cos (5\omega t + 11.3°) - 0.09 \cos (7\omega t + 8.0°) + \ldots$

4-9. $$i(t) = \frac{2}{\pi}\left[1 + \frac{1}{2}\cos \omega t + \frac{2}{3}\cos 2\omega t - \frac{2}{15}\cos 4\omega t + \frac{2}{35}\cos 6\omega t + \ldots\right]$$

4-11. **(a)** 0.06 sec
(b) 150 Hz
600 Hz

Chapter 5

5-1. 0.288 μV
0.299 μV

5-3. $R_n = 1\ k\Omega$
$I_n = 0.288$ nA

5-5. **(a)** 1.33 MHz
(b) 10.25 μV

5-7. **(a)** 109.32 dB
(b) 7.78 dB

5-9. **(a)** 66.58 °K
(b) 8.04×10^{-11} watts

5-11. 12 MΩ
100 μV

5-13. 4.98 μV

5-15. 1.727×10^{-14} watts

Chapter 6

6-3. $1.2 + j\,0.3\ \Omega$

6-5. 1 Ω

Chapter 7

7-5. **(a)** $H(j\omega) = \dfrac{1 + j\omega R_2C}{1 + j\omega c(R_1 + R_2)}$

(b) $\omega_2 = \dfrac{1}{R_2C}$

$\omega_1 = \dfrac{1}{C(R_1 + R_2)}$

7-7. $C = 1\mu\text{F}$
$R_1 = 394\ K\Omega$
$F_2 = 710\ \Omega$

Chapter 8

8-1. $e_{\text{AM}}(t) = 10\,[1 + 0.25 \sin 2000\,\pi t] \sin (2 \times 10^6\,\pi t)$
$m = 0.25$

8-3. $P_C = 2000$ watts
$P_{\text{SB}} = 160$ watts

8-5. 0.803

8-9. $\dfrac{(\text{S/N})_{\text{AM}}}{(\text{S/N})_{\text{DSB - SC.}}} = \dfrac{m^2}{2 + m^2}$

8-13. **(a)** $P_C/P_t = \dfrac{1}{1 + \dfrac{m^2}{2}}$

$$\frac{P_{SB}}{P_T} = \frac{m^2}{2 + m^2}$$

(b) $P_{C(peak)} = v^2/R$

$$P_{SB(peak)} = \frac{m^2 v^2}{4R}$$

8-15. **(a)** 0.3
(b) 0.6
(c) 1.0

8-17. $P_C = 339$ watts
$P_{SB} = 61$ watts

8-21. **(a)** 20.0015 MHz
19.9985 MHz
(b) 0.15 watts; 2.74 volts rms
1.095 watts; 7.4 volts rms
(c) 0.447 amps
0.450 amps

Chapter 9

9-1. **(a)** 525 rad/s
(b) 3.55 rad/s
(c) 42 rad/s

9-5. $10 + 60t$ rad/s

9-7. $BW_{FM} = 21 \times BW_{AM}$

9-11. f_{IF} 10.7 MHz
$\Delta f = \pm 75$ kHz

9-13. **(a)** 10 watts
(b) 6 watts

9-15. 2.618 kHz

9-17. 5

9-19. 48 kHz

9-21. $\Delta f_2 = 1.3\ \Delta f_1$

Chapter 10

10-1. 13 to 18 MHz

10-3. −96 dBm

10-5. −154 dBm
0.00524 μV

10-7. 3.5 dB

10-9. **(a)** 26.67 kHz
(b) 3.375 kHz

Chapter 11

11-1. 5.7 μsec
4.4 kHz
88 kHz

11-3. 80 sec

11-7. **(a)** 12 MHz
(b) 10 kHz
(c) 40 kHz

11-4. 768 kHz

Chapter 12

12-1. **(a)** 10100
(b) 00110
(c) 00010
(d) 11011

12-3. RMS (peak) = 0.0093 volts
S/N = 40.6 dB

12-5. 0.3 Mbits

12-9. **(a)** yes
(b) 1.13 mvolt

12-11. $F_0 = 8.0 + j0$

$F_1 = 0.5 + 21.5$
$F_2 = -3 + j0$
$F_3 = -3 + j2$
$f_1 = 250$ kHz

$y(t) = \frac{1}{4}\{8 + 1.58 \cos \omega_1 t + 71.56° + 3 \cos (2\omega_1 t + 180°) + 3.6$

$\cos (3\omega_1 t + 156.4°)\}$

Chapter 13

13-1. +87 dBrn

13-3. 16,000 bits/sec
39,868 bits/sec

13-5.

1 0 1 0 0 1 1 0	1
	0
↑	1
DATA	1
↓	0
	0
1 1 1 0 0 1 1 0	0
1 1 0 0 0 1 0 1	

13-7. **(a)** $\Delta_f = 500$ Hz
Baud rate = bit rate = 1200 bits/sec
(b) $m_f = 0.416$

13-9. −90 dBm

13-11. 8.12 dBm

13-13.

Power (W)	dBm	dBrn
10^1	40	130
10^0	30	120
10^{-3}	0	90
10^{-6}	−30	60
10^{-12}	−90	0

13-15. **(a)** 1.544 Mbits/sec
(b) 5.568 MHz

13-17. 29,904 bits/sec

Chapter 14

14-1. **(a)** 80.6°
(b) 1014 m
(c) 69 ns

14-3. 8052 modes

14-5. 30

14-7. −9.17 dB

14-9. 12.4 dB/km
1.9 dB/km

14-11. **(a)** 9.96°
(b) 0.173

14.13 0.768

Chapter 15

15-1. 1.1 nwatts
7.52 in.

15-7. 1,103,389.3 miles
after

15-9. 290°K

15-11. 28.4 dB

15-13. 15 mm
100

Chapter 16

16-1. 0.33
−0.33

16-3. $31.67 \angle -2.3° \ \Omega$
$0.005 + j0.126$
0.005 n/mi
0.126 rad/mi

16-5. $Z_a = \infty$
$Z_b = Z_0$

16-7. $l_1 = 0.373\lambda$ $l_1' = 0.143\lambda$
$l_2 = 0.337\lambda$ $l_2' = 0.088\lambda$

16-9. $l = 0.165\lambda$

16-11. $\Gamma_R = -0.2$
SWR = 1.5

16-13. 2500 watts

16-15. $27 - j32\ \Omega$

16-17. 0.003λ

Chapter 17

17-3. −162.01 dBw

17-5. **(a)** 0.0194 $\mu W/m^2$
(b) 7.178×10^{-6} A/m

17-7. 0.546 volts/meter

17-9. 178.9 Ω

17-11. **(a)** 34.82 ft.
(c) 470.9 ft.
(d) 0 dB

17-13. **(a)** 38.73 mi 27.39 mi
(b) 12.24 mi 8.66 mi

17-15. 39.03 dB

17-17. 24.14 mi

17-19. 4.106 kwatts

17-21. 223.6 per cent

INDEX